T0182018

QUANTUM MEASUREMENT AND CONTROL

The control of individual quantum systems promises a new technology for the twenty-first century – quantum technology. This book is the first comprehensive treatment of modern quantum measurement and measurement-based quantum control, which are vital elements for realizing quantum technology.

Readers are introduced to key experiments and technologies through dozens of recent experiments in cavity QED, quantum optics, mesoscopic electronics and trapped particles, several of which are analysed in detail. Nearly 300 exercises help build understanding, and prepare readers for research in these exciting areas.

This important book will interest graduate students and researchers in quantum information, quantum metrology, quantum control and related fields. Novel topics covered include adaptive measurement; realistic detector models; mesoscopic current detection; Markovian, state-based and optimal feedback; and applications to quantum information processing.

HOWARD M. WISEMAN is Director of the Centre for Quantum Dynamics at Griffith University, Australia. He has worked in quantum measurement and control theory since 1992, and is a Fellow of the Australian Academy of Science (AAS). He has received the Bragg Medal of the Australian Institute of Physics, the Pawsey Medal of the AAS and the Malcolm Macintosh Medal of the Federal Science Ministry.

GERARD J. MILBURN is an Australian Research Council Federation Fellow at the University of Queensland, Australia. He has written three previous books, on quantum optics, quantum technology and quantum computing. He has been awarded the Boas Medal of the Australian Institute of Physics and is a Fellow of the Australian Academy of Science and the American Physical Society.

An outstanding introduction, at the advanced graduate level, to the mathematical description of quantum measurements, parameter estimation in quantum mechanics, and open quantum systems, with attention to how the theory applies in a variety of physical settings. Once assembled, these mathematical tools are used to formulate the theory of quantum feedback control. Highly recommended for the physicist who wants to understand the application of control theory to quantum systems and for the control theorist who is curious about how to use control theory in a quantum context.

Carlton Caves, University of New Mexico

A comprehensive and elegant presentation at the interface of quantum optics and quantum measurement theory. Essential reading for students and practitioners, both, in the growing quantum technologies revolution.

Howard Carmichael, The University of Auckland

Quantum Measurement and Control provides a comprehensive and pedagogical introduction to critical new engineering methodology for emerging applications in quantum and nano-scale technology. By presenting fundamental topics first in a classical setting and then with quantum generalizations, Wiseman and Milburn manage not only to provide a lucid guide to the contemporary toolbox of quantum measurement and control but also to clarify important underlying connections between quantum and classical probability theory. The level of presentation is suitable for a broad audience, including both physicists and engineers, and recommendations for further reading are provided in each chapter. It would make a fine textbook for graduate-level coursework.

Hideo Mabuchi, Stanford University

This book present a unique summary of the theory of quantum measurements and control by pioneers in the field. The clarity of presentation and the varied selection of examples and exercises guide the reader through the exciting development from the earliest foundation of measurements in quantum mechanics to the most recent fundamental and practical developments within the theory of quantum measurements and control. The ideal blend of precise mathematical arguments and physical explanations and examples reflects the authors' affection for the topic to which they have themselves made pioneering contributions.

Klaus Mølmer, University of Aarhus

QUANTUM MEASUREMENT
AND CONTROL

HOWARD M. WISEMAN
Griffith University

GERARD J. MILBURN
University of Queensland

CAMBRIDGE
UNIVERSITY PRESS

CAMBRIDGE
UNIVERSITY PRESS

University Printing House, Cambridge CB2 8BS, United Kingdom

Published in the United States of America by Cambridge University Press, New York

Cambridge University Press is part of the University of Cambridge.

It furthers the University's mission by disseminating knowledge in the pursuit of education, learning and research at the highest international levels of excellence.

www.cambridge.org
Information on this title: www.cambridge.org/9781107424159

First published 2010
First paperback edition 2014

A catalogue record for this publication is available from the British Library

Library of Congress Cataloguing in Publication data
Wiseman, H. M. (Howard M.)
Quantum measurement and control / Howard M. Wiseman, Gerard J. Milburn.
p.   cm.
Includes bibliographical references and index.
ISBN 978-0-521-80442-4 (hardback)
1. Quantum measure theory.   I. Milburn, G. J. (Gerard J.)   II. Title.
QC174.17.M4W57   2009
530.1201'51542 – dc22   2009034266

ISBN   978-0-521-80442-4   Hardback
ISBN   978-1-107-42415-9   Paperback

To our boys: Tom & Andy, Finlay & Bailey,
who were much smaller when we began.

Contents

Preface

The twenty-first century is seeing the emergence of the first truly quantum technologies; that is, technologies that rely on the counter-intuitive properties of individual quantum systems and can often outperform any conventional technology. Examples include quantum computing, which promises to be much faster than conventional computing for certain problems, and quantum metrology, which promises much more sensitive parameter estimation than that offered by conventional techniques. To realize these promises, it is necessary to understand the measurement and control of quantum systems. This book serves as an introduction to quantum measurement and control, including some of the latest developments in both theory and experiment.

Scope and aims

To begin, we should make clear that the title of this book is best taken as short-hand for 'Quantum measurements with applications, principally to quantum control'. That is, the reader should be aware that (i) a considerable part of the book concerns quantum measurement theory, and applications other than quantum control; and (ii) the sort of quantum control with which we are concerned is that in which measurement plays an essential role, namely feedback (or feedforward) control of quantum systems.[1]

Even with this somewhat restricted scope, our book cannot hope to be comprehensive. We aim to teach the reader the fundamental theory in quantum measurement and control, and to delve more deeply into some particular topics, in both theory and experiment.

Much of the material in this book is new, published in the last few years, with some material in Chapter 6 yet to be published elsewhere. Other material, such as the basic quantum mechanics, is old, dating back a lifetime or more. However, the way we present the material, being informed by new fields such as quantum information and quantum control, is often quite unlike that in older text-books. We have also ensured that our book is relevant to current developments by discussing in detail numerous experimental examples of quantum measurement and control.

---

[1] In using the term 'feedback' or 'feedforward' we are assuming that a measurement step intervenes – see Section 5.8.1 for further discussion.

We have not attempted to give a full review of research in the field. The following section of this preface goes some way towards redressing this. The 'further reading' section which concludes each chapter also helps. Our selection of material is naturally biassed towards our own work, and we ask the forbearance of the many workers in the field, past or present, whom we have overlooked.

We have also not attempted to write an introduction to quantum mechanics suitable for those who have no previous knowledge in this area. We do cover all of the fundamentals in Chapter 1 and Appendix A, but formal knowledge is no substitute for the familiarity which comes with working through exercises and gradually coming to grips with new concepts through an introductory course or text-book.

Our book is therefore aimed at two groups wishing to do research in, or make practical use of, quantum measurement and control theory. The first is physicists, for whom we provide the necessary introduction to concepts in classical control theory. The second is control engineers who have already been introduced to quantum mechanics, or who are introducing themselves to it in parallel with reading our book.

In all but a few cases, the results we present are derived in the text, with small gaps to be filled in by the reader as exercises. The substantial appendices will help the reader less familiar with quantum mechanics (especially quantum mechanics in phase space) and stochastic calculus. However, we keep the level of mathematical sophistication to a minimum, with an emphasis on building intuition. This is necessarily done at the expense of rigour; ours is not a book that is likely to appeal to mathematicians.

Historical background

Quantum measurement theory provides the essential link between the quantum formalism and the familiar classical world of macroscopic apparatuses. Given that, it is surprising how much of quantum mechanics was developed in the absence of formal quantum measurement theory – the structure of atoms and molecules, scattering theory, quantized fields, spontaneous emission etc. Heisenberg [Hei30] introduced the 'reduction of the wavepacket', but it was Dirac [Dir30] who first set out quantum measurement theory in a reasonably rigorous and general fashion. Shortly afterwards von Neumann [vN32] added a mathematician's rigour to Dirac's idea. A minor correction of von Neumann's projection postulate by Lüders [Lüd51] gave the theory of projective measurements that is still used today.

After its formalization by von Neumann, quantum measurement theory ceased to be of interest to most quantum physicists, except perhaps in debates about the interpretation of quantum mechanics [Sch49]. In most experiments, measurements were either made on a large ensemble of quantum particles, or, if they were made on an individual particle, they effectively destroyed that particle by detecting it. Thus a theory of how the state of an individual quantum system changed upon measurement was unnecessary. However, some mathematical physicists concerned themselves with generalizing quantum measurement theory to describe non-ideal measurements, a programme that was completed in the 1970s by Davies [Dav76] and Kraus [Kra83]. Davies in particular showed how the new formalism could describe a continuously monitored quantum system, specifically for the case of quantum jumps [SD81].

By this time, experimental techniques had developed to the point where it was possible to make quantum-limited measurements on an individual quantum system. The prediction [CK85] and observation [NSD86, BHIW86] of quantum jumps in a single trapped ion was a watershed in making physicists (in quantum optics at least) realize that there was more to quantum measurement theory than was contained in von Neumann's formalization. This led to a second watershed in the early 1990s when it was realized that quantum jumps could be described by a stochastic dynamical equation for the quantum state, giving a new numerical simulation method for open quantum systems [DCM92, GPZ92, Car93]. Carmichael [Car93] coined the term 'quantum trajectory' to describe this stochastic evolution of the quantum state. He emphasized the relation of this work to the theory of photodetection, and generalized the equations to include quantum diffusion, relating to homodyne detection.

Curiously, quantum diffusion equations had independently, and somewhat earlier, been derived in other branches of physics [Bel02]. In the mathematical-physics literature, Belavkin [Bel88, BS92] had made use of quantum stochastic calculus to derive quantum diffusion equations, and Barchielli [Bar90, Bar93] had generalized this to include quantum-jump equations. Belavkin had drawn upon the classical control theory of how a probability distribution could be continuously (in time) conditioned upon noisy measurements, a process called filtering. He thus used the term quantum filtering equations for the quantum analogue. Meanwhile, in the quantum-foundations literature, several workers also derived these sorts of equations as attempts to solve the quantum-measurement problem by incorporating an objective collapse of the wavefunction [Gis89, Dió88, Pea89, GP92a, GP92b].

In this book we are not concerned with the quantum measurement problem. By contrast, Belavkin's idea of making an analogy with classical control theory is very important for this book. In particular, Belavkin showed how quantum filtering equations can be applied to the problem of feedback control of quantum systems [Bel83, Bel88, Bel99]. A simpler version of this basic idea was developed independently by the present authors [WM93c, Wis94]. Quantum feedback experiments (in quantum optics) actually date back to the mid 1980s [WJ85a, MY86]. However, only in recent years have sophisticated experiments been performed in which the quantum trajectory (quantum filtering equation) has been essential to the design of the quantum control algorithm [AAS+02, SRO+02].

At this point we should clarify exactly what we mean by 'quantum control'. Control is, very roughly, making a device work well under adverse conditions such as (i) uncertainties in parameters and/or initial conditions; (ii) complicated dynamics; (iii) noise in the dynamics; (iv) incomplete measurements; and (v) resource constraints. Quantum control is control for which the design requires knowledge of quantum mechanics. That is, it does not mean that the whole control process must be treated quantum mechanically. Typically only a small part (the 'system') is treated quantum mechanically, while the measurement device, amplifiers, collators, computers, signal generators and modulators are all treated classically.

As stated above, we are primarily concerned in this book with quantum feedback control. However, there are other sorts of quantum control in which measurement theory does not play a central role. Here we briefly discuss a few of these; see Ref. [MK05] for a fuller review of types of quantum control and Ref. [MMW05] for a recent sample of the field. First, *open-loop control* means applying control theory to manipulate the dynamics of systems in the absence of measurement [HTC83, D'A07]. The first models of quantum computing were all based upon open-loop control [Pre98]. It has been applied to good effect in finite quantum systems in which the Hamiltonian is known to great precision and real-time measurement is impossible, such as in nuclear magnetic resonance [KGB02, KLRG03]. Second, there is *learning control*, which applies to systems in which the Hamiltonian is not known well and real-time measurement is again impossible, such as chemical reactions [PDR88]. Here the idea is to try some control strategy with many free parameters, see what results, adjust these parameters, and try again. Over time, an automated learning procedure can lead to significant improvements in the performance of the control strategy [RdVRMK00]. Finally, general mathematical techniques developed by control theorists, such as semi-definite programming and model reduction, have found application in quantum information theory. Examples include distinguishing separable and entangled states [DPS02] and determining the performance of quantum codes [RDM02], respectively.

The structure of this book

The structure of this book is shown in Fig. 1. It is not a linear structure; for example, the reader interested in Chapter 7 could skip most of the material in Chapters 2, 3 and 6. Note that the reliance relation (indicated by a solid arrow) is meant to be transitive. That is, if Chapter C is indicated to rely upon Chapter B, and likewise Chapter B upon Chapter A, then Chapter C may also rely directly upon Chapter A. (This convention avoids a proliferation of arrows.) Not shown in the diagram are the two Appendices. Material in the first, an introduction to quantum mechanics and phase space, is used from Chapter 1 onwards. Material in the second, on stochastic differential equations, is used from Chapter 3 onwards.

For the benefit of readers who wish to skip chapters, we will explain the meaning of each of the dashed arrows. The dashed arrow from Chapter 2 to Chapter 7 is for Section 2.5 on adaptive measurements, which is used in Section 7.9. That from Chapter 3 to Chapter 4 is for Section 3.6, on the Lindblad form of the master equation, and Section 3.11, on the Heisenberg picture dynamics. That from Chapter 3 to Chapter 6 is for Section 3.8 on preferred ensembles. That from Chapter 5 to Chapter 6 is for Section 5.5 on homodyne-based Markovian feedback. Finally, that from Chapter 6 to Chapter 7 is for the concept of an optimal quantum filter, which is introduced in Section 6.5. Of course, there are other links between various sections of different chapters, but these are the most important.

Our book is probably too long to be covered in a single graduate course. However, selected chapters (or selected topics within chapters) could be used as the basis of such a course, and the above diagram should aid a course organizer in the selection of material. Here are some examples. Chapters 1, 3 and 4 could be the text for a course on open quantum

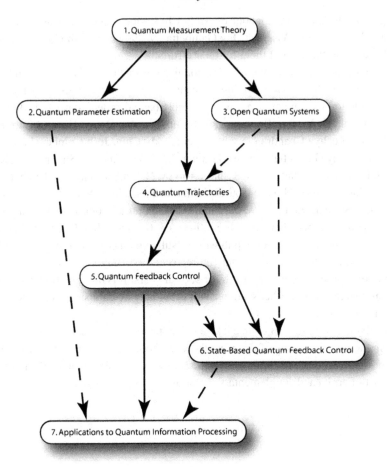

Fig. 1 The structure of this book. A solid arrow from one chapter to another indicates that the latter relies on the former. A dashed arrow indicates a partial reliance.

systems. Chapters 1, 4 and 6 (plus selected other sections) could be the text for a course on state-based quantum control. Chapters 1 and 2 could be the text for a course on quantum measurement theory.

Acknowledgements

This book has benefited from the input of many people over the years, both direct (from reading chapter drafts) and indirect (in conversation with the authors). Andy Chia deserves particular thanks for his very thorough reading of several chapters and his helpful suggestions for improvements and additions. Prahlad Warszawski also deserves mention for reading and commenting upon early drafts of the early chapters. The quality of the figures is due largely to the meticulous labour of Andy Chia. Much other painstaking work was assisted greatly by Nadine Wiseman, Jay Gambetta, Nisha Khan, Josh Combes and Ruth Forrest. We are grateful to all of them for their efforts.

In compiling a list of all those who deserve thanks, it is inevitable that some will be inadvertently omitted. At the risk of offending these people, we would also like to acknowledge scientific interaction over many years with (in alphabetical order) Slava Belavkin, Andy Berglund, Dominic Berry, Luc Bouten, Sam Braunstein, Zoe Brady, Howard Carmichael, Carlton Caves, Anushya Chandran, Yanbei Chen, Andy Chia, Josh Combes, Lajos Diósi, Andrew Doherty, Jay Gambetta, Crispin Gardiner, J. M. Geremia, Hsi-Sheng Goan, Ramon van Handel, Kurt Jacobs, Matt James, Sasha Korotkov, Navin Khaneja, P. S. Krishnaprasad, Erik Lucero, Hans Maasen, Hideo Mabuchi, John Martinis, Ahsan Nazir, Luis Orozco, Neil Oxtoby, Mohan Sarovar, Keith Schwab, John Stockton, Laura Thomsen and Stuart Wilson.

Chapter 6 requires a special discussion. Sections 6.3–6.6, while building on Ref. [WD05], contain a large number of hitherto unpublished results obtained by one of us (H.M.W) in collaboration with Andrew Doherty and (more recently) Andy Chia. This material has circulated in the community in draft form for several years. It is our intention that much of this material, together with further sections, will eventually be published as a review article by Wiseman, Doherty and Chia.

The contributions of others of course take away none of the responsibility of the authors for errors in the text. In a book of this size, there are bound to be very many. Readers are invited to post corrections and comments on the following website, which will also contain an official list of errata and supplementary material:

www.quantum-measurement-and-control.org

1

Quantum measurement theory

1.1 Classical measurement theory

1.1.1 Basic concepts

Although this chapter, and indeed this book, is concerned with quantum measurements, we begin with a discussion of some elementary notions of measurement for classical systems. These are systems that operate at a level where quantum effects are not apparent. The purpose of this discussion is to introduce some ideas, which carry over to quantum theory, concerning states, conditional and non-conditional distributions, and stochastic processes. It will also make the distinct features of quantum measurements plainer.

A classical system can be described by a set of *system variables*, which we will call the system *configuration*. For example, for a system of N interacting particles these could be the N position and momentum vectors of the particles. The possible values of these variables form the *configuration space* $\mathbb{S}$ for the system. In the above dynamical example, the configuration space would be $\mathbb{R}^{6N}$, where $\mathbb{R}$ is the real line.[1] Alternatively, to take the simplest possible example, there may be a single system variable X that takes just two values, $X = 0$ or $X = 1$, so that the configuration space would be $\{0, 1\}$. Physically, this binary variable could represent a coin on a table, with $X = 0$ and $X = 1$ corresponding to heads and tails, respectively.

We define the *state* of a classical system to be a probability distribution on configuration space. Say (as in the example of the coin) that there is a single system variable $X \in \mathbb{S}$ that is discrete. Then we write the probability that X has the value x as $\Pr[X = x]$. Here, in general, $\Pr[E]$ is the probability of an event E. When no confusion is likely to arise, we write $\Pr[X = x]$ simply as $\wp(x)$. Here we are following the convention of representing variables by upper-case letters and the corresponding arguments in probability distributions by the corresponding lower-case letters. If X is a continuous variable, then we define a probability density $\wp(x)$ by $\wp(x)\mathrm{d}x = \Pr[X \in (x, x + \mathrm{d}x)]$. In either case, the state of the system is represented by the function $\wp(x)$ for all values of x. When we choose to be more careful, we write this as $\{\wp(x)\colon x \in \mathbb{S}\}$, or as $\{\wp(x)\colon x\}$. We will use these conventions

---

[1] This space is often called 'phase space', with 'configuration space' referring only to the space of positions. We will not use 'configuration space' with this meaning.

conscientiously for the first two chapters, but in subsequent chapters we will become more relaxed about such issues in order to avoid undue notational complexity.

The system state, as we have defined it, represents an observer's knowledge about the system variables. Unless the probability distribution is non-zero only for a single configuration, we say that it represents a state of *uncertainty* or incomplete knowledge. That is, in this book we adopt the position that probabilities are *subjective*: they represent degrees of certainty rather than objective properties of the world. This point of view may be unfamiliar and lead to uncomfortable ideas. For example, different observers, with different knowledge about a system, would in general assign different states to the same system. This is not a problem for these observers, as long as the different states are *consistent*. This is the case as long as their supports on configuration space are not disjoint (that is, as long as they all assign a non-zero probability to at least one set of values for the system variables). This guarantees that there is at least one state of complete knowledge (that is, one configuration) that all observers agree is a possible state.

We now consider measurement of a classical system. With a perfect measurement of X, the observer would simply find out its value, say x'. The system state would then be a state of complete knowledge about this variable. For discrete variables this is represented by the Kronecker δ-function $\wp(x) = \delta_{x,x'}$, whereas for a continuous variable it is represented by the Dirac δ-function $\wp(x) = \delta(x - x')$. For comparison with the quantum case (in following sections), it is more enlightening to consider imperfect measurements. Suppose that one only has access to the values of the system variables indirectly, through an *apparatus* variable Y. The state of the apparatus is also specified by a probability distribution $\wp(y)$. By some physical process, the apparatus variable becomes statistically dependent on the system variable. That is, the configuration of the apparatus is correlated (perhaps imperfectly) with the configuration of the system. If the apparatus variable is observed, $\{\wp(y): y\}$ is simply the probability distribution of measurement outcomes.

One way of thinking about the system–apparatus correlation is illustrated in Fig. 1.1. The correlation is defined by a functional relationship among the readout variable, Y, the system variable, X, before the measurement, and a random variable, Ξ, which represents extra noise in the measurement outcome. We can specify this by a function

$$Y = G(X, \Xi), \tag{1.1}$$

together with a probability distribution $\wp(\xi)$ for the noise. Here, the noise is assumed to be independent of the system, and is assumed not to affect the system. That is, we restrict our consideration for the moment to *non-disturbing measurements*, for which X after the measurement is the same as X before the measurement.

1.1.2 Example: binary variables

To illustrate the above theory, consider the case of binary variables. As we will see, this is relevant in the quantum setting also. For this case, the state of the system $\wp(x)$ is completely specified by the probability $\wp(x := 0)$, since $\wp(x := 1) = 1 - \wp(x := 0)$. Here

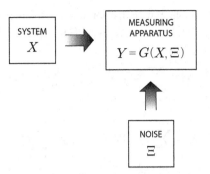

Fig. 1.1 System–apparatus correlation in a typical classical measurement.

we have introduced another abuse of notation, namely that $\wp(x := a)$ means $\wp(x)$ evaluated at $x = a$, where a is any number or variable. In other words, it is another way of writing $\Pr[X = a]$ (for the case of discrete variables). The convenience of this notation will become evident.

We assume that the apparatus and noise are also described by binary variables with values 0 and 1. We take the output variable Y to be the *binary addition* (that is, addition modulo 2) of the system variable X and the noise variable Ξ. In the language of binary logic, this is called the 'exclusive or' (XOR) of these two variables, and is written as

$$Y = X \oplus \Xi. \tag{1.2}$$

We specify the noise by $\wp(\xi := 0) = \mu$.

Equation 1.2 implies that the readout variable Y will reproduce the system variable X if $\Xi = 0$. If $\Xi = 1$, the readout variable is (in the language of logic) the negation of the system variable. That is to say, the readout has undergone a *bit-flip* error, so that $Y = 1$ when $X = 0$ and vice versa. We can thus interpret μ as the probability that the readout variable is 'correct'. If no noise is added by the measurement apparatus, so that $\wp(\xi := 0) = 1$, we call the measurement *ideal*.

It should be intuitively clear that the apparatus state (i.e. the readout distribution) $\wp(y)$ is determined by the function G together with the noise probability $\wp(\xi)$ and the system state before the measurement, $\wp(x)$. This last state is called the a-priori state, or prior state. In the example above we find that

$$\wp(y := 1) = \mu\wp(x := 1) + (1 - \mu)\wp(x := 0), \tag{1.3}$$
$$\wp(y := 0) = \mu\wp(x := 0) + (1 - \mu)\wp(x := 1). \tag{1.4}$$

This may be written more succinctly by inverting Eq. (1.2) to obtain $\Xi = X \oplus Y$ and writing

$$\wp(y) = \sum_{x=0}^{1} \wp(\xi := x \oplus y)\wp(x). \tag{1.5}$$

In the case of a binary variable X with distribution $\wp(x)$, it is easy to verify that the mean is given by $E[X] = \wp(x := 1)$. Here we are using E to represent 'expectation of'. That is, in the general case,

$$E[X] = \sum_x x \Pr[X = x] = \sum_x x \wp(x). \tag{1.6}$$

More generally,

$$E[f(X)] = \sum_x f(x) \Pr[X = x] = \sum_x f(x) \wp(x). \tag{1.7}$$

Using this notation, we define the variance of a variable as

$$\mathrm{Var}[X] \equiv E[X^2] - (E[X])^2. \tag{1.8}$$

Exercise 1.1 *Show that*

$$E[Y] = (1 - \mu) + (2\mu - 1)E[X], \tag{1.9}$$

$$\mathrm{Var}[Y] = \mu(1 - \mu) + (2\mu - 1)^2 \mathrm{Var}[X]. \tag{1.10}$$

Equation (1.9) shows that the average measurement result is the system variable mean, scaled by a factor of $2\mu - 1$, plus a constant off-set of $1 - \mu$. The scaling factor also appears in the variance equation (1.10), together with a constant (the first term) due to the noise added by the measurement process. When the measurement is ideal ($\mu = 1$), the mean and variance of the readout variable directly reflect the statistics of the measured system state.

1.1.3 Bayesian inference

We stated above that we are considering, at present, non-disturbing measurements, in which the system variable X is unaffected by the measurement. However, this does not mean that the system *state* is unaffected by the measurement. Recall that the state represents the observer's incomplete knowledge of the system, and the point of making a measurement is (usually) to obtain more knowledge. Thus we *should* expect the state to change *given* that a certain readout is obtained.

The concept we are introducing here is the *conditional* state of the system, also known as the state conditioned on the readout. This state is sometimes called the a-posteriori state, or posterior state. The key to finding the conditioned state is to use Bayesian inference. Here, inference means that one infers information about the system from the readout, and Bayesian inference means doing this using Bayes' theorem. This theorem is an elementary consequence of basic probability theory, via the double application of the conditional-probability definition

$$\Pr(A|B) = \Pr(A \cap B)/\Pr(B), \tag{1.11}$$

where A and B are events, $A \cap B$ is their intersection and $A|B$ is to be read as 'A given B'. In an obvious generalization of this notation from events to the values of system variables,

Bayes' theorem says that the conditional system state may be written in terms of the a-priori (or prior) system state $\wp(x)$ as

$$\wp'(x|y) = \frac{\wp(y|x)\wp(x)}{\wp(y)}. \tag{1.12}$$

Here the prime emphasizes that this is an a-posteriori state, and (sticking to the discrete case as usual)

$$\wp(y) = \sum_x \wp(y|x)\wp(x), \tag{1.13}$$

as required for the conditional state to be normalized.

The crucial part of Bayesian inference is the known conditional probability $\wp(y|x)$, also known as the 'forward probability'. This is related to the measurement noise and the function G as follows:

$$\wp(y|x) = \sum_\xi \wp(y|x,\xi)\wp(\xi) = \sum_\xi \delta_{y,G(x,\xi)}\wp(\xi). \tag{1.14}$$

Here $\wp(y|x,\xi)$ means the state of y given the values of x and ξ. If the output function $Y = G(X,\Xi)$ is invertible in the sense that there is a function G^{-1} such that $\Xi = G^{-1}(X,Y)$, then we can further simplify this as

$$\wp(y|x) = \sum_\xi \delta_{\xi,G^{-1}(x,y)}\wp(\xi) = \wp(\xi := G^{-1}(x,y)). \tag{1.15}$$

Thus we obtain finally for the conditional system state

$$\wp'(x|y) = \frac{\wp(\xi := G^{-1}(x,y))\wp(x)}{\wp(y)}. \tag{1.16}$$

Exercise 1.2 *If you are unfamiliar with probability theory, derive the first equality in Eq. (1.14).*

As well as defining the conditional post-measurement system state, we can define an unconditional posterior state by averaging over the possible measurement results:

$$\wp'(x) = \sum_y \wp'(x|y)\wp(y) = \sum_y \wp(\xi := G^{-1}(x,y))\wp(x). \tag{1.17}$$

The terms conditional and unconditional are sometimes replaced by the terms selective and non-selective, respectively. In this case of a non-disturbing measurement, it is clear that

$$\wp'(x) = \wp(x). \tag{1.18}$$

That is, the unconditional posterior state is always the same as the prior state. This is the counterpart of the statement that the system variable X is unaffected by the measurement.

Exercise 1.3 *Determine the posterior conditional states $\wp'(x|y)$ in the above binary example for the two cases $y = 0$ and $y = 1$. Show that, in the limit $\mu \to 1$, $\wp'(x|y) \to \delta_{x,y}$ (assuming that $\wp(x := y) \neq 0$), whereas, in the case $\mu = 1/2$, $\wp'(x|y) = \wp(x)$. Interpret these results.*

1.1.4 Example: continuous variables

We now turn to the case of continuous state variables. Suppose one is interested in determining the position of a particle on the real line. Let the a-priori state of this system be the probability density $\wp(x)$. As explained earlier, this means that, if an ideal measurement of the position X is made, then the probability that a value between x and $x + dx$ will be obtained is $\wp(x)dx$.

As in the binary case, we introduce an apparatus with configuration Y and a noise variable Ξ, both real numbers. To define the measurement we specify the output function, $G(X, \Xi)$, for example

$$Y = X + \Xi, \tag{1.19}$$

so that $\Xi = G^{-1}(X, Y) = Y - X$. We must also specify the probability density for the noise variable ξ, and a common choice is a zero-mean Gaussian with a variance Δ^2:

$$\wp(\xi) = (2\pi \Delta^2)^{-1/2} e^{-\xi^2/(2\Delta^2)}. \tag{1.20}$$

The post-measurement apparatus state is given by the continuous analogue of Eq. (1.13),

$$\wp(y) = \int_{-\infty}^{\infty} \wp(y|x)\wp(x)dx. \tag{1.21}$$

Exercise 1.4 *Show that the mean and variance of the state $\wp(y)$ are $E[X]$ and $\mathrm{Var}[X] + \Delta^2$, respectively. This clearly shows the effect of the noise.*

Finding the conditional states in this case is difficult in general. However, it is greatly simplified if the a-priori system state is Gaussian:

$$\wp(x) = (2\pi\sigma^2)^{-1/2} \exp\left(-\frac{(x - \bar{x})^2}{2\sigma^2}\right), \tag{1.22}$$

because then the conditional states are still Gaussian.

Exercise 1.5 *Verify this, and show that the conditional mean and variance given a result y are, respectively,*

$$\bar{x}' = \frac{\sigma^2 y + \Delta^2 \bar{x}}{\Delta^2 + \sigma^2}, \qquad (\sigma')^2 = \frac{\sigma^2 \Delta^2}{\Delta^2 + \sigma^2}. \tag{1.23}$$

Hence show that, in the limit $\Delta \to 0$, the conditional state $\wp'(x|y)$ converges to $\delta(x - y)$, and an ideal measurement is recovered.

1.1.5 Most general formulation of classical measurements

As stated above, so far we have considered only non-disturbing classical measurements; that is, measurements with no *back-action* on the system. However, it is easy to consider classical measurements that do have a back-action on the system. For example, one could measure whether or not a can has petrol fumes in it by dropping a lit match inside. The result (nothing, or flames) will certainly reveal whether or not there *was* petrol inside the can, but the final state of the system after the measurement will have no petrol fumes inside in either case.

We can generalize Bayes' theorem to deal with this case by allowing a state-changing operation to act upon the state after applying Bayes' theorem. Say the system state is $\wp(x)$. For simplicity we will take X to be a discrete random variable, with the configuration space being $\{0, 1, \ldots, n - 1\}$. Say Y is the result of the measurement as usual. Then this state-changing operation is described by an $n \times n$ matrix $\mathcal{B}_y$, whose element $\mathcal{B}_y(x|x')$ is the probability that the measurement will cause the system to make a transition, from the state in which $X = x'$ to the state in which $X = x$, given that the result $Y = y$ was obtained. Thus, for all x' and all y,

$$\mathcal{B}_y(x|x') \geq 0, \qquad \sum_x \mathcal{B}_y(x|x') = 1. \tag{1.24}$$

The posterior system state is then given by

$$\wp'(x|y) = \frac{\sum_{x'} \mathcal{B}_y(x|x')\wp(y|x')\wp(x')}{\wp(y)}, \tag{1.25}$$

where the expression for $\wp(y)$ is unchanged from before.

We can unify the Bayesian part and the back-action part of the above expression by defining a new $n \times n$ matrix $\mathcal{O}_y$ with elements

$$\mathcal{O}_y(x|x') = \mathcal{B}_y(x|x')\wp(y|x'), \tag{1.26}$$

which maps a normalized probability distribution $\wp(x)$ onto an unnormalized probability distribution:

$$\tilde{\wp}'(x|y) = \sum_{x'} \mathcal{O}_y(x|x')\wp(x'). \tag{1.27}$$

Here we are introducing the convention of using a tilde to indicate an unnormalized state, with a norm of less than unity. This norm is equal to

$$\wp(y) = \sum_x \sum_{x'} \mathcal{O}_y(x|x')\wp(x'), \tag{1.28}$$

the probability of obtaining the result $Y = y$. Maps that take states to (possibly unnormalized) states are known as *positive maps*. The normalized conditional system state is

$$\wp'(x|y) = \sum_{x'} \mathcal{O}_y(x|x')\wp(x')/\wp(y). \tag{1.29}$$

From the properties of $\mathcal{O}_y$, it follows that it is possible to find an n-vector E_y with positive elements $E_y(x)$, such that the probability formula simplifies:

$$\sum_x \sum_{x'} \mathcal{O}_y(x|x')\wp(x') = \sum_x E_y(x)\wp(x). \tag{1.30}$$

Specifically, in terms of Eq. (1.26),

$$E_y(x) = \wp(y|x). \tag{1.31}$$

This satisfies the completeness condition

$$\forall x, \sum_y E_y(x) = 1. \tag{1.32}$$

This is the only mathematical restriction on $\{\mathcal{O}_y: y\}$ (apart from requiring that it be a positive map).

Exercise 1.6 *Formulate the match-in-the-tin measurement technique described above. Let the states $X = 1$ and $X = 0$ correspond to petrol fumes and no petrol fumes, respectively. Let the results $Y = 1$ and $Y = 0$ correspond to flames and no flames, respectively. Determine the two matrices $\mathcal{O}_{y:=0}$ and $\mathcal{O}_{y:=1}$ (each of which is a 2×2 matrix).*

The unconditional system state after the measurement is

$$\wp'(x) = \sum_y \sum_{x'} \mathcal{O}_y(x|x')\wp(x') = \sum_{x'} \mathcal{O}(x|x')\wp(x'). \tag{1.33}$$

Here the unconditional evolution map $\mathcal{O}$ is

$$\mathcal{O} = \sum_y \mathcal{O}_y. \tag{1.34}$$

Exercise 1.7 *Show that $\mathcal{O}$ is the identity if and only if there is no back-action.*

1.2 Quantum measurement theory

1.2.1 Probability and quantum mechanics

As we have discussed, with a classical system an ideal measurement can determine with certainty the values of all of the system variables. In this situation of complete knowledge, all subsequent ideal measurement results are determined with certainty. In consequence, measurement and probability do not play a significant role in the foundation of classical mechanics (although they do play a very significant role in practical applications of classical mechanics, where noise is inevitable [Whi96]).

The situation is very different in quantum mechanics. Here, for any sort of measurement, there are systems about which one has maximal knowledge, but for which the result of the measurement is not determined. The best one can do is to give the probability distributions for measurement outcomes. From this it might be inferred that a state of maximal knowledge about a quantum system is not a state of complete knowledge. That is, that there are 'hidden' variables about which one has incomplete knowledge, even when one has maximal knowledge, and these hidden variables determine the measurement outcomes.

Although it is possible to build a perfectly consistent interpretation of quantum mechanics based on this idea (see for example Ref. [BH93]), most physicists reject the idea. Probably the chief reason for this rejection is that in 1964 John Bell showed that any such deterministic hidden-variables theory must be nonlocal (that is, it must violate local causality) [Bel64]. That is, in such a theory, an agent with control over some local macroscopic parameters (such as the orientation of a magnet) can, under particular circumstances, instantaneously affect the hidden variables at an arbitrarily distant point in space. It can be shown that this effect cannot allow faster-than-light signalling, and hence does not lead to causal paradoxes within Einstein's theory of relativity. Nevertheless, it is clearly against the spirit of relativity theory. It should be noted, however, that this result is not restricted to hidden-variable interpretations; any interpretation of quantum mechanics that allows the concept of local causality to be formulated will be found to violate it – see Refs. [Bel87] (p. 172) and [Wis06].

Another, perhaps better, justification for ignoring hidden-variables theories is that there are infinitely many of them (see Ref. [GW04] and references therein). While some are more natural than others [Wis07], there is, at this stage, no compelling reason to choose one over all of the others. Thus one would be forced to make a somewhat arbitrary choice as to which hidden-variables interpretation to adopt, and each interpretation would have its own unique explanation as to the nature of quantum-mechanical uncertainty.

Rather than grappling with these difficulties, in this book we take an *operational* approach. That is, we treat quantum mechanics as simply an algorithm for calculating what one expects to happen when one performs a measurement. We treat uncertainty about future measurement outcomes as a primitive in the theory, rather than ascribing it to lack of knowledge about existing hidden variables.

We will still talk of a quantum state as representing our knowledge about a system, even though strictly it is our knowledge about the outcomes of our future measurements on that system. Also it is still useful, in many cases, to think of a quantum state of maximal knowledge as being like a classical state of incomplete knowledge about a system. Very crudely, this is the idea of 'quantum noise'. The reader who is not familiar with basic quantum mechanics (pure states, mixed states, time-evolution, entanglement etc.) should consult Appendix A for a summary of this material. However, before moving to quantum measurements, we note an important point of terminology. The matrix representation ρ of a mixed quantum state is usually called (for historical reasons) the density operator, or density matrix. We will call it the *state matrix*, because it generalizes the *state vector* for pure states.

Finally, just as in the classical case, observers with different knowledge may assign different states simultaneously to a single system. The most natural way to extend the concept of consistency to the quantum case is to replace the common state of maximal knowledge with a common pure state. That is, the condition for the consistency of a collection of states $\{\rho_j\}$ from different observers is that there exists a positive ϵ and a ket $|\psi\rangle$ such that, for all j, $\rho_j - \epsilon|\psi\rangle\langle\psi|$ is a positive operator. In other words, each observer's state ρ_j can be written as a mixture of the pure state $|\psi\rangle\langle\psi|$ and some other states.

Exercise 1.8 *Show, from this definition, that two different pure states $\hat{\pi}_1$ and $\hat{\pi}_2$ cannot be consistent states for any system.*
Hint: *Consider the operator $\hat{\sigma}_j = \hat{\pi}_j - \epsilon|\psi\rangle\langle\psi|$. Assuming $\hat{\pi}_j \neq |\psi\rangle\langle\psi|$, show that $\mathrm{Tr}[\sigma_j^2] > (\mathrm{Tr}[\sigma_j])^2$ and hence deduce the result.*

1.2.2 Projective measurements

The traditional description of measurement in quantum mechanics is in terms of projective measurements, as follows. Consider a measurement of the physical quantity Λ. First we note that the associated operator $\hat{\Lambda}$ (often called an *observable*) can be diagonalized as

$$\hat{\Lambda} = \sum_\lambda \lambda \hat{\Pi}_\lambda, \tag{1.35}$$

where $\{\lambda\}$ are the eigenvalues of $\hat{\Lambda}$ which are real and which we have assumed for convenience are discrete. $\hat{\Pi}_\lambda$ is called the projection operator, or projector, onto the subspace of eigenstates of $\hat{\Lambda}$ with eigenvalue λ. If the spectrum (set of eigenvalues $\{\lambda\}$) is non-degenerate, then the projector would simply be the rank-1 projector $\hat{\pi}_\lambda = |\lambda\rangle\langle\lambda|$. We will call this special case *von Neumann measurements*.

In the more general case, where the eigenvalues of $\hat{\Lambda}$ are N_λ-fold degenerate, $\hat{\Pi}_\lambda$ is a rank-N_λ projector, and can be written as $\sum_{j=1}^{N_\lambda} |\lambda, j\rangle\langle\lambda, j|$. For example, in the simplest model of the hydrogen atom, if Λ is the energy then λ would be the principal quantum number n and j would code for the angular-momentum and spin quantum numbers l, m and s of states with the same energy. The projectors are orthonormal, obeying

$$\hat{\Pi}_\lambda \hat{\Pi}_{\lambda'} = \delta_{\lambda,\lambda'} \hat{\Pi}_\lambda. \tag{1.36}$$

The existence of this orthonormal basis is a consequence of the spectral theorem (see Box 1.1).

When one measures Λ, the result one obtains is one of the eigenvalues λ. Say the measurement begins at time t and takes a time T. Assuming that the system does not evolve significantly from other causes during the measurement, the probability for obtaining that particular eigenvalue is

$$\Pr[\Lambda(t) = \lambda] = \wp_\lambda = \mathrm{Tr}[\rho(t)\hat{\Pi}_\lambda]. \tag{1.37}$$

After the measurement, the conditional (a-posteriori) state of the system given the result λ is

$$\rho_\lambda(t + T) = \frac{\hat{\Pi}_\lambda \rho(t) \hat{\Pi}_\lambda}{\Pr[\Lambda(t) = \lambda]}. \tag{1.38}$$

That is to say, the final state has been projected by $\hat{\Pi}_\lambda$ into the corresponding subspace of the total Hilbert space. This is known as the *projection postulate*, or sometimes as *state collapse*, or *state reduction*. The last term is best avoided, since it invites confusion with the reduced state of a bipartite system as discussed in Section A.2.2 of Appendix A. This process

---

Box 1.1 Spectral theorem

The spectral theorem states that any *normal operator* $\hat{N}$ has a complete set of eigenstates that are orthonormal. A normal operator is an operator such that $[\hat{N}, \hat{N}^\dagger] = 0$. That is, for every such $\hat{N}$ there is a basis $\{|v, j\rangle: v, j\}$ for the Hilbert space such that, with $v \in \mathbb{C}$,

$$\hat{N}|v, j\rangle = v|v, j\rangle.$$

Here the extra index j is necessary because the eigenvalues v may be degenerate. The general diagonal form of $\hat{N}$ is

$$\hat{N} = \sum_v v\hat{\Pi}_v,$$

where the $\hat{\Pi}_v = \sum_j |v, j\rangle\langle v, j|$ form a set of orthogonal projectors obeying

$$\hat{\Pi}_v\hat{\Pi}_{v'} = \delta_{v,v'}\hat{\Pi}_v.$$

Hermitian operators (for which $\hat{N} = \hat{N}^\dagger$) are a special class of normal operators. It can be shown that a normal operator that is not Hermitian can be written in the form $\hat{N} = \hat{R} + i\hat{H}$, where $\hat{R}$ and $\hat{H}$ are commuting Hermitian operators. For any two operators that commute, there is some complete basis comprising states that are eigenstates of both operators, which in this case will be a basis $|v, j\rangle$ diagonalizing $\hat{N}$. As will be discussed in Section 1.2.2, operators that share eigenstates are simultaneously measurable.

Thus it is apparent that a non-Hermitian normal operator is really just a compact way to represent two simultaneously observable quantities by having eigenvalues v in the complex plane rather than the real line. By considering vectors or other multi-component objects, any number of commuting operators can be combined to represent the corresponding simultaneously observable quantities. This demonstrates that, for projective quantum measurement theory, the important thing is not an operator representing the observables, but rather the projector $\hat{\Pi}_r$ corresponding to a result r.

---

should be compared to the classical Bayesian update rule, Eq. (1.12). A consequence of this postulate is that, if the measurement is immediately repeated, then

$$\Pr[\Lambda(t + T) = \lambda'|\Lambda(t) = \lambda] = \mathrm{Tr}[\rho_\lambda(t + T)\hat{\Pi}_{\lambda'}] = \delta_{\lambda',\lambda}. \tag{1.39}$$

That is to say, the same result is guaranteed. Moreover, the system state will not be changed by the second measurement. For a deeper understanding of the above theory, see Box 1.2.

For pure states, $\rho(t) = |\psi(t)\rangle\langle\psi(t)|$, the formulae (1.37) and (1.38) can be more simply expressed as

$$\Pr[\Lambda(t) = \lambda] = \wp_\lambda = \langle\psi(t)|\hat{\Pi}_\lambda|\psi(t)\rangle \tag{1.40}$$

Box 1.2 Gleason's theorem

It is interesting to ask how much of quantum measurement theory one can derive from assuming that quantum measurements are described by a complete set of projectors, one for each result r. Obviously there must be some rule for obtaining a probability $\wp_r$ from a projector $\hat{\Pi}_r$, such that $\sum_r \wp_r = 1$. Gleason [Gle57] proved that, if one considers measurements with at least three outcomes (requiring a Hilbert-space dimension of at least three), then it follows that there exists a non-negative operator ρ of unit trace such that

$$\wp_r = \mathrm{Tr}[\hat{\Pi}_r \rho].$$

That is, the probability rule (1.37) can be derived, not assumed.

It must be noted, however, that Gleason required an additional assumption: *non-contextuality*. This means that $\wp_r$ depends only upon $\hat{\Pi}_r$, being independent of the other projectors which complete the set. That is, if two measurements each have one outcome represented by the same projector $\hat{\Pi}_r$, the probabilities for those outcomes are necessarily the same, even if the measurements cannot be performed simultaneously.

Gleason's theorem shows that the state matrix ρ is a consequence of the structure of Hilbert space, if we require probabilities to be assigned to projection operators. It suggests that, rather than introducing pure states and then generalizing to mixed states, the state matrix ρ can be taken as fundamental.

---

and

$$|\psi_\lambda(t+T)\rangle = \hat{\Pi}_\lambda |\psi(t)\rangle / \sqrt{\wp_\lambda}. \tag{1.41}$$

However, if one wishes to describe the unconditional state of the system (that is, the state *if one makes the measurement, but ignores the result*) then one must use the state matrix:

$$\rho(t+T) = \sum_\lambda \Pr[\Lambda(t) = \lambda]\rho_\lambda(t+T) = \sum_\lambda \hat{\Pi}_\lambda \rho(t)\hat{\Pi}_\lambda. \tag{1.42}$$

Thus, if the state were pure at time t, and we make a measurement, but ignore the result, then in general the state at time $t + T$ will be mixed. That is, projective measurement, unlike unitary evolution,[2] is generally an entropy-increasing process unless one keeps track of the measurement results. This is in contrast to non-disturbing measurements in classical mechanics, where (as we have seen) the unconditional a-posteriori state is identical to the a-priori state (1.17).

Exercise 1.9 *Show that a projective measurement of Λ decreases the purity $\mathrm{Tr}\left[\rho^2\right]$ of the unconditional state unless the a-priori state $\rho(t)$ can be diagonalized in the same basis as can $\hat{\Lambda}$.*

---

[2] Of course unitary evolution *can* change the entropy of a *sub*system, as we will discuss in Chapter 3.

Hint: *Let* $p_{\lambda\lambda'} = \text{Tr}[\hat{\Pi}_\lambda \rho(t)\hat{\Pi}_{\lambda'}\rho(t)]$ *and show that* $\forall \lambda, \lambda',\ p_{\lambda\lambda'} \geq 0$. *Then express* $\text{Tr}[\rho(t)^2]$ *and* $\text{Tr}[\rho(t+T)^2]$ *in terms of these* $p_{\lambda\lambda'}$.

From the above measurement theory, it is simple to show that the mean value for the result Λ is

$$
\begin{aligned}
\langle \Lambda \rangle &= \sum_\lambda \Pr[\Lambda = \lambda]\lambda \\
&= \sum_\lambda \text{Tr}[\rho \hat{\Pi}_\lambda]\lambda \\
&= \text{Tr}\left[\rho \left(\sum_\lambda \lambda \hat{\Pi}_\lambda\right)\right] = \text{Tr}[\rho \hat{\Lambda}].
\end{aligned} \tag{1.43}
$$

Here we are using angle brackets as an alternative notation for expectation value when dealing with quantum observables.

Exercise 1.10 *Using the same technique, show that*

$$
\langle \Lambda^2 \rangle = \sum_\lambda \lambda^2 \Pr[\Lambda = \lambda] = \text{Tr}\left[\rho \hat{\Lambda}^2\right]. \tag{1.44}
$$

Thus the mean value (A.6) and variance (A.8) can be derived rather than postulated, provided that they are interpreted in terms of the moments of the results of a projective measurement of Λ.

Continuous spectra. The above results can easily be generalized to treat physical quantities with a continuous spectrum, such as the position $\hat{X}$ of a particle on a line. Considering this non-degenerate case for simplicity, the spectral theorem becomes

$$
\hat{X} = \int_{-\infty}^{\infty} x \hat{\Pi}(x)\mathrm{d}x = \int_{-\infty}^{\infty} x|x\rangle\langle x|\mathrm{d}x. \tag{1.45}
$$

Note that $\hat{\Pi}(x)$ is not strictly a projector, but a projector density, since the orthogonality conditions are

$$
\hat{\Pi}(x)\hat{\Pi}(x') = \delta(x - x')\hat{\Pi}(x), \tag{1.46}
$$

or, in terms of the unnormalizable *improper states* $|x\rangle$,

$$
\langle x|x'\rangle = \delta(x - x'). \tag{1.47}
$$

These states are discussed in more detail in Appendix A.

The measurement outcomes are likewise described by probability densities. For example, if the system is in a pure state $|\psi(t)\rangle$, the probability that an ideal measurement of position gives a result between x and $x + \mathrm{d}x$ is

$$
\wp(x)\mathrm{d}x = \text{Tr}\left[|\psi(t)\rangle\langle\psi(t)|\hat{\Pi}(x)\right]\mathrm{d}x = |\psi(x,t)|^2\,\mathrm{d}x. \tag{1.48}
$$

Here we have defined the wavefunction $\psi(x,t) = \langle x|\psi(t)\rangle$. Unfortunately it is not possible to assign a proper a-posteriori state to the system, since $|x\rangle$ is unnormalizable. This problem can be avoided by considering an approximate measurement of position with finite accuracy Δ, as will always be the case in practice. This can still be described as a projective measurement, for example by using the (now discrete) set of projectors

$$\hat{\Pi}_j = \int_{x_j}^{x_{j+1}} \hat{\Pi}(x)\mathrm{d}x, \tag{1.49}$$

where, for all j, $x_{j+1} = x_j + \Delta$.

Exercise 1.11 *Show that the $\hat{\Pi}_j$ defined here form an orthonormal set.*

Simultaneous measurements. Heisenberg's uncertainty relation (see Exercise A.3) shows that it is impossible for both the position and the momentum of a particle to be known exactly (have zero variance). This is often used as an argument for saying that it is impossible simultaneously to measure position and momentum. We will see in the following sections that this is not strictly true. Nevertheless, it is the case that it is impossible to carry out a simultaneous *projective* measurement of position and momentum, and this is the case of interest here.

For two quantities A and B to be measurable simultaneously, it is sufficient (and necessary) for them to be measurable consecutively, such that the joint probability of the results a and b does not depend on the order of the measurement. Considering a system in a pure state for simplicity, we thus require for all a and b and all $|\psi\rangle$ that

$$\hat{\Pi}_a\hat{\Pi}_b|\psi\rangle = \mathrm{e}^{\mathrm{i}\theta}\hat{\Pi}_b\hat{\Pi}_a|\psi\rangle, \tag{1.50}$$

for some θ. By considering the norm of these two vectors (which must be equal) it can be seen that $\mathrm{e}^{\mathrm{i}\theta}$ must equal unity.

Exercise 1.12 *Prove this, and hence that*

$$\forall a, b, \ [\hat{\Pi}_a, \hat{\Pi}_b] = 0. \tag{1.51}$$

This is equivalent to the condition that $[\hat{A}, \hat{B}] = 0$, and means that there is a basis, say $\{|\phi_k\rangle\}$, in which both $\hat{A}$ and $\hat{B}$ are diagonal. That is,

$$\hat{A} = \sum_k a_k|\phi_k\rangle\langle\phi_k|, \qquad \hat{B} = \sum_k b_k|\phi_k\rangle\langle\phi_k|, \tag{1.52}$$

where the eigenvalues a_k may be degenerate (that is, there may exist k and k' such that $a_k = a_{k'}$) and similarly for b_k. Thus, one way of making a simultaneous measurement of A and B is to make a measurement of $\hat{K} = \sum_k k|\phi_k\rangle\langle\phi_k|$, and from the result k determine the appropriate values a_k and b_k for A and B.

1.2.3 Systems and meters

The standard (projective) presentation of quantum measurements is inadequate for a number of reasons. A prosaic, but very practical, reason is that very few measurements can be made in such a way that the apparatus adds no classical noise to the measurement result. A more interesting reason is that there are many measurements in which the a-posteriori conditional system state is clearly not left in the eigenstate of the measured quantity corresponding to the measurement result. For example, in photon counting by a photodetector, at the end of the measurements all photons have been absorbed, so that the system (e.g. the cavity that originally contained the photons) is left in the vacuum state, not a state containing the number n of photons counted. Another interesting reason is that non-projective measurements allow far greater flexibility than do projective measurements. For example, the simultaneous measurement of position and momentum is a perfectly acceptable idea, so long as the respective accuracies do not violate the Heisenberg uncertainty principle, as we will discuss below.

The fundamental reason why projective measurements are inadequate for describing real measurements is that experimenters never directly measure the system of interest. Rather, the system of interest (such as an atom) interacts with its environment (the continuum of electromagnetic field modes), and the experimenter observes the effect of the system on the environment (the radiated field). Of course, one could argue that the experimenter does not observe the radiated field, but rather that the field interacts with a photodetector, which triggers a current in a circuit, which is coupled to a display panel, which radiates more photons, which interact with the experimenter's retina, and so on. Such a chain of systems is known as a von Neumann chain [vN32]. The point is that, at some stage before reaching the mind of the observer, one has to cut the chain by applying the projection postulate. This cut, known as Heisenberg's cut [Hei30], is the point at which one considers the measurement as having been made.

If one were to apply a projection postulate directly to the atom, one would obtain wrong predictions. However, assuming a projective measurement of the field will yield results negligibly different from those obtained assuming a projective measurement at any later stage. This is because of the rapid decoherence of macroscopic material objects such as photodetectors (see Chapter 3). For this reason, it is sufficient to consider the field to be measured projectively. Because the field has interacted with the system, their quantum states are correlated (indeed, they are entangled, provided that their initial states are pure enough). The projective measurement of the field is then effectively a measurement of the atom. The latter measurement, however, is not projective, and we need a more general formalism to describe it.

Let the initial system state vector be $|\psi(t)\rangle$, and say that there is a second quantum system, which we will call the *meter*, or *apparatus*, with the initial state $|\theta(t)\rangle$. Thus the initial (unentangled) combined state is

$$|\Psi(t)\rangle = |\theta(t)\rangle|\psi(t)\rangle. \tag{1.53}$$

Let these two systems be coupled together for a time T_1 by a unitary evolution operator $\hat{U}(t + T_1, t)$, which we will write as $\hat{U}(T_1)$. Thus the combined system–meter state after this coupling is

$$|\Psi(t + T_1)\rangle = \hat{U}(T_1)|\theta(t)\rangle|\psi(t)\rangle. \tag{1.54}$$

This cannot in general be written in the factorized form of Eq. (1.53).

Now let the meter be measured projectively over a time interval T_2, and say $T = T_1 + T_2$. We assume that the evolution of the system and meter over the time T_2 is negligible (this could be either because $T_2 \ll T_1$, or because the coupling Hamiltonian is time-dependent). Let the projection operators for the meter be rank-1 operators, so that $\hat{\Pi}_r = \hat{\pi}_r \otimes \hat{1}$. The order of the tensor product is meter then system, as in Eq. (1.53), and $\hat{\pi}_r = |r\rangle\langle r|$. Here r denotes the value of the observed quantity R. The set $\{|r\rangle\}$ forms an orthonormal basis for the meter Hilbert space. Then the final combined state is

$$|\Psi_r(t + T)\rangle = \frac{|r\rangle\langle r|\hat{U}(T_1)|\theta(t)\rangle|\psi(t)\rangle}{\sqrt{\wp_r}}, \tag{1.55}$$

where the probability of obtaining the value r for the result R is

$$\Pr[R = r] = \wp_r = \langle\psi(t)|\langle\theta(t)|\hat{U}^\dagger(T_1)[|r\rangle\langle r| \otimes \hat{1}]\hat{U}(T_1)|\theta(t)\rangle|\psi(t)\rangle. \tag{1.56}$$

The measurement on the meter disentangles the system and the meter, so that the final state (1.55) can be written as

$$|\Psi_r(t + T)\rangle = |r\rangle\frac{\hat{M}_r|\psi(t)\rangle}{\sqrt{\wp_r}}, \tag{1.57}$$

where $\hat{M}_r$ is an operator that acts only in the system Hilbert space, defined by

$$\hat{M}_r = \langle r|\hat{U}(T_1)|\theta(t)\rangle. \tag{1.58}$$

We call it a *measurement operator*. The probability distribution (1.56) for R can similarly be written as

$$\wp_r = \langle\psi(t)|\hat{M}_r^\dagger\hat{M}_r|\psi(t)\rangle. \tag{1.59}$$

1.2.4 Example: binary measurement

To understand the ideas just introduced, it is helpful to consider a specific example. We choose one analogous to the classical discrete binary measurement discussed in Section 1.1.2. The quantum analogue of a system with a single binary system variable is a quantum system in a two-dimensional Hilbert space. Let $\{|x\rangle: x = 0, 1\}$ be an orthonormal basis for this Hilbert space. An obvious physical realization is a spin-half particle. The spin in any direction is restricted to one of two possible values, $\pm\hbar/2$. These correspond to the spin being up $(+)$ or down $(-)$ with respect to the given direction. Choosing a particular direction (z is conventional), we label these states as $|0\rangle$ and $|1\rangle$, respectively. Other physical realizations include an atom with only two relevant levels, or a single electromagnetic cavity mode containing no photon or one photon. The latter two examples will be discussed in detail in Section 1.5.

Now consider a measured system S and a measurement apparatus A, both described by two-dimensional Hilbert spaces. We will use x for the system states and y for the apparatus states. Following the above formalism, we assume that initially both systems are in pure states, so that the joint state of the system at time t is

$$|\Psi(t)\rangle = |\theta(t)\rangle|\psi(t)\rangle = \sum_{x,\xi} s_x a_\xi |y := \xi\rangle|x\rangle. \tag{1.60}$$

Note that we have used an analogous notation to the classical case, so that $|y := \xi\rangle$ is the apparatus state $|y\rangle$ with y taking the value ξ. To make a measurement, the system and apparatus states must become correlated. We will discuss how this may take place physically in Section 1.5. For now we simply postulate that, as a result of the unitary interaction between the system and the apparatus, we have

$$|\Psi(t + T_1)\rangle = \hat{G}|\Psi(t)\rangle = \sum_{x,\xi} a_\xi s_x |y := G(x, \xi)\rangle|x\rangle, \tag{1.61}$$

where $\hat{G}$ is a unitary operator defined by

$$\hat{G}|y := \xi\rangle|x\rangle = |y := G(x, \xi)\rangle|x\rangle. \tag{1.62}$$

Note that the interaction between the system and the apparatus has been specified by reference to a particular basis for the system and apparatus, $\{|y\rangle|x\rangle\}$. We will refer to this (for the system, or apparatus, or both together) as the *measurement basis*.

Exercise 1.13 *Show that $\hat{G}$ as defined is unitary if there exists an inverse function G^{-1} in the sense that, for all y, $y = G(x, G^{-1}(x, y))$.*
Hint: *Show that $\hat{G}^\dagger \hat{G} = \hat{1} = \hat{G}\hat{G}^\dagger$ using the matrix representation in the measurement basis.*

The invertibility condition is the same as we used in Section 1.1.3 for the classical binary measurement model.

As an example, consider $G(x, \xi) = x \oplus \xi$, as in the classical case, where again this indicates binary addition. In this case $\hat{G} = \hat{G}^{-1}$. The system state is unknown and is thus arbitrary. However, the apparatus is assumed to be under our control and can be prepared in a *fiducial* state. This means a standard state for the purpose of measurement. Often the fiducial state is a particular state in the measurement basis, and we will assume that it is $|y := 0\rangle$, so that $a_\xi = \delta_{\xi,0}$. In this case the state after the interaction is

$$|\Psi(t + T_1)\rangle = \sum_x s_x |y := x\rangle|x\rangle \tag{1.63}$$

and there is a perfect correlation between the system and the apparatus. Let us say a projective measurement (of duration T_2) of the apparatus state in the measurement basis is made. This will give the result y with probability $|s_y|^2$, that is, with exactly the probability that a projective measurement directly on the system in the measurement basis would have given. Moreover, the conditioned system state at time $t + T$ (where $T = T_1 + T_2$ as above), given the result y, is

$$|\psi_y(t + T)\rangle = |x := y\rangle. \tag{1.64}$$

Again, this is as would have occurred with the appropriate projective measurement of duration T on the system, as in Eq. (1.41).

This example is a special case of a model introduced by von Neumann. It would appear to be simply a more complicated version of the description of standard projective measurements. However, as we now show, it enables us to describe a more general class of measurements in which extra noise appears in the result due to the measurement apparatus.

Suppose that for some reason it is not possible to prepare the apparatus in one of the measurement basis states. In that case we must use the general result given in Eq. (1.61). Using Eq. (1.58), we find

$$
\begin{aligned}
\hat{M}_y|\psi(t)\rangle &= \langle y|\Psi(t+T_1)\rangle \\
&= \sum_{x,\xi} \delta_{y,G(x,\xi)} a_\xi s_x |x\rangle \\
&= \sum_x a_{G^{-1}(x,y)} s_x |x\rangle \\
&= \sum_{x'} a_{G^{-1}(x',y)} |x'\rangle\langle x'| \sum_x s_x |x\rangle.
\end{aligned}
\tag{1.65}
$$

Thus we have the measurement operator

$$
\hat{M}_y = \sum_x a_{G^{-1}(x,y)} |x\rangle\langle x|.
\tag{1.66}
$$

For the particular case $G^{-1}(x, y) = x \oplus y$, this simplifies to

$$
\hat{M}_y = \sum_\xi a_\xi |x := y \oplus \xi\rangle\langle x := y \oplus \xi|.
\tag{1.67}
$$

Returning to the more general form of Eq. (1.65), we find that the probability for the result y is

$$
\wp(y) = \langle\psi(t)|\hat{M}_y^\dagger \hat{M}_y|\psi(t)\rangle = \sum_x |s_x|^2 |a_{G^{-1}(x,y)}|^2.
\tag{1.68}
$$

If we define

$$
\wp(\xi) = |a_\xi|^2,
\tag{1.69}
$$

$$
\wp(x) = |s_x|^2 = \mathrm{Tr}[\rho(t)|x\rangle\langle x|],
\tag{1.70}
$$

where $\rho(t) = |\psi(t)\rangle\langle\psi(t)|$ is the system state matrix, then the probability distribution for measurement results may then be written as

$$
\wp(y) = \sum_x \wp(\xi := G^{-1}(x, y))\wp(x).
\tag{1.71}
$$

This is the same form as for the classical binary measurement scheme; see Eq. (1.13) and Eq. (1.15). Here the noise distribution arises from *quantum noise* associated with the fiducial (purposefully prepared) apparatus state. It is quantum noise because the initial apparatus state is still a pure state. The noise arises from the fact that it is not prepared in one of the measurement basis states. Of course, the apparatus may be prepared in a mixed state, in which case the noise added to the measurement result may have a classical origin. This is discussed below in Section 1.4.

The system state conditioned on the result y is

$$|\psi_y(t+T)\rangle = \hat{M}_y|\psi(t)\rangle/\sqrt{\wp(y)} = \sum_x a_{G^{-1}(x,y)} s_x |x\rangle/\sqrt{\wp(y)}. \tag{1.72}$$

If, from this, we calculate the probability $|\langle x|\psi_y(t+T)\rangle|^2$ for the system to have $X = x$ after the measurement giving the result y, we find this probability to be given by

$$\wp'(x|y) = \frac{\wp(y|x)\wp(x)}{\wp(y)}. \tag{1.73}$$

Again, this is the same as the classical result derived using Bayes' theorem. The interesting point is that the projection postulate does that work for us in the quantum case. Moreover, it gives us the full a-posteriori conditional state, from which the expectation value of any observable (not just X) can be calculated. The quantum measurement here is thus more than simply a reproduction of the classical measurement, since the conditional state (1.72) cannot be derived from Bayes' theorem.

Exercise 1.14 *Consider two infinite-dimensional Hilbert spaces describing a system and a meter. Show that the operator $\hat{G}$, defined in the joint position basis $|y\rangle|x\rangle$ by*

$$\hat{G}|y := \xi\rangle|x\rangle = |y := \xi + x\rangle|x\rangle, \tag{1.74}$$

is unitary. Let the fiducial apparatus state be

$$|\theta\rangle = \int_{-\infty}^{\infty} d\xi \left[(2\pi\Delta^2)^{-1/2} \exp(-\xi^2/(2\Delta^2)) \right]^{1/2} |y := \xi\rangle. \tag{1.75}$$

Following the example of this subsection, show that, insofar as the statistics of $\hat{X}$ are concerned, this measurement is equivalent to the classical measurement analysed in Section 1.1.4.

1.2.5 Measurement operators and effects

As discussed in Section 1.2.3, the system and apparatus are no longer entangled at the end of the measurement. Thus it is not necessary to continue to include the meter in our description of the measurement. Rather we can specify the measurement completely in terms of the measurement operators $\hat{M}_r$. The conditional state of the system, given that the result R has the value r, after a measurement of duration T, is

$$|\psi_r(t+T)\rangle = \frac{\hat{M}_r|\psi(t)\rangle}{\sqrt{\wp_r}}. \tag{1.76}$$

As seen above, the probabilities are given by the expectation of another operator, defined in terms of the measurement operators by

$$\hat{E}_r = \hat{M}_r^\dagger \hat{M}_r. \tag{1.77}$$

These operators are known as *probability operators*, or *effects*. The fact that $\sum_r \wp_r$ must equal unity for all initial states gives a *completeness condition* on the measurement operators:

$$\sum_r \hat{E}_r = \hat{1}_S. \tag{1.78}$$

This restriction, that $\{\hat{E}_r: r\}$ be a *resolution of the identity* for the system Hilbert space, is the only restriction on the set of measurement operators (apart from the fact that they must be positive, of course).

The set of all effects $\{\hat{E}_r: r\}$ constitutes an *effect-valued measure* more commonly known as a *probability-operator-valued measure* (POM[3]) on the space of results r. This simply means that, rather than a probability distribution (or probability-valued measure) over the space of results, we have a probability-operator-valued measure. Note that we have left behind the notion of 'observables' in this formulation of measurement. The possible measurement results r are not the eigenvalues of an Hermitian operator representing an observable; they are simply labels representing possible results. Depending on the circumstances, it might be convenient to represent the result R by an integer, a real number, a complex number, or an even more exotic quantity.

If one were making only a single measurement, then the conditioned state $|\psi_r\rangle$ would be irrelevant. However, one often wishes to consider a sequence of measurements, in which case the conditioned system state is vital. In terms of the state matrix ρ, which allows the possibility of mixed initial states, the conditioned state is

$$\rho_r(t + T) = \frac{\mathcal{J}[\hat{M}_r]\rho(t)}{\wp_r}, \tag{1.79}$$

where $\wp_r = \mathrm{Tr}[\rho(t)\hat{E}_r]$ and, for arbitrary operators A and B,

$$\mathcal{J}[\hat{A}]\hat{B} \equiv \hat{A}\hat{B}\hat{A}^\dagger. \tag{1.80}$$

The *superoperator*

$$\mathcal{O}_r = \mathcal{J}[\hat{M}_r] \tag{1.81}$$

is known as the *operation* for r. It is called a superoperator because it takes an operator (here ρ) to another operator. Operations can be identified with the class of superoperators that take physical states to physical states. (See Box 1.3.) This very important class is also known as *completely positive maps*.

---

[3] The abbreviation POVM is used also, and, in both cases, PO is sometimes understood to denote 'positive operator' rather than 'probability operator'.

Box 1.3 Superoperators and operations

A superoperator $\mathcal{S}$ is an operator on the space of Hilbert-space operators:

$$\hat{A} \rightarrow \hat{A}' = \mathcal{S}\hat{A}. \tag{1.82}$$

A superoperator $\mathcal{S}$ must satisfy three conditions in order to correspond to a physical processes (such as measurement or dynamics).

1. $\mathcal{S}$ is trace-preserving or decreasing. That is, $0 \leq \mathrm{Tr}[\mathcal{S}\rho] \leq 1$ for any state ρ. Moreover, $\mathrm{Tr}[\mathcal{S}\rho]$ is the probability that the process occurs.
2. $\mathcal{S}$ is a convex linear map on operators. That is, for probabilities $\wp_j$, we have that $\mathcal{S}\sum_j \wp_j \rho_j = \sum_j \wp_j \mathcal{S}\rho_j$.
3. $\mathcal{S}$ is *completely positive*. That is, not only does $\mathcal{S}$ map positive operators to positive operators for the system of interest S, but so does $(\mathcal{I} \otimes \mathcal{S})$. Here $\mathcal{I}$ is the identity superoperator for an arbitrary second system R.

The final property deserves some comment. It might have been thought that positivity of a superoperator would be sufficient to represent a physical process. However, it is always possible that a system S is entangled with another system R before the physical process represented by $\mathcal{S}$ acts on system S. It must still be the case that the total state of both systems remains a physical state with a positive state matrix. This gives condition 3.

If a superoperator satisfies these three properties then it is called an *operation*, and has the Kraus representation [Kra83], or *operator sum representation*,

$$\mathcal{S}(\rho) = \sum_j \hat{K}_j \rho \hat{K}_j^\dagger \tag{1.83}$$

for some set of operators $\hat{K}_j$ satisfying

$$\hat{1} - \sum_j \hat{K}_j^\dagger \hat{K}_j \geq 0. \tag{1.84}$$

There is another important representation theorem for operations, which follows from the Gelfand–Naimark–Segal theorem [Con90]. Consider, as above, an apparatus or ancilla system A in addition to the quantum system of interest S. Then there is a pure state $|\theta\rangle_A$ of A and some unitary evolution, $\hat{U}_{SA}$, describing the coupling of system S to system A, such that

$$\mathcal{S}\rho_S = \mathrm{Tr}_A\left[(\hat{1}_S \otimes \hat{\Pi}_A)\hat{U}_{SA}(\rho_S \otimes |\theta\rangle_A\langle\theta|)\hat{U}_{SA}^\dagger\right], \tag{1.85}$$

where $\hat{\Pi}_A$ is some projector for the ancilla system A. This is essentially the converse of the construction of operations for measurements from a system–apparatus coupling in Section 1.2.3.

If the measurement were performed but the result R ignored, the final state of the system would be

$$\rho(t+T) = \sum_r \wp_r \rho_r(t+T) = \sum_r \mathcal{J}[\hat{M}_r]\rho(t) \equiv \mathcal{O}\rho(t). \tag{1.86}$$

Here $\mathcal{O}$ is also an operation, and is trace-preserving.

For non-projective measurements, there is no guarantee that repeating the measurement will yield the same result. In fact, the final state of the system may be completely unrelated to either the initial state of the system or the result obtained. This is best illustrated by an example.

Example 1. Consider the set of measurement operators $\{\hat{M}_r\}$ defined by $\hat{M}_r = |0\rangle\langle r|$, where $r \in \{0, 1, 2, \ldots\}$ and $\{|r\rangle\}$ is a complete basis for the system Hilbert space. Then the effects for the measurements are projectors $\hat{E}_r = \hat{\Pi}_r = |r\rangle\langle r|$, which obviously obey the completeness condition (1.78). The probability of obtaining $R = r$ is just $\langle r|\rho(t)|r\rangle$. However, the final state of the system, regardless of the result r, is $\rho_r(t+T) = |0\rangle\langle 0|$. Lest it be thought that this is an artificial example, it in fact arises very naturally from counting photons. There R is the number of photons, and, because photons are typically absorbed in order to be counted, the number of photons left after the measurement has finished is zero.

In the above example, the effects are still projection operators. However, there are other measurements in which this is not the case.

Example 2. Consider a two-dimensional Hilbert space with the basis $|0\rangle$, $|1\rangle$. Consider a *continuous* measurement result ϕ that can take values between 0 and 2π. We define the measurement operators $\hat{M}_\phi = |\phi\rangle\langle\phi|/\sqrt{\pi}$, where $|\phi\rangle$ is defined by

$$|\phi\rangle = [|0\rangle + \exp(i\phi)|1\rangle]/\sqrt{2}. \tag{1.87}$$

In this case the effects are

$$\hat{E}_\phi = \frac{1}{\pi}|\phi\rangle\langle\phi|, \tag{1.88}$$

and the completeness condition, which is easy to verify, is

$$\int_0^{2\pi} \mathrm{d}\phi\, \hat{E}_\phi = |0\rangle\langle 0| + |1\rangle\langle 1| = \hat{1}. \tag{1.89}$$

Although $\hat{E}_\phi$ is proportional to a projection operator it is not equal to one. It does not square to itself: $(\hat{E}_\phi\, \mathrm{d}\phi)^2 = \hat{E}_\phi\, \mathrm{d}\phi(\mathrm{d}\phi/\pi)$. Neither are different effects orthogonal in general: $\hat{E}_\phi \hat{E}_{\phi'} \neq 0$ unless $\phi' = \phi + \pi$. Thus, even if the system is initially in the state $|\phi\rangle$, there is a finite probability for any result to be obtained except $\phi + \pi$.

The effects $\hat{E}_r$ need not even be proportional to projectors, as the next example shows.

Example 3. Consider again an infinite-dimensional Hilbert space, but now use the continuous basis $|x\rangle$ (see Section 1.2.2 and Appendix A), for which $\langle x|x'\rangle = \delta(x - x')$. Define an effect

$$\hat{E}_y = \int_{-\infty}^{\infty} dx (2\pi \Delta^2)^{-1/2} \exp[-(y - x)^2/(2\Delta^2)]|x\rangle\langle x|. \qquad (1.90)$$

This describes an imprecise measurement of position. It is easy to verify that the effects are not proportional to projectors by showing that $\hat{E}_y^2$ is not proportional to $\hat{E}_y$. Nevertheless, they are positive operators and obey the completeness relation

$$\int_{-\infty}^{\infty} dy \, \hat{E}_y = \hat{1}. \qquad (1.91)$$

Exercise 1.15 *Verify Eq. (1.91). Also show that these effects can be derived from the measurement model introduced in Exercise 1.14.*

The previous examples indicate some of the flexibility that arises from not requiring the effects to be projectors. As mentioned above, another example of the power offered by generalized measurements is the simultaneous measurement of position $\hat{X}$ and momentum $\hat{P}$. This is possible provided that the two measurement results have a certain amount of error. A simple model for this was first described by Arthurs and Kelly [AK65]. A more abstract description directly in terms of the resulting projection valued measure was given by Holevo [Hol82]. The description given below is based on the discussion in [SM01].

Example 4. The model of Arthurs and Kelly consists of two meters that are allowed to interact instantaneously with the system. The interaction couples one of the meters to position and the other to momentum, encoding the results of the measurement in the final states of the meters. Projective measurements are then made on each of the meter states separately. These measurements can be carried out simultaneously since operators for distinct meters commute. For appropriate meter states, this measurement forces the conditional state of the system into a Gaussian state (defined below). We assume some appropriate length scale such that the positions and momenta for the system are dimensionless, and satisfy $[\hat{X}, \hat{P}] = i$.

The appropriate unitary interaction is

$$\hat{U} = \exp\left[-i\left(\hat{X}\hat{P}_1 + \hat{P}\hat{P}_2\right)\right]. \qquad (1.92)$$

Here the subscripts refer to the two detectors, which are initially in minimum-uncertainty states (see Appendix A) $|d_1\rangle$ and $|d_2\rangle$, respectively. Specifically, we choose the wavefunctions in the position representation to be

$$\langle x_j|d_j\rangle = (\pi/2)^{-1/4} e^{-x_j^2}. \qquad (1.93)$$

After the interaction, the detectors are measured in the position basis. The measurement result is thus the pair of numbers (X_1, X_2). Following the theory given above, the *measurement operator* for this result is

$$\hat{M}(x_1, x_2) = \langle x_1 | \langle x_2 | \hat{U} | d_2 \rangle | d_1 \rangle. \tag{1.94}$$

With a little effort it is possible to show that $\hat{M}(x_1, x_2)$ is proportional to a projection operator:

$$\hat{M}(x_1, x_2) = \frac{1}{\sqrt{2\pi}} |(x_1, x_2)\rangle \langle (x_1, x_2)|. \tag{1.95}$$

Here the state $|(x_1, x_2)\rangle$ is a minimum-uncertainty state for the system, with a position probability amplitude distribution

$$\langle x | (x_1, x_2) \rangle = (\pi)^{-1/4} \exp\left[i x x_2 - \frac{1}{2}(x - x_1)^2 \right]. \tag{1.96}$$

From Appendix A, this is a state with mean position and momentum given by x_1 and x_2, respectively, and with the variances in position and momentum equal to $1/2$.

Exercise 1.16 *Verify Eq. (1.95).*

The corresponding probability density for the observed values, (x_1, x_2), is found from the effect density

$$\hat{E}(x_1, x_2) dx_1\, dx_2 = \frac{1}{2\pi} |(x_1, x_2)\rangle \langle (x_1, x_2)| dx_1\, dx_2. \tag{1.97}$$

Exercise 1.17 *Show that*

$$\int_{-\infty}^{\infty} dx_1 \int_{-\infty}^{\infty} dx_2\, \hat{E}(x_1, x_2) = \hat{1}. \tag{1.98}$$

From this POM we can show that

$$E[X_1] = \langle \hat{X} \rangle, \qquad E[X_1^2] = \langle \hat{X}^2 \rangle + \frac{1}{2}, \tag{1.99}$$

$$E[X_2] = \langle \hat{P} \rangle, \qquad E[X_2^2] = \langle \hat{P}^2 \rangle + \frac{1}{2}, \tag{1.100}$$

where $\langle \hat{A} \rangle = \text{Tr}[\hat{A}\rho]$ is the quantum expectation, while E is a classical average computed by evaluating an integral over the probability density $\wp(x_1, x_2)$. Thus the readout variables X_1 and X_2 give, respectively, the position and momentum of the system with additional noise.

It is more conventional to denote the state $|(x_1, x_2)\rangle$ by $|\alpha\rangle$, where the single complex parameter α is given by $\alpha = (x_1 + ix_2)/\sqrt{2}$. In this form the states are known as coherent states (see Section A.4). The corresponding effect density is $\hat{F}(\alpha) = |\alpha\rangle\langle\alpha|/\pi$ and the resulting probability density $\wp(\alpha)d^2\alpha = \text{Tr}\left[\hat{F}(\alpha)\rho\right] d^2\alpha$. This is known as the Q-function in quantum optics – see Section A.5. For a general choice of initial pure states for the detectors, the probability density for observed results is known as the Husimi function [Hus40].

1.2.6 Non-selective evolution and choice of basis

Recall that in the analysis above, using system and meter states, the combined state prior to the measurement of the meter was

$$|\Psi(t + T_1)\rangle = \hat{U}(T_1)|\theta(t)\rangle|\psi(t)\rangle. \qquad (1.101)$$

As explained there, it is not possible to assign a state vector to the system at time $t + T_1$, because it is entangled with the meter. However, it *is* possible to assign a state matrix to the system. This state matrix is found by taking the partial trace over the meter:

$$\rho(t + T_1) = \mathrm{Tr}_A[|\Psi(t + T_1)\rangle\langle\Psi(t + T_1)|]$$

$$\equiv \sum_j {}_A\langle\phi_j|\Psi(t + T_1)\rangle\langle\Psi(t + T_1)|\phi_j\rangle_A, \qquad (1.102)$$

where $\{|\phi_j\rangle_A: j\}$ is an arbitrary set of basis states for the meter. But this basis can of course be the basis $\{|r\rangle: r\}$ appropriate for a measurement of R on the meter. Thus the reduced system state $\rho(t + T_1)$ is the same as the average system state $\rho(t + T)$ (for $T \geq T_1$) of Eq. (1.86), which is obtained by averaging over the measurement results. That is, the non-selective system state after the measurement does not depend on the basis in which the meter is measured.

Different measurement bases for the meter can be related by a unitary transformation thus:

$$|r\rangle = \sum_s U^*_{r,s}|s\rangle, \qquad (1.103)$$

where U is a c-number matrix satisfying $\sum_r U_{r,s}U^*_{r,q} = \delta_{s,q}$. In terms of the measurement operators $\hat{M}_s$, this amounts to a unitary rearrangement to $\hat{M}_r$ defined by

$$\hat{M}_r = \sum_s U_{r,s}\hat{M}_s. \qquad (1.104)$$

Exercise 1.18 *Verify that the unconditional final state under the new measurement operators $\{\hat{M}_s\}$ is the same as that under the old measurement operators $\{\hat{M}_r\}$.*

The binary example. Although the unconditional system state is the same regardless of how the meter is measured, the conditional system states are quite different. This can be illustrated using the binary measurement example of Section 1.2.4. Consider the simple case in which the fiducial apparatus state is the measurement basis state $|0\rangle_A = |y := 0\rangle$. The measurement basis states are eigenstates of the apparatus operator

$$\hat{Y} = \sum_{y=0}^{1} y|y\rangle\langle y|. \qquad (1.105)$$

Then, if the apparatus is measured in the measurement basis, the measurement operators are

$$\hat{M}_y = {}_A\langle y|\hat{G}|0\rangle_A = |x := y\rangle\langle x := y|. \qquad (1.106)$$

As stated before, these simply project or 'collapse' the system into its measurement basis, the eigenstates of

$$\hat{X} = \sum_{x=0}^{1} x|x\rangle\langle x|. \tag{1.107}$$

Now consider an alternative orthonormal basis for the apparatus, namely the eigenstates of the complementary operator

$$\hat{P}_A = \sum_{p=0}^{1} p|p\rangle_A\langle p|. \tag{1.108}$$

Here the eigenstates for the apparatus are

$$|p\rangle_A = 2^{-1/2}(|y := 0\rangle + e^{i\pi p}|y := 1\rangle), \tag{1.109}$$

and $\hat{X}$ and $\hat{P}$ are complementary in the sense that $\hat{X}$ is maximally uncertain for a system in a $\hat{P}$-eigenstate, and vice versa. In this case the measurement operators are, in the measurement (x) basis,

$$\hat{M}_p = 2^{-1/2}(|0\rangle\langle 0| + e^{-i\pi p}|1\rangle\langle 1|). \tag{1.110}$$

Exercise 1.19 *Verify that the non-selective evolution is the same under these two different measurements, and that it always turns the system into a mixture diagonal in the measurement basis.*

Clearly, measurement of the apparatus in the complementary basis does not collapse the system into a pure state in the measurement basis. In fact, it does not change the occupation probabilities for the measurement basis states at all. This is because the measurement yields no information about the system, since the probabilities for the two results are independent of the system:

$$\Pr[P_A = p] = \langle\psi(t)|\hat{M}_p^\dagger\hat{M}_p|\psi(t)\rangle = 1/2. \tag{1.111}$$

This 'measurement' merely changes the relative phase of these states by π if and only if $p = 1$:

$$\frac{\hat{M}_p \sum_x s_x|x\rangle}{\sqrt{\Pr[P_A = p]}} = \sum_x s_x e^{-i\pi px}|x\rangle. \tag{1.112}$$

That is to say, with probability 1/2, the relative phase of the system states is flipped. In this guise, the interaction between the system and the apparatus is seen not to collapse the system into a measurement eigenstate, but to introduce noise into a complementary system property: the relative phase.

This dual interpretation of an interaction between a system and another system (the meter) is very common. The non-selective evolution reduces the system to a mixture diagonal in some basis. One interpretation (realized by measuring the meter in an appropriate way) is that the system is collapsed into a particular state in that basis, but an equally valid interpretation (realized by measuring the meter in a complementary way) is that the meter is merely adding noise into the relative phases of the system components in this basis. In the following section, we will see how both of these interpretations can be seen simultaneously in the Heisenberg picture.

Exercise 1.20 *Consider the quantum position-measurement model introduced in Exercise 1.14. Show that, if the apparatus is measured in the momentum basis (see Appendix A), then the measurement operators are*

$$\hat{M}_p = \left(\frac{2\Delta^2}{\pi}\right)^{1/4} \exp(-p^2\Delta^2)\exp(-ip\hat{X}). \tag{1.113}$$

Show also that the non-selective evolution is the same as in Exercise 1.14, and that the selective evolution in this exercise can be described in terms of random momentum kicks on the system.

1.3 Representing outcomes as operators

1.3.1 'Correlations without correlata'

We have already met the idea that an operator can represent an outcome in Sections A.1 and 1.2.2, where it was shown that

$$\langle f(\Lambda) \rangle = \text{Tr}[f(\hat{\Lambda})\rho]. \tag{1.114}$$

That is, if an operator $\hat{\Lambda}$ represents an observable Λ, then any function of the result of a measurement of Λ is represented by that function of the operator $\hat{\Lambda}$. Here ρ is the state of the system at the time of the measurements. Clearly, if ρ evolves after the measurement has finished, then the formula (1.114) using this new ρ might no longer give the correct expectation values for the results that had been obtained.

This problem can be circumvented by using the system–meter model of measurement we have presented. Let us assume an entangled system–meter state of the form

$$|\Psi(t + T_1)\rangle = \sum_\lambda |\lambda\rangle_A \hat{\Pi}_\lambda |\psi(t)\rangle_S, \tag{1.115}$$

where $\{|\lambda\rangle_A: \lambda\}$ is an orthonormal set of apparatus states and $\{\hat{\Pi}_\lambda: \lambda\}$ is the set of eigen-projectors of the system observable Λ_S. This is the ideal correlation for the apparatus to 'measure' Λ_S. The *apparatus* observable represented by

$$\hat{\Lambda}_A = \sum_\lambda \lambda |\lambda\rangle_A \langle\lambda| \tag{1.116}$$

has identical moments to the system observable Λ_S for the original system state $|\psi(t)\rangle$, or indeed for the (mixed) system state at time $t + T_1$ derived from Eq. (1.115).

Exercise 1.21 *Show this.*

What has been gained by introducing the meter is that $\hat{\Lambda}_A$ will continue to represent the result Λ of the measurement made at time t, for all times in the future, regardless of the system evolution. We require only that the statistics of $\hat{\Lambda}_A$ do not change after the measurement; that is, that Λ_A be a so-called QND (*quantum non-demolition*) observable. Since meter operators by definition commute with system operators, the meter operator is a classical quantity insofar as the system is concerned – a c-number rather than a q-number. For instance, one could consider a Hamiltonian, acting some time after the measurement, of the form

$$\hat{H} = \hat{\Lambda}_A \otimes \hat{F}_S, \tag{1.117}$$

where $\hat{F}_S$ is an Hermitian system operator, and not have to worry about the operator ordering. In fact, insofar as the system is concerned, this Hamiltonian is equivalent to the Hamiltonian

$$\hat{H} = \Lambda \hat{F}_S, \tag{1.118}$$

where here Λ is the measurement result (a random variable) obtained in the projective measurement of the system at time t.

Exercise 1.22 *Convince yourself of this.*

The action of Hamiltonians such as these (a form of feedback) will be considered in greater detail in later chapters.

This idea of representing measurement results by meter operators is not limited to projective measurements of the system. Say one has the entangled state between system and meter

$$|\Psi(t + T_1)\rangle = \hat{U}(T_1)|\theta\rangle_A |\psi(t)\rangle_S, \tag{1.119}$$

and one measures the meter in the (assumed non-degenerate) eigenbasis $\{|r\rangle_A\}$ of the operator

$$\hat{R}_A = \sum_r r|r\rangle_A \langle r|. \tag{1.120}$$

Then the operator $\hat{R}_A$ represents the outcome of the measurement that, for the system, is described using the measurement operators $\hat{M}_r = \langle r|\hat{U}(T_1)|\theta\rangle$. Recall that the results r are just labels, which need not be real numbers, so $\hat{R}_A$ is not necessarily an Hermitian operator. If the result R is a complex number, then $\hat{R}_A$ is a *normal* operator (see Box 1.1). If R is a real vector, then $\hat{R}_A$ is a vector of commuting Hermitian operators.

It is important to note that $\hat{R}_A$ represents the measurement outcome *whether or not* the projective measurement of the apparatus is made. That is, it is possible to represent a measurement outcome simply by modelling the apparatus, without including the extra step of apparatus state collapse. In this sense, the von Neumann chain can be avoided, not by placing the Heisenberg cut between apparatus and higher links (towards the observer's consciousness), but by ignoring these higher links altogether. The price to be paid for this parsimony is a high one: the loss of any notion of *actual* outcomes. The measurement result R remains a random variable (represented by the operator $\hat{R}_A$) that never takes any particular one of its 'possible' values r. Within this philosophical viewpoint one denies the existence of events, but nevertheless calculates their statistics; in other words, 'correlations without correlata' [Mer98].

1.3.2 Measurement in the Heisenberg picture

The 'measurement without collapse' formulation outlined above is obviously the ideal one for working in the Heisenberg picture. Recall from Section A.1.3 that, in the Heisenberg picture, the state vector or state matrix is constant, while operators evolve in time. However, this was formulated only for unitary evolution; if one wishes to describe a measurement for which some particular result is obtained, this can be done only by invoking state collapse. That is, one must still allow the state to change, even though one is working in the Heisenberg picture. But if one is content to describe a measurement simply as the coupling of the system to the meter, with the result being represented by a meter operator, then state collapse never occurs. Consequently, it is possible to describe all evolution, including measurement, in terms of changing operators. Of course, to do this, one needs to consider system and apparatus operators, not just system operators.

The necessity of using apparatus operators might not be obvious to the reader. After all, when considering unitary evolution of the system alone, we can use essentially the same transformation in the Schrödinger and Heisenberg pictures: $\rho \to \hat{U}\rho\hat{U}^\dagger$ and $\hat{O} \to \hat{U}^\dagger\hat{O}\hat{U}$, respectively. This suggests that for measurement the analogue of $\rho \to \rho'_r = \hat{M}_r\rho\hat{M}_r^\dagger$ would be $\hat{O} \to \hat{O}'_r = \hat{M}_r^\dagger\hat{O}\hat{M}_r$. However, this construction does not work when one considers operator products. The correct post-measurement expectation for $\hat{A}\hat{B}$, weighted by the probability for outcome r, is

$$\mathrm{Tr}\big[\hat{A}\hat{B}\rho'_r\big] = \mathrm{Tr}\big[\hat{M}_r^\dagger\hat{A}\hat{B}\hat{M}_r\rho\big]. \tag{1.121}$$

In general, this is quite different from

$$\mathrm{Tr}\big[\hat{A}'_r\hat{B}'_r\rho\big] = \mathrm{Tr}\big[\hat{M}_r^\dagger\hat{A}\hat{M}_r\hat{M}_r^\dagger\hat{B}\hat{M}_r\rho\big], \tag{1.122}$$

because, in general, $\hat{M}_r$ is not unitary.

The correct Heisenberg formulation of measurement is as follows. The total state (of system plus apparatus) remains equal to the initial state, which is usually taken to factorize as

$$\rho_{\text{total}}(t) = \rho_S \otimes \rho_A. \tag{1.123}$$

The measurement outcome is described, as above, by the apparatus operator

$$\hat{R}_A(t) = \sum_r r|r\rangle_A\langle r|, \tag{1.124}$$

which here is for time t, before the measurement interaction between system and apparatus. This interaction, of duration T_1, changes $\hat{R}_A$ to

$$\hat{R}_A(t + T) = \hat{U}^\dagger(T_1)\left[\hat{R}_A(t) \otimes \hat{1}_S\right]\hat{U}(T_1) \tag{1.125}$$

$$= \sum_r r\, \hat{U}^\dagger(T_1)(|r\rangle_A\langle r| \otimes \hat{1}_S)\hat{U}(T_1). \tag{1.126}$$

Here T is any time greater than or equal to T_1, since we are assuming that the measurement interaction ceases at time $t + T_1$ and that $\hat{R}_A$ is a QND observable for all subsequent evolution of the meter.

It follows trivially from the analysis of Section A.1.3 that the Heisenberg-picture operator $\hat{R}_A(t + T)$ with respect to $\rho_{\text{total}}(t)$ has the same statistics as does the Schrödinger-picture operator with respect to $\rho_{\text{total}}(t + T)$, evolved according to the measurement interaction. Hence, if the initial apparatus state is pure,

$$\rho_A = |\theta\rangle_A\langle\theta|, \tag{1.127}$$

as we assumed, then these statistics are identical to those of the random variable R_A, the result of a measurement on the system with measurement operators $\{\hat{M}_r\}$.

Being an apparatus operator, $\hat{R}_A(s)$ commutes with system operators at all times s. For $s \leq t$ (that is, before the system and apparatus interact), it is also uncorrelated with all system operators. That is, for $s \leq t$, expectation values factorize:

$$\langle\hat{O}_S(t)f(\hat{R}_A(t))\rangle = \langle\hat{O}_S(t)\rangle\langle f(\hat{R}_A(t))\rangle. \tag{1.128}$$

Here $\hat{O}_S$ is an arbitrary system operator and f is an arbitrary function. For $s > t$, this is no longer true. In particular, for $s = t + T$ the correlation with the system is the same as one would calculate using state collapse, namely

$$\langle\hat{O}_S(t + T)f(\hat{R}_A(t + T))\rangle = \sum_r \wp_r f(r)\mathrm{Tr}\left[\hat{O}_S\,\rho_r(t + T)\right], \tag{1.129}$$

where $\rho_r(t + T)$ is the a-posteriori conditioned system state.

Exercise 1.23 *Convince yourself of this.*

It should be noted that these two descriptions of measurement, in terms of changing operators or changing states, have classical equivalents. They are descriptions in terms of changing system variables or changing probability distributions for these variables. We have already used these two descriptions in Section 1.1. Specifically, we began with the 'Heisenberg' description, with correlations arising between system and apparatus variables, and then moved to the complementary 'Schrödinger' description with system state collapse derived using Bayes' theorem.

In the Heisenberg picture, an important difference between quantum and classical measurement stands out. The back-action of the measurement on the system is seen in changes in the system operators, rather than changes in the system state. Classically, a non-disturbing measurement does not introduce any noise into the system. Hence classically there may be no change in the system variables, but in the quantum case any measurement will necessarily cause changes to the system operators. This quantum back-action is best illustrated by example, as we will do in the next subsection. The same distinction between quantum and classical mechanics is also present in the Schrödinger picture, but only in the weaker form given in Exercise 1.9.

1.3.3 The binary example

To illustrate the description of measurement in the Heisenberg picture, we use again the example of a binary measurement. Rather than using $\hat{X}$ for the system and $\hat{Y}$ for the apparatus, we use $\hat{X}_S$ and $\hat{X}_A$. Similarly, we use $\hat{P}_S$ and $\hat{P}_A$ for the complementary operators. These are defined by the relation between the eigenstates $|x\rangle$ and $|p\rangle$ defined in Section 1.2.6. The operators $\hat{X}$ and $\hat{P}$ each act as a displacement in the complementary basis, by which we mean that, for binary variables k and n,

$$\exp(i\pi k \hat{X})|p\rangle = |p \oplus k\rangle, \tag{1.130}$$

$$\exp(i\pi n \hat{P})|x\rangle = |x \oplus n\rangle. \tag{1.131}$$

It is now easy to see that the measurement interaction between the system and the apparatus may be realized by

$$\hat{G} = \exp(i\pi \hat{X}_S \otimes \hat{P}_A). \tag{1.132}$$

Exercise 1.24 *Show that this does produce Eq. (1.63).*

In the Heisenberg picture, this unitary operator transforms the operators according to $\hat{O}(t + T_1) = \hat{G}^\dagger \hat{O}(t)\hat{G}$, where $\hat{O}$ is an arbitrary operator. Thus we find

$$\hat{X}_S(t + T_1) = \hat{X}_S(t), \tag{1.133}$$

$$\hat{P}_S(t + T_1) = \hat{P}_S(t) \oplus \hat{P}_A(t), \tag{1.134}$$

$$\hat{X}_A(t + T_1) = \hat{X}_S(t) \oplus \hat{X}_A(t), \tag{1.135}$$

$$\hat{P}_A(t + T_1) = \hat{P}_A(t). \tag{1.136}$$

The binary addition $\oplus$ is defined for operators by, for example,

$$\hat{X}_S \oplus \hat{X}_A = \sum_{x,y} (x \oplus y)|x\rangle_S \langle x| \otimes |y\rangle_A \langle y|. \tag{1.137}$$

If we make the identifications

$$X = X_S, \qquad Y = X_A(t + T_1), \qquad \Xi = X_A(t), \tag{1.138}$$

then Eq. (1.135) is identical in form and content to the classical Eq. (1.2). The noise term is seen to arise from the initial apparatus state. Note that $\hat{X}_S$ is unchanged by the interaction. This quantity is a QND variable and the measurement interaction realizes a QND measurement of $\hat{X}_S$. However, unlike in the classical case, the system is affected by the measurement. This quantum back-action is seen in the change in the complementary system quantity, $\hat{P}_S$, in Eq. (1.134). The 'quantum noise' added to the system here is $\hat{P}_A$, which is another QND variable. Clearly, if one were to measure $\hat{P}_A$, one would gain no information about the system. (Indeed, one gains most information about the system by measuring the apparatus in the $\hat{X}_A$ basis, which is a basis complementary to the $\hat{P}_A$ basis). However, by measuring $\hat{P}_A$, one directly finds out the noise that has affected the system, as discussed in Section 1.2.6. We see now that, in the Heisenberg picture, both interpretations of the interaction, namely in terms of gaining information about the system and in terms of adding noise to the system, can be seen simultaneously.

Exercise 1.25 *Analyse the case of a generalized position measurement from Exercise 1.14 in the same manner as the binary example of this subsection.*
Hint: *First show that $\hat{P}$ generates displacements of position $\hat{X}$ and vice versa. That is, $e^{iq\hat{P}}|x\rangle = |x+q\rangle$ and $e^{-ik\hat{X}}|p\rangle = |p+k\rangle$ (see Section A.3). Note that, unlike in the binary example, there is no π in the exponential.*

1.4 Most general formulation of quantum measurements

1.4.1 Operations and effects

The theory of measurements we have presented thus far is not quite the most general, but can easily be generalized to make it so. This generalization is necessary to deal with some cases in which the initial meter state is not pure, or the measurement on the meter is not a von Neumann measurement. In such cases the conditioned system state may be impure, even if the initial system state was pure. We call these *inefficient measurements*.

To give the most general formulation[4] we must dispense with the measurement operators $\hat{M}_r$ and use only operations and effects. The operation $\mathcal{O}_r$ for the result r is a completely positive superoperator (see Box 1.3), not restricted to the form of Eq. (1.81). It can nevertheless be shown that an operation can always be written as

$$\mathcal{O}_r = \sum_j \mathcal{J}[\hat{\Omega}_{r,j}], \qquad (1.139)$$

for some set of operators $\{\hat{\Omega}_{r,j}: j\}$.

For a given operation $\mathcal{O}_r$, the set $\{\hat{\Omega}_{r,j}: j\}$ is not unique. For this reason it would be wrong to think of the operators $\hat{\Omega}_{r,j}$ as measurement operators. Rather, the operation is the basic element in this theory, which takes the a-priori system state to the conditioned a-posteriori state:

$$\tilde{\rho}_r(t+T) = \mathcal{O}_r\rho(t). \qquad (1.140)$$

---

[4] It is possible to be even more general by allowing the apparatus to be initially correlated with the system. We do not consider this situation because it removes an essential distinction between apparatus and system, namely that the former is in a fiducial state known to the experimenter, while the latter can be in an arbitrary state (perhaps known to a different experimenter). If the two are initially correlated they should be considered jointly as the system.

The state in Eq. (1.140) is unnormalized. Its norm is the probability $\wp_r$ for obtaining the result $R = r$,

$$\wp_r = \text{Tr}[\mathcal{O}_r \rho(t)], \tag{1.141}$$

so that the normalized state is

$$\rho_r(t + T) = \mathcal{O}_r \rho(t)/\wp_r. \tag{1.142}$$

As for efficient measurements in Section 1.2.5, it is possible to define a *probability operator*, or *effect*, $\hat{E}_r$, such that, for all ρ,

$$\text{Tr}[\mathcal{O}_r \rho] = \text{Tr}[\rho \hat{E}_r]. \tag{1.143}$$

It is easy to verify that

$$\hat{E}_r = \sum_j \hat{\Omega}_{r,j}^\dagger \hat{\Omega}_{r,j}, \tag{1.144}$$

which is obviously Hermitian and positive. The completeness condition

$$\sum_r \hat{E}_r = \hat{1} \tag{1.145}$$

is the only mathematical restriction on the set of operations $\mathcal{O}_r$.

The unconditional system state after the measurement is

$$\rho(t + T) = \sum_r \mathcal{O}_r \rho(t) = \mathcal{O}\rho(t). \tag{1.146}$$

Here the non-selective operation can be written

$$\mathcal{O} = \sum_{r,j} \mathcal{J}[\hat{\Omega}_{r,j}]. \tag{1.147}$$

In terms of the unitary operator $\hat{U}(T_1)$ coupling system to apparatus, this operation can also be defined by

$$\mathcal{O}\rho \equiv \text{Tr}_A[\hat{U}(T_1)(\rho \otimes \rho_A)\hat{U}^\dagger(T_1)], \tag{1.148}$$

where ρ_A is the initial apparatus state matrix.

Exercise 1.26 *By decomposing ρ_A into an ensemble of pure states, and considering an apparatus basis $\{|r\rangle\}$, derive an expression for $\hat{\Omega}_{r,j}$. Also show the non-uniqueness of the set $\{\hat{\Omega}_{r,j}: j\}$.*

This completes our formal description of quantum measurement theory. Note that the above formulae, from Eq. (1.139) to Eq. (1.147), are exact analogues of the classical formulae from Eq. (1.26) to Eq. (1.34). The most general formulation of classical measurement was achieved simply by adding back-action to Bayes' theorem. The most general formulation of quantum measurement should thus be regarded as the quantum generalization of

Table 1.1. *Quantum measurement theory as generalized Bayesian analysis*

Concept	Quantum formula	Bayesian formula
Initial state	$\rho(t)$, a positive operator	$\wp(t)$, a positive vector
such that	$\text{Tr}[\rho(t)] = 1$	$\sum_x \wp(x; t) = 1$
Measurement result	R, a random variable	R, a random variable
For each r define	an operation $\mathcal{O}_r$	a matrix $\mathcal{O}_r$
such that	$\tilde{\rho}_r(t + T) = \mathcal{O}_r \rho(t) \geq 0.$	$\tilde{\wp}_r(t + T) = \mathcal{O}_r \wp(t) \geq 0$
$\text{Pr}[R = r]$	$\wp(r) = \text{Tr}[\tilde{\rho}_r(t + T)]$	$\wp(r) = \sum_x \tilde{\wp}_r(x; t + T)$
can be written as	$\wp(r) = \text{Tr}[\rho(t)\hat{E}_r]$	$\wp(r) = \sum_x \wp(x; t) E_r(x)$
where	$\sum_r \hat{E}_r = \hat{1}$	$\forall x, \sum_r E_r(x) = 1$
Conditioned state	$\rho_r(t + T) = \tilde{\rho}_r(t + T)/\wp(r)$	$\wp_r(t + T) = \tilde{\wp}_r(t + T)/\wp(r)$
Interpretation	a matter of debate!	Bayes' rule: $E_r(x) = \wp(r|x)$

Bayes' theorem, in which back-action is an inseparable part of the measurement. This difference arises simply from the fact that a quantum state is represented by a positive matrix, whereas a classical state is represented by a positive vector (i.e. a vector of probabilities). This analogy is summarized in Table 1.1.

We now give a final example to show how generalized measurements such as these arise in practice, and why the terminology *inefficient* is appropriate for those measurements for which measurement operators cannot be employed. It is based on Example 1 in Section 1.2.5, which is a description of *efficient* photon counting if $|n\rangle$ is interpreted as the state with n photons.

Say one has an inefficient photon detector, which has only a probability η of detecting each photon. If the perfect detector would detect n photons, then, from the binomial expansion, the imperfect detector would detect r photons with probability

$$\wp(r|n) = \eta^r (1 - \eta)^{n-r} \binom{n}{r}. \tag{1.149}$$

Thus, if r photons are counted at the end of the measurement, the probability that n photons 'would have been' counted by the perfect detector is, by Bayes' theorem,

$$\wp(n|r) = \frac{\wp(r|n)\langle n|\rho(t)|n\rangle}{\sum_m \wp(r|m)\langle m|\rho(t)|m\rangle}. \tag{1.150}$$

Hence, the conditioned system state is the mixture

$$\rho_r(t + T) = \sum_n \wp(n|r) \frac{\mathcal{J}[|0\rangle\langle n|]\rho(t)}{\langle n|\rho(t)|n\rangle} \tag{1.151}$$

$$= \sum_n \frac{\wp(r|n)\mathcal{J}[|0\rangle\langle n|]\rho(t)}{\sum_m \wp(r|m)\langle m|\rho(t)|m\rangle} \tag{1.152}$$

$$= \frac{\mathcal{O}_r \rho(t)}{\text{Tr}[\rho(t)\hat{E}_r]}, \tag{1.153}$$

Table 1.2. *Eight useful or interesting classes of quantum measurements*

Symbol	Name	Definition
E	Efficient	$\forall r, \exists \hat{M}_r, \; \mathcal{O}_r = \mathcal{J}[\hat{M}_r]$
C	Complete	$\forall \rho, \forall r, \; \mathcal{O}_r \rho \propto \mathcal{O}_r \hat{1}$
S	Sharp	$\forall r, \; \text{rank}(\hat{E}_r) = 1$
O	Of an observable X	$\forall r, \; \hat{E}_r = E_r(\hat{X})$
BAE	Back-action-evading	O with $\forall \rho, \forall x \in \lambda(\hat{X}), \; \text{Tr}[\hat{\Pi}_x \rho] = \text{Tr}[\hat{\Pi}_x \mathcal{O} \rho]$
MD	Minimally disturbing	E with $\forall r, \; \hat{M}_r = \hat{M}_r^\dagger$
P	Projective	MD and O
VN	von Neumann	P and S

where the operations and effects are, respectively,

$$\mathcal{O}_r = \sum_n \eta^r (1-\eta)^{n-r} \binom{n}{r} \mathcal{J}[|0\rangle\langle n|], \tag{1.154}$$

$$\hat{E}_r = \sum_n \eta^r (1-\eta)^{n-r} \binom{n}{r} |n\rangle\langle n|. \tag{1.155}$$

1.4.2 Classification of measurements

The formalism of operations and effects encompasses an enormous, even bewildering, variety of measurements. By placing restrictions on the operations, different classes of measurements may be defined. In this section, we review some of these classes and their relation to one another. We restrict our consideration to eight classes, identified and defined in Table 1.2. Their complicated inter-relations are defined graphically by the Venn diagram in Fig. 1.2.

Some classes of measurement are characterized by the disturbance imposed on the system by the measurement ('efficient', 'complete', and 'minimally disturbing'). Others are characterized by the sort of information the measurement yields ('sharp' and 'of an observable X'), and so can be defined using the effects only. The remainder are characterized by both the sort of information obtained and the disturbance of the system ('back-action-evading', 'projective', and 'von Neumann').

Some of these classes are well known (such as back-action-evading measurements) while others are not (such as complete measurements). Below, we briefly discuss each of the eight. This also allows us to discuss various concepts relevant to quantum measurement theory.

[E]: Efficient measurements. As already discussed, efficient measurements are ones for which each operation is defined in terms of a measurement operator: $\mathcal{O}_r = \mathcal{J}[\hat{M}_r]$. These measurements take pure states to pure states. Any noise in efficient measurements can be interpreted as quantum noise. The complementary set is that of *inefficient* measurements, which introduce classical noise or uncertainty into the measurement.

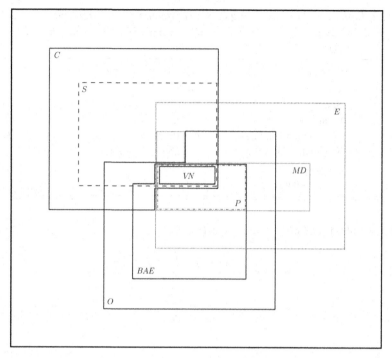

Fig. 1.2 A Venn diagram for the eight classes of quantum measurements described in Table 1.2.

It is only for the class of efficient measurements that one can derive the following powerful theorem [Nie01, FJ01]:

$$H[\rho(t)] \geq \sum_r \wp_r H[\rho_r(t+T)]. \tag{1.156}$$

Here, $H[\rho]$ is any measure of the mixedness of ρ that is invariant under unitary transformations of ρ and satisfies

$$H[w_1\rho_1 + w_2\rho_2] \geq w_1 H[\rho_1] + w_2 H[\rho_2] \tag{1.157}$$

for arbitrary state matrices ρ_j and positive weights w_j summing to unity. Examples of such measures are the entropy $-\mathrm{Tr}[\rho \log \rho]$ and the 'linear entropy' $1 - \mathrm{Tr}[\rho^2]$.[5] The interpretation of this theorem is that, as long as no classical noise is introduced in the measurement, the a-posteriori conditional state is *on average* less mixed than (or just as mixed as) the a-priori state. That is, the measurement refines one's knowledge of the system, as one would hope. Note that it is *not* true that the conditional a-posteriori state is *always* less mixed than the a-priori state.

---

[5] An even stronger version of this theorem, using majorization to classify the relative mixedness of two states, has also been proven [Nie01, FJ01].

Exercise 1.27 *Prove the foregoing statement, by finding an example for a binary system.*
Hint: *This is a classical phenomenon, so the measurement operators and state matrix in the example can all be diagonal in the same basis.*

[C]: Complete measurements. The definition of complete measurements in Table 1.2 implies that, for all results r, the conditioned a-posteriori state

$$\rho_r(t + T) = \mathcal{O}_r \rho(t)/\text{Tr}[\hat{E}_r \rho(t)] \tag{1.158}$$

is independent of $\rho(t)$. In other words, at the end of the measurement, no information remains in the system about its initial state. This is the sense in which the measurement is *complete*: no further measurements could yield any more information about the initial system state.

The definition of complete measurements implies that the operations must be of the form

$$\mathcal{O}_r = \sum_{j,k} \mathcal{J}\big[|\theta_{rk}\rangle\langle\phi_{rj}|\big], \tag{1.159}$$

where θ and ϕ denote (possibly unnormalized) system states. From this, it is easy to see that the conditioned state, independently of $\rho(t)$, is

$$\rho_r(t + T) = \frac{\sum_k |\theta_{rk}\rangle\langle\theta_{rk}|}{\sum_k \langle\theta_{rk}|\theta_{rk}\rangle}. \tag{1.160}$$

The concept of complete measurements (or, more particularly, 'incomplete measurements') will be seen to be very useful when discussing *adaptive measurements* in Section 2.5.

[S]: Sharp measurements. The definition of sharp measurements in Table 1.2 implies that the effects are rank-1 positive operators. That is to say, each effect is of the form $\hat{E}_r = |\phi_r\rangle\langle\phi_r|$, for some (possibly unnormalized) state $|\phi_r\rangle$. This implies that the operations must be of the form

$$\mathcal{O}_r = \sum_k \mathcal{J}[|\theta_{rk}\rangle\langle\phi_r|]. \tag{1.161}$$

From this it is apparent that sharp measurements are a subclass of complete measurements. Also, it is apparent that, for efficient measurements, sharpness and completeness are identical properties.

The significance of sharpness is that a sharp measurement cannot be an unsharp version of a different measurement [MdM90a, MdM90b]. That is, the results of a sharp measurement cannot be generated by making a different measurement and then rendering it 'unsharp' by classically processing the results. Mathematically, a sharp measurement $\{\hat{E}_r\}$ is one for which there is no other measurement $\{\hat{E}'_s: s\}$ such that

$$\hat{E}_r = \sum_s w_{r|s} \hat{E}'_s, \tag{1.162}$$

where $w_{r|s}$ is the probability that r is reported as the measurement result when the second measurement result is s. The object $\{w_{r|s}\}$ is sometimes called a stochastic map from $\{s\}$ to $\{r\}$. We also require that this stochastic map $\{w_{r|s}\}$ be nontrivial. A trivial stochastic map is a deterministic one for which $w_{r|s}^2 = w_{r|s}$ for all r and s, which simply relabels measurement results.

Another fact about sharp measurements is that it is always possible to prepare the system in a state such that a given result r *cannot* be obtained. Note, however, that there is no requirement that the effects be orthogonal, so it is not necessarily possible to prepare the system such that a given result r is guaranteed.

[O]: Measurements of an observable. If an effect $\hat{E}_r$ is a function of an Hermitian operator $\hat{X}$, then the probability of obtaining the result r is given by

$$\wp_r = \text{Tr}\left[E_r(\hat{X})\rho(t)\right] = \sum_x E_r(x)\text{Tr}\left[\hat{\Pi}_x \rho(t)\right], \tag{1.163}$$

where $\{x\}$ are the (assumed discrete for simplicity) eigenvalues of $\hat{X}$ and $\hat{\Pi}_x$ the corresponding projectors. If all of the effects are functions of the same operator $\hat{X}$, then it is evident that the measurement is equivalent to a (possibly unsharp) measurement of the observable X. That is, the result R could be obtained by making a projective measurement of X and then processing the result. Note that this definition places no restriction on the state of the system after the measurement.

The class labelled O in Fig. 1.2 should be understood to be the class of measurements that are measurements of some observable X. Note that, by virtue of the definition here, a measurement in this class may be a measurement of more than one observable. For example, it is obvious from the above definition that any measurement of X^2 is also a measurement of X. However, if $\hat{X}$ has eigenvalues of equal magnitude but opposite sign, then the converse is not true. This is because, for example, it is not possible to write the effects for a projective measurement of $\hat{X}$, which are

$$\hat{E}_x = |x\rangle\langle x| = \delta_{\hat{X},x}, \tag{1.164}$$

as a function of $\hat{S} = \hat{X}^2$. This is the case even though the projectors for the latter *are* functions of $\hat{X}$:

$$\hat{E}_s = \sum_x \delta_{x^2,s}|x\rangle\langle x| = \delta_{\hat{X}^2,s}. \tag{1.165}$$

By binning results (corresponding to values of X with the same magnitude), one can convert the measurement of X into a measurement of X^2. However, it is not permissible to allow such binning in the above definition, because then every measurement would be a measurement of any observable; simply binning all the results together gives a single $\hat{E} = \hat{1}$, which can be written as a (trivial) function of any observable.

[BAE]: Back-action-evading measurements. Consider a measurement of an observable X according to the above definition. A hypothetical projective measurement of X before this measurement will not affect the results of this measurement, because the effects are

a function of $\hat{X}$. However, the converse is not necessarily true. Because the definition of 'measurement of an observable X' is formulated in terms of the effects alone, it takes no account of the disturbance or *back-action* of the measurement on the system. A back-action-evading (BAE) measurement of X is one for which a projective measurement of X *after* the measurement will have the same statistics as one *before*. If the total (i.e. non-selective) operation for the measurement in question is $\mathcal{O} = \sum_r \mathcal{O}_r$, then the requirement is that, for all ρ and all eigenvalues x of $\hat{X}$,

$$\mathrm{Tr}[\hat{\Pi}_x \rho] = \mathrm{Tr}[\hat{\Pi}_x \mathcal{O}\rho]. \tag{1.166}$$

This is the condition in Table 1.2, where we use $\lambda(\hat{X})$ to denote the set of eigenvalues of $\hat{X}$.

A concept closely related to BAE measurement is QND measurement. Recall from Section 1.3 that X is a QND (quantum non-demolition) observable if the operator $\hat{X}$ is a constant of motion (in the Heisenberg picture). Thus, we can talk of a QND measurement of $\hat{X}$ if the effects are functions of $\hat{X}$ and

$$\hat{X} = \hat{U}^\dagger(T_1) \hat{X} \hat{U}(T_1), \tag{1.167}$$

where $\hat{U}(T_1)$ is the unitary operator describing the coupling of the system to the meter, as in Section 1.3.2, so that $\hat{X}$ is to be understood as $\hat{X}_S \otimes \hat{1}_A$.

The condition for a back-action-evading measurement (1.166) is implied by (and hence is weaker than) that for a quantum non-demolition measurement. To see this, first note that a unitary transformation preserves eigenvalues, so that Eq. (1.167) implies that, for all x,

$$\hat{\Pi}_x \otimes \hat{1}_A = \hat{U}^\dagger(T_1)(\hat{\Pi}_x \otimes \hat{1}_A)\hat{U}(T_1). \tag{1.168}$$

Now post-multiply both sides of Eq. (1.168) by $\rho \otimes \rho_A$, where ρ_A is the initial apparatus state. This gives

$$(\hat{\Pi}_x \rho) \otimes \rho_A = \hat{U}^\dagger(T_1)(\hat{\Pi}_x \otimes \hat{1}_A)\hat{U}(T_1)(\rho \otimes \rho_A). \tag{1.169}$$

Now pre- and post-multiply by $\hat{U}(T_1)$ and $\hat{U}^\dagger(T_1)$, respectively. This gives

$$\hat{U}(T_1)[(\hat{\Pi}_x \rho) \otimes \rho_A]\hat{U}^\dagger(T_1) = (\hat{\Pi}_x \otimes \hat{1}_A)\hat{U}(T_1)(\rho \otimes \rho_A)\hat{U}^\dagger(T_1). \tag{1.170}$$

Taking the total trace of both sides then yields Eq. (1.166), from the result in Eq. (1.148).

Often the terms back-action-evading (BAE) measurement and quantum non-demolition (QND) measurement are used interchangeably, and indeed the authors are not aware of any proposal for a BAE measurement that is not also a QND measurement. The advantage of the BAE definition given above is that it is formulated in terms of the operations and effects, as we required.

It is important not to confuse the non-selective and selective a-posteriori states. The motivating definition (1.166) is formulated in terms of the non-selective total operation $\mathcal{O}$. The definition would be silly if we were to replace this by the selective operation $\mathcal{O}_r$ (even if an appropriate normalizing factor were included). That is because, if the system were prepared in a state with a non-zero variance in X, then the measurement would in general collapse the state of the system into a new state with a smaller variance for X.

That is, the statistics of X would not remain the same. The actual definition ensures that *on average* (that is, ignoring the measurement results) the statistics for X are the same after the measurement as before.

[MD]: Minimally disturbing measurements. Minimally disturbing measurements are a subclass of efficient measurements. The *polar decomposition theorem* says that an arbitrary operator, such as the measurement operator $\hat{M}_r$, can be decomposed as

$$\hat{M}_r = \hat{U}_r \hat{V}_r, \tag{1.171}$$

where $\hat{U}_r$ is unitary and $\hat{V}_r = \sqrt{\hat{E}_r}$ is Hermitian and positive. We can interpret these two operators as follows. The Hermitian $\hat{V}_r$ is responsible for generating the necessary back-action (the 'state collapse') associated with the information gained in obtaining the result r (since the statistics of the results are determined solely by $\hat{E}_r$, and hence solely by $\hat{V}_r$). The unitary $\hat{U}_r$ represents surplus back-action: an extra unitary transformation independent of the state.

A minimally disturbing measurement is one for which $\hat{U}_r$ is (up to an irrelevant phase factor) the identity. That is,

$$\hat{M}_r = \sqrt{\hat{E}_r}, \tag{1.172}$$

so that the only disturbance of the system is the necessary back-action determined by the probability operators $\hat{E}_r$. The name 'minimally disturbing' can be justified rigorously as follows. The fidelity between an a-priori state of maximal knowledge $|\psi\rangle$ and the a-posteriori state $\tilde{\rho}_r = \mathcal{O}_r|\psi\rangle\langle\psi|$, averaged over r and ψ, is

$$F_{\text{average}} = \int d\mu_{\text{Haar}}(\psi) \sum_r \langle\psi|\tilde{\rho}_r|\psi\rangle. \tag{1.173}$$

Here $d\mu_{\text{Haar}}(\psi)$ is the Haar measure over pure states, the unique measure which is invariant under unitary transformations. For a given POM $\{\hat{E}_r\}$, this is maximized for efficient measurements with measurement operators given by Eq. (1.172) [Ban01].

Exercise 1.28 *Show that, for a given POM and a* particular *initial state ψ, a minimally disturbing measurement (as defined here) is in general* not *the one which maximizes the fidelity between a-priori and a-posteriori states.*
Hint: *Consider a QND measurement of $\hat{\sigma}_z$ on a state $|\psi\rangle = \alpha|\sigma_z := -1\rangle + \beta|\sigma_z := 1\rangle$. Compare this with the non-QND measurement of $\hat{\sigma}_z$ with measurement operators $|\psi\rangle\langle\sigma_z := -1|$ and $|\psi\rangle\langle\sigma_z := 1|$.*

For minimally disturbing measurements, it is possible to complement the relation (1.156) by the following equally powerful theorem:

$$H[\rho(t + T)] \geq H[\rho(t)], \tag{1.174}$$

where $\rho(t + T) = \sum_r \wp_r \rho_r(t + T)$. That is, the unconditional a-posteriori state is at least as mixed as the a-priori state – if one does not take note of the measurement result, one's information about the system can only decrease. This does not hold for measurements in general; for the measurement in Example 1, the a-posteriori state is the pure state $|0\rangle$ regardless of the a-priori state. However, it does hold for a slightly broader class than minimally disturbing measurements, namely measurements in which the surplus back-action U_r in Eq. (1.171) is the same for all r. These can be thought of as minimally disturbing measurements followed by a period of unitary evolution.

A minimally disturbing measurement of an observable X is a BAE measurement of that observable, but, of course, minimally disturbing measurements are not restricted to measurements of observables. Finally, it is an interesting fact that the class of minimally disturbing measurements does not have the property of closure. Closure of a class means that, if an arbitrary measurement in a class is followed by another measurement from the same class, the 'total' measurement (with a two-fold result) is guaranteed to be still a member of that class.

Exercise 1.29 *Find an example that illustrates the lack of closure for the MD class.*

[P]: Projective measurements. These are the measurements with which we began our discussion of quantum measurements in Section 1.2.2. They are sometimes referred to as orthodox measurements, and as Type I measurements (all other measurements being Type II) [Pau80]. From the definition that they are minimally disturbing and a measurement of an observable, it follows that the measurement operators $\hat{M}_r$ and effects $\hat{E}_r$ are identical and equal to projectors $\hat{\Pi}_r$.

[VN]: Von Neumann measurements. Sometimes the term 'von Neumann measurement' is used synonymously with the term 'projective measurements'. We reserve the term for sharp projective measurements (that is, those with rank-1 projectors). This is because von Neumann actually got the projection postulate wrong for projectors of rank greater than 1, as was pointed out (and corrected) by Lüders [Lüd51]. Von Neumann measurements are the only measurements which are members of all of the above classes.

1.4.3 Classification exercise

Appreciating the relations among the above classes of measurements requires a careful study of Fig. 1.2. To assist the reader in this study, we here provide a prolonged exercise. The Venn diagram in Fig. 1.2 has 17 disjoint regions. If there were no relations among the eight classes, there would be 2^8, that is 256, regions. Thus the fact that there are only 17 testifies to the many inter-relationships among classes.

Below, we have listed 17 different measurements, defined by their set of operations $\{\mathcal{O}_r\}$. Each measurement belongs in a distinct region of 'measurement space' in Fig. 1.2. The object of the exercise is to number the 17 regions in this figure with the number (from 1 to 17) corresponding to the appropriate measurement in the list below.

All of the measurements are on an infinite-dimensional system, with basis states $\{|n\rangle: n = 0, 1, 2, \ldots\}$, called number states. Any ket containing n or m indicates a number state. Any ket containing a complex number $\pm\alpha$, β or γ indicates a *coherent state*, defined as

$$|\alpha\rangle = e^{-|\alpha|^2/2} \sum_{n=0}^{\infty} \frac{\alpha^n}{\sqrt{n!}} |n\rangle. \tag{1.175}$$

See also Appendix A. It is also useful to define sets $\mathbb{E}$ and $\mathbb{O}$, the even and odd counting numbers, respectively. If the result r is denoted n, then the resolution of the identity is $\sum_{n=0}^{\infty} \hat{E}_n$. If it is denoted α then it is $\int d^2\alpha \, \hat{E}_\alpha$. If denoted $\mathbb{E}$, $\mathbb{O}$ then it is $\hat{E}_\mathbb{E} + \hat{E}_\mathbb{O}$.

We also use the following operators in the list below. The operator $\hat{D}_\beta$ denotes a *displacement operator* defined by how it affects a coherent state:

$$\hat{D}_\beta |\alpha\rangle = |\alpha + \beta\rangle, \tag{1.176}$$

for some non-zero complex number β. The number operator $\hat{N}$ has the number states as its eigenstates. The two operators $\hat{\Pi}_\mathbb{E}$ and $\hat{\Pi}_\mathbb{O}$ are defined by

$$\hat{\Pi}_{\mathbb{E},\mathbb{O}} = \sum_{n\in\mathbb{E},\mathbb{O}} |n\rangle\langle n|. \tag{1.177}$$

Finally, $\wp_\eta(n|m)$ is as defined in Eq. (1.149) for some $0 < \eta < 1$.

Here is the list.

1. $\mathcal{O}_\alpha = \pi^{-1}\mathcal{J}[|\alpha\rangle\langle\alpha|]$
2. $\mathcal{O}_\alpha = \int d^2\beta \, \pi^{-2} e^{-|\alpha-\beta|^2} \mathcal{J}[|\beta\rangle\langle\beta|]$
3. $\mathcal{O}_\alpha = \pi^{-1}\mathcal{J}[|0\rangle\langle\alpha|]$
4. $\mathcal{O}_\alpha = \int d^2\gamma \, \pi^{-2} e^{-|\gamma|^2} \mathcal{J}[|\gamma\rangle\langle\alpha|]$
5. $\mathcal{O}_\alpha = \int d^2\gamma \, \pi^{-1} e^{-|\gamma|^2} \int d^2\beta \, \pi^{-2} e^{-|\alpha-\beta|^2} \mathcal{J}[|\gamma\rangle\langle\beta|]$
6. $\mathcal{O}_\alpha = \mathcal{J}[\hat{E}_\alpha^{1/2}]$, $\hat{E}_\alpha = (2\pi)^{-1}(|\alpha\rangle\langle\alpha| + |-\alpha\rangle\langle-\alpha|)$
7. $\mathcal{O}_\alpha = \mathcal{J}[\hat{D}_\beta \hat{E}_\alpha^{1/2}]$, $\hat{E}_\alpha = (2\pi)^{-1}(|\alpha\rangle\langle\alpha| + |-\alpha\rangle\langle-\alpha|)$
8. $\mathcal{O}_n = \mathcal{J}[|n\rangle\langle n|]$
9. $\mathcal{O}_n = \sum_{m=0}^{\infty} \wp_\eta(n|m)\mathcal{J}[|m\rangle\langle m|]$
10. $\mathcal{O}_n = \sum_{m=0}^{\infty} \wp_\eta(n|m)\mathcal{J}[\hat{D}_\beta|m\rangle\langle m|]$
11. $\mathcal{O}_n = \mathcal{J}[|0\rangle\langle n|]$
12. $\mathcal{O}_n = \sum_{m=0}^{\infty} 2^{-(m+1)}\mathcal{J}[|m\rangle\langle n|]$
13. $\mathcal{O}_{\mathbb{E},\mathbb{O}} = \mathcal{J}[\hat{\Pi}_{\mathbb{E},\mathbb{O}}]$
14. $\mathcal{O}_{\mathbb{E},\mathbb{O}} = \mathcal{J}[\exp(i\pi\hat{N})\hat{\Pi}_{\mathbb{E},\mathbb{O}}]$
15. $\mathcal{O}_\mathbb{E} = \sum_{n\in\mathbb{E}} \mathcal{J}[|0\rangle\langle n|]$, $\mathcal{O}_\mathbb{O} = \sum_{n\in\mathbb{O}} \mathcal{J}[|1\rangle\langle n|]$
16. $\mathcal{O}_{\mathbb{E},\mathbb{O}} = \mathcal{J}[\hat{D}_\beta \hat{\Pi}_{\mathbb{E},\mathbb{O}}]$
17. $\mathcal{O}_{\mathbb{E},\mathbb{O}} = \sum_{n\in\mathbb{E},\mathbb{O}} \mathcal{J}[|0\rangle\langle n|]$

1.5 Measuring a single photon

In this section we give an experimental example of the quantum measurement of a binary variable, as introduced in Section 1.2.4. This experiment was realized as a 'cavity QED' system, a term used to denote the interaction between a discrete-level atomic system and a small number of electromagnetic field modes, which are also treated as quantum systems. In the experiment performed by the Haroche group in Paris in 1999 [NRO⁺99], the measured system was the state of an electromagnetic field in a microwave cavity. Apart from small imperfections, the preparation procedure produced a pure state containing no more than a single photon. Thus the state of the cavity field may be written as $|\psi\rangle = c_0|0\rangle + c_1|1\rangle$. The measured variable is the photon number with result 0 or 1. The apparatus was an atom with three levels: ground state $|g\rangle$, excited state $|e\rangle$, and an auxiliary state $|i\rangle$. The final readout on the apparatus determines whether the atom is in state $|g\rangle$ by a selective ionization process, which we will describe below. This final readout is not ideal and thus we will need to add an extra classical noise to the description of the measurement.

We begin with a brief description of the interaction between the cavity field and a single two-level atom in order to specify how the correlation between the system and the apparatus is established. If, through frequency or polarization mismatching, the cavity mode does not couple to the auxiliary level $|i\rangle$, then we can define the atomic lowering operator by $\hat\sigma = |g\rangle\langle e|$. The field annihilation operator is $\hat a$ (see Section A.4). The relevant parts of the total Hamiltonian are

$$\hat H = \omega_c \hat a^\dagger \hat a + \omega_g |g\rangle\langle g| + \omega_e |e\rangle\langle e| + \omega_i |i\rangle\langle i| + (i\Omega/2)(\hat\sigma^\dagger + \hat\sigma)(\hat a - \hat a^\dagger), \quad (1.178)$$

where Ω is known as the single-photon Rabi frequency and is proportional to the dipole moment of the atom and inversely proportional to the square root of the volume of the cavity mode. We work in the interaction frame (see Section A.1.3) with the free Hamiltonian

$$\hat H_0 = \omega_c \hat a^\dagger \hat a + \omega_g |g\rangle\langle g| + (\omega_g + \omega_c)|e\rangle\langle e| + (\omega_g + \omega_d)|i\rangle\langle i|, \quad (1.179)$$

where ω_d is the frequency of a 'driving field', a classical microwave field (to be discussed later). The 'interaction Hamiltonian' $\hat V = \hat H - \hat H_0$ becomes the time-dependent Hamiltonian $\hat V_{\mathrm{IF}}(t)$ in the interaction frame. However, the evolution it generates is well approximated by the time-*independent* Hamiltonian

$$\hat V_{\mathrm{IF}} = \Omega(i\hat\sigma^\dagger \hat a - i\hat\sigma \hat a^\dagger)/2 + \Delta\hat\sigma^\dagger\hat\sigma + \delta|i\rangle\langle i|, \quad (1.180)$$

where Δ is the detuning $\omega_e - \omega_g - \omega_c$ of the $|e\rangle \leftrightarrow |g\rangle$ transition from the cavity resonance, and $\delta = \omega_i - \omega_g - \omega_d$ is that of the $|i\rangle \leftrightarrow |g\rangle$ transition from the classical driving field. The necessary approximation (called the *rotating-wave approximation*) is to drop terms rotating (in the complex plane) at high frequencies $\sim \omega_c \gg \Delta, \delta, \Omega$. This is justified because they average to zero over the time-scale on which evolution occurs in the interaction frame.

Exercise 1.30 *Derive Eq. (1.180) and convince yourself of the validity of the rotating-wave approximation.*

Hint: *Finding $\hat V_{\mathrm{IF}}(t)$ is the same as solving for this operator in the Heisenberg picture with the Hamiltonian $\hat H_0$. Since $\hat H_0$ splits into a part acting on the atom and a part acting on the field, the Heisenberg equations of motion for the atom and field operators in $\hat V_{\mathrm{IF}}$ can be solved separately.*

Let us now assume that the atom is resonant with the cavity ($\Delta = 0$), in which case the Hamiltonian (1.180) (apart from the final term) is known as the Jaynes-Cummings Hamiltonian. If this Hamiltonian acts for a time τ on an initial state $|1, g\rangle$, the final state is

$$\exp(-i\hat{V}_{\mathrm{IF}}\tau)|1, g\rangle = \cos(\Omega\tau/2)|1, g\rangle + \sin(\Omega\tau/2)|0, e\rangle, \tag{1.181}$$

where $|n, g\rangle \equiv |n\rangle|g\rangle$ and $|n, e\rangle \equiv |n\rangle|e\rangle$.

Exercise 1.31 *Show this and the analogous result for the initial state $|0, e\rangle$.*
Hint: *Derive the equations of motion for the coefficients of the state vector in the above basis.*

If the total interaction time $\tau = 2\pi/\Omega$, then the probability that the atom is in the ground state again is unity, but the quantum state has acquired an overall phase. That is to say, for this interaction time, the state changes as $|1, g\rangle \rightarrow -|1, g\rangle$. However, if the field is initially in the vacuum state, there is no change: $|0, g\rangle \rightarrow |0, g\rangle$.

Exercise 1.32 *Show this by verifying that $|0, g\rangle$ is an eigenstate of $\hat{V}_{\mathrm{IF}}$ with eigenvalue zero.*

This sign difference in the evolution of states $|0\rangle$ and $|1\rangle$ provides the essential correlation between the system and the apparatus that is used to build a measurement. If the field is in a superposition of vacuum and one photon, the interaction with the atom produces the 'conditional' transformation

$$(c_0|0\rangle + c_1|1\rangle) \otimes |g\rangle \overset{C}{\rightarrow} (c_0|0\rangle - c_1|1\rangle) \otimes |g\rangle. \tag{1.182}$$

It is called conditional because the sign of the state is flipped if and only if there is one photon present. Note that we are not using the term here in the context of a measurement occurring.

As it stands this is not of the form of a binary quantum measurement discussed in Section 1.2.4 since the meter state (the atom) does not change at all. In order to configure this interaction as a measurement, we need to find a way to measure the relative phase shift introduced by the interaction between the field and the atom. This is done using the 'auxiliary' electronic level, $|i\rangle$, which does not interact with the cavity mode and cannot undergo a conditional phase shift. We begin by using a classical microwave pulse R_1 of frequency ω_d, to prepare the atom in a superposition of the auxiliary state and the ground state: $|g\rangle \rightarrow (|g\rangle + |i\rangle)/\sqrt{2}$. For the moment, we assume that this is resonant, so that $\delta = 0$ in Eq. (1.180). After the conditional interaction, C, between the atom and the cavity field, another microwave pulse R_2 of frequency ω_d again mixes the states $|g\rangle$ and $|i\rangle$. It reverses the action of R_1, taking $|g\rangle \rightarrow (|g\rangle - |i\rangle)/\sqrt{2}$ and $|i\rangle \rightarrow (|g\rangle + |i\rangle)/\sqrt{2}$.

Exercise 1.33 *Show that this transformation is unitary and reverses R_1.*

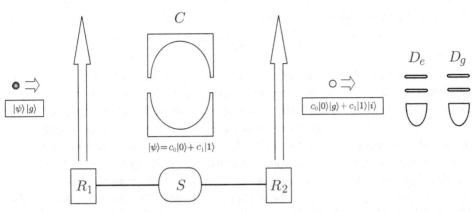

Fig. 1.3 A schematic diagram of the Haroche single-photon measurement [NRO$^+$99]. A single atom traverses three microwave fields R_1, C and R_2, the middle one being described by a single-mode cavity field. It then encounters two ionization detectors, D_e and D_g, which detect whether the atom is in the excited state or ground state, respectively. The driving fields R_1 and R_2 are produced by the same microwave source, which locks their relative phase. Adapted by permission from Macmillan Publishers Ltd: G. Nogues *et al.*, *Nature* **400**, 239–242 (1999), (Fig. 1), copyright Macmillan Magazines Ltd 1999.

Finally, a projective readout of the ground state $|g\rangle$ is made, as shown in Fig. 1.3. The full measurement protocol can now be described:

$$(c_0|0\rangle + c_1|1\rangle)|g\rangle \overset{R_1}{\to} (c_0|0\rangle + c_1|1\rangle)\frac{1}{\sqrt{2}}(|i\rangle + |g\rangle)$$

$$\overset{C}{\to} \frac{1}{\sqrt{2}}(c_0|0\rangle(|i\rangle + |g\rangle) + c_1|1\rangle(|i\rangle - |g\rangle))$$

$$\overset{R_2}{\to} c_0|0\rangle|g\rangle + c_1|1\rangle|i\rangle. \tag{1.183}$$

An ideal measurement of the ground state of the atom gives a yes (no) result with probability $|c_0|^2$ ($|c_1|^2$), and a measurement of the photon number has been made without absorbing the photon.

To compare this with the binary measurement discussed in Section 1.2.4, we use the apparatus state encoding $|g\rangle \leftrightarrow |0\rangle_A$, $|i\rangle \leftrightarrow |1\rangle_A$. The overall interaction ($R_2 \circ C \circ R_1$) between the system and the apparatus is then defined by Eq. (1.62). We can then specify the apparatus operators $\hat{X}_A$ and $\hat{P}_A$ used in Section 1.2.6,

$$\hat{X}_A = |i\rangle\langle i|, \tag{1.184}$$

$$\hat{P}_A = \frac{1}{2}(|g\rangle - |i\rangle)(\langle g| - \langle i|). \tag{1.185}$$

Likewise the equivalent operators for the system can be defined in the photon-number basis, $\hat{X}_S = |1\rangle\langle 1|$, $\hat{P}_S = (|0\rangle - |1\rangle)(\langle 0| - \langle 1|)/2$. Provided that the atom is initially restricted to the subspace spanned by $\{|g\rangle, |i\rangle\}$, the action of $R_2 \circ C \circ R_1$ can be represented in terms of these operators by the unitary operator

$$\hat{U}_{R_2 \circ C \circ R_1} = \exp[i\pi \hat{X}_S \hat{P}_A]. \tag{1.186}$$

Certain aspects of the Paris experiment highlight the kinds of considerations that distinguish an actual measurement from simple theoretical models. To begin, it is necessary to prepare the states of the apparatus (the atoms) appropriately. Rubidium atoms from a thermal beam are first prepared by laser-induced optical pumping into the circular Rydberg states with principal quantum numbers 50 (for level g) or 51 (for level e). The e → g transition is resonant with a cavity field at 51.1 GHz. The auxiliary level, i, corresponds to a principal quantum number of 49 and the i → g transition is resonant at 54.3 GHz.

Next it is necessary to control the duration of the interaction between the system and the apparatus in order to establish the appropriate correlation. To do this, the atoms transiting the cavity field must have a velocity carefully matched to the cavity length. The optical-pumping lasers controlling the circular states are pulsed, generating at a preset time an atomic sample with on average 0.3–0.6 atoms. Together with velocity selection, this determines the atomic position at any time within ±1 mm. The single-photon Rabi frequency at the cavity centre is $\Omega/(2\pi) = 47$ kHz. The selected atomic velocity is 503 m s^{-1} and the beam waist inside the cavity is 6 mm, giving an effective interaction time τ such that $\Omega\tau = 2\pi$. Finally, a small external electric field Stark-shifts the atomic frequency out of resonance with the cavity. This gives rise to an adjustable detuning Δ in Eq. (1.180), which allows fine control of the effective interaction.

The experiment is designed to detect the presence or absence of a single photon. Thus it is necessary to prepare the cavity field in such a way as to ensure that such a state is typical. The cavity is cooled to below 1.2 K, at which temperature the average thermal excitation of photon number $\bar{n}$ in the cavity mode is 0.15. The thermal state of a cavity field is a mixed state of the form

$$\rho_c = (1 + \bar{n})^{-1} \sum_{n=0}^{\infty} e^{-n\beta} |n\rangle \langle n|, \tag{1.187}$$

where $\beta = \hbar\omega_c/(k_B T)$. At these temperatures, $\beta \ll 1$ and we can assume that the cavity field is essentially in the vacuum state $|0\rangle$. The small components of higher photon number lead to experimental errors.

In order to generate an average photon number large enough for one to see a single-photon signal, it is necessary to excite a small field coherently. This is done by injecting a 'preparatory' atom in the excited state, $|e\rangle$, and arranging the interaction time so that the atom-plus-cavity state is $|0, e\rangle + |1, g\rangle$. The state of this atom is then measured after the interaction. If it is found to be $|g\rangle$, then a single photon has been injected into the cavity field mode. If it is found to be $|e\rangle$, the cavity field mode is still the vacuum. Thus each run consists of randomly preparing either a zero- or a one-photon state and measuring it. Over many runs the results are accumulated, and binned according to what initial field state was prepared. The statistics over many runs are then used to generate the

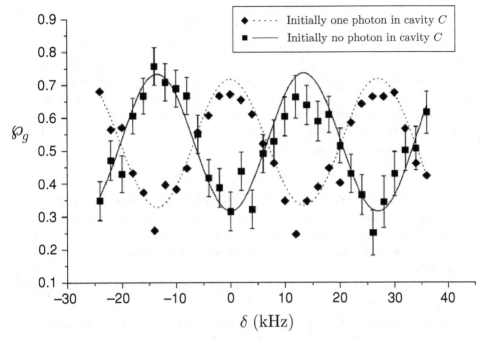

Fig. 1.4 The experimental results of the Paris single-photon experiment, showing the probability of measuring the atom in the ground state versus detuning of the cavity field. The dashed line corresponds to an initial field with a single photon, whereas the solid line is for an initial vacuum field state. Reprinted by permission from Macmillan Publishers Ltd: G. Nogues *et al.*, *Nature* **400**, 239–242 (1999), (Fig. 2), copyright Macmillan Magazines Ltd 1999.

conditional probability of finding the atom in the state $|g\rangle$ when there is one photon in the cavity.

Another refinement of the experiment is to use the detuning δ of the fields R_1 and R_2 to vary the quality of the measurement. This is a standard technique in atomic physics known as *Ramsey fringe interferometry*, or just Ramsey interferometry. This is explained in Box 1.4, where $|e\rangle$ plays the role of $|i\rangle$ in the present discussion. The extra Hamiltonian $\delta|i\rangle\langle i|$ causes free evolution of the atomic dipole. Its net effect is to introduce an extra phase factor δT, proportional to the time T between applications of each of these fields. The probability of finding the atom in state $|g\rangle$ at the end of measurement is then given by

$$\wp_g = \wp_0\mu + \wp_1(1-\mu), \tag{1.188}$$

where $\wp_0$ and $\wp_1$ are the probabilities that the cavity contains no or one photon, respectively, and $\mu = \cos^2(\delta T)$. If $\wp_0 = 1$ or 0 at the start of the measurement, then $\wp_g$ is an oscillatory function of the detuning δ, and the phase of the oscillation distinguishes the two cases.

In Fig. 1.4 we show the experimental results from the Paris experiment. Two cases are shown: in one case (dashed line) the initial state of the field was prepared in a one-photon state (the preparatory atom exited in the ground state), whereas in the second case (solid

Box 1.4 Ramsey fringe interferometry

Ramsey interferometry was developed to measure accurately the frequency ω_{eg} of an atomic transition. It works by producing a signal that depends on the difference δ between the unknown frequency and a known frequency ω_d. It is a standard technique for atomic frequency standards with application to time standards [Maj98]. We here give a simplified treatment of the essential physics behind the technique.

Consider a two-level atom, with ground and excited states $|g\rangle$ and $|e\rangle$, described in the interaction frame with respect to

$$\hat{H}_0 = \omega_g |g\rangle\langle g| + (\omega_g + \omega_d)|e\rangle\langle e|. \tag{1.189}$$

The atom is prepared in the ground state and injected through a classical field R_1 with frequency ω_d that differs from the atomic resonance frequency ω_{eg} by a small detuning δ. The atomic velocity is chosen so that the atom interacts with the field for a precise time τ. The interaction induces a superposition between the ground and excited states of the form

$$|g\rangle \rightarrow \alpha|g\rangle + \beta|e\rangle, \tag{1.190}$$

where the coefficients depend on τ and δ and the Rabi frequency for the transition. (The Rabi frequency is roughly the dot product of the classical electric field with the electric dipole moment of the atomic transition, divided by $\hbar$. It also equals the single-photon Rabi frequency times the square root of the mean number of photons in the field. For a classical field in a mode with a large mode volume (as here), the former is very small and the latter very large, giving a finite product.) If the detuning δ is small enough, one can arrange to obtain $\alpha = \beta = 1/\sqrt{2}$.

The atom then evolves freely for a time T during which the Hamiltonian in the interaction frame is $\hat{V}_{\rm IF} = \delta|e\rangle\langle e|$. This changes β to $\beta e^{-i\delta T}$. After this it interacts with another classical field, R_2, of the same frequency, which undoes the transformation R_1. This means that we have to adjust T and/or the phase of R_2 so that, if $\delta = 0$, all atoms emerge in the ground state. Then the state of the atom after the second field is

$$\cos(\delta T/2)|g\rangle - i\sin(\delta T/2)|e\rangle. \tag{1.191}$$

The probability that an atom will emerge in the excited state when $\delta \neq 0$ is thus

$$\wp_e(\delta) = \sin^2(\delta T/2). \tag{1.192}$$

By varying the frequency ω_d of the driving fields R_1 and R_2, and sampling this probability by repeated measurement, we produce interference fringes with a spacing proportional to T^{-1}. A complicating effect is that, for large detuning, the coefficients α and β are not exactly $1/\sqrt{2}$, but also depend on the detuning δ in both amplitude and phase, and this causes the interference-fringe visibility to decrease.

line) the field was prepared in a zero-photon state (the preparatory atom exited in the excited state). In both cases the probability of finding the apparatus atom in the ground state, $|g\rangle$, is plotted as a function of the detuning of the R_1 and R_2 fields. Note that the two cases are π out of phase, as expected.

It is quite apparent from the data that the measurement is far from perfect. The probabilities do not vary from zero to unity, so the contrast or visibility, defined as $(\wp_{max} - \wp_{min})/(\wp_{max} + \wp_{min})$, is not unity. A primary source of error is the efficiency of the ionization detectors, which is as low as 30%. Also, the interaction that correlates the field and the apparatus is not perfect, and there is a 20% residual probability for the apparatus atom to absorb the photon, rather than induce a π conditional phase shift. Other sources of error are imperfections of the $\pi/2$ Ramsey pulses, samples containing two atoms in the cavity, the residual thermal field in the cavity and the possibility that the injected photon will escape from the cavity before the detection atom enters.

Exercise 1.34 *What is the effect of imperfect ionization detection on the readout? Calculate the mean and variance of the readout variable Y in terms of the system (X) mean and variance for the binary asymmetric measurement defined by the conditioned probabilities $\wp(y|x)$ below:*

$$\wp(1|1) = \eta = 1 - \wp(0|1),$$

$$\wp(1|0) = \epsilon = 1 - \wp(0|0).$$

Here η is the detection efficiency, while ϵ is related to the rate of so-called dark-counts.

1.6 Further reading

Bayesian inference is named after its discoverer, the Reverend Thomas Bayes, who published the idea in 1764. For a modern treatment, see *Bayesian Theory* by Bernardo and Smith [BS94]. It is also covered in *Introduction to Control Theory* by Jacobs [Jac93], and *Optimal Control Theory* by Whittle [Whi96], which are useful references for later chapters as well.

For good introductions to quantum mechanics, see the textbooks by Peres [Per95] and Jauch [Jau68]. Both are notable for discussion of the interpretation of Gleason's theorem. An on-line (in 2009) text that contains a good introduction to the quantum formalism using the Dirac notation is that of Preskill [Pre97].

The theory of generalized measurement has a long history. Early work by Davies and Kraus is summarized in *Quantum Theory of Open Systems* [Dav76], and *States, Effects, Operations: Fundamental Notions of Quantum Theory* [Kra83], respectively. However, both these sources are rather mathematical in style. A more readable book (for physicists) is *Quantum Measurement* by Braginsky and Khalili [BK92]. This is particularly good at helping to build intuition based on simple ideas such as the Heisenberg uncertainty principle. It also contains a detailed discussion of quantum non-demolition measurements.

A different approach to obtaining a formal analogy between classical measurement theory and quantum measurement theory has been developed by Hardy in Ref. [Har02]. In this he derives, for systems with a finite number of distinguishable states, both quantum theory and classical theory from a small number of plausible axioms. The extra axiom that leads to quantum theory rather than classical theory is that there is a *continuous* reversible transformation between any two pure states (pure states having been operationally defined). As suggested by our presentation, the essential difference that follows is one of dimensionality. In the classical case, K, the number of probabilities needed to specify the state, is equal to N, the maximum number of states that can be reliably distinguished in a single shot. In the quantum case, K is equal to N^2, reflecting the fact that quantum states are represented most naturally as matrices. Note, however, that, in the quantum case, Hardy's map from the K probabilities defining the state to the probabilities of a particular measurement outcome is *not* a positive map – another difference from the classical case.

The consistency of states, both classical and quantum, is much more involved than the treatment we have given (which is equivalent to that in Ref. [BFM02]). The definitions we have given can be viewed as simply one definition in a hierarchy of operational definitions. Interestingly, the hierarchy has a different structure in the quantum and classical cases. This is discussed in detail by Caves, Fuchs and Schack in Ref. [CFS02a]. In this and other papers [CFS02b], these authors strongly argue for interpreting quantum theory using the analogy with classical Bayesian inference. A related question is that of how two or more observers can pool their states of knowledge. For a review of work in this area, and some special results in the quantum case, see Ref. [SW07].

Finally, we note that the work on QND photon detection has advanced far beyond that of Ref. [NRO$^+$99] described above. See for example Refs. [GKG$^+$07, GBD$^+$07], also from the laboratory of Haroche, in which back-action-evading measurements of photon number are performed. This enabled the experimentalists to distinguish several different photon numbers with great accuracy, and to observe quantum jumps between them.

2

Quantum parameter estimation

2.1 Quantum limits to parameter estimation

2.1.1 Introduction

Many experiments can be thought of as comprising two steps: (i) a preparation procedure in which the system to be measured is isolated and prepared, and the apparatus is initialized; and (ii) a measurement step in which the system is coupled to an apparatus and the measurement result recorded. The preparation procedure can be specified by a set of classical parameters, or settings of a physical device. The measurement results are random classical variables that will be correlated with the preparation procedure. In this chapter we are concerned with the case in which the classical parameters specifying the preparation of the state are imperfectly known. Then, assuming that the physical system is well understood, these correlations allow the unknown parameters to be estimated from the measurement results.

As we saw in the last chapter, in quantum mechanics the results of measurements are generally statistical, even when one has complete knowledge of the preparation procedure. A single preparation step and measurement step might not be sufficient to estimate a parameter well. Thus it is common to repeat the two steps of preparation and measurement on a large number of systems, either all at one time or sequentially. Whether measuring one quantum system or many, one is faced with a number of questions. How should one prepare the system state? What sort of measurement should one make on the system? What is the optimum way to extract the parameter from the measurement results? These questions generally fall under the heading of quantum parameter estimation, the topic of this chapter.

From the above, it should be clear that the 'quantum' in 'quantum parameter estimation' is there simply because it is a quantum system that mediates the transfer of information from the classical parameters of the preparation procedure to the classical measurement results. If the system were classical then in principle the parameters could be estimated with perfect accuracy from a single measurement. The role of quantum mechanics is therefore to set *limits* on the accuracy of the estimation of these parameters. Since all of Nature is ultimately quantum mechanical, these are of course fundamental limits. In the following section, we show how a simple expression for quantum limits to parameter estimation can be easily derived.

51

2.1.2 The Helstrom–Holevo lower bound

The classic studies of Holevo [Hol82] and Helstrom [Hel76] laid the foundations for a study of quantum parameter estimation. They did so in the context of quantum communication theory. This fits within the above paradigm of parameter estimation as follows. The message or signal is the set of parameters used by the sender to prepare the state of a quantum system. This system is the physical medium for the message. The receiver of the system must make a measurement upon it, to try to recover the message. In this context we need to answer questions such as the following. What are the physical limitations on encoding the message (i.e. in preparing a sequence of quantum systems)? Given these, what is the optimal way to encode the information? What are the physical limitations on decoding the message (i.e. in measuring a sequence of quantum systems)? Given these, what is the optimal way to decode the information?

The work of Helstrom and Holevo established an ultimate quantum limit for the error in parameter estimation. We refer to this as the Helstrom–Holevo *lower bound*. A special case can be derived simply as follows. We assume that the sender has access to a quantum system in a *fiducial state* ρ_0. The only further preparation the sender can do is to transform the state:

$$\rho_0 \rightarrow \rho_X = e^{-iX\hat{G}}\rho e^{iX\hat{G}}. \tag{2.1}$$

Here $\hat{G}$ is an Hermitian operator known as the *generator*, and X is a real parameter. The aim of the receiver is to estimate X. The receiver does this by measuring a quantity X_{est}, which is the receiver's best estimate for X. We assume that X_{est} is represented by an Hermitian operator $\hat{X}_{\text{est}}$.

The simplest way to characterize the quality of the estimate is the mean-square deviation

$$\langle(X_{\text{est}} - X)^2\rangle_X = \langle(\Delta X_{\text{est}})^2\rangle_X + [b(X)]^2. \tag{2.2}$$

Here this is decomposed into the variance of the estimator (in the transformed state),

$$\langle(\Delta X_{\text{est}})^2\rangle_X = \text{Tr}\big[(\hat{X}_{\text{est}} - \langle X_{\text{est}}\rangle_X)^2\rho_X\big], \tag{2.3}$$

plus the square of the *bias* of the estimator X_{est},

$$b(X) = \langle X_{\text{est}}\rangle_X - X, \tag{2.4}$$

that is, how different the mean of the estimator $\langle X_{\text{est}}\rangle_X = \text{Tr}\big[\hat{X}_{\text{est}}\rho_X\big]$ is from the true value of X.

We now derive an inequality for the mean-square deviation of the estimate. First we note that, from Eq. (2.1),

$$\frac{d\langle X_{\text{est}}\rangle_X}{dX} = -i\,\text{Tr}\big[[\hat{X}_{\text{est}}, \hat{G}]\rho_X\big]. \tag{2.5}$$

Using the general Heisenberg uncertainty principle (A.9),

$$\langle(\Delta X_{\text{est}})^2\rangle_X\langle(\Delta G)^2\rangle_X \geq \frac{1}{4}\big|\text{Tr}\big[[\hat{X}_{\text{est}}, \hat{G}]\rho_X\big]\big|^2, \tag{2.6}$$

we find that

$$\langle(\Delta X_{\text{est}})^2\rangle_X \langle(\Delta G)^2\rangle_X \geq \frac{1}{4}\left|\frac{\mathrm{d}\langle X_{\text{est}}\rangle_X}{\mathrm{d}X}\right|^2. \tag{2.7}$$

This inequality then sets a lower bound for the mean-square deviation,

$$\langle(X_{\text{est}} - X)^2\rangle_X \geq \frac{[1 + b'(X)]^2}{4\langle(\Delta G)^2\rangle_X} + b^2(X). \tag{2.8}$$

If there is no systematic error in the estimator, then $\langle X_{\text{est}}\rangle_X = X$ and the bias is zero, $b(X) = 0$. In this case, the lower bound to the mean-square deviation of the estimate is

$$\langle(\Delta X_{\text{est}})^2\rangle_X \geq \frac{1}{4\langle(\Delta G)^2\rangle_0}. \tag{2.9}$$

Here we have set $X = 0$ on the right-hand side using the fact that $\hat{G}$ commutes with the unitary parameter transformation, so that moments of the generator do not depend on the shift parameter.

Consider the case in which $\hat{X}_{\text{est}}$ and $\hat{G}$ are conjugate operators, by which we mean

$$[\hat{G}, \hat{X}_{\text{est}}] = -\mathrm{i}. \tag{2.10}$$

Assume also that $\text{Tr}\left[\hat{X}_{\text{est}}\rho_0\right] = 0$, so that $\langle X_{\text{est}}\rangle_X = X$.

Exercise 2.1 *Show this, by considering the Taylor-series expansion for $e^{\mathrm{i}X\hat{G}}\hat{X}_{\text{est}}e^{-\mathrm{i}X\hat{G}}$ and using Eq. (2.10).*

In this case the parameter-estimation uncertainty principle, given in Eq. (2.9), follows directly from the general Heisenberg uncertainty relation (2.6) on using the commutation relations (2.10).

The obvious example of canonically conjugate operators is position and momentum. Let the unitary parameter transformation be $\exp(-\mathrm{i}X\hat{P})$, where $\hat{P}$ is the momentum operator. Let $\hat{Q}$ be the canonically conjugate position operator, defined by the inner product $\langle q|p\rangle = \exp(\mathrm{i}pq)/\sqrt{2\pi}$. Then $[\hat{Q}, \hat{P}] = \mathrm{i}$ (see Appendix A). Choosing $\hat{X}_{\text{est}} = \hat{Q} - \text{Tr}\left[\hat{Q}\rho_0\right]$ gives

$$\langle(X_{\text{est}} - X)^2\rangle \geq \frac{1}{4\langle(\Delta P)^2\rangle_0}. \tag{2.11}$$

In general, $\hat{X}_{\text{est}}$ need not be canonically conjugate to the generator $\hat{G}$. Indeed, in general $\hat{X}_{\text{est}}$ need not be an operator in the Hilbert space of the system at all. This means that the Holevo–Helstrom lower bound on the mean-square deviation in the estimate applies not only to projective measurements on the system. It also applies to generalized measurements described by effects. This is because, as we have seen in Section 1.3.2, a generalized measurement on the system is equivalent to a projective measurement of a joint observable on system and meter, namely the unitarily evolved meter readout observable $\hat{R}_A(t + T)$. In fact, it turns out that in many cases the *optimal measurement* is just such a generalized measurement. This is one of the most important results to come out of the work by Helstrom and Holevo.

In the above we have talked of optimal measurement without defining what we mean by optimal. Typically we assume that the receiver knows the fiducial state ρ_0 and the generator $\hat{G}$, but has no information about the parameter X. The aim of the receiver is then to minimize some *cost function* associated with the error in X_{est}. The Helstrom–Holevo lower bound relates to a particular cost function, the mean-square error $\langle (X_{\text{est}} - X)^2 \rangle$. An alternative cost function is $-\delta(X_{\text{est}} - X)$, which when minimized yields the *maximum-likelihood estimate*. However, the singular nature of this function makes working with it difficult. We will consider other alternatives later in this chapter, and in the next section we treat in detail optimality defined in terms of Fisher information.

2.2 Optimality using Fisher information

In some contexts, it makes sense to define optimality in terms other than minimizing the mean-square error $\langle (X_{\text{est}} - X)^2 \rangle$. In particular, for measurements repeated an asymptotically large number of times, what matters asymptotically is not the mean-square error, but the *distinguishability*. That is, how well two slightly different values of X can be distinguished on the basis of a set of M measurement results derived from an ensemble of M systems, each prepared in state ρ_X. This notion of distinguishability is quantified by the *Fisher information* [KJ93], defined as

$$F(X) = \int d\xi \, \wp(\xi|X) \left(\frac{d \ln \wp(\xi|X)}{dX} \right)^2 . \tag{2.12}$$

Here ξ is the result of a measurement of X_{est} on a single copy of the system.

As we will show, not only can this quantity be used instead of the mean-square error, but also it actually provides a stronger lower bound for it than does the Holevo–Helstrom bound:

$$\langle (\delta X_{\text{est}})^2 \rangle_X \geq \frac{1}{M F(X)} \geq \frac{1}{M 4 \langle (\Delta G)^2 \rangle_0} . \tag{2.13}$$

Here M is the number of copies of the system used to obtain the estimate X_{est}. The deviation δX_{est} is not $X_{\text{est}} - X$, but rather

$$\delta X_{\text{est}} = \frac{X_{\text{est}}}{|d \langle X_{\text{est}} \rangle_X / dX|} - X . \tag{2.14}$$

This is necessary to compensate for any bias in the estimate, and, for an unbiased estimate, $\langle (\delta X_{\text{est}})^2 \rangle_X$ reduces to the mean-square error. Equation (2.13) will be derived later in this section, but first we show how the Fisher information arises from consideration of distinguishability.

2.2.1 Distinguishability and Fisher information

As noted in Box 1.4, Ramsey interferometry is a way to measure the passage of time using measurements on a large ensemble of two-level atoms. The probability for an atom to be found in the ground state by the final measurement is $\wp_g = \cos^2\theta$, where $\theta = \delta T/2$. Here δ is an adjustable detuning and T is the time interval to be measured.

Clearly a measurement on a single atom would not tell us very much about the parameter θ. If we could actually measure the probability $\wp_g$ then we could easily determine θ. However, the best we can do is to measure whether the atom is in the ground state on a large number M of atoms prepared in the same way. For a finite sample we will then obtain an estimate f_g of $\wp_g$, equal to the observed frequency of the ground-state outcome. Owing to statistical fluctuations, this estimate will not be exactly the same as the actual probability.

In a sample of size M, the probability of obtaining m_g outcomes in the ground state is given by the binomial distribution

$$\wp^{(M)}(m_g) = \binom{M}{m_g} \wp_g(\theta)^{m_g} (1 - \wp_g(\theta))^{M-m_g}. \tag{2.15}$$

The mean and variance for the fraction $f_g = m_g/M$ are

$$\langle f_g \rangle_\theta = \wp_g(\theta), \tag{2.16}$$

$$\langle (\Delta f_g)^2 \rangle_\theta = \wp_g(\theta)(1 - \wp_g(\theta))/M. \tag{2.17}$$

It is then easy to see that the error in estimating θ by estimating the probability $\wp_g(\theta)$ in a finite sample is

$$\delta\theta = \left| \frac{\mathrm{d}\wp_g}{\mathrm{d}\theta} \right|^{-1} \delta\wp_g = \left| \frac{\mathrm{d}\wp_g}{\mathrm{d}\theta} \right|^{-1} \left[\frac{\wp_g(1 - \wp_g)}{M} \right]^{1/2}. \tag{2.18}$$

In order to be able to measure a small shift, $\Delta\theta = \theta' - \theta$, in the parameter from some fiducial setting, θ, the shift must be larger than this error: $\Delta\theta \geq \delta\theta$.

Since $\delta\theta$ is the minimum distance in parameter space between two distinguishable distributions, we can characterize the *statistical distance* between two distributions as the number of distinguishable distributions that can fit between them, along a line joining them in parameter space. This idea was first applied to quantum measurement by Wootters [Woo81]. Because $\delta\theta$ varies inversely with the square root of the the sample size M, we define the statistical distance between two distributions with close parameters θ and θ' as

$$\Delta s = \lim_{M \to \infty} \frac{1}{\sqrt{M}} \frac{\Delta\theta}{\delta\theta}. \tag{2.19}$$

Strictly, for any finite difference $\Delta\theta$ we should use the integral form

$$\Delta s(\theta, \theta') = \lim_{M \to \infty} \frac{1}{\sqrt{M}} \int_\theta^{\theta'} \frac{\mathrm{d}\theta}{\delta\theta}. \tag{2.20}$$

Exercise 2.2 *Show that, for this case of Ramsey interferometry, $\delta\theta = 1/2\sqrt{M}$, independently of θ, so that $\Delta s(\theta, \theta') = 2|\theta' - \theta|$.*

The result in Eq. (2.20) is a special case of a more general result for a probability distribution for a measurement with K outcomes. Let $\wp_k$ be the probability for the outcome

k. It can be shown that the infinitesimal statistical distance ds between two distributions, $\wp_k$ and $\wp_k + \mathrm{d}\wp_k$, is best defined by [CT06]

$$(\mathrm{d}s)^2 = \sum_{k=1}^{K} \frac{(\mathrm{d}\wp_k)^2}{\wp_k} = \sum_{k=1}^{K} \wp_k (\mathrm{d}\ln\wp_k)^2. \tag{2.21}$$

If we assume that the distribution depends on a single parameter X, then we have

$$\left(\frac{\mathrm{d}s}{\mathrm{d}X}\right)^2 = \sum_{k} \wp_k \left(\frac{\mathrm{d}\ln\wp_k(X)}{\mathrm{d}X}\right)^2 \equiv F(X), \tag{2.22}$$

where this quantity is known as the *Fisher information*. The generalization for continuous readout results was already given in Eq. (2.12).

Clearly the Fisher information has the same dimensions as X^{-2}. From the above arguments we see that the reciprocal square root of $MF(X)$ is a measure of the change ΔX in the parameter X that can be detected reliably by M trials. It can be proven that $(\Delta X)^2$ is a lower bound to the mean of the square of the 'debiased' error (2.14) in the estimate X_{est} of X from the set of M measurement results. That is,

$$\langle(\delta X_{\mathrm{est}})^2\rangle_X \geq \frac{1}{MF(X)}. \tag{2.23}$$

This, the first half of Eq. (2.13), is known as the *Cramér–Rao lower bound*.

Exercise 2.3 *Show that, for the Ramsey-interferometry example, $F(\theta) = 4$, independently of θ. If, for $M = 1$, one estimates θ as 0 if the atom is found in the ground state and $\pi/2$ if it is found in the excited state, show that*

$$\langle(\delta\theta_{\mathrm{est}})^2\rangle_\theta = \theta^2 \cos^2\theta + (\theta - |\csc(2\theta)|)^2 \sin^2\theta. \tag{2.24}$$

Verify numerically that the inequality Eq. (2.23) is always satisfied, and is saturated at discrete points, $\theta \approx 0, 1.1656, 1.8366, \ldots$.

Estimators that saturate the Cramér–Rao lower bound at *all* parameter values are known in the statistical literature as *efficient*. We will not use that term, because we use it with a very different meaning for quantum measurements. Instead we will call such estimators Cramér–Rao optimal (CR optimal).

Exercise 2.4 *Show that, if $\wp(\xi|X)$ is a Gaussian of mean X, then $X_{\mathrm{est}} = \xi$ is a Cramér–Rao-optimal estimate of X.*

2.2.2 Quantum statistical distance

The Cramér–Rao lower bound involves properties of the probability distribution $\wp(\xi|X)$ for a single measurement on a system parameterized by X. Quantum measurement theory so far has only entered in that the system is taken to be a quantum system, so that $\wp(\xi|X)$

is generated by some POM:

$$\wp(\xi|X) = \text{Tr}\big[\rho_X \hat{E}_\xi\big]. \tag{2.25}$$

By contrast, the third expression in Eq. (2.13) involves only the properties of $\rho(X) = \exp(-i\hat{G}X)\rho(0)\exp(i\hat{G}X)$. Thus, to prove the second inequality in Eq. (2.13), we must seek an upper bound on the Fisher information over the set of all possible POMs $\{\hat{E}_\xi : \xi\}$. Recall that the Fisher information is related to the squared statistical distance between two distributions $\wp(\xi|X)$ and $\wp(\xi|X + dX)$:

$$(ds)^2 = (dX)^2 F(X). \tag{2.26}$$

What we want is a measure of the squared distance between two states ρ_X and ρ_{X+dX} that generate the distributions:

$$(ds_Q)^2 = (dX)^2 \max_{\{\hat{E}_\xi : \xi\}} F(X). \tag{2.27}$$

Here we use the notation ds_Q to denote the infinitesimal *quantum statistical distance* between two states. Clearly $(ds_Q)^2/(dX)^2$ will be the sought upper bound on $F(X)$.

We now present a heuristic (rather than rigorous) derivation of an explicit expression for $(ds_Q)^2/(dX)^2$. The classical statistical distance in Eq. (2.21) can be rewritten appealingly as

$$(ds)^2 = 4 \sum_{k=1}^{K} (da_k)^2, \tag{2.28}$$

where $a_k = \sqrt{\wp_k}$.

Exercise 2.5 *Verify this.*

That is, the statistical distance between probability distributions is the Euclidean distance between the *probability amplitude* distributions. Now consider two quantum states, $|\psi\rangle$ and $|\psi\rangle + d|\psi\rangle$. In terms of distinguishing these states, any change in the global phase is irrelevant; the only relevant change is the part $|d\psi_\perp\rangle = (1 - |\psi\rangle\langle\psi|)d|\psi\rangle$ which is orthogonal to the first state. That is, without loss of generality, we can take the second state to be $|\psi\rangle + |d\psi_\perp\rangle$.

From the above it is apparent that, regardless of the dimensionality of the system, there are only two relevant basis states for the problem, $|1\rangle \equiv |\psi\rangle$ and $|2\rangle \equiv |d\psi_\perp\rangle/||d\psi_\perp\rangle|$. That is, the system effectively reduces to a two-dimensional system. It is intuitively clear that the relevant eigenstates for the optimal measurement observable will be linear combinations of $|1\rangle$ and $|2\rangle$. If we choose to measure in the basis $|1\rangle$ and $|2\rangle$, then the result $k = 2$ will imply that the state is definitely $|\psi\rangle + d|\psi\rangle$, whereas the result 1 leaves us uncertain as to the state. In this case, we have $(da_1)^2 = 0$, and $(da_2)^2 = \langle d\psi_\perp|d\psi_\perp\rangle$. Thus, Eq. (2.27) evaluates to

$$(ds)_Q^2 = 4\langle d\psi_\perp|d\psi_\perp\rangle. \tag{2.29}$$

Exercise 2.6 *Show that this expression holds for* any *projective measurement in the two-dimensional Hilbert space spanned by* $|1\rangle$ *and* $|2\rangle$. *(This follows from the result of Exercise 2.2.)*

For the case of single-parameter estimation with $|\psi_X\rangle = \exp(-i\hat{G}X)|\psi_0\rangle$, it is clear that

$$|d\psi_\perp\rangle = -i(\hat{G} - \langle G\rangle_X)|\psi_X\rangle dX. \tag{2.30}$$

Thus we get

$$\left(\frac{ds_Q}{dX}\right)^2 = 4\langle(\Delta G)^2\rangle_0, \tag{2.31}$$

proving the second lower bound in Eq. (2.13) for the pure-state case.

The case of mixed states is considerably more difficult. The explicit form for the quantum statistical distance turns out to be

$$(ds)_Q^2 = \text{Tr}[d\rho\,\mathcal{L}[\rho]d\rho]. \tag{2.32}$$

Here $\mathcal{L}[\rho]$ is a superoperator taking ρ as its argument. If ρ has the diagonal representation $\rho = \sum_j p_j|j\rangle\langle j|$, then

$$\mathcal{L}[\rho]\hat{A} = \sum_{j,k}{}' \frac{2}{p_j + p_k} A_{jk}|j\rangle\langle k|, \tag{2.33}$$

where the prime on the sum means that it excludes the terms for which $p_j + p_k = 0$. If ρ has all non-zero eigenvalues, then $\mathcal{L}[\rho]$ can be defined more elegantly as

$$\mathcal{L}[\rho] = \mathcal{R}^{-1}[\rho], \tag{2.34}$$

where the action of the superoperator $\mathcal{R}[\rho]$ on an arbibtrary operator $\hat{A}$ is defined as

$$\mathcal{R}[\rho]\hat{A} = (\rho\hat{A} + \hat{A}\rho)/2. \tag{2.35}$$

It is clear that $\mathcal{L}[\rho]$ is a superoperator version of the reciprocal of ρ. With this understanding, Eq. (2.32) also looks like a quantum version of the classical statistical distance (2.21).

Exercise 2.7 *Show that, for the pure-state case, Eq. (2.32) reduces to Eq. (2.29), by using the basis* $|1\rangle$, $|2\rangle$ *defined above.*

Now consider again the case of a unitary transformation as X varies, so that

$$d\rho = -i[\hat{G}, \rho]dX. \tag{2.36}$$

To find $(ds)_Q^2$ from Eq. (2.32), we first need to find the operator $\hat{A} = \mathcal{R}^{-1}[\rho]d\rho$. From Eq. (2.35), this must satisfy

$$(\rho\hat{A} + \hat{A}\rho) = -2i[\hat{G}, \rho]dX. \tag{2.37}$$

If $\rho = \hat{\pi}$ (a pure state satisfying $\hat{\pi}^2 = \hat{\pi}$), then $\hat{A} = -2i[\hat{G}, \hat{\pi}]dX = 2\,d\hat{\pi}$ is a solution of this equation. That gives

$$\left(\frac{ds_Q}{dX}\right)^2 = -2\,\mathrm{Tr}\big[[\hat{G}, \hat{\pi}]^2\big] = 4\langle(\Delta G)^2\rangle_0\,, \tag{2.38}$$

as found above. If ρ is not pure then it can be shown that

$$\left(\frac{ds_Q}{dX}\right)^2 \leq 4\langle(\Delta G)^2\rangle_0. \tag{2.39}$$

Putting all of the above results together, we have now three inequalities:

$$M\langle(\delta X_{\mathrm{est}})^2\rangle_X \geq \frac{1}{F(X)} \geq \left(\frac{dX}{ds_Q}\right)^2 \geq \frac{1}{4\langle(\Delta G)^2\rangle_0}. \tag{2.40}$$

The first of these is the classical Cramér–Rao inequality. The second we will call the Braunstein–Caves inequality. It applies even if the transformation of $\rho(X)$ as X varies is non-unitary. That is, even if there is no $\hat{G}$ that generates the transformation. The final inequality obviously applies only if there is such a generator. In the case of pure states, it can be replaced by an equality.

Omitting the second term (the classical Fisher information) gives what we will call the Helstrom–Holevo inequality. Like the Cramér–Rao inequality, this cannot always be saturated for a given set $\{\rho_X\}_X$. The advantage of the Braunstein–Caves inequality is that it can always be saturated, as is clear from the definition of the quantum statistical distance in Eq. (2.27). If there is a unitary transformation generated by $\hat{G}$, omitting both the second and the third term gives the inequality (2.9) for the special case of unbiased estimates with $M = 1$. As is apparent, the inequality (2.9) was derived much more easily than those in Eq. (2.40), but the advantage of generality and saturability offered by Eq. (2.40) should also now be apparent.

2.2.3 Achieving Braunstein–Caves optimality

For the case of pure states $|\psi_X\rangle = \exp(-i\hat{G}X)|\psi_0\rangle$, the inequality

$$F(X) = 4\langle(\Delta G)^2\rangle_0 \tag{2.41}$$

can always be saturated, and the POM $\{E_\xi\}_\xi$ that achieves this for all X we will call Braunstein–Caves optimal (BC optimal). Clearly this optimality is for a given fiducial state $|\psi_0\rangle$. We will follow the reasoning of Ref. [BCM96] in seeking this optimality.

Requiring the bound to be achieved for all X suggests considering distributions $\wp(\xi|X)$ that are functions of $\xi - X$ only. Since

$$\wp(\xi|X) = \langle\psi_0|e^{i\hat{G}X}\hat{E}(\xi)e^{-i\hat{G}X}|\psi_0\rangle, \tag{2.42}$$

this implies that

$$e^{iX\hat{G}}\hat{E}(\xi)e^{-iX\hat{G}} = \hat{E}(\xi - X). \tag{2.43}$$

Such measurements are called *covariant* by Holevo [Hol82]. In addition we will posit that the optimal POM is a multiple of a projection operator:

$$\hat{E}(\xi)d\xi = \mu|\xi\rangle\langle\xi|d\xi \tag{2.44}$$

for μ a real constant. It is important to note that we do *not* require the states $\{|\xi\rangle\}$ to be orthogonal.

Since the POM is independent of a change of phase for the states $|\xi\rangle$, we can choose with no loss of generality

$$e^{-iX\hat{G}}|\xi\rangle = |\xi + X\rangle. \tag{2.45}$$

This means that

$$\langle\xi|e^{-iX\hat{G}}|\psi\rangle = \langle\xi - X|\psi\rangle = \exp\left(-X\frac{\partial}{\partial\xi}\right)\langle\xi|\psi\rangle. \tag{2.46}$$

In other words the generator $\hat{G}$ is a displacement operator in the $|\xi\rangle$ representation:

$$\hat{G} \equiv -i\frac{\partial}{\partial\xi}. \tag{2.47}$$

Moreover, the probability distribution $\wp(\xi|X)$ is simply expressed in the ξ representation as

$$\wp(\xi|X) = |\psi_X(\xi)|^2 = |\psi_0(\xi - X)|^2 \equiv \wp_0(\xi - X), \tag{2.48}$$

where $\psi(\xi) \equiv \langle\xi|\psi\rangle/\sqrt{\mu}$.

For the POM to be optimal, it must maximize the Fisher information at $F = 4\langle(\Delta G)^2\rangle_0$. For a covariant measurement, the Fisher information takes the form

$$F = \int d\xi \frac{[\wp_0'(\xi)]^2}{\wp_0(\xi)}, \tag{2.49}$$

where the prime here denotes differentiation with respect to the argument. Note that the conditioning on the true value X has been dropped, because for a covariant measurement F is independent of X. Braunstein and Caves have shown [BC94] that F is maximized if and only if the wavefunction of the fiducial state is, up to an overall phase, given by

$$\psi_0(\xi) = \sqrt{\wp_0(\xi)}e^{i\langle G\rangle_0\xi}. \tag{2.50}$$

To see this we can calculate the mean and variance of $\hat{G}$ in the $|\xi\rangle$ representation for the state $\sqrt{\wp_0(\xi)}e^{i\Theta(\xi)}$,

$$\langle G\rangle_0 = \int d\xi \, \psi_0^*(\xi)\left(-i\frac{\partial}{\partial\xi}\right)\psi_0(\xi) \tag{2.51}$$

$$= \int d\xi \, \wp_0(\xi)\Theta'(\xi), \tag{2.52}$$

$$\langle(\Delta G)^2\rangle_0 = \int d\xi \, \psi_0^*(\xi)\left(-i\frac{\partial}{\partial\xi} - \langle\hat{G}\rangle_0\right)^2 \psi_0(\xi) \tag{2.53}$$

$$= \frac{1}{4}\int d\xi \frac{[\wp_0'(\xi)]^2}{\wp_0(\xi)} + \int d\xi \, \wp_0(\xi)[\Theta'(\xi) - \langle G\rangle_0]^2. \tag{2.54}$$

Exercise 2.8 *Verify these results.*

Clearly, for the Fisher information to attain its maximum value, the phase $\Theta(\xi)$ must be linear in ξ with slope $\langle G \rangle_0$.

We now address the conditions under which a BC-optimal measurement is also a CR-optimal measurement. That is, when it also saturates the first inequality in Eq. (2.40). The mean and variance of the measurement result Ξ are

$$\langle \Xi \rangle_X = \int d\xi \, \wp_0(\xi - X)\xi = X + \langle \Xi \rangle_0, \qquad (2.55)$$

$$\langle (\Delta \Xi)^2 \rangle_X = \int d\xi \, \wp_0(\xi - X)(\xi - \langle \Xi \rangle_X)^2 = \langle (\Delta \Xi)^2 \rangle_0. \qquad (2.56)$$

Thus there may be a global bias in the mean of Ξ, but the variance of Ξ is independent of X. Suppose now we make M measurements and form the unbiased estimator

$$X_{\text{est}} = \frac{1}{M} \sum_{j=1}^{M} (\Xi_j - \langle \Xi \rangle_0). \qquad (2.57)$$

The deviation is $\delta X_{\text{est}} = X_{\text{est}} - X$ and

$$\langle (\delta X_{\text{est}})^2 \rangle = \langle (\Delta \Xi)^2 \rangle_0 / M. \qquad (2.58)$$

Thus, to be a CR-optimal measurement for any M, the POM must saturate the Cramér–Rao bound for $M = 1$. It can be shown that this requires that $\wp_0(\xi)$ be a Gaussian. (Recall Exercise 2.4.) It is very important to remember, however, that, for a given generator $\hat{G}$, physical restrictions on the form of the wavefunctions may make Gaussian states impossible. Thus there may be no states that achieve the Cramér–Rao lower bound for a BC-optimal measurement. Moreover, if we choose estimators other than the sample mean, the fiducial wavefunction that achieves the lower bound need not be Gaussian. In particular, for $M \to \infty$, a maximum-likelihood estimate of X will be CR optimal for any wavefunction of the form (2.50).

The relation

$$\langle (\Delta \Xi)^2 \rangle \langle (\Delta G)^2 \rangle \geq 1/4 \qquad (2.59)$$

looks like the Heisenberg uncertainty relation of the usual form, since $\hat{G} = -i \, \partial/\partial\xi$ in the ξ representation. However, nothing in our derivation assumed that the states $|\xi\rangle$ were the eigenstates of an Hermitian operator. Indeed, as we shall see, there are many examples for which the BC-optimal measurement is described by a POM with non-orthogonal elements. This is an important reason for introducing generalized measurements, as we did in Chapter 1. One further technical point should be made. In order to find the BC-optimal measurement, we must carefully consider the states for which the generator $\hat{G}$ is a displacement operator. If $\hat{G}$ has a degenerate spectrum then it is not possible to find a BC-optimal measurement in terms of a POM described by a single real number ξ. Further details may be found in [BCM96].

2.3 Examples of BC-optimal parameter estimation

2.3.1 Spatial displacement

The generator of spatial displacements is the momentum, $\hat{P}$, so we consider families of states defined by

$$|\psi_X\rangle = e^{-iX\hat{P}}|\psi_0\rangle. \tag{2.60}$$

The uncertainty relation Eq. (2.59) then becomes

$$\langle(\delta X_{\text{est}})^2\rangle_X\langle(\Delta P)^2\rangle \geq \frac{1}{4M}. \tag{2.61}$$

The BC-optimal POM $\{\hat{E}(\xi)d\xi\}_\xi$ is of the form $|\xi\rangle\langle\xi|$, with $\hat{P} = -i\,\partial/\partial\xi$. This is satisfied for

$$|\xi\rangle = \frac{1}{\sqrt{2\pi}}\int_{-\infty}^{\infty} dp|p\rangle e^{-i\xi p}e^{if(p)}, \tag{2.62}$$

where $|p\rangle$ are the canonical delta-function-normalized eigenstates of $\hat{P}$ (see Appendix A) and $f(p)$ is an arbitrary real function.

Exercise 2.9 *Show that* $\exp(-i\hat{P}X)|\xi\rangle = |\xi + X\rangle$ *regardless of* $f(p)$.

This illustrates an important point: the conjugate basis to momentum is not unique. The fiducial states $|\xi\rangle$ can be written as

$$|\xi\rangle = \exp[if(\hat{P})]|q := \xi\rangle, \tag{2.63}$$

where $|q\rangle$ is a *canonical* position state, defined by Eq. (2.62) with $\xi = q$ and $f(p) \equiv 0$. (See Appendix A.)

In this case the states $|\xi\rangle$ *are* eigenstates of an Hermitian operator, namely

$$\hat{\Xi} = \hat{Q} + f'(\hat{P}), \tag{2.64}$$

where $\hat{Q}$ is the *canonical* position operator with $|q\rangle$ as its eigenstates.

Exercise 2.10 *Show this, using the fact that* $\hat{Q} = i\,\partial/\partial p$ *in the momentum basis.*

The condition for BC optimality is that the position wavefunction of the fiducial state have the form

$$\langle\xi|\psi_0\rangle = \psi_0(\xi) = r(\xi)e^{i\langle P\rangle_0\xi}, \tag{2.65}$$

for $r(\xi)$ real. In the momentum basis, this becomes

$$\langle p|\psi_0\rangle = e^{if(p)}\tilde{r}(p - \langle P\rangle_0), \tag{2.66}$$

where $\tilde{r}$, the Fourier transform of r, is a skew-symmetric function (that is, $\tilde{r}(k) = \tilde{r}^*(-k)$). Thus, if any $f(p)$ is allowed, the condition on $|\psi_0\rangle$ for achieving BC optimality is just that $|\langle p|\psi_0\rangle|^2$ be symmetric in p about $p_0 \equiv \langle P\rangle_0$. If we allow only *canonical* position measurements, with $f(p) = 0$, then Eq. (2.66) implies that

$$\forall k, \ \langle p := p_0 + k | \psi_0 \rangle = \langle \psi_0 | p := p_0 - k \rangle. \tag{2.67}$$

2.3.2 Spatial displacement of a squeezed state

It is instructive to illustrate these ideas by considering a fiducial state that does not satisfy Eq. (2.67), but that does have a symmetric momentum distribution and does achieve the Cramér–Rao lower bound. We use the special class of fiducial states, the *squeezed vacuum state* [Sch86], defined by

$$|\psi_0\rangle = |r, \phi\rangle = \exp\left[r\left(e^{-2i\phi}\hat{a}^2 - e^{2i\phi}\hat{a}^{\dagger 2}\right)/2\right]|0\rangle. \tag{2.68}$$

As discussed in Appendix A, this squeezed state is in fact a zero-amplitude coherent state for rotated and rescaled canonical coordinates, $\hat{Q}'$ and $\hat{P}'$, defined by

$$\hat{Q} + i\hat{P} = (\hat{Q}'e^r + i\hat{P}'e^{-r})e^{i\phi}. \tag{2.69}$$

If we graphically represent a vacuum state as a circle in phase space with the parametric equation

$$Q^2 + P^2 = \langle 0|(\hat{Q}^2 + \hat{P}^2)|0\rangle = 1, \tag{2.70}$$

then the squeezed vacuum state can be represented by an ellipse in phase space with the parametric equation

$$Q'^2 + P'^2 = \langle \psi_0|(\hat{Q}'^2 + \hat{P}'^2)|\psi_0\rangle = 1. \tag{2.71}$$

This ellipse, oriented at angle ϕ, is shown in Fig. 2.1. These curves can also be thought of as contours for the Wigner or Q function – see Section A.5.

The momentum wavefunction for this fiducial state can be shown to be

$$\langle p|\psi_0\rangle \propto \exp\left(-\frac{p^2}{2\gamma}\right), \tag{2.72}$$

where γ is a complex parameter

$$\gamma = \frac{\cosh r + e^{2i\phi}\sinh r}{\cosh r - e^{2i\phi}\sinh r}. \tag{2.73}$$

The condition for BC optimality is Eq. (2.66). In this case, since $\langle P \rangle_0 = 0$, it reduces to

$$\langle p|\psi_0\rangle e^{-if(p)} = \langle -p|\psi_0\rangle^* e^{+if(-p)}. \tag{2.74}$$

From Eq. (2.72), this will be the case if

$$f(p) = \frac{p^2 \operatorname{Im}(\gamma)}{2|\gamma|^2}. \tag{2.75}$$

That is, the CR-optimal measurement is a measurement of

$$\hat{\Xi} = \hat{Q} - \operatorname{Im}(\gamma^{-1})\hat{P}. \tag{2.76}$$

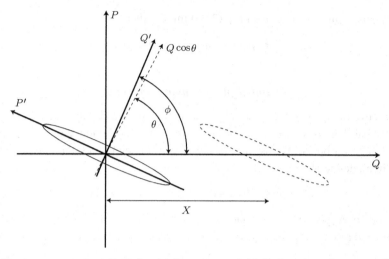

Fig. 2.1 Phase-space representation of optimal measurement of displacement of a squeezed vacuum state. The solid ellipse represents (see the text) the squeezed vacuum state and the dashed-line ellipse that of the state displaced by X in the Q direction. The principal axes of the ellipse are labelled Q' and P', rotated by ϕ relative to Q and P. The BC-optimal measurement for estimating X is to measure the observable $\hat{\Xi} = (\hat{Q} \cos\theta + \hat{P} \sin\theta)/\cos\theta$. Its eigenstates are displaced along the dashed line labelled $Q \cos\theta$, oriented at an angle θ (determined by ϕ and the degree of squeezing). This measurement can be understood as a compromise between the maximal 'signal', which would be obtained by measuring Q, and the minimal 'noise', which would be obtained by measuring Q'. Adapted from *Annals of Physics* **247**, S. L. Braunstein *et al.*, *Generalized uncertainty relations: Theory, examples, and Lorentz invariance*, 135, Copyright (1996), with permission from Elsevier.

In other words, if we use the squeezed coherent state as a fiducial state, then the optimal measurement is a *linear combination* of $\hat{Q}$ and $\hat{P}$. In this case the $\hat{\Xi}$ representation of the fiducial state is a Gaussian state as expected, with the probability density

$$\wp_0(\xi) = \left(\pi \, \mathrm{Re}[\gamma^{-1}]\right)^{-1/2} \exp\left(-\frac{\xi^2}{\mathrm{Re}[\gamma^{-1}]}\right). \tag{2.77}$$

This has a mean of zero (indicating that Ξ is an unbiased estimator) and a variance of

$$\langle \psi_0|(\Delta \Xi)^2|\psi_0\rangle = \frac{1}{2} \mathrm{Re}(\gamma^{-1}) = \frac{1}{4\langle \psi_0|(\Delta P)^2|\psi_0\rangle}. \tag{2.78}$$

Since the probability density is Gaussian, we do not need to appeal to the large-M limit to achieve the Cramér–Rao lower bound. The sample mean of Ξ provides an efficient unbiased estimator of X for all values of M.

The optimal observable $\hat{\Xi}$ can be written as

$$\hat{\Xi} = \frac{\hat{Q} \cos\theta + \hat{P} \sin\theta}{\cos\theta}, \tag{2.79}$$

where $\theta = \mathrm{artan}[-\mathrm{Im}(\gamma^{-1})]$. This operator can be thought of as a modified position operator arising from the rotation in the phase plane by an angle θ followed by a rescaling by $1/\cos\theta$.

The rescaling means that displacement by X produces the same change in $\hat{\Xi}$ as it does in the canonical position operator $\hat{Q}$. Note that the optimal rotation angle θ is not the same as ϕ, the rotation angle that defines the major and minor axes of the squeezed state. Figure 2.1 and its caption give an intuitive explanation for this optimal measurement.

2.3.3 Harmonic oscillator phase

The question of how to give a quantum description for the phase coordinate of a simple harmonic oscillator goes back to the beginning of quantum mechanics [Dir27] and has given rise to a very large number of attempted answers. Many of these answers involve using an appropriate operator to represent the phase (see Ref. [BP86] for what is probably the most successful approach). Quantum parameter estimation avoids the need to find an operator to represent the parameter.

In appropriately scaled units, the energy of a simple harmonic oscillator of angular frequency ω is given by

$$\hat{H} = \frac{\omega}{2}(\hat{Q}^2 + \hat{P}^2) = \omega(\hat{N} + 1/2). \tag{2.80}$$

Here $\hat{N}$ is the number operator (see Section A.4). In a time τ the phase of a local oscillator changes by $\Theta = \omega\tau$. Thus the unitary operator for a phase shift Θ is

$$\exp(-i\hat{H}\tau) = \exp(-i\hat{N}\Theta). \tag{2.81}$$

Here we have removed the constant vacuum energy by redefining $\hat{H}$ as $\omega\hat{N}$.

To find a BC-optimal measurement we seek a POM of the form

$$\hat{E}(\phi)d\phi = \mu|\phi\rangle\langle\phi|d\phi, \tag{2.82}$$

such that, following Eq. (2.45), $|\phi\rangle$ is a state for which $\hat{N}$ generates displacement:

$$\exp(-i\hat{N}\Theta)|\phi\rangle = |\phi + \Theta\rangle. \tag{2.83}$$

This implies that $|\phi\rangle$ is of the form

$$|\phi\rangle = \sum_{n=0}^{\infty} e^{-i\phi n + if(n)}|n\rangle. \tag{2.84}$$

The canonical choice is $f(n) \equiv 0$. This will be appropriate if the fiducial state is of the form

$$|\psi_0\rangle = \sum \sqrt{\wp_n} e^{i\theta_0 n}|n\rangle. \tag{2.85}$$

This is the case for many commonly produced states, such as the coherent states (see Section A.4).

Since $\hat{E}(\phi)$ is periodic with period 2π, we have to restrict the range of results, for example to the interval $-\pi \leq \phi < \pi$. Normalizing the canonical phase POM then gives

$$\hat{E}(\phi)\mathrm{d}\phi = \frac{1}{2\pi}|\phi\rangle\langle\phi|\mathrm{d}\phi \tag{2.86}$$

with

$$|\phi\rangle = \sum_{n=0}^{\infty} \mathrm{e}^{-\mathrm{i}\phi n}|n\rangle. \tag{2.87}$$

These are the Susskind–Glogower phase states [SG64], which are not orthogonal:

$$\langle\phi|\phi'\rangle = \pi\delta(\phi - \phi') - \frac{\mathrm{i}}{2}\cot\left(\frac{\phi - \phi'}{2}\right) + \frac{1}{2}. \tag{2.88}$$

They are overcomplete and are not the eigenstates of any Hermitian operator.

As mooted earlier, this example illustrates the important feature of quantum parameter estimation, namely that it does not restrict us to measuring system observables, but rather allows general POMs. The phase states are in fact eigenstates of a *non-unitary* operator

$$\widehat{\mathrm{e}^{\mathrm{i}\Phi}} = (\hat{N} + 1)^{-1/2}\hat{a} = \hat{a}\hat{N}^{-1/2} = \sum_{n=1}^{\infty}|n - 1\rangle\langle n|, \tag{2.89}$$

such that

$$\widehat{\mathrm{e}^{\mathrm{i}\Phi}}|\phi\rangle = \mathrm{e}^{\mathrm{i}\phi}|\phi\rangle. \tag{2.90}$$

The ϕ and n representations of the state $|\psi\rangle$ are related by

$$\langle n|\psi\rangle = \frac{1}{2\pi}\int_{-\pi}^{\pi}\mathrm{d}\phi\,\mathrm{e}^{-\mathrm{i}n\phi}\langle\phi|\psi\rangle. \tag{2.91}$$

The condition on the fiducial state for the measurement to be BC optimal is

$$\langle n := \langle N\rangle + u|\psi_0\rangle = \langle n := \langle N\rangle - u|\psi_0\rangle^*. \tag{2.92}$$

This can be satisfied only for a limited class of states because n is discrete and bounded below by zero. Specifically, choosing $\theta_0 = 0$, it is satisfied by states of the form

$$|\psi_0\rangle = \sum_{n=0}^{2\mu}\sqrt{\wp_n}|n\rangle; \quad \wp_n = \wp_{2\mu-n}, \tag{2.93}$$

where μ (integer or half-integer) is the mean photon number. States for which this is satisfied achieve BC optimality:

$$F(\Theta) = 4\langle(N - \mu)^2\rangle. \tag{2.94}$$

However, because $\wp_n$ has finite support (that is, it is zero outside a finite range of ns), $\wp_0(\phi) = |\langle\phi|\psi_0\rangle|^2$ cannot be a Gaussian. Thus, even assuming that Θ is restricted to

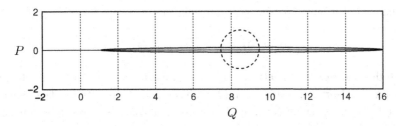

Fig. 2.2 A heuristic phase-space representation of a coherent state (dashed) with amplitude $\alpha = 6$ and a phase-squeezed state (solid) with amplitude $\alpha = 6$ and squeezing parameters $r = 2$ and $\phi = \pi/2$. The contours are defined by parametric equations like Eq. (2.70) and Eq. (2.71).

$[-\pi, \pi)$, the CR lower bound cannot be achieved and we have a strict inequality

$$\langle(\delta\Theta_{\text{est}})^2\rangle\langle(\Delta N)^2\rangle > 1/(4M), \tag{2.95}$$

where M is the number of ϕ measurements contributing to the estimate Θ_{est}, as usual.

It is difficult to produce states with an exact upper bound on their photon number. However, there are many states that can satisfy the BC-optimality condition approximately. For example a state with large mean $\hat{N}$, so that the lower bound at $n = 0$ is not important, and a broad spread in the number distribution that is approximately symmetric about the mean will do. The easiest to produce is a coherent state $|\alpha\rangle$ (see Section A.4). These have a Poissonian number distribution, for which the mean and variance are both equal to $|\alpha|^2$. Thus for coherent states (see Fig. 2.2) we have

$$\langle(\delta\Theta_{\text{est}})^2\rangle > \frac{1}{MF(\Theta)} > \frac{1}{4M\mu}, \tag{2.96}$$

where the inequalities can be approximately satisfied for large μ. The μ^{-1} scaling is known as the *standard quantum limit*. Here 'standard' arises simply because coherent states are the easiest suitable states to produce.

One could beat the standard quantum limit with another fiducial state, such as a phase-squeezed state of a simple harmonic oscillator. We have already met a class of squeezed state, defined in Eq. (2.68). The phase-squeezed state is this state displaced in phase space in the direction orthogonal to the direction of squeezing (see Fig. 2.2).

The *ultimate quantum limit* arises from choosing a BC-optimal state with the largest number variance for a fixed mean number μ. Clearly this is a state of the form

$$\sqrt{2}|\psi_0\rangle = |0\rangle + |2\mu\rangle, \tag{2.97}$$

which has a variance of μ^2. Thus, for a fixed mean photon number, the ultimate limit is

$$F(\Theta) = 4\mu^2. \tag{2.98}$$

The probability distribution for the measurement result ϕ is

$$\wp(\phi|\Theta) = \frac{1}{\pi} \cos^2[\mu(\phi + \Theta)]. \tag{2.99}$$

Although this satisfies Eq. (2.98), for $\mu \geq 1$ it is clear that ϕ is useless for finding an estimate for $\Theta \in [-\pi, \pi)$, because Eq. (2.99) has a periodicity of π/μ. The explanation is that the Fisher information quantifies how well small changes in Θ can be detected, not how well an unknown Θ can be estimated.

Exercise 2.11 *Show Eq. (2.99), and verify Eq. (2.98) by calculating the Fisher information directly from Eq. (2.99).*

2.4 Interferometry – other optimality conditions

As the immediately preceding discussion shows, although BC optimality captures the optimal states and measurements for detecting small shifts in parameters from multiple measurements, it does not necessarily guarantee states or measurements that are good for estimating a completely unknown parameter from one measurement. For this we need some other conditions for optimality. As discussed in Section 2.1.2, Helstrom and Holevo consider optimality in terms of minimizing a cost function. It turns out that, for many problems, any reasonable cost function is minimized for a measurement constructed in the manner described above. That is, for a generator $\hat{G}$, the effects are proportional to rank-1 projectors (Eq. (2.44)), such that $\hat{G}$ generates displacements in the effect basis (Eq. (2.45)). However, the states that minimize the cost may be very different from the states that maximize the Fisher information. (To obtain a finite optimal state in both cases it may be necessary to apply some constraint, such as a fixed mean energy, as we considered above.) In this section we investigate this difference in the context of phase-difference estimation.

2.4.1 The standard quantum limit for interferometry

We have already met interferometry via the Ramsey technique in Section 2.2.1. Interferometry is the basis of many high-precision measurements of time, distance and other physical quantities. The ultimate limit to the precision is, of course, set by quantum mechanics. It is easiest to investigate this limit using a device known as a Mach–Zehnder interferometer (MZI) (see Fig. 2.3). This is built from two optical beam-splitters and two mirrors. The input of the device consists of two modes of a bosonic field, such as the electromagnetic field. The beam-splitter mixes them coherently and turns them into a new pair of modes that form the two arms of the interferometer. In one arm there is an element that introduces a *phase difference* Θ, which is to be estimated. The second beam-splitter coherently mixes the modes of the two arms and gives the two output modes. The outputs of this device are then measured to yield an estimate $\check{\Theta}$ of the phase difference Θ between the two arms of

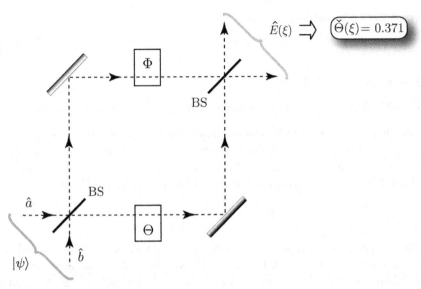

Fig. 2.3 The Mach–Zehnder interferometer. The unknown phase to be estimated is Θ. Both beam-splitters (BS) are 50 : 50. The final measurement is described by a POM $\hat{E}(\xi)$ whose outcome ξ is used to obtain a phase estimate $\breve{\Theta}$. (The value shown for this was chosen arbitrarily.) Figure 1 adapted with permission from D. W. Berry *et al.*, *Phys. Rev. A* **63**, 053804, (2001). Copyrighted by the American Physical Society.

the interferometer. We use $\breve{\Theta}$ rather than Θ_{est} because in this case the estimate $\breve{\Theta}$ is made from a *single* measurement.

The quantum description of the MZI requires a two-mode Hilbert space, with annihilation operators $\hat{a}$ and $\hat{b}$ obeying $[\hat{a}, \hat{a}^{\dagger}] = [\hat{b}, \hat{b}^{\dagger}] = 1$, with all other commutators in $\hat{a}$, $\hat{a}^{\dagger}$, $\hat{b}$ and $\hat{b}^{\dagger}$ being zero. For convenience we will use nomenclature appropriate for the electromagnetic field and call the eigenstates of $\hat{a}^{\dagger}\hat{a}$ and $\hat{b}^{\dagger}\hat{b}$ photon number states, since they have integer eigenvalues (see Section A.4). It is useful to define the following operators:

$$\hat{J}_x = (\hat{a}^{\dagger}\hat{b} + \hat{a}\hat{b}^{\dagger})/2, \tag{2.100}$$

$$\hat{J}_y = (\hat{a}^{\dagger}\hat{b} - \hat{a}\hat{b}^{\dagger})/(2i), \tag{2.101}$$

$$\hat{J}_z = (\hat{a}^{\dagger}\hat{a} - \hat{b}^{\dagger}\hat{b})/2, \tag{2.102}$$

$$\hat{J}^2 = \hat{J}_x^2 + \hat{J}_y^2 + \hat{J}_z^2 = \hat{j}(\hat{j} + 1), \tag{2.103}$$

where

$$\hat{j} = (\hat{a}^{\dagger}\hat{a} + \hat{b}^{\dagger}\hat{b})/2 \tag{2.104}$$

has integer and half-integer eigenvalues. This is known as the Schwinger representation of angular momentum, because the operators obey the usual angular-momentum *operator algebra*. A set of operators is said to form an operator algebra if all the commutators are members of that set, as in this case:

$$[\hat{J}_x, \hat{J}_y] = \mathrm{i}\hat{J}_z, \tag{2.105}$$

with cyclic permutations of x, y, z.

Exercise 2.12 *Show this from the commutation relations for the mode operators.*

For simplicity we consider states that are eigenstates of $\hat{j}$. That is, states with an exact total number of photons $\hat{a}^\dagger \hat{a} + \hat{b}^\dagger \hat{b} = 2j$. The MZI elements preserve photon number, so we can always work using the angular-momentum algebra appropriate to a particle of spin j. A balanced (50/50) beam-splitter can be described by the unitary operator

$$\hat{B}_\pm = \exp(\pm \mathrm{i}\pi \hat{J}_x/2), \tag{2.106}$$

where the $\pm$ represents two choices for a phase convention. For convenience we will take the first beam-splitter to be described by $\hat{B}_+$ and the second by $\hat{B}_-$. Thus, in the absence of a phase shift in one of the arms, the nett effect of the MZI is nothing: $\hat{B}_- \hat{B}_+ = \hat{I}$ and the beams a and b come out in the same state as that in which they entered. The choice of $\hat{B}_\pm$ is a convention, rather than a physically determinable fact, because in optics the distances in the interferometer are not usually measured to wavelength scale except by using interferometry. Thus an experimenter would set up an interferometer with the unknown phase Θ set to zero, and then adjust the arms until the desired output (no change) is achieved.

The effect of the unknown phase shift in the lower arm of the interferometer is described by the unitary operator $\hat{U}(\Theta) = \exp(\mathrm{i}\Theta \hat{a}^\dagger \hat{a})$. The operator $\hat{a}$ (rather than $\hat{b}$) appears here because the input beam $\hat{a}$ is identified with the *transmitted* (i.e. straight through) beam. Because $\hat{j}$ is a constant with value j, we can add $\exp(-\mathrm{i}\Theta j)$ to this unitary operator with no physical effect, and rewrite it as $\hat{U}(\Theta) = \exp(\mathrm{i}\Theta \hat{J}_z)$. If we also include a *known* phase shift Φ in the other arm of the MZI, as shown in Fig. 2.3 (this will be motivated later), then we have between the beam-splitters

$$\hat{U}(\Theta - \Phi) = \exp[\mathrm{i}(\Theta - \Phi)\hat{J}_z]. \tag{2.107}$$

The total unitary operator for the MZI is thus

$$\hat{I}(\Theta - \Phi) = \hat{B}_- \hat{U}(\Theta - \Phi)\hat{B}_+ = \exp[-\mathrm{i}(\Theta - \Phi)\hat{J}_y]. \tag{2.108}$$

Exercise 2.13 *Show this. First show the following theorem for arbitrary operators $\hat{R}$ and $\hat{S}$:*

$$\mathrm{e}^{\xi \hat{R}} \hat{S} \mathrm{e}^{-\xi \hat{R}} = \hat{S} + \xi[\hat{R}, \hat{S}] + \frac{\xi^2}{2!}[\hat{R}, [\hat{R}, \hat{S}]] + \cdots . \tag{2.109}$$

Then use the commutation relations for the $\hat{J}$s to show that $\hat{B}_- \hat{J}_z \hat{B}_+ = -\hat{J}_y$. Use this to show that $\hat{B}_- f(\hat{J}_z)\hat{B}_+ = f(-\hat{J}_y)$ for an arbitrary function f.

The MZI unitary operator $\hat{I}(\Theta - \Phi)$ transforms the photon-number difference operator from the input $2\hat{J}_z$ to the output

$$(2\hat{J}_z)_{\text{out}} = \cos(\Theta - \Phi)2\hat{J}_z + \sin(\Theta - \Phi)2\hat{J}_x. \tag{2.110}$$

Here the subscript 'out' refers to an output operator (that is, a Heisenberg-picture operator for a time after the pulse has traversed the MZI). An output operator is related to the corresponding input operator (that is, the Heisenberg-picture operator for a time before the pulse has met the MZI) by

$$\hat{O}_{\text{out}} = \hat{I}(\Theta - \Phi)\hat{O}\hat{I}(\Theta - \Phi)^\dagger. \tag{2.111}$$

Exercise 2.14 *Show Eq. (2.110) using similar techniques to those in Exercise 2.13 above.*

We can use this expression to derive the standard quantum limit (SQL) to interferometry. As defined in Section 2.3.3 above, the SQL is smply the limit that can be obtained using an easily prepared state and a simple measurement scheme. The easily prepared state is a state with all photons in one input, say the a mode.[1] That is, the input state is a $\hat{J}_z$ eigenstate with eigenvalue j. If Θ is approximately known already, we can choose $\Phi \approx \Theta + \pi/2$. Then the SQL is achieved simply by measuring the output photon-number difference operator

$$(2\hat{J}_z)_{\text{out}} = \sin(\Theta + \pi/2 - \Phi)2\hat{J}_z - \cos(\Theta + \pi/2 - \Phi)2\hat{J}_x \tag{2.112}$$

$$\simeq (\Theta + \pi/2 - \Phi)2\hat{J}_z - 2\hat{J}_x \tag{2.113}$$

$$= (\Theta + \pi/2 - \Phi)2j - 2\hat{J}_x. \tag{2.114}$$

This operator can be measured simply by counting the numbers of photons in the two output modes and subtracting one number from the other. We can use this to obtain an estimate via

$$\check{\Theta} = (2J_z)_{\text{out}}/(2j) - (\pi/2 - \Phi), \tag{2.115}$$

where $(2J_z)_{\text{out}}$ is the result of the measurement. It is easy to verify that for a $\hat{J}_z$ eigenstate $\langle J_x \rangle = 0$, so that the mean of the estimate is approximately Θ, as desired. From Eq. (2.114) the variance is

$$\left\langle \check{\Theta}^2 - \langle \check{\Theta} \rangle^2 \right\rangle \approx \left\langle J_x^2 \right\rangle / j^2. \tag{2.116}$$

For the state $J_z = j$, we have $\langle J_z^2 \rangle = j^2$, while by symmetry $\langle J_x^2 \rangle = \langle J_y^2 \rangle$. Since the sum of these three squared operators is $j(j+1)$, it follows that $\langle J_x^2 \rangle = j/2$. Thus we get

$$\left\langle \check{\Theta}^2 - \langle \check{\Theta} \rangle^2 \right\rangle \approx 1/(2j). \tag{2.117}$$

---

[1] Actually it is not easy experimentally to prepare a state with a definite number of photons in one mode. However, it is easy to prepare a state with an indefinite number of photons in one mode, and then to measure the photon number in each output beam (as discussed below). Since the total number of photons is preserved by the MZI, the experimental results are exactly the same as if a photon-number state, containing the measured number of photons, had been prepared.

That is, provided that the unknown phase is approximately known already, the SQL for the variance in the estimate is equal to the reciprocal of the photon number.

In fact, this SQL can be obtained without the restriction that $\Phi \approx \Theta + \pi/2$, provided that one uses a more sophisticated technique for estimating the phase from the data. This can be understood from the fact that, with all $2j$ photons entering one port, the action of the MZI is equivalent to that of the Ramsey interferometer (Section 2.2.1) repeated $2j$ times.

Exercise 2.15 *Convince yourself of this fact. Note that the parameter θ in the Ramsey-interferometry example is analogous to $\Theta/2$ in the MZI example.*

As shown in Exercise 2.2, the Fisher information implies that the minimum detectable phase shift is independent of the true phase. Indeed, with $M = 2j$ repetitions we get $(2\,\delta\theta)^2 = 1(2j)$, which is exactly the same as the SQL found above for the MZI.

Note, however, that using the Fisher information to define the SQL has problems, as discussed previously. In the current situation, it is apparent from Eq. (2.112) that the same measurement statistics will result if $\Theta + \pi/2 - \Phi$ is replaced by $\Phi + \pi/2 - \Theta$.

Exercise 2.16 *Convince yourself of this. Remember that $\hat{J}_x$ is pure noise.*

That is, the results make it impossible to distinguish Θ from $2\Phi - \Theta$. Thus, it is still necessary to have prior knowledge, restricting Θ to half of its range, say $[0, \pi)$.

More importantly, if one tries to go beyond the SQL by using states entangled across both input ports (as will be considered in Section 2.4.3) then the equivalence between the MZI and Ramsey interferometry breaks down. In such cases, the simple measurement scheme of counting photons in the output ports will *not* enable an estimate of Θ with accuracy independent of Θ. Rather, one finds that one does need to be able to set $\Phi \approx \Theta + \pi/2$ in order to obtain a good estimate of Θ. To get around the restriction (of having to know Θ before one tries to estimate it), it is necessary to consider measurement schemes that go beyond simply counting photons in the output ports. It is to this topic that we now turn.

2.4.2 Canonical phase-difference measurements

The optimal measurement scheme is of course the BC-optimal measurement, as defined in Section 2.2.3. It follows from Eq. (2.108) that we seek a continuum of states for which $\hat{J}_y$ generates displacements. First we introduce the $\hat{J}_y$ eigenstates $|j, \mu\rangle^y$, satisfying $\hat{J}_y|j, \mu\rangle^y = \mu|j, \mu\rangle^y$, with $-j \leq \mu \leq j$. Then we define unnormalized phase states

$$|j\xi\rangle = \sum_{\mu=-j}^{j} \mathrm{e}^{-\mathrm{i}\mu\xi}|j, \mu\rangle^y, \qquad (2.118)$$

with ξ an angle variable. We could have included an additional exponential term $\mathrm{e}^{\mathrm{i}f(\mu)}$ for an arbitrary function f, analogously to Eq. (2.62). By choosing $f \equiv 0$ we are defining *canonical* phase states.

The canonical POM using these phase states is

$$\hat{E}(\xi)\mathrm{d}\xi = |j\xi\rangle\langle j\xi|\mathrm{d}\xi/(2\pi). \tag{2.119}$$

In terms of the $\hat{J}_y$ eigenstates,

$$\hat{E}(\xi)\mathrm{d}\xi = \frac{1}{2\pi}\sum_{\mu,\nu=-j}^{j} \mathrm{e}^{-\mathrm{i}(\mu-\nu)\xi}|j,\mu\rangle^y\langle j,\nu|\mathrm{d}\xi. \tag{2.120}$$

A canonical phase-difference measurement is appropriate for a fiducial state $|\psi_0\rangle$ that satisfies

$$^y\langle j,\mu|\psi_0\rangle = {}^y\langle j,2j-\mu|\psi_0\rangle^*. \tag{2.121}$$

We also want Ξ to be an unbiased estimate of Θ. For cyclic variables, the appropriate sense of unbiasedness is that

$$\arg\langle \mathrm{e}^{\mathrm{i}\Xi}\rangle = \arg\int \langle\psi_0|\hat{I}(\Theta)^\dagger \hat{E}(\xi)\hat{I}(\Theta)|\psi_0\rangle \mathrm{e}^{\mathrm{i}\xi}\,\mathrm{d}\xi = \Theta. \tag{2.122}$$

Exercise 2.17 *Show that this will be the case, if we make the coefficients $^y\langle j,\mu|\psi_0\rangle$ real and positive.*

Since we are going to optimize over the input states, we can impose these restrictions without loss of generality. Similarly, there is no need to consider the auxiliary phase shift Φ.

The fiducial state in the $|j,\mu\rangle^y$ basis is not easily physically interpretable. We would prefer to have it in the $|j,\mu\rangle^z$ basis, which is equivalent to the photon-number basis for the two input modes:

$$|j,\mu\rangle^z = |n_a := j+\mu\rangle|n_b := j-\mu\rangle. \tag{2.123}$$

It can be shown [SM95] that the two angular-momentum bases are related by

$$^y\langle j\mu|j\nu\rangle^z = \mathrm{e}^{\mathrm{i}(\pi/2)(\nu-\mu)} I^j_{\mu\nu}(\pi/2), \tag{2.124}$$

where $I^j_{\mu\nu}(\pi/2)$ are the interferometer matrix elements in the $|j,\mu\rangle^z$ basis given by

$$I^j_{\mu\nu}(\pi/2) = 2^{-\mu}\left[\frac{(j-\mu)!\,(j+\mu)!}{(j-\nu)!\,(j+\nu)!}\right]^{1/2} P^{(\mu-\nu,\mu+\nu)}_{j-\mu}(0),$$
$$\text{for } \mu-\nu > -1, \quad \mu+\nu > -1, \tag{2.125}$$

where $P^{(\alpha,\beta)}_n(x)$ are the Jacobi polynomials, and the other matrix elements are obtained using the symmetry relations

$$I^j_{\mu\nu}(\Theta) = (-1)^{\mu-\nu} I^j_{\nu\mu}(\Theta) = I^j_{-\nu,-\mu}(\Theta). \tag{2.126}$$

2.4.3 Optimal states for interferometry

We expect that, with a canonical measurement and optimized input states, the interferometer should perform quadratically better than the standard quantum limit of $(\Delta \check{\Theta})^2 \simeq 1/(2j)$. This expectation is based on an analogy with measurement of the phase of a single mode, treated in Section 2.3.3. There the SQL was achieved with a canonical measurement and coherent states, which gave a Fisher information equal to 4μ, where μ was half the maximum photon number. By contrast, the ultimate quantum limit was a Fisher information scaling as $4\mu^2$.

In order to prove rigorously that there is a quadratic improvement, we need to use a better measure for spread than the variance, because this is strictly infinite for cyclic variables, and depends upon θ_0 if the range is restricted to $[\theta_0, \theta_0 + 2\pi)$. We could use the Fisher information, but, as discussed above, this is not necessarily appropriate if Θ is completely unknown in a range of 2π. Instead we choose the natural measure of spread for a cyclic variable [Hol84], which we will call the Holevo variance:

$$\mathrm{HV} \equiv S^{-2} - 1, \tag{2.127}$$

where $S \in [0, 1]$ we call the *sharpness* of the phase distribution, defined as

$$S \equiv |\langle e^{i\Phi} \rangle| \equiv \int_0^{2\pi} d\phi \, \wp(\phi) e^{i(\phi - \bar{\phi})}, \tag{2.128}$$

where the 'mean phase' $\bar{\phi}$ is here defined by the requirement that S is real and non-negative. If the Holevo variance is small then it can be shown that

$$\mathrm{HV} \simeq \int_{-\pi}^{\pi} 4 \sin^2 \left(\frac{\phi - \bar{\phi}}{2} \right) \wp(\phi) d\phi. \tag{2.129}$$

Exercise 2.18 *Verify this.*

From this it is apparent that, provided that there is no significant contribution to the variance from $\wp(\phi)$ far from $\bar{\phi}$, this definition of the variance is equivalent to the usual definition. Note that (unlike the usual variance) the Holevo phase variance approaches infinity in the limit of a phase distribution that is flat on $[\theta_0, \theta_0 + 2\pi)$.

We now assume as above that the $c_\mu = \langle \psi_0 | j, \mu \rangle^y$ are positive. Then, from Eq. (2.120), the sharpness of the distribution $\wp(\xi)$ for the fiducial state $|\psi_0\rangle$ is

$$S = \sum_{\mu=-j}^{j-1} c_\mu c_{\mu+1}. \tag{2.130}$$

We wish to maximize this subject to the constraint $\sum_\mu |c_\mu|^2 = 1$. From linear algebra the solution can be shown to be [BWB01]

$$S_{\mathrm{max}} = \cos \left(\frac{\pi}{2j+2} \right) \tag{2.131}$$

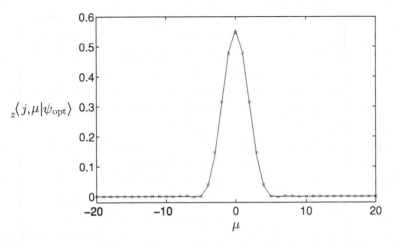

Fig. 2.4 The coefficients $_z\langle j, \mu | \psi_{\text{opt}} \rangle$ for the state optimized for minimum phase variance under ideal measurements. All coefficients for a photon number of $2j = 40$ are shown as the continuous line, and those near $\mu = 0$ for a photon number of $2j = 1200$ as crosses. Figure 2 adapted with permission from D. W. Berry *et al.*, *Phys. Rev.* A **63**, 053804, (2001). Copyrighted by the American Physical Society.

for

$$c_\mu = \frac{1}{\sqrt{j+1}} \sin\left[\frac{(\mu + j + 1)\pi}{2j + 2}\right]. \tag{2.132}$$

The minimum Holevo variance is thus

$$\text{HV} = \tan^2\left(\frac{\pi}{2j+2}\right) = \frac{\pi^2}{(2j)^2} + O(j^{-3}). \tag{2.133}$$

This is known as the Heisenberg limit and is indeed quadratically improved over the SQL. Note that the coefficients (2.132) are symmetric about the mean, so these states are also BC optimal. However, they are very different from the states that maximize the Fisher information, which, following the argument in Section 2.3.3, would have only two non-zero coefficients, $c_{\pm j} = 1/\sqrt{2}$.

Using Eq. (2.124), the state in terms of the eigenstates of $\hat{J}_z$ is

$$|\psi_{\text{opt}}\rangle = \frac{1}{\sqrt{j+1}} \sum_{\mu,\nu=-j}^{j} \sin\left[\frac{(\mu + j + 1)\pi}{2j + 2}\right] e^{i(\pi/2)(\mu-\nu)} I_{\mu\nu}^j(\pi/2) |j\nu\rangle_z. \tag{2.134}$$

An example of this state for 40 photons is plotted in Fig. 2.4. This state contains contributions from all the $\hat{J}_z$ eigenstates, but the only significant contributions are from 9 or 10 states near $\mu = 0$. The distribution near the centre is fairly independent of photon number. To demonstrate this, the distribution near the centre for 1200 photons is also shown in Fig. 2.4. In Ref. [YMK86] a practical scheme for generating a combination of two states near $\mu = 0$ was proposed. Since the optimum states described here have significant contributions

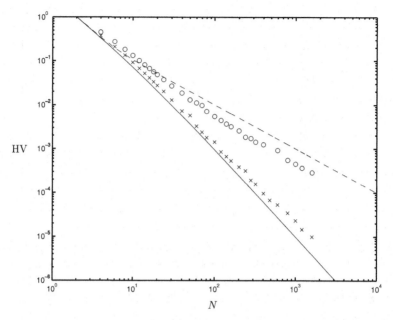

HV

N

Fig. 2.5 Variances in the phase estimate versus input photon number $N = 2j$. The lines are exact results for canonical measurements on optimized states $|\psi_{\text{opt}}\rangle$ (continuous line) and on states with all photons incident on one input port $|jj\rangle_z$ (dashed line). The crosses are the numerical results for the adaptive phase-measurement scheme on $|\psi_{\text{opt}}\rangle$. The circles are numerical results for a non-adaptive phase-measurement scheme on $|\psi_{\text{opt}}\rangle$. Figure 3 adapted with permission from D. W. Berry *et al.*, *Phys. Rev. A* **63**, 053804, (2001). Copyrighted by the American Physical Society.

from a small number of states near $\mu = 0$, it should be possible to produce a reasonable approximation of these states using a similar method to that in Ref. [YMK86].

The Holevo phase variance (2.133) for the optimal state is plotted in Fig. 2.5. The exact Holevo phase variance of the state for which all the photons are incident on one port, $|jj\rangle_z$, is also shown for comparison. This is the state used in Section 2.4.1 to obtain the SQL $\langle \breve{\Theta}^2 - \langle \breve{\Theta} \rangle^2 \rangle \approx 1/(2j)$ under a simple measurement scheme for an approximately known phase shift. As Fig. 2.5 shows, exactly the same result is achieved here asymptotically for the Holevo variance (that is, HV $\sim 1/(2j)$). The difference is that the canonical measurement, as used here, does not require Θ to be approximately known before the measurement begins. Other results are also shown in this figure, which are to be discussed later.

2.5 Interferometry – adaptive parameter estimation

2.5.1 Constrained measurements

The theory of parameter estimation we have presented above is guaranteed to find the BC-optimal measurement scheme (at least for the simple case of a single parameter with generator $\hat{G}$ having a non-degenerate spectrum). In practice, this may be of limited use, because

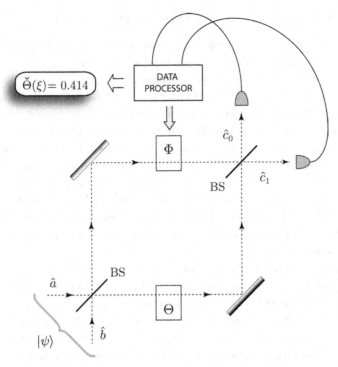

Fig. 2.6 The adaptive Mach–Zehnder interferometer, allowing for feedback to control the phase Φ. Figure 1 adapted with permission from D. W. Berry *et al.*, *Phys. Rev. A*, **63**, 053804, (2001). Copyrighted by the American Physical Society.

it might not be possible to perform the optimal measurement with available experimental techniques. For this reason, it is often necessary to consider measurements constrained by some practical consideration. It is easiest to understand this idea in a specific context, so in this section we again consider the case of interferometric phase measurements.

As we explained in Section 2.4.1, the standard way to do quantum-limited interferometry is simply to count the total number of photons exiting at each output port. For a $2j$-photon input state, the total number of output photons is fixed also, so all of the information from the measurement is contained in the operator

$$(\hat{J}_z)_{\text{out}} = (\hat{a}^\dagger_{\text{out}}\hat{a}_{\text{out}} - \hat{b}^\dagger_{\text{out}}\hat{b}_{\text{out}})/2. \tag{2.135}$$

In Fig. 2.6 we use the new notation

$$\hat{c}_0 = \hat{a}_{\text{out}}, \qquad \hat{c}_1 = \hat{b}_{\text{out}}. \tag{2.136}$$

These output annihilation operators are given by

$$\hat{c}_u(\Theta, \Phi) = \hat{b}\sin\left(\frac{\Theta - \Phi + u\pi}{2}\right) + \hat{a}\cos\left(\frac{\Theta - \Phi + u\pi}{2}\right). \tag{2.137}$$

Exercise 2.19 *Show this, again using the technique of Exercise 2.13.*

As noted above, for an arbitrary input state, measuring $(\hat{J}_z)_{\text{out}}$ gives a good estimate of Θ only if $\Phi \simeq \Theta + \pi/2$. This of course requires prior knowledge of Θ, which is not always true. However, it is possible to perform a measurement, still constrained to be realized by photon counting, whose accuracy is close to that of the canonical measurement and independent of Θ. In this respect it is like the canonical measurement, but its accuracy will typically be worse than that of the canonical measurement. In order to realize the measurement we are referring to, it is necessary to make the auxiliary phase Φ *time-varying*. That is, it will be adjusted *over the course of a single measurement*. For example, it could be changed after each detection. This requires breaking down the measurement into individual detections, rather than counting only the total number of detections at one detector minus the total at the other. The measurement operators which describe individual detections are in fact just proportional to the output annihilation operators defined above, as we will now show.

Let us denote the result u from the mth detection as u_m (which is 0 or 1 according to whether the photon is detected in mode c_0 or c_1, respectively) and the measurement record up to and including the mth detection as the binary string $r_m \equiv u_m \ldots u_2 u_1$. The state of the two-mode field after m detections will be a function of the measurement record and we denote it as $|\psi(r_m)\rangle$. Denoting the null string by r_0, the state before any detections is $|\psi\rangle = |\psi(r_0)\rangle$. Since we are considering demolition detection, the state after the $(m-1)$th detection will be a two-mode state containing exactly $2j + 1 - m$ photons.

Define measurement operators corresponding to the two outcomes resulting from the mth photodetection:

$$\hat{M}_{u_m}^{(m)} = \frac{\hat{c}_{u_m}}{\sqrt{2j + 1 - m}}. \tag{2.138}$$

From Eq. (2.137), the effects

$$\hat{E}_{u_1}^{(m)} = \frac{\hat{c}_{u_m}^\dagger \hat{c}_{u_m}}{2j + 1 - m} \tag{2.139}$$

satisfy

$$\hat{E}_1^{(m)} + \hat{E}_0^{(m)} = \frac{\hat{a}^\dagger \hat{a} + \hat{b}^\dagger \hat{b}}{2j + 1 - m}. \tag{2.140}$$

On the two-mode subspace of states having exactly $2j + 1 - m$ photons, this is an identity operator. Thus, these effects do satisfy the completeness condition (1.78) for all states on which they act. Moreover, it is clear that the action of either measurement operator is to reduce the number of photons in the state by one (see Section A.4), as stated above.

The probability for a complete measurement record r_{2j} is

$$\Pr[R_{2j} = u_{2j} u_{2j-1} \ldots u_2 u_1] = \langle \psi | (\hat{M}_{u_1}^{(1)})^\dagger \ldots (\hat{M}_{u_{2j}}^{(2j)})^\dagger \hat{M}_{u_{2j}}^{(2j)} \ldots \hat{M}_{u_1}^{(1)} | \psi \rangle. \tag{2.141}$$

Now, if Φ is fixed, the $\hat{M}_u^{(m)}$ are independent of m (apart from a constant). Moreover, $\hat{M}_1^m$ and $\hat{M}_0^{m'}$ commute, because $\hat{a}_{\text{out}}$ and $\hat{b}_{\text{out}}$ commute for Φ fixed. Thus we obtain

$$\Pr[R_{2j} = u_{2j}u_{2j-1}\ldots u_2u_1] = \frac{1}{(2j)!}\langle\psi|(\hat{a}^\dagger)_{\text{out}}^{n_a}(\hat{b}^\dagger)_{\text{out}}^{n_b}\hat{b}_{\text{out}}^{n_b}\hat{a}_{\text{out}}^{n_a}|\psi\rangle, \qquad (2.142)$$

where

$$n_a = 2j - n_b = \sum_{m=1}^{2j} u_m. \qquad (2.143)$$

Exercise 2.20 *Show this.*

That is to say, the probability for the record does not depend at all upon the order of the results, only upon the total number of detections n_a in mode a_{out}. Thus, for Φ fixed we recover the result that photon counting measures $(\hat{J}_z)_{\text{out}}$, with result $(n_a - n_b)/2$.

If Φ is not fixed, but is made to change during the course of the measurement, then more general measurements can be made. The only relevant values of Φ will be those pertaining to the times at which detections occur, which we will denote Φ_m. Obviously, if Φ_m depends upon m, the results $u_m = 0, 1$ have different significance for different m. Thus the order of the bits in r_{2j} will be important. In general the measurement operators will not commute, and it will not be possible to collapse the probability as in Eq. (2.142).

In the next subsection we will consider the case of adaptive measurements, for which Φ_m depends upon previous results r_{m-1}. However, before considering that, we note that an adjustable second phase Φ is of use even without feedback [HMP+96]. By setting

$$\Phi_m = \Phi_0 + \frac{m\pi}{2j}, \qquad (2.144)$$

where Φ_0 is chosen randomly, we vary the total phase shift $\Theta - \Phi$ by a half-cycle over the course of the measurement. (A full cycle is not necessary because an additional phase shift of π merely swaps the operators $\hat{M}_1$ and $\hat{M}_0$.) This means that an estimate $\check{\Theta}$ of Θ can be made with an accuracy independent of Θ. However, as we will show, this phase estimate always has a variance scaling as $O(j^{-1})$, which is much worse than the optimal limit of $O(j^{-2})$ from a canonical measurement.

2.5.2 Adaptive measurements

Before turning to adaptive interferometric measurements, it is worth making a few remarks about what constitutes an adaptive measurement in general.

In Section 1.4.2 we introduced the idea of a *complete* measurement as one for which the conditioned state of the system ρ'_r after the measurement depended only on the result r, not upon the initial state. Clearly no further measurements on this system will yield any more information about its initial state. The complementary class of measurements, *incomplete* measurements, consists of ones for which further measurement of ρ may yield more information about the initial state.

If an incomplete measurement is followed by another measurement, then the results of the two measurements can be taken together, so as to constitute a greater measurement. Say the set of operations for the first measurement is $\{\mathcal{O}_p\colon p\}$ and that for the second is $\{\mathcal{O}'_q\colon q\}$. Then the operation for the greater measurement is simply the second operation acting after the first:

$$\mathcal{O}_r\rho = \mathcal{O}'_q(\mathcal{O}_p\rho), \qquad (2.145)$$

where $r = (q, p)$. Depending upon what sort of information one wishes to obtain, it may be advantageous to choose a different second measurement depending on the result of the first measurement. That is, the measurement $\{\mathcal{O}'_q\colon q\}$ will depend upon p. This is the idea of an *adaptive measurement*.

By making a measurement adaptive, the greater measurement may more closely approach the ideal measurement one would like to make. As long as the greater measurement remains incomplete, one may continue to add to it by making more adaptive measurements. Obviously it only makes sense to consider adaptive measurements in the context of measurements that are constrained in some way. For unconstrained measurements, one would simply make the ideal measurement one wishes to make.

It is worth emphasizing again that when we say adaptive measurements we mean measurements on a single system. Another concept of adaptive measurement is to make a (perhaps complete) measurement on the system, and use the result to determine what sort of measurement to make on a second identical copy of the system, and so on. This could be incorporated into our definition of adaptive measurements by considering the system to consist of the original system plus the set of all copies.

The earliest example of using adaptive measurements to make a better constrained measurement is due to Dolinar [Dol73] (see also Ref. [Hel76], p. 163). The *Dolinar receiver* was proposed in the context of trying to discriminate between two non-orthogonal (coherent) states by photodetection, and has recently been realized experimentally [CMG07]. Adaptive measurements have also been found to be useful in estimating the phase (relative to a phase reference called a local oscillator) of a single-mode field, with the measurement again constrained to be realized by photodetection [Wis95]. An experimental demonstration of this will be discussed in the following section. Meanwhile we will illustrate adaptive detection by a similar application: estimating the phase difference in an interferometer as introduced in Ref. [BW00] and studied in more detail in Ref. [BWB01].

Unconstrained interferometric measurements were considered in Section 2.4.2, and constrained interferometric measurements in Section 2.5.1. Here we consider again constrained measurements, where all one can do is detect photons in the output ports, but we allow the measurement to be adaptive, by making the auxiliary phase Φ depend upon the counts so far. Using the notation of Section 2.5.1, the phase Φ_m, before the detection of the mth photon, depends upon the record $r_{m-1} = u_{m-1} \cdots u_1$ of detections (where $u_k = 0$ or 1 denotes a detection in detector 0 or 1, respectively). The question is, how should Φ_m depend upon r_{m-1}?

We will assume that the two-mode $2j$-photon input state $|\psi\rangle$ is known by the experimenter – only the phase Θ is unknown. The state after m detections will be a function of the measurement record r_m and Θ, and we denote it as $|\tilde{\psi}(r_m, \Theta)\rangle$. It is determined by the initial condition $|\tilde{\psi}(r_0, \Theta)\rangle = |\psi\rangle$ and the recurrence relation

$$|\tilde{\psi}(u_m r_{m-1}, \Theta)\rangle = \hat{M}_{u_m}^{(m)}(\Theta, \Phi_m)|\tilde{\psi}(r_{m-1}, \Theta)\rangle. \tag{2.146}$$

These states are unnormalized, and the norm of the state matrix represents the probability for the record r_m, given Θ:

$$\wp(r_m|\Theta) = \langle\tilde{\psi}(r_m, \Theta)|\tilde{\psi}(r_m, \Theta)\rangle. \tag{2.147}$$

Thus the probability of obtaining the result u_m at the mth measurement, given the previous results r_{m-1}, is

$$\wp(u_m|\Theta, r_{m-1}) = \frac{\langle\tilde{\psi}(u_m r_{m-1}, \Theta)|\tilde{\psi}(u_m r_{m-1}, \Theta)\rangle}{\langle\tilde{\psi}(r_{m-1}, \Theta)|\tilde{\psi}(r_{m-1}, \Theta)\rangle}. \tag{2.148}$$

Also, the posterior probability distribution for Θ is

$$\wp(\Theta|r_m) = N_m(r_m)\langle\tilde{\psi}(r_m, \Theta)|\tilde{\psi}(r_m, \Theta)\rangle, \tag{2.149}$$

where $N(r_m)$ is a normalization factor. To obtain this we have used Bayes' theorem assuming a flat prior distribution for Θ (that is, an initially unknown phase). A Bayesian approach to interferometry was realized experimentally in Ref. [HMP+96], but only with non-adaptive measurements.

With this background, we can now specify the adaptive algorithm for Φ_m. The sharpness of the distribution after the mth detection is given by

$$S(u_m r_{m-1}) = \left| \int_0^{2\pi} \wp(\theta|u_m r_{m-1})e^{i\theta}\, d\theta \right|. \tag{2.150}$$

A reasonable (not necessarily optimal) choice for the feedback phase before the mth detection, Φ_m, is the one that will maximize the sharpness after the mth detection. Since we do not know u_m beforehand, we weight the sharpnesses for the two alternative results by their probabilities of occurring on the basis of the previous measurement record. Therefore the expression we wish to maximize is

$$M(\Phi_m|r_m) = \sum_{u_m=0,1} \wp(u_m|r_{m-1})S(u_m r_{m-1}). \tag{2.151}$$

Using Eqs. (2.148), (2.149) and (2.150), and ignoring the constant $N_m(r_m)$, the maximand can be rewritten as

$$\sum_{u_m=0,1} \left| \int_0^{2\pi} \langle\tilde{\psi}(u_m r_{m-1}, \theta)|\tilde{\psi}(u_m r_{m-1}, \theta)\rangle e^{i\theta}\, d\theta \right|. \tag{2.152}$$

The controlled phase Φ_m appears implicitly in Eq. (2.152) through the recurrence relation (2.146), since the measurement operator $\hat{M}_{u_m}^{(m)}$ in Eq. (2.138) is defined in terms of $\hat{c}_{u_m}(\Theta, \Phi_m)$ in Eq. (2.137). The maximizing solution Φ_m can be found analytically [BWB01], but we will not exhibit it here.

The final part of the adaptive scheme is choosing the phase estimate $\check{\Theta}$ of Θ from the complete data set r_{2j}. For cyclic variables, the analogue to minimizing the mean-square error is to maximize

$$\langle \cos(\check{\Theta} - \Theta) \rangle. \tag{2.153}$$

To achieve this, $\check{\Theta}$ is chosen to be the appropriate mean of the posterior distribution $\wp(\theta|r_{2j})$, which from Eq. (2.149) is

$$\check{\Theta} = \arg \int_0^{2\pi} \langle \tilde{\psi}(r_{2j}, \theta) | \tilde{\psi}(r_{2j}, \Theta) \rangle e^{i\theta} \, d\theta. \tag{2.154}$$

This completes the formal description of the algorithm. Its effectiveness can be determined numerically, by generating the measurement results randomly with probabilities determined using $\Theta = 0$, and the final estimate $\check{\Theta}$ determined as above. From Eq. (2.127), an ensemble $\{\check{\Theta}_\mu\}_{\mu=1}^M$ of M final estimates allows the Holevo phase variance to be approximated by

$$\text{HV} \simeq -1 + \left| M^{-1} \sum_{\mu=1}^M e^{i\check{\Theta}_\mu} \right|^{-2}. \tag{2.155}$$

It is also possible to determine the phase variance exactly by systematically going through all the possible measurement records and averaging over Φ_1 (the auxiliary phase before the first detection). However, this method is feasible only for photon numbers up to about 30.

The results of using this adaptive phase-measurement scheme on the optimal input states determined above are shown in Fig. 2.5. The phase variance is very close to the phase variance for ideal measurements, with scaling very close to j^{-2}. The phase variances do differ relatively more from the ideal values for larger photon numbers, however, indicating a scaling slightly worse than j^{-2}. For comparison, we also show the variance from the non-adaptive phase measurement defined by Eq. (2.144). As is apparent, this has a variance scaling as j^{-1}. Evidently, an adaptive measurement has an enormous advantage over a non-adaptive measurement, at least for the optimal input state.

We can sum up the results of this section as follows. Constrained non-adaptive measurements are often far inferior to constrained adaptive measurements, which are often almost as good as unconstrained measurements. That is, a measurement constrained by some requirement of experimental feasibility typically reaches only the standard quantum limit of parameter estimation. This may be much worse than the Heisenberg limit, which can be achieved by the optimal unconstrained measurement. However, if the experiment is made just a little more complex, by allowing adaptive measurements, then most of the difference can be made up. Note, however, that achieving the Heisenberg limit, whether by adaptive or unconstrained measurements, typically requires preparation of an optimal (i.e. non-standard) input state.

2.6 Experimental results for adaptive phase estimation

The above adaptive interferometric phase-estimation scheme has recently been achieved experimentally [HBB$^+$07] for the case of $|jj\rangle_z$ input, although there is a twist in the tale (see the final paragraph of Section 7.10). However, it was preceded some years earlier by the closely related single-mode adaptive phase estimation referred to above, in work done by the group of Mabuchi [AAS$^+$02]. In the single-mode case, an unknown phase shift is imprinted upon a single mode, and this mode is made to interfere with an optical local oscillator (that is, an effectively classical mode) before detection. We call this form of measurement *dyne* detection, for reasons that will become obvious. It allows one to estimate the phase of the system relative to that of the local oscillator. The level of mathematics required to analyse this single-mode case is considerably higher than that for the interferometric case (although it yields asymptotic analytical solutions more easily). We will therefore not present the theory, which is contained in Refs. [WK97, WK98]. However, we will present a particularly simple case [Wis95] in Section 7.9.

In a single-shot adaptive phase measurement, the aim is to make a good estimate of the phase of a single pulse of light relative to the optical local oscillator. In the experiment of Armen *et al.* [AAS$^+$02], each pulse was prepared (approximately) in a coherent state of mean photon number $\bar{n}$, with a randomly assigned phase. The best possible phase measurement would be a canonical phase measurement, as described in Section 2.3.3. From Eq. (2.96) with $M = 1$ (for a single-shot measurement), the canonical phase variance is, for $\bar{n}$ large, close to the Helstrom–Holevo lower bound:

$$\langle (\delta \breve{\Theta}_{\text{can}})^2 \rangle \simeq 1/(4\bar{n}). \tag{2.156}$$

As in the interferometric case, if the phase to be estimated was known approximately before the measurement, then a simple scheme would allow the phase to be estimated with an uncertainty close to the canonical limit. This is the technique of *homodyne* detection, so called because the local oscillator frequency is the same as that of the signal. But, in a communication context, the phase would be completely unknown. Since canonical measurements are not feasible, the usual alternative is *heterodyne* detection. This involves a local oscillator, which is detuned (i.e. at slightly different frequency from the system) so that it cycles over all possible relative phases with the system. That is, it is analogous to the non-adaptive interferometric phase measurement introduced in Section 2.5.1. Again, as in the interferometric case, this technique introduces noise scaling as $1/\bar{n}$. Specifically, the heterodyne limit to phase measurements on a coherent state is twice the canonical limit [WK97]:

$$\langle (\delta \breve{\Theta}_{\text{het}})^2 \rangle \simeq 1/(2\bar{n}). \tag{2.157}$$

The aim of the experiments by Armen *et al.* was to realize an adaptive measurement that can beat the standard limit of heterodyne detection. As in the interferometric case, this involves real-time feedback to control an auxiliary phase Φ, here that of the local oscillator. Since each optical pulse has some temporal extent, the measurement signal generated by the leading edge of a given pulse can be used to form a preliminary estimate of its phase.

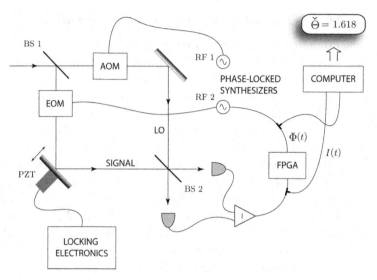

Fig. 2.7 Apparatus used to perform both adaptive homodyne and heterodyne measurements (see the text) in the experiment of Armen *et al.* Solid lines denote optical paths, and dashed lines denote electrical paths. 'PZT' indicates a piezoelectric transducer. Figure 2(a) adapted with permission from M. A. Armen *et al.*, *Phys. Rev. Lett.*, **89**, 133602, (2002). Copyrighted by the American Physical Society.

This can then be used to adjust the local oscillator phase in a sensible way before the next part of the pulse is detected, and so on. Detailed theoretical analyses of such adaptive 'dyne' schemes [WK97, WK98] show that they are very close to canonical measurements. Specifically, for coherent state inputs the difference is negligible even for mean photon numbers of order 10.

Accurately assessing the performance of a single-shot measurement requires many repetitions of the measurement under controlled conditions. Figure 2.7 shows a schematic diagram of the experimental apparatus [AAS+02]. Light from a single-mode (continuous-wave) laser enters the Mach–Zehnder interferometer at beam-splitter 1 (BS 1), thereby creating two beams with well-defined relative phase. The local oscillator (LO) is generated using an acousto-optic modulator (AOM) driven by a radio-frequency (RF) synthesizer (RF 1 in Fig. 2.7). The signal whose phase is to be measured is a weak *sideband* to the carrier (local oscillator). That is, it is created from the local oscillator by an electro-optic modulator (EOM) driven by a RF synthesizer (RF 2) that is phase-locked to RF 1. A pair of photodetectors is used to collect the light emerging from the two output ports of the final 50 : 50 beam-splitter (BS 2). Balanced detection is used: the difference of their photocurrents provides the basic signal used for either heterodyne or adaptive phase estimation. The measurements were performed on optical pulses of duration 50 µs.

In this experimental configuration, the adaptive measurement was performed by feedback control of the phase of RF 2, which sets the relative phase between the signal and the LO. The real-time electronic signal processing required in order to implement the feedback algorithm was performed by a field-programmable gate array (FPGA) that can execute complex computations with high bandwidth and short delays. The feedback and phase-estimation procedure corresponded to the 'Mark II' scheme of Ref. [WK97], in which the photocurrent is integrated with time-dependent gain to determine the instantaneous feedback signal. When performing heterodyne measurements, RF 2 was simply detuned from RF 1 by 1.8 MHz. For both types of measurement, both the photocurrent, $I(t)$, and the feedback signal, $\Phi(t)$, were stored on a computer for post-processing. This is required because the final phase estimate in the 'Mark II' scheme of Ref. [WK97] is not simply the estimate used in the feedback loop, but rather depends upon the full history of the photocurrent and feedback signal. (This estimate is also the optimal one for heterodyne detection.)

The data plotted in Fig. 2.8(a) demonstrate the superiority of an adaptive homodyne measurement procedure over the standard heterodyne measurement procedure. Also plotted is the theoretical prediction for the variance of ideal heterodyne measurement (2.157), both with (thin solid line) and without (dotted line) correction for a small amount of excess electronic noise in the balanced photocurrent. The excellent agreement between the heterodyne data and theory indicates that there is no excess phase noise in the coherent signal states. In the range of 10–300 photons per pulse, most of the adaptive data lie below the absolute theoretical limit for heterodyne measurement (dotted line), and all of them lie below the curve that has been corrected for excess electronic noise (which also has a detrimental effect on the adaptive data).

For signals with large mean photon number, the adaptive estimation scheme used in the experiment was inferior to heterodyne detection, because of technical noise in the feedback loop. At the other end of the scale (very low photon numbers), the intrinsic phase uncertainty of coherent states becomes large and the relative differences among the expected variances for adaptive, heterodyne and ideal estimation become small. Accordingly, Armen *et al.* were unable to beat the heterodyne limit for the mean-square error in the phase estimates for mean photon numbers less than about 8.

However, Armen *et al.* were able to show that the estimator distribution for adaptive homodyne detection remains narrower than that for heterodyne detection even for pulses with mean photon number down to $\bar{n} \approx 0.8$. This is shown in Fig. 2.8(b), which plots the adaptive and heterodyne phase-estimator distributions for $\bar{n} \approx 2.5$. Note that the distributions are plotted on a logarithmic scale. The adaptive phase distribution has a narrower peak than the heterodyne distribution, but exhibits rather high tails. These features agree qualitatively with the numerical and analytical predictions of Ref. [WK98]. It can be partly explained by the fact that the feedback loop occasionally locks on to a phase that is wrong by π.

2.7 Quantum state discrimination

So far in this chapter we have considered the parameter to be estimated as having a continuous spectrum. However, it is quite natural, especially in the context of communication, to

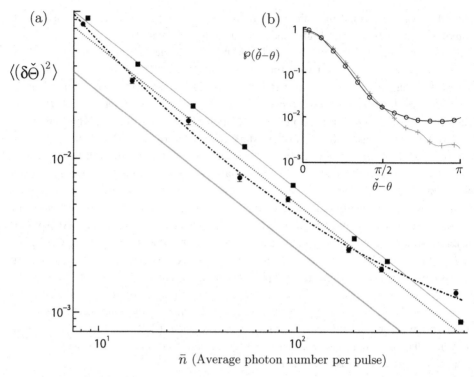

$\bar{n}$ (Average photon number per pulse)

Fig. 2.8 Experimental results from the adaptive and heterodyne measurements. (a) Adaptive (circles) and heterodyne (crosses) phase-estimate variance versus mean photon number per pulse. The dash–dotted line is a second-order curve through the adaptive data, to guide the eye. The thin lines are the theoretical curves for heterodyne detection with (solid) and without (dotted) corrections for detector electronic noise. The thick solid line denotes the fundamental quantum uncertainty limit, given the overall photodetection efficiency. (b) Probability distributions for the error in the phase estimate for adaptive (circles) and heterodyne (crosses) measurements, for pulses with mean photon number of about 2.5. Figure 1 adapted with permission from M. A. Armen *et al.*, *Phys. Rev. Lett.* **89**, 133602, (2002). Copyrighted by the American Physical Society.

consider the case in which the parameter can take values in a finite discrete set. In this case the best estimate should be one of the values in this set, and the problem is really that of deciding which one. This is known as a *quantum decision* or *quantum state-discrimination* problem.

2.7.1 Minimizing the error probability

The simplest example is a communication system that transmits binary information ($s = 0$ or 1) encoded as two states of a physical system ($|\psi_1\rangle$ or $|\psi_0\rangle$). For example, they could be states of light with different intensities. A sequence of systems is directed to a receiver where a measurement is made on each element in the sequence to try to determine which state, and thus which symbol, was transmitted in each case. This is an interesting problem for the case of non-orthogonal states, $\langle \psi_1 | \psi_0 \rangle = \mu \neq 0$. We will suppose that the prior probability distributions for the source are π_0 and π_1 for the states $|\psi_0\rangle$ and $|\psi_1\rangle$, respectively. The measurements are described by some POM $\{\hat{E}_a: a\}$. If the states were orthogonal, they could be discriminated without error, but otherwise there is some finite probability that an error will be made. An error is when the receiver assigns a value 1 when the state transmitted was in fact 0, or conversely. We want to find the optimal POM, which will minimize the effect of errors in a suitable sense to be discussed below. Pioneering work on this question was done by Helstrom [Hel76]. We will follow the presentation of Fuchs [Fuc96].

For a given POM, the above problem also arises in classical decision problems. We need consider only the two probability distributions $\wp(a|s)$ for $s = 0$ and $s = 1$. Classically, such probability distributions arise from noisy transmission and noisy measurement. In order to distinguish $s = 0$ and $s = 1$, the receiver must try to discriminate between the two distributions $\wp(a|s)$ on the basis of a single measurement. If the distributions overlap then this will result in errors, and one should minimize the probability of making an error.[2] Another approach is to relax the requirement that the decision is conclusive; that is, to allow for the possibility of three decisions, yes, no and inconclusive. It may then be possible to make a decision without error, at the expense of a finite probability of not being able to make a decision at all. We will return to this approach in Section 2.7.3.

In each trial of the measurement there are n possible results $\{a\}$, but there are only two possible decision outcomes, 0 and 1. A decision function δ must then take one of the n results of the measurement and give a binary number; $\delta: \{1, \ldots, n\} \to \{0, 1\}$. The probability that this decision is wrong is then

$$\wp_e(\delta) = \pi_0 \wp(\delta := 1 | s := 0) + \pi_1 \wp(\delta := 0 | s := 1). \tag{2.158}$$

The optimal strategy (for minimizing the error probability) is a *Bayesian decision function*, defined as follows. The posterior conditional probability distributions for the two states, given a particular outcome a, are

$$\wp(s|a) = \frac{\wp(a|s)\pi_s}{\wp(a)}, \tag{2.159}$$

where $\wp(a) = \pi_0 \wp(a|s := 0) + \pi_1 \wp(a|s := 1)$ is the total probability for outcome a in the measurement. Then the optimal decision function is

$$\delta_B(a) = \begin{cases} 0 & \text{if } \wp(s := 0|a) > \wp(s := 1|a), \\ 1 & \text{if } \wp(s := 1|a) > \wp(s := 0|a), \\ \text{either} & \text{if } \wp(s := 0|a) = \wp(s := 1|a). \end{cases} \tag{2.160}$$

---

[2] A more sophisticated strategy is to minimize some cost associated with making the wrong decision. The simplest cost function is one that is the same for any wrong decision. That is, we care as much about wrongly guessing $s = 1$ as we do about wrongly guessing $s = 0$. This leads back simply to minimizing the probability of error. However, there are many situations for which other cost functions may be more appropriate, such as in weather prediction, where $s = 1$ indicates a cyclone and $s = 0$ indicates none.

The (minimal) probability of error under this strategy is

$$\wp_e = \sum_{a=1}^{n} \wp(a)(1 - \max\{\wp(s := 0|a), \wp(s := 1|a)\}) \tag{2.161}$$

$$= \sum_{a=1}^{n} \min\{\pi_0 \wp(a|s := 0), \pi_1 \wp(a|s := 1)\}. \tag{2.162}$$

Exercise 2.21 *Show this.*

Hint: *For a given outcome a, the probability of a correct decision is* $\max\{\wp(s := 0|a), \wp(s := 1|a)\}$.

In the quantum case, the probability of measurement outcome a depends both on the states of the system and on the particular measurement we make on the system: $\wp(a|s) = \text{Tr}[\rho_s \hat{E}_a]$. Thus the probability of error is

$$\wp_e = \sum_a \min\{\pi_0 \text{Tr}[\rho_0 \hat{E}_a], \pi_1 \text{Tr}[\rho_1 \hat{E}_a]\}. \tag{2.163}$$

In this case we wish to minimize the error over all POMs. Because we sum over all the results a when making our decision, we really need consider only a binary POM with outcomes $\{0, 1\}$ corresponding to the decision δ introduced above. Then the probability of error becomes

$$\wp_e = \pi_0 \text{Tr}[\rho_0 \hat{E}_1] + \pi_1 \text{Tr}[\rho_1 \hat{E}_0]. \tag{2.164}$$

Exercise 2.22 *Show this.*

In this case the optimum (i.e. minimum) probability of error is achieved by minimization over all POMs. Because $\hat{E}_0 + \hat{E}_1 = \hat{1}$, we can rewrite Eq. (2.164) as

$$\wp_e = \pi_0 + \text{Tr}[(\pi_1 \rho_1 - \pi_0 \rho_0)\hat{E}_0] = \pi_0 + \text{Tr}[\hat{\Gamma} \hat{E}_0], \tag{2.165}$$

where $\hat{\Gamma} = \pi_1 \rho_1 - \pi_0 \rho_0$. The optimization problem is thus one of finding the minimum of $\text{Tr}[\hat{\Gamma} \hat{E}_0]$ over all Hermitian operators $0 \le \hat{E}_0 \le \hat{1}$.

Let us write $\hat{\Gamma}$ in terms of its eigenvalues, which may be positive or negative:

$$\hat{\Gamma} = \sum_j \gamma_j |j\rangle\langle j|. \tag{2.166}$$

It is not difficult to see that $\text{Tr}[\hat{\Gamma} \hat{E}_0]$ will be minimized if we choose

$$\hat{E}_0^{\text{opt}} = \sum_{j:\gamma_j < 0} |j\rangle\langle j|. \tag{2.167}$$

Exercise 2.23 *Convince yourself of this.*

The minimum error probability is thus

$$\wp_e = \pi_0 + \sum_{j:\gamma_j < 0} \gamma_j, \tag{2.168}$$

and this is known as the *Helstrom lower bound*.

The optimal measurement $\{\hat{E}_0^{\text{opt}}, \hat{E}_1^{\text{opt}}\}$ with $\hat{E}_1 = \hat{1} - \hat{E}_0^{\text{opt}}$ can be performed by making a measurement of the operator $\hat{\Gamma}$, and sorting the results into the outcomes $s = 0$, $s = 1$ or either, according to whether they correspond to positive, negative or zero eigenvalues, respectively. This is exactly as given in Eq. (2.163), where a plays the role of γ. This shows that the Helstrom lower bound can be achieved using a projective measurement.

We now restrict the discussion to pure states $\rho_s = |\psi_s\rangle\langle\psi_s|$, in which case $\hat{\Gamma} = \pi_1|\psi_1\rangle\langle\psi_1| - \pi_o|\psi_0\rangle\langle\psi_0|$. The eigenvalues are given by

$$\gamma_\pm = -\frac{\pi_0 - \pi_1}{2} \pm \frac{1}{2}\sqrt{1 - 4\pi_0\pi_1|\mu|^2}. \tag{2.169}$$

Exercise 2.24 *Show this.*

Hint: *Only two basis states are needed in order to express $\hat{\Gamma}$ as a matrix.*

Thus we find the well-known Helstrom lower bound for the error probability for descriminating two pure states,

$$\wp_e^{\min} = \frac{1}{2}\left(1 - \sqrt{1 - 4\pi_0\pi_1|\mu|^2}\right), \tag{2.170}$$

2.7.2 Experimental demonstration of the Helstrom bound

Barnett and Riis [BR97] performed an experiment that realizes the Helstrom lower bound when trying to discriminate between non-orthogonal polarization states of a single photon. The polarization state of a single photon is described in a two-dimensional Hilbert space with basis states corresponding to horizontal (H) and vertical (V) polarization. Barnett and Riis set up the experiment to prepare either of the two states,

$$|\psi_0\rangle = \cos\theta|H\rangle + \sin\theta|V\rangle, \tag{2.171}$$

$$|\psi_1\rangle = \cos\theta|H\rangle - \sin\theta|V\rangle, \tag{2.172}$$

with prior probabilities π_0 and π_1, respectively, and a measurement of the projector

$$\hat{E}_\phi = |\phi\rangle\langle\phi|, \tag{2.173}$$

with $|\phi\rangle = \cos\phi|H\rangle + \sin\phi|V\rangle$, for which an outcome of 1 indicates a polarization in the direction ϕ with respect to the horizontal, while an outcome of 0 indicates a polarization in the direction $\pi/2 + \phi$ with respect to the horizontal. The Helstrom lower bound is attained for

$$\tan(2\phi^{\text{opt}}) = \frac{\tan(2\theta)}{\pi_0 - \pi_1}. \tag{2.174}$$

Exercise 2.25 *Show this.*

If the prepared states are equally likely then we have $\phi^{\text{opt}} = \pi/4$ and the Helstrom lower bound for the error probability is

$$\wp_e = \frac{1}{2}[1 - \sin(2\theta)]. \tag{2.175}$$

In order to perform this experiment as described above we would need a reliable source of single-photon states, suitably polarized. However, deterministic single-photon sources do not yet exist, though they are in active development. Instead Barnett and Riis used an attenuated coherent state from a pulsed source. A coherent-state pulse has a non-determinate photon number, with a Poissonian distribution (see Section A.4). In the experiment light from a mode-locked laser produced a sequence of pulses, which were heavily attenuated (with a neutral-density filter) so that on average each pulse contained about 0.1 photons. For such a low-intensity field, only one in 200 pulses will have more than one photon and most will have none. The laser was operated at a wavelength of 790 nm and had a pulse-repetition rate of 80.3 MHz. The output was linearly polarized in the horizontal plane. Each pulse was then passed through a Glan–Thompson polarizer set at θ or $-\theta$ to produce either of the two prescribed input states.

The polarization measurement was accomplished by passing the pulse through a polarizing beam-splitter, set at an angle $\pi/4$ to the horizontal. This device transmits light polarized in this direction while reflecting light polarized in the orthogonal direction. If the pulse was transmitted, the measurement was said to give an outcome of 1, whereas if it was reflected, the outcome was 0. A 'right' output is when the outcome $a = 0$ or 1 agreed with the prepared state $|\psi_0\rangle$, or $|\psi_1\rangle$, respectively, and a 'wrong' output occurs when it did not. The pulses were directed to photodiodes and the photocurrent integrated, this being simply proportional to the probability of detecting a single photon.

In the experiment the probability of error was determined by repeating the experiment for many photons; that is, simply by running it continually. Call the integrated output from the 'wrong' output I_W, and that from the 'right' output I_R, in arbitrary units. If the Glan–Thompson polarizer is set to θ then the error probability is given by the quantity

$$\wp_e^0 = \frac{I_W}{I_R + I_W} = \left(2 + \frac{I_R - I_W}{I_W}\right)^{-1}. \tag{2.176}$$

If this polarizer is rotated from θ to $-\theta$, the corresponding error probability $\wp_e^1$ is determined similarly. The mean probability of these errors is then taken as an experimental determination of the error probability

$$\wp_e = \frac{1}{2}(\wp_e^0 + \wp_e^1). \tag{2.177}$$

Barnett and Riis determined this quantity as a function of θ over the range 0 to $\pi/4$. The results are shown in Fig. 2.9. Good agreement with the Helstrom lower bound was found.

2.7.3 Inconclusive state discrimination

Thus far, we have discussed an optimal protocol for discriminating two non-orthogonal states that is conclusive, but likely to result in an error. A different protocol, first introduced by Ivanovic [Iva87], requires that the discrimination be *unambiguous* (that is, the probability

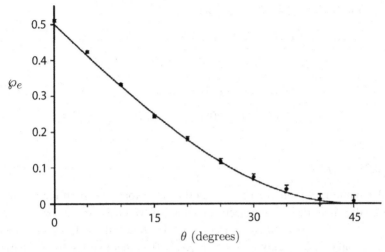

Fig. 2.9 The experimental results of Barnett and Riis. The measured error probability is plotted as a function of the half-angle between the two linear polarization states. The solid curve is the Helstrom bound. Figure 2 adapted with permission from S. M. Barnett and E. Riis, *Experimental demonstration of polarization discrimination at the Helstrom bound, Journal of Modern Optics* **44**, 1061, (1997), Taylor & Francis Ltd, http://www.informaworld.com, reprinted by permission of the publisher.

of making an error must be zero), but at the expense of a non-zero probability of an *inconclusive* outcome (that is, an outcome for which no decision can be made). The idea was subsequently elaborated by Dieks [Die88] and Peres [Per88]. We shall refer to the optimal protocol, which minimizes the probability of an inconclusive result, while never making an error, as realizing the IDP lower bound. Unlike the case of the Helstrom lower bound, the IDP bound cannot be achieved using projective measurements on the system. Instead, an ancilla is introduced, so as to make a generalized measurement on the system.

Consider the simple case in which we seek to discriminate between two pure states $|\psi_\pm\rangle$ that have equal prior probability. We can always choose a two-dimensional Hilbert space to represent these states as

$$|\psi_\pm\rangle = \cos\alpha|1\rangle \pm \sin\alpha|0\rangle, \qquad (2.178)$$

where $|0\rangle$ and $|1\rangle$ constitute an orthonormal basis, and without loss of generality we can take $0 \le \alpha \le \pi/4$. The first step in the protocol is to couple the system to an ancilla two-level system in an appropriate way, so that in the full four-dimensional tensor-product space we can have at least two mutually exclusive outcomes (and hence at most two inconclusive outcomes). Let the initial state of the ancilla be $|0\rangle$. The coupling is the 'exchange coupling' and performs a rotation in the two-dimensional subspace of the tensor-product space spanned by $\{|0\rangle \otimes |1\rangle, |1\rangle \otimes |0\rangle\}$. The states $|0\rangle \otimes |0\rangle$ and $|1\rangle \otimes |1\rangle$ are invariant. On writing the unitary operator for the exchange as $\hat{U}$, and parameterizing it by

θ, the total initial state $|\psi_{\pm}\rangle \otimes |0\rangle$ transforms to

$$|\Psi_{\pm}\rangle = \hat{U}(\theta)[(\cos\alpha|1\rangle \pm \sin\alpha|0\rangle)] \otimes |0\rangle \tag{2.179}$$

$$= \cos\alpha[\cos\theta|1\rangle \otimes |0\rangle + \sin\theta|0\rangle \otimes |1\rangle] \pm \sin\alpha|0\rangle \otimes |0\rangle.$$

If we choose $\cos\theta = \tan\alpha$ this state may be written as

$$|\Psi_{\pm}\rangle = (1-\nu)^{1/2}(|1\rangle \pm |0\rangle) \otimes |0\rangle + (2\nu-1)^{1/2}|0\rangle \otimes |1\rangle, \tag{2.180}$$

where $\nu = \cos^2\alpha$.

Exercise 2.26 *Verify this.*

Since the amplitudes in the first term are orthogonal for the two different input states, they may be discriminated by a projective readout. If we measure the operators $\hat{\sigma}_x \otimes \hat{\sigma}_z$, where $\hat{\sigma}_z = |1\rangle\langle1| - |0\rangle\langle0|$ and $\hat{\sigma}_x = |1\rangle\langle0| + |0\rangle\langle1|$, the results will be $(1, -1)$ only for the state $|\psi_+\rangle$ and $(-1, -1)$ only for the state $|\psi_-\rangle$. The other two results could arise from either state and indicate an inconclusive result. The probability of an inconclusive result is easily seen to be

$$\wp_i = |\langle\psi_-|\psi_+\rangle| = \cos(2\alpha). \tag{2.181}$$

This is the optimal result, the IDP bound. Of course, the unitary interaction between ancilla and system followed by a projective measurement is equivalent to a generalized measurement on the system alone, as explained in Section 1.2.3.

Exercise 2.27 *Determine the effects $\hat{E}_-$, $\hat{E}_+$ and $\hat{E}_i$ (operators in the two-dimensional system Hilbert space) corresponding to the three measurement outcomes.*

2.7.4 Experimental demonstration of the IDP bound

Experiments achieving the IDP bound have been performed by Huttner *et al.* [HMG+96] and Clarke *et al.* [CCBR01]. Here we present results from the latter. A simplified schematic diagram of the experiment is shown in Fig. 2.10. The two-dimensional state space for the system is the polarization degree of freedom for a single photon in a fixed momentum mode, initially taken as the a direction. Thus

$$|1\rangle = |H\rangle_a, \tag{2.182}$$

$$|0\rangle = |V\rangle_a, \tag{2.183}$$

where $|H\rangle_a$ means a single-photon state in the horizontal polarization of the a momentum mode, and $|V\rangle_a$ similarly for the vertical polarization. The first step is to separate out the horizontal and vertical polarization so that we may conditionally couple to an ancilla mode, labelled c, which is initially in the vacuum state. This is easily achieved by changing the momentum mode of the H-polarized photon so that it is now travelling in a different direction. This may be done using the polarizing beam-splitter, PBS 1 in Fig. 2.10, to

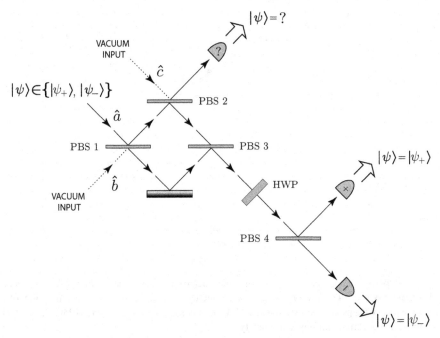

Fig. 2.10 A schematic representation of the optimal state-discrimination experiment of Clarke *et al.* [CCBR01]. PBS indicates a polarizing beam-splitter and PD indicates a photon detector. Note that the device is essentially a polarization Mach–Zehnder interferometer.

transmit vertically polarized light and reflect horizontally polarized light. The total state of the system after PBS 1 is

$$|\psi_{\pm}^{(1)}\rangle = (\cos\alpha|0_a\rangle|H\rangle_b \pm \sin\alpha|V\rangle_a|0_b\rangle)|0_c\rangle, \tag{2.184}$$

where we have dispensed with tensor-product symbols.

The next step is to couple modes b and c using a beam-splitter with a variable transmittivity, PBS 2, with amplitude transmittivity given by $\cos\theta$. For single-photon states, a beam-splitter is equivalent to the exchange interaction. Thus at PBS 2 we have the transformation

$$|H\rangle_b|0\rangle_c \rightarrow \cos\theta|H\rangle_b|0\rangle_c + \sin\theta|0\rangle_b|H\rangle_c, \tag{2.185}$$

where we assume that the transmitted photon does not change its polarization. Thus, just after PBS 2 we can write the total state as

$$|\psi_{\pm}^{(2)}\rangle = (1-v)^{1/2}\left[|0_a\rangle|H\rangle_b \pm |V\rangle_a|0_b\rangle\right]|0\rangle_c + (2v-1)^{1/2}|0_a\rangle|0\rangle_b|H\rangle_c, \tag{2.186}$$

where $v = \cos^2\alpha$ and we have taken $\cos\theta = \tan\alpha$ as in Section 2.7.3.

A photon detector at the output of PBS 2 will now determine whether there is a photon in mode c, which is simply the output for an inconclusive result, labelled '$|\psi\rangle =$?' in Fig. 2.10. The final step is to make projective measurements to distinguish the two possible

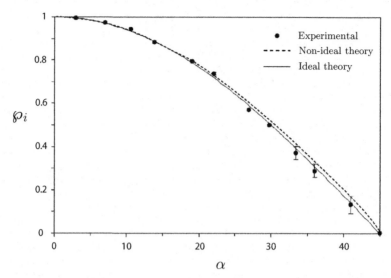

Fig. 2.11 Experiment results for an IDP state-discrimination measurement for the states $\cos\alpha|H\rangle_b \pm \sin\alpha|V\rangle_a$. The probability of an inconclusive result is plotted as a function of α. Figure 3 adapted with permission from R. B. M. Clarke *et al.*, *Phys. Rev.* A **63**, 040305 (R), (2001). Copyrighted by the American Physical Society.

orthogonal states that occur in the first term. This may be done by using a polarizing beam-splitter to put both photons back into the same momentum mode. Thus, after PBS 3 the total state is

$$|\psi_{\pm}^{(3)}\rangle = (1 - v)^{1/2} \left[|H\rangle_a \pm |V\rangle_a\right]|0\rangle_b|0\rangle_c + (2v - 1)^{1/2}|0\rangle_a|0\rangle_b|H\rangle_c. \qquad (2.187)$$

It now suffices to measure the polarization of mode a in the diagonal basis $|H\rangle_a \pm |V\rangle_a$. This is done by using a sequence of polarizer and polarizing beam-splitter followed by two photon detectors, as indicated in Fig. 2.10.

In Fig. 2.11 we show the results of Clarke *et al.* for the probability of inconclusive results, at the optimal configuration, as the angle between the two states is varied, as a function of α. Within experimental error the agreement between the IDP bound and the experiment is very good.

2.8 Further reading

2.8.1 Quantum tomography

In this chapter we have been concerned mainly with the problem in which the state ρ being measured depends (smoothly) upon a single unknown parameter λ. Of course, it is possible to consider the case in which it depends upon more than one parameter. For a D-dimensional system, if there are $D^2 - 1$ parameters upon which ρ depends, then it is possible that the complete space of all possible ρs can be generated by varying the parameter values (an

Hermitian matrix has D^2 independent elements, but the condition $\text{Tr}[\rho] = 1$ removes one of them.) This situation, in which one is effectively trying to identify a completely unknown ρ from measurements on multiple copies, is known as *quantum state estimation* or *quantum tomography*.

A good review of this topic for D-dimensional systems is Ref. [DPS03]. Note that, in order to measure $D^2 - 1$ parameters using *projective* measurements, it is necessary to consider $D + 1$ different projective measurements, that is, measuring $D + 1$ different observables. Such a set of observables is called a *quorum* [Fan57]. That a quorum size of $D + 1$ observables is *necessary* is obvious from the fact that measuring one observable repeatedly can give only $D - 1$ parameters: the D probabilities (one for each outcome) minus one because they must sum to unity. That $D + 1$ is a *sufficient* size for a (suitably chosen) quorum was first proven by Fano [Fan57].

The term *quantum tomography* was coined in quantum optics for estimating the state of an electromagnetic field mode. Here the quorum consists of a collection of quadrature operators $\hat{X}_\theta = \cos\theta\,\hat{Q} + \sin\theta\,\hat{P}$ for different values of θ. This is analogous to the reconstruction of a two-dimensional tissue density distribution from measurements of density profiles along various directions in the plane in medical tomography. Of course, the quantum harmonic oscillator has an infinite Hilbert-space dimension D, so strictly an infinite quorum (infinitely many θs) must be used. In practice, it is possible to reconstruct an arbitrary ρ with a finite quorum, and a finite number of measurements for each observable, if certain assumptions or approximations are made. We refer the reader to the review [PR04] for a discussion of these techniques, including maximum-likelihood estimation.

2.8.2 Other work

A general treatment of the resources (state preparation, and measurement) required to attain the Heisenberg limit in parameter estimation is given by Giovanneti, Lloyd and Maccone [GLM06]. This is done using the language of quantum information processing, which is treated by us in Chapter 7. See the discussion of recent theory and experiment [HBB$^+$07] in optical phase estimation in Section 7.10. The theory in this work is closely related to the adaptive algorithm of Section 2.5.2.

Understanding the quantum limits to parameter estimation has also thrown light on the controversial issue of the time–energy uncertainty relation. It has long been recognized that the time–energy uncertainty relation is of a different character from the position–momentum uncertainty relation, since there is no time operator in quantum mechanics; see for example Ref. [AB61]. However, if $\hat{H}$ is taken as the generator $\hat{G}$ of the unitary, and t as the parameter X to be estimated, then the Holevo upper bound (2.9) gives a precise meaning to the time–energy uncertainty relation.

Developing physical systems to estimate time translations as accurately as possible is, of course, the business of the time-standards laboratories. Modern time standards are based on the oscillations of atomic dipoles, and their quantum description is very similar to that given in Section 2.2.1. However, in practice, clocks are limited in their performance by their *instability*, which can be thought of as the relative variation in the time interval between ticks of the clock. In terms of the general theory of parameter estimation, it is as if the generator $\hat{G}$ itself varied randomly by a tiny amount from shot to shot. For a model based on a two-level atomic transition the instability is the ratio of the variation in the frequency of the transition $\delta\omega$ to the mean frequency ω. Currently, the best clocks have an instability $\delta\omega/\omega$ at or below 10^{-17} [HOW$^+$05].

The topic of discriminating non-orthogonal quantum states has been reviewed by Chefles [Che00]; see also Ref. [PR04]. More recently, Jacobs [Jac07] has considered this problem in the context of continuous adaptive measurements similar to that discussed in Section 2.6 above. Jacobs shows that, by using an adaptive technique, one can increase the rate at which the information regarding the initial preparation is obtained. However, in the long-time limit, such an adaptive measurement actually *reduces* the total amount of information obtained, compared with the non-adaptive measurement that reproduces (in the long-time limit) the optimal projective measurement discussed in Section 2.7.1. That is, the adaptive measurement fails to attain the Helstrom lower bound. This is an instructive example of the fact that locally (in time) optimizing the rate of increase in some desired quantity (such as the information about the initial state preparation) does not necessarily lead to the globally optimal scheme. There is thus no reason to expect the adaptive measurement schemes discussed in Sections 2.5 and 2.6 to be optimal.

3

Open quantum systems

3.1 Introduction

As discussed in Chapter 1, to understand the general evolution, conditioned and unconditioned, of a quantum system, it is necessary to consider coupling it to a second quantum system. In the case in which the second system is much larger than the first, it is often referred to as a *bath*, *reservoir* or *environment*, and the first system is called an *open system*. The study of open quantum systems is important to quantum measurement for two reasons.

First, all real systems are open to some extent, and the larger a system is, the more important its coupling to its environment will be. For a macroscopic system, such coupling leads to very rapid *decoherence*. Roughly, this term means the irreversible loss of quantum coherence, that is the conversion of a quantum superposition into a classical mixture. This process is central to understanding the emergence of classical behaviour and ameliorating, if not solving, the so-called *quantum measurement problem*.

The second reason why open quantum systems are important is in the context of generalized quantum measurement theory as introduced in Chapter 1. Recall from there that, by coupling a quantum system to an 'apparatus' (a second quantum system) and then measuring the apparatus, a generalized measurement on the system is realized. For an open quantum system, the coupling to the environment is typically continuous (present at all times). In some cases it is possible to *monitor* (i.e. continuously measure) the environment so as to realize a continuous generalized measurement on the system.

In this chapter we are concerned with introducing open quantum systems, and with discussing the first point, decoherence. We introduced the decoherence of a macroscopic apparatus in Section 1.2.3, in the context of the von Neumann chain and Heisenberg's cut. To reiterate that discussion, direct projective measurements on a quantum system do not adequately describe realistic measurements. Rather, one must consider making measurements on an apparatus that has been coupled to the system. But how does one make a direct observation on the apparatus? Should one introduce yet another system to model the readout of the meter coupled to the actual system of study, and so on with meters upon meters *ad infinitum*? This is the von Neumann chain [vN32]. To obtain a finite theory, the experimental result must be considered to have been recorded definitely at some point: Heisenberg's cut [Hei30].

The quantum measurement problem is that there is no physical basis for inserting a cut at any particular point. However, there is a physical basis for determining the point in the chain *after which* the cut may be placed without affecting any theoretical predictions. This point is the point at which, for all practical purposes, the meter can be treated as a classical, rather than a quantum, object. That such a point exists is due to decoherence brought about by the environment of the apparatus.

Consider, for example, the single-photon measurement discussed in Section 1.5. The system of study was the electromagnetic field of a single-mode microwave cavity. The meter was an atomic system, suitably prepared. This meter clearly still behaves as a quantum system; however, as other experiments by the same group have shown [RBH01], the atomic 'meter' is in turn measured by ionization detectors. These detectors are, of course, rather complicated physical systems involving electrical fields, solid-state components and sophisticated electronics. Should we include these as quantum systems in our description? No, for two reasons.

First, it is too hard. Quantum systems with many degrees of freedom are generally intractable. This is due to the exponential increase in the dimension of the Hilbert space with the number of components for multi-partite systems, as discussed in Section A.2. Except for cases in which the Hamiltonian has an exceptionally simple structure, numerical solutions are necessary for the quantum many-body problem.

Exercise 3.1 *For the special case of a Hamiltonian that is invariant under particle permutations show that the dimension of the total Hilbert space increases only linearly in the number of particles.*

However, even on today's supercomputers, numerical solutions are intractable for 100 particles or more. Detectors typically have far more particles than this, and, more importantly, they typically interact strongly with other systems in their environment.

Second, it is unnecessary. Detectors are not arbitrary many-body systems. They are designed for a particular purpose: to be a detector. This means that, despite its being coupled to a large environment, there are certain properties of the detector that, if initially well defined, remain well defined over time. These classical-like properties are those that are *robust* in the face of decoherence, as we will discuss in Section 3.7. Moreover, in an ideal detector, one of these properties is precisely the one which becomes correlated with the quantum system and apparatus, and so constitutes the *measurement result*. As we will discuss in Section 4.8, sometimes it may be necessary to treat the detector dynamics in greater detail in order to understand precisely what information the experimenter has obtained about the system of study from the measurement result. However, in this case it is still unnecessary to treat the detector as a quantum system; a classical model is sufficient.

The remainder of this chapter is organized as follows. In Section 3.2 we introduce the simplest approach to modelling the evolution of open quantum systems: the *master equation* derived in the Born–Markov approximations. In Section 3.3 we apply this to the simplest (and historically first) example: radiative damping of a two-level atom. In the same section we also describe damping of an optical cavity; this treatment is very similar, insofar as both involve a rotating-wave approximation. In Section 3.4 we consider systems in which the rotating-wave approximation cannot be made: the spin–boson model and Brownian motion. In all of these examples so far, the reservoir consists of harmonic oscillators, modes of a bosonic field (such as the electromagnetic field). In Section 3.5 we treat a rather different sort of reservoir, consisting of a fermionic (electron) field, coupled to a single-electron system.

In Section 3.6 we turn to more formal results: the mathematical conditions that a Markovian theory of open quantum systems should satisfy. Armed with these examples and this theory, we tackle the issue of decoherence and its relation to the quantum measurement problem in Section 3.7, using the example of Brownian motion. Section 3.8 develops this idea in the direction of continuous measurement (which will be considered in later chapters), using the examples of the spin–boson model, and the damped and driven atom. The ground-breaking decoherence experiment from the group of Haroche is analysed in Section 3.9 using the previously introduced damped-cavity model. In Section 3.10 we discuss two more open systems of considerable experimental interest: a quantum electromechanical oscillator and a superconducting qubit. Finally (apart from the further reading), we present in Section 3.11 a Heisenberg-picture description of the dynamics of open quantum systems, and relate it to the descriptions in earlier sections.

3.2 The Born–Markov master equation

In this section we derive a general expression for the evolution of an open quantum system in the Born and Markov approximations. This will then be applied to particular cases in subsequent sections. The essential idea is that the system couples weakly to a very large environment. The weakness of the coupling ensures that the environment is not much affected by the system: this is the Born approximation. The largeness of the environment (strictly, the closeness of its energy levels) ensures that from one moment to the next the system effectively interacts with a different part of the environment: this is the Markov approximation.

Although the environment is relatively unaffected by the system, the system is profoundly affected by the environment. Specifically, it typically becomes entangled with the environment. For this reason, it cannot be described by a pure state, even if it is initially in a pure state. Rather, as shown in Section A.2.2, it must be described by a mixed state ρ. The aim of the Born–Markov approximation is to derive a differential equation for ρ. That is, rather than having to use a quantum state for the system and environment, we can find the approximate evolution of the system by solving an equation for the system state alone. For historical reasons, this is called a *master equation*.

The dynamics of the state ρ_{tot} for the system plus environment is given in the Schrödinger picture by

$$\dot{\rho}_{\text{tot}}(t) = -\mathrm{i}[\hat{H}_S + \hat{H}_E + \hat{V}, \rho_{\text{tot}}(t)]. \tag{3.1}$$

Here $\hat{H}_S$ is the Hamiltonian for the system (that is, it acts as the identity on the environment Hilbert space), $\hat{H}_E$ is that for the environment, and $\hat{V}$ includes the coupling between the two. Following the formalism in Section A.1.3, it is convenient to move into an interaction frame with free Hamiltonian $\hat{H}_0 = \hat{H}_S + \hat{H}_E$. That is, instead of $\hat{H}_{\text{tot}} = \hat{H}_0 + \hat{V}$, we use

$$\hat{V}_{\text{IF}}(t) = e^{i\hat{H}_0 t} \hat{V} e^{-i\hat{H}_0 t}. \tag{3.2}$$

In this *frame*, the Schrödinger-*picture* equation is

$$\dot{\rho}_{\text{tot;IF}}(t) = -i[\hat{V}_{\text{IF}}(t), \rho_{\text{tot;IF}}(t)], \tag{3.3}$$

where the original solution to Eq. (3.1) is found as

$$\rho_{\text{tot}}(t) = e^{-i\hat{H}_0 t} \rho_{\text{tot;IF}} e^{i\hat{H}_0 t}. \tag{3.4}$$

The equations below are all in the interaction frame, but for ease of notation we drop the IF subscripts. That is, $\hat{V}$ will now denote $\hat{V}_{\text{IF}}(t)$, etc.

Since the interaction is assumed to be weak, the differential equation Eq. (3.3) may be solved as a perturbative expansion. We solve Eq. (3.3) implicitly to get

$$\rho_{\text{tot}}(t) = \rho_{\text{tot}}(0) - i \int_0^t dt_1 [\hat{V}(t_1), \rho_{\text{tot}}(t_1)]. \tag{3.5}$$

We then substitute this solution back into Eq. (3.3) to yield

$$\dot{\rho}_{\text{tot}}(t) = -i[\hat{V}(t), \rho_{\text{tot}}(0)] - \int_0^t dt_1 [\hat{V}(t), [\hat{V}(t_1), \rho_{\text{tot}}(t_1)]]. \tag{3.6}$$

Since we are interested here only in the evolution of the system, we trace over the environment to get an equation for $\rho \equiv \rho_S = \text{Tr}_E[\rho_{\text{tot}}]$ as follows:

$$\dot{\rho}(t) = -i \, \text{Tr}_E \left([\hat{V}(t), \rho_{\text{tot}}(0)] \right)$$

$$- \int_0^{t_1} dt_1 \, \text{Tr}_E \left([\hat{V}(t), [\hat{V}(t_1), \rho_{\text{tot}}(t_1)]] \right). \tag{3.7}$$

This is still an exact equation but is also still implicit because of the presence of $\rho_{\text{tot}}(t_1)$ inside the integral. However, it can be made explicit by making some approximations, as we will see. It might be asked why we carry the expansion to second order in V, rather than use the first-order equation (3.3), or some higher-order equation. The answer is simply that second order is the lowest order which generally gives a non-vanishing contribution to the final master equation.

We now assume that at $t = 0$ there are no correlations between the system and its environment:

$$\rho_{\text{tot}}(0) = \rho(0) \otimes \rho_E(0). \tag{3.8}$$

This assumption may be physically unreasonable for some interactions between the system and its environment [HR85]. However, for weakly interacting systems it is a reasonable approximation. We also split $\hat{V}$ (which, it must be remembered, denotes the Hamiltonian in the interaction frame) into two parts:

$$\hat{V}(t) = \hat{V}_S(t) + \hat{V}_{SE}(t), \tag{3.9}$$

where $\hat{V}_S(t)$ acts nontrivially only on the system Hilbert space, and where $\text{Tr}[\hat{V}_{SE}(t)\rho_{\text{tot}}(0)] = 0$.

Exercise 3.2 *Show that this can be done, irrespective of the initial system state $\rho(0)$, by making a judicious choice of $\hat{H}_0$.*

We now make a very important assumption, namely that the system only weakly affects the bath so that in the last term of Eq. (3.7) it is permissible to replace $\rho_{\text{tot}}(t_1)$ by $\rho(t_1) \otimes \rho_E(0)$. This is known as the *Born approximation*, or the weak-coupling approximation. Under this assumption, the evolution becomes

$$\dot{\rho}(t) = -i[\hat{V}_S(t), \rho(t)] - \int_0^t dt_1 \, \text{Tr}_E\left([\hat{V}_{SE}(t), [\hat{V}_{SE}(t_1), \rho(t_1) \otimes \rho_E(0)]]\right). \tag{3.10}$$

Note that this assumption is not saying that $\rho_{\text{tot}}(t_1)$ is well approximated by $\rho(t_1) \otimes \rho_E(0)$ for all purposes, and indeed this is not the case; the coupling between the system and the environment in general entangles them. This is why the system becomes mixed, and why measuring the environment can reveal information about the system, as will be considered in later chapters, but this factorization assumption is a good one for the purposes of deriving the evolution of the system alone.

The equation (3.10) is an integro-differential equation for the system state matrix ρ. Because it is nonlocal in time (it contains a convolution), it is still rather difficult to solve. We seek instead a local-in-time differential equation, sometimes called a time-convolutionless master equation, that is, an equation in which the rate of change of $\rho(t)$ depends only upon $\rho(t)$ and t. This can be justified if the integrand in Eq. (3.10) is small except in the region $t_1 \approx t$. Since the modulus of $\rho(t_1)$ does not depend upon t_1, this property must arise from the physics of the bath. As we will show in the next section, it typically arises when the system couples roughly equally to many energy levels of the bath (eigenstates of $\hat{H}_E$) that are close together in energy. Under this approximation it is permissible to replace $\rho(t_1)$ in the integrand by $\rho(t)$, yielding

$$\dot{\rho}(t) = -i[\hat{V}_S(t), \rho(t)] - \int_0^t dt_1 \, \text{Tr}_E\left([\hat{V}(t), [\hat{V}(t_1), \rho(t) \otimes \rho_E(0)]]\right). \tag{3.11}$$

This is sometimes called the Redfield equation [Red57].

Even though the approximation of replacing $\rho(t_1)$ by $\rho(t)$ is sometimes referred to as a Markov approximation [Car99, GZ04], the resulting master equation (3.11) is not strictly Markovian. That is because it has time-dependent coefficients, as will be discussed in Section 3.4. In fact, it can be argued [BP02] that this additional approximation is not really an additional approximation at all: the original Born master equation Eq. (3.10) would not be expected to be more accurate than the Redfield equation Eq. (3.11).

To obtain a true Markovian master equation, an autonomous differential equation for $\rho(t)$, it is necessary to make a more substantial Markov approximation. This consists of again appealing to the sharpness of the integrand at $t_1 \approx t$, this time to replace the lower limit of the integral in Eq. (3.11) by $-\infty$. In that way we get finally the Born–Markov master equation for the system in the interaction frame:

$$\dot{\rho}(t) = -i[\hat{V}_S(t), \rho(t)] - \int_{-\infty}^t dt_1 \, \text{Tr}_E\left([\hat{V}(t), [\hat{V}(t_1), \rho(t) \otimes \rho_E(0)]]\right). \tag{3.12}$$

We will see in examples below how, for physically reasonable properties of the bath, this gives a master equation with time-independent coefficients, as required. In particular, we require $\hat{H}_E$ to have a continuum spectrum in the relevant energy range, and we require

$\rho_E(0)$ to commute with $\hat{H}_E$. In practice, the latter condition is often relaxed in order to yield an equation in which $V_S(t)$ may be time-dependent, but the second term in Eq. (3.12) is still required to be time-independent.

3.3 The radiative-damping master equation

In this section we repeat the derivation of the Born–Markov master equation for a specific case: radiative damping of quantum optical systems (a two-level atom and a cavity mode). This provides more insight into the Born and Markov approximations made above.

3.3.1 Spontaneous emission

Historically, the irreversible dynamics of spontaneous emission were introduced by Bohr [Boh13] and, more quantitatively, by Einstein [Ein17], before quantum theory had been developed fully. It was Wigner and Weisskopf [WW30] who showed in 1930 how the radiative decay of an atom from the excited to the ground state could be explained within quantum theory. This was possible only after Dirac's quantization of the electromagnetic field, since it is the infinite (or at least arbitrarily large) number of electromagnetic field modes which forms the environment or bath into which the atom radiates. The theory of spontaneous emission is described in numerous recent texts [GZ04, Mil93], so our treatment will just highlight key features.

As discussed in Section A.4, the free Hamiltonian for a mode of the electromagnetic field is that of a harmonic oscillator. The total Hamiltonian for the bath is thus

$$\hat{H}_E = \sum_k \omega_k \hat{b}_k^\dagger \hat{b}_k, \tag{3.13}$$

where the integer k codes all of the information specifying the mode: its frequency, direction, transverse structure and polarization. The mode structure incorporates the effect of bulk materials with a linear refractive index (such as mirrors) and the like, so this is all described by the Hamiltonian $\hat{H}_E$. The annihilation and creation operators for each mode are independent and they obey the *bosonic* commutation relations

$$[\hat{b}_k, \hat{b}_l^\dagger] = \delta_{kl}. \tag{3.14}$$

We will assume that only two energy levels of the atom are relevant to the problem, so the free Hamiltonian for the atom is

$$\hat{H}_a = \frac{\omega_a}{2} \hat{\sigma}_z. \tag{3.15}$$

Here ω_a is the energy (or frequency) difference between the ground $|g\rangle$ and excited $|e\rangle$ states, and $\hat{\sigma}_z = |e\rangle\langle e| - |g\rangle\langle g|$ is the inversion operator for the atom. (See Box 3.1.) The coupling of the electromagnetic field to an atom can be described by the so-called dipole-coupling Hamiltonian

$$\hat{V} = \sum_k (g_k \hat{b}_k + g_k \hat{b}_k^\dagger)(\hat{\sigma}_+ + \hat{\sigma}_-). \tag{3.16}$$

Box 3.1 The Bloch representation

Consider a two-level system with basis states $|0\rangle$ and $|1\rangle$. The three *Pauli* operators for the system are defined as

$$\hat{\sigma}_x = |0\rangle\langle 1| + |1\rangle\langle 0|, \tag{3.17}$$

$$\hat{\sigma}_y = i|0\rangle\langle 1| - i|1\rangle\langle 0|, \tag{3.18}$$

$$\hat{\sigma}_z = |1\rangle\langle 1| - |0\rangle\langle 0|. \tag{3.19}$$

These obey the following product relations:

$$\hat{\sigma}_j\hat{\sigma}_k = \delta_{jk}\hat{1} + i\epsilon_{jkl}\hat{\sigma}_l. \tag{3.20}$$

Here the subscripts stand for x, y or z, while $\hat{1}$ is the 2×2 unit matrix, i is the unit imaginary and ϵ_{jkl} is the completely antisymmetric tensor (that is, transposing any two subscripts changes its sign) satisfying $\epsilon_{xyz} = 1$. From this commutation relations like $[\hat{\sigma}_x, \hat{\sigma}_y] = 2i\hat{\sigma}_z$ and anticommutation relations like $\hat{\sigma}_x\hat{\sigma}_y + \hat{\sigma}_y\hat{\sigma}_x = 0$ are easily derived.

The state matrix for a two-level system can be written using these operators as

$$\rho(t) = \tfrac{1}{2}[\hat{1} + x(t)\hat{\sigma}_x + y(t)\hat{\sigma}_y + z(t)\hat{\sigma}_z], \tag{3.21}$$

where x, y, z are the averages of the Pauli operators. That is, $x = \text{Tr}[\hat{\sigma}_x\rho]$ *et cetera*. Recall that $\text{Tr}[\rho^2] \leq 1$, with equality for and only for pure states. This translates to

$$x^2 + y^2 + z^2 \leq 1, \tag{3.22}$$

again with equality iff the system is pure. Thus, the system state can be represented by a 3-vector inside (on) the unit sphere for a mixed (pure) state. The vector is called the *Bloch vector* and the sphere the *Bloch sphere*.

For a two-level atom, it is conventional to identify $|1\rangle$ and $|0\rangle$ with the ground $|g\rangle$ and excited $|e\rangle$ states. Then z is called the atomic inversion, because it is positive iff the atom is inverted, that is, has a higher probability of being in the excited state than in the ground state. The other components, y and x, are called the atomic coherences, or components of the atomic dipole.

Another two-level system is a spin-half particle. Here 'spin-half' means that the maximum angular momentum contained in the intrinsic spin of the particle is $\hbar/2$. The operator for the spin angular momentum (a 3-vector) is $(\hbar/2) \times (\hat{\sigma}_x, \hat{\sigma}_y, \hat{\sigma}_z)$. That is, in this case the Bloch vector (x, y, z) has a meaning in ordinary three-dimensional space, as the mean spin angular momentum, divided by $\hbar/2$.

Nowadays it is common to study a two-level quantum system without any particular physical representation in mind. In this context, it is appropriate to use the term *qubit* – a quantum bit.

Here $\hat{\sigma}_+ = (\hat{\sigma}_-)^\dagger = |e\rangle\langle g|$ is the raising operator for the atom. The coefficient g_k (which can be assumed real without loss of generality) is proportional to the dipole matrix element for the transition (which we will assume is non-zero) and depends on the structure of mode k. In particular, it varies as $V_k^{-1/2}$, where V_k is the physical volume of mode k.

It turns out that the rate γ of radiative decay for an atom in free space is of order $10^8\,\mathrm{s}^{-1}$ or smaller. This is much smaller than the typical frequency ω_a for an optical transition, which is of order $10^{15}\,\mathrm{s}^{-1}$ or greater. Since γ is due to the interaction Hamiltonian $\hat{V}$, it seems reasonable to treat $\hat{V}$ as being small compared with $\hat{H}_0 = \hat{H}_a + \hat{H}_E$. Thus we are justified in following the method of Section 3.2. We begin by calculating $\hat{V}$ in the interaction frame:

$$\hat{V}_{\mathrm{IF}}(t) = \sum_k (g_k \hat{b}_k e^{-i\omega_k t} + g_k \hat{b}_k^\dagger e^{i\omega_k t})(\hat{\sigma}_+ e^{+i\omega_a t} + \hat{\sigma}_- e^{-i\omega_a t}). \tag{3.23}$$

Exercise 3.3 *Show this, using the same technique as in Exercise 1.30.*

The first approximation we make is to remove the terms in $\hat{V}_{\mathrm{IF}}(t)$ that rotate (in the complex plane) at frequency $\omega_a + \omega_k$ for all k, yielding

$$\hat{V}_{\mathrm{IF}}(t) = \sum_k (g_k \hat{b}_k \hat{\sigma}_+ e^{-i(\omega_k - \omega_a)t} + g_k \hat{b}_k^\dagger \hat{\sigma}_- e^{i(\omega_k - \omega_a)t}). \tag{3.24}$$

As discussed in Section 1.5, this is known as the rotating-wave approximation (RWA). It is justified on the grounds that these terms rotate so fast ($\sim 10^{15}\,\mathrm{s}^{-1}$) that they will average to zero over the time-scale of radiative decay ($\sim 10^{-8}$ s) and hence not contribute to this process.[1] This approximation leads to significant simplifications.

Now substitute Eq. (3.24) into the exact equation (3.7) for the system state $\rho(t)$ in Section 3.2. To proceed we need to specify the initial state of the field, which we take to be the vacuum state (see Appendix A). The first term in Eq. (3.7) is then exactly zero.

Exercise 3.4 *Show this, and show that it holds also for a field state in a thermal state* $\rho_E \propto \exp[-\hat{H}_E/(k_B T)]$.
Hint: *Expand ρ_E in the number basis.*

For this choice of ρ_E, we have $\hat{V}_S = 0$; later, we will relax this assumption.

For convenience, we now drop the IF subscripts, while still working in the interaction frame. Under the Born approximation, the equation for $\rho(t)$ becomes

$$\dot{\rho} = -\int_0^t dt_1 \{\Gamma(t - t_1)[\hat{\sigma}_+ \hat{\sigma}_- \rho(t_1) - \hat{\sigma}_- \rho(t_1)\hat{\sigma}_+] + \mathrm{H.c.}\}, \tag{3.25}$$

where H.c. stands for the Hermitian conjugate term, and

$$\Gamma(\tau) = \sum_k g_k^2 e^{-i(\omega_k - \omega_a)\tau}. \tag{3.26}$$

---

[1] Terms like these are, however, important for a proper calculation of the Lamb frequency shift $\Delta\omega_a$, but that is beyond the scope of this treatment.

Exercise 3.5 *Show this, using the properties of the vacuum state and the field operators.*

Next, we wish to make the *Markov* approximation. This can be justified by considering the reservoir correlation function (3.26). For an atom in free space, there is an infinite number of modes, each of which is infinite in volume, so the modulus squared of the coupling coefficients is infinitesimal. Thus we can justify replacing the sum in Eq. (3.26) by an integral,

$$\Gamma(\tau) = \int_0^\infty d\omega \, \rho(\omega)g(\omega)^2 e^{i(\omega_a - \omega)\tau}. \tag{3.27}$$

Here $\rho(\omega)$ is the density of field modes as a function of frequency. This is infinite but the product $\rho(\omega)g(\omega)^2$ is finite. Moreover, $\rho(\omega)g(\omega)^2$ is a smoothly varying function of frequency for ω in the vicinity of ω_a. This means that the reservoir correlation function, $\Gamma(\tau)$, is sharply peaked at $\tau = 0$.

Exercise 3.6 *Convince yourself of this by considering a toy model in which $\rho(\omega)g(\omega)^2$ is independent of ω in the range $(0, 2\omega_a)$ and zero elsewhere.*

Thus we can apply the Markov approximation to obtain the master equation

$$\dot{\rho} = -i\frac{\Delta\omega_a}{2}[\hat{\sigma}_z, \rho] + \gamma\mathcal{D}[\hat{\sigma}_-]\rho. \tag{3.28}$$

Here the superoperator $\mathcal{D}[\hat{A}]$ is defined for an arbitrary operator $\hat{A}$ by

$$\mathcal{D}[\hat{A}]\rho \equiv \hat{A}\rho\hat{A}^\dagger - \frac{1}{2}(\hat{A}^\dagger\hat{A}\rho + \rho\hat{A}^\dagger\hat{A}). \tag{3.29}$$

The real parameters $\Delta\omega_a$ (the frequency shift) and γ (the radiative decay rate) are defined as

$$\Delta\omega_a - i\frac{\gamma}{2} = -i\int_0^\infty \Gamma(\tau)d\tau. \tag{3.30}$$

Exercise 3.7 *Derive Eq. (3.28)*

In practice the frequency shift (called the Lamb shift) due to the atom coupling to the electromagnetic vacuum is small, but can be calculated properly only by using renormalization theory and relativistic quantum mechanics.

The solution of Eq. (3.28) at any time $t > 0$ depends only on the initial state at time $t = 0$; there is no memory effect. The evolution is non-unitary because of the $\mathcal{D}$ term, which represents radiative decay. This can be seen by considering the Bloch representaton of the atomic state, as discussed in Box 3.1.

Exercise 3.8 *Familiarize yourself with the Bloch sphere by finding the points on it corresponding to the eigenstates of the Pauli matrices, and the point corresponding to the maximally mixed state.*

For example, the equation of motion for the inversion can be calculated as $\dot{z} = \text{Tr}[\hat{\sigma}_z\dot{\rho}]$, and re-expressing the right-hand side in terms of x, y and z. In this case we find simply

$\dot{z} = -\gamma(z+1)$, so the inversion decays towards the ground state ($z = -1$) exponentially at rate γ. Thus we can equate γ to the A coefficient of Einstein's theory [Ein17], and $1/\gamma$ to the atomic lifetime. The energy lost by the atom is radiated into the field, hence the term radiative decay. The final state here is pure, but, if it starts in the excited state, then the atom will become mixed before it becomes pure again. This mixing is due to entanglement between the atom and the field: the total state is a superposition of excited atom and vacuum-state field, and ground-state atom and field containing one photon of frequency ω_0. This process is called spontaneous emission because it occurs even if there are initially no photons in the field.

Exercise 3.9 *Show that, if the atom is prepared in the excited state at time $t = 0$, the Bloch vector at time t is $(0, 0, 2\mathrm{e}^{-\gamma t} - 1)$. At what time is the entanglement between the atom and the radiated field maximal?*

Strictly, the frequency of the emitted photon has a probability distribution centred on ω_0 with a full width-at-half-maximum height of γ. Thus a finite lifetime of the atomic state leads to an uncertainty in the energy of the emitted photon, which can be interpreted as an uncertainty in the energy separation of the atomic transition. The reciprocal relation between the lifetime $1/\gamma$ and the energy uncertainty γ is sometimes referred to as an example of the time–energy uncertainty relation. It should be noted that its meaning is quite different from that of the Heisenberg uncertainty relations mentioned in Section 1.2.1, since time is not a system property represented by an operator; it is merely an external parameter. Nevertheless, this relation is of great value heuristically, as we will see.

We note two important generalizations. Firstly, the atom may be driven coherently by a classical field. As long as the system Hamiltonian which describes this driving is weak compared with $\hat{H}_a$, it will have negligible effect on the derivation of the master equation in the interaction frame, and can simply be added at the end. Alternatively, this situation can be modelled quantum mechanically by taking the bath to be initially in a coherent state, which will make $\hat{V}_S(t)$ non-zero, and indeed time-dependent in general (this is discussed in Section 3.11.2 below). In any case, the effect of driving is simply to add another Hamiltonian evolution term to the final master equation (3.12) in the interaction frame. If the frequency of oscillation of the driving field is $\omega_0 \approx \omega_a$, then it is most convenient to work in an interaction frame using $\hat{H}_a = \omega_0 \hat{\sigma}_z/2$, rather than $\hat{H}_a = \omega_a \hat{\sigma}_z/2$. This is because, on moving to the interaction frame, it makes the total effective Hamiltonian for the atom time-independent:

$$\hat{H}_{\mathrm{drive}} = \frac{\Omega}{2}\hat{\sigma}_x + \frac{\Delta}{2}\hat{\sigma}_z. \tag{3.31}$$

Here $\Delta = \omega_a + \Delta\omega_a - \omega_0$ is the effective detuning of the atom, while Ω, the *Rabi frequency*, is proportional to the amplitude of the driving field and the atomic dipole moment. Here the phase of the driving field acts as a reference to define the in-phase (x) and in-quadrature (y) parts of the atomic dipole relative to the imposed force. The master equation for a resonantly driven, damped atom is known as the resonance fluorescence master equation.

Exercise 3.10 *(a) Show that the* Bloch equations *for resonance fluorescence are*

$$\dot{x} = -\Delta y - \frac{\gamma}{2}x, \tag{3.32}$$

$$\dot{y} = -\Omega z + \Delta x - \frac{\gamma}{2}y, \tag{3.33}$$

$$\dot{z} = +\Omega y - \gamma(z+1), \tag{3.34}$$

and that the stationary solution is

$$\begin{pmatrix} x \\ y \\ z \end{pmatrix}_{ss} = \begin{pmatrix} -4\Delta\Omega \\ 2\Omega\gamma \\ -\gamma^2 - 4\Delta^2 \end{pmatrix} (\gamma^2 + 2\Omega^2 + 4\Delta^2)^{-1}. \tag{3.35}$$

(b) Compare $\theta = \arctan(y_{ss}/x_{ss})$ and $A = \sqrt{x_{ss}^2 + y_{ss}^2}$ with the phase and amplitude of the long-time response of a classical, lightly damped, harmonic oscillator to an applied periodic force with magnitude proportional to Ω and detuning Δ. In what regime does the two-level atom behave like the harmonic oscillator?

Hint: *First, define interaction-frame phase and amplitude variables for the classical oscillator; that is, variables that would be constant in the absence of driving and damping.*

The second generalization is that the field need not be in a vacuum state, but rather (for example) may be in a thermal state (i.e. with a Planck distribution of photon numbers [GZ04]). This gives rise to *stimulated* emission and absorption of photons. In that case, the total master equation in the Markov approximation becomes

$$\dot{\rho} = -i\left[\frac{\Omega}{2}\hat{\sigma}_x + \frac{\Delta}{2}\hat{\sigma}_z, \rho\right] + \gamma(\bar{n}+1)\mathcal{D}[\hat{\sigma}_-]\rho + \gamma\bar{n}\mathcal{D}[\hat{\sigma}_+]\rho, \tag{3.36}$$

where $\bar{n} = \{\exp[\hbar\omega_a/(k_B T)] - 1\}^{-1}$ is the thermal mean photon number evaluated at the atomic frequency (we have here restored $\hbar$). This describes the (spontaneous and stimulated) emission of photons at a rate proportional to $\gamma(\bar{n}+1)$, and (stimulated) absorption of photons at a rate proportional to $\gamma\bar{n}$.

3.3.2 Cavity emission

Another system that undergoes radiative damping is a mode of the electromagnetic field in an optical cavity. In quantum optics the term 'cavity' is used for any structure (typically made of dielectric materials) that will store electromagnetic energy at discrete frequencies. The simplest sort of cavity is a pair of convex mirrors facing each other, but no mirrors are perfectly reflecting, and the stored energy will decay because of transmission through the mirrors.

Strictly speaking, a mode of the electromagnetic field should be a stationary solution of Maxwell's equations [CRG89] and so should not suffer a decaying amplitude. However, it is often convenient to treat *pseudomodes*, such as those that are localized within a cavity, as if they were modes, and to treat the amplitude decay as radiative damping due to coupling to the (pseudo-)modes that are localized outside the cavity [GZ04]. This is a good approximation, provided that the coupling is weak; that is, that the transmission at the mirrors is small.

The simplest case to consider is a single mode (of frequency ω_c) of a one-dimensional cavity with one slightly lossy mirror and one perfect mirror. We use $\hat{a}$ for the annihilation operator for the cavity mode of interest and $\hat{b}_k$ for those of the bath as before. The total Hamiltonian for system plus environment, in the RWA, is [WM94a]

$$\hat{H} = \omega_c \hat{a}^\dagger \hat{a} + \sum_k \omega_k \hat{b}_k^\dagger \hat{b}_k + \sum_k g_k(\hat{a}^\dagger \hat{b}_k + \hat{a} \hat{b}_k^\dagger). \tag{3.37}$$

The first term represents the free energy of the cavity mode of interest, the second is for the free energy of the many-mode field outside the cavity, and the last term represents the dominant terms in the coupling of the two for optical frequencies.

For weak coupling the Born–Markov approximations are justified just as for spontaneous emission. Following the same procedure leads to a very similar master equation for the cavity field, in the interaction frame:

$$\dot{\rho} = \gamma(\bar{n} + 1)\mathcal{D}[\hat{a}]\rho + \gamma\bar{n}\mathcal{D}[\hat{a}^\dagger]\rho. \tag{3.38}$$

Here $\bar{n}$ is the mean thermal photon number of the external field evaluated at the cavity frequency ω_c. We have ignored any environment-induced frequency shift, since this simply redefines the cavity resonance ω_c.

The first irreversible term in Eq. (3.38) represents emission of photons from the cavity. The second irreversible term represents an incoherent excitation of the cavity due to thermal photons in the external field.

Exercise 3.11 *Show that the rate of change of the average photon number in the cavity is given by*

$$\frac{d\langle \hat{a}^\dagger \hat{a}\rangle}{dt} = -\gamma\langle \hat{a}^\dagger \hat{a}\rangle + \gamma\bar{n}. \tag{3.39}$$

Note that here (and often from here on) we are relaxing our convention on angle brackets established in Section 1.2.1. That is, we may indicate the average of a property for a quantum system by angle brackets around the corresponding operator.

From Eq. (3.39) it is apparent that γ is the decay rate for the energy in the cavity. Assuming that $\rho(\omega)g(\omega)^2$ is slowly varying with frequency, we can evaluate this decay rate to be

$$\gamma \simeq 2\pi\rho(\omega_c)g(\omega_c)^2. \tag{3.40}$$

Exercise 3.12 *Show this explicitly using the example of Exercise 3.6.*
Note: *This result can be obtained more simply by replacing $\int_{-\infty}^{0} d\tau\, e^{-i\omega\tau}$ by $\pi\delta(\omega)$, which is permissible when it appears in an ω-integral with a flat integrand.*

In more physical terms, if the mirror transmits a proportion $T \ll 1$ of the energy in the cavity on each reflection, and the round-trip time for light in the cavity is τ, then $\gamma = T/\tau$.

As in the atomic case, we can include other dynamical processes by simply adding an appropriate Hamiltonian term to the interaction-frame master equation (3.38), as long as the added Hamiltonian is (in some sense) small compared with $\hat{H}_0$. In particular, we can include a coherent driving term, to represent the excitation of the cavity mode by an external laser of frequency ω_c, by adding the following driving Hamiltonian [WM94a]:

$$\hat{H}_{\text{drive}} = i\epsilon\hat{a}^\dagger - i\epsilon^*\hat{a}. \tag{3.41}$$

Exercise 3.13 *Show that, in the zero-temperature limit, the stationary state for the driven, damped cavity is a coherent state $|\alpha\rangle$ with $\alpha = 2\epsilon/\gamma$.*
Hint: *Make the substitution $\hat{a} = 2\epsilon/\gamma + \hat{a}_0$, and show that the solution of the master equation is the vacuum state for $\hat{a}_0$.*

3.4 Irreversibility without the rotating-wave approximation

In the previous examples of radiative decay of an atom and a cavity, the system Hamiltonian $\hat{H}_S$ produced oscillatory motion in the system with characteristic frequencies (ω_a and ω_c, respectively) much larger than the rate of decay. This allowed us to make a RWA in describing the system–environment coupling Hamiltonian as $\sum_k g_k(\hat{s}\hat{b}_k^\dagger + \hat{s}^\dagger\hat{b}_k)$, where $\hat{s}$ is a system lowering operator. That is, the coupling describes the transfer of quanta of excitation of the oscillation between the system and the bath. When there is no such large characteristic frequency, it is not possible to make such an approximation. In this section we discuss two examples of this, the spin–boson model and quantum Brownian motion. We will, however, retain the model for the bath as a collection of harmonic oscillators and the assumption that the interaction between system and environment is weak in order to derive a master equation perturbatively.

3.4.1 The spin–boson model

Consider a two-level system, coupled to a reservoir of harmonic oscillators, such that the total Hamiltonian is

$$\hat{H} = \frac{\Delta}{2}\hat{\sigma}_x + \sum_k \left(\frac{\hat{p}_k^2}{2m_k} + \frac{m_k\omega_k^2\hat{q}_k^2}{2} \right) + \hat{\sigma}_z \sum_k g_k\hat{q}_k, \tag{3.42}$$

where $\hat{q}_k$ are the coordinates of each of the environmental oscillators. This could describe a spin-half particle (see Box 3.1), in the interaction frame with respect to a Hamiltonian proportional to $\hat{\sigma}_z$. Such a Hamiltonian would describe a static magnetic field in the z ('longitudinal') direction. Then the first term would describe resonant driving (as in the two-level atom case) by a RF magnetic field in the x–y ('transverse') plane, and the last term would describe fluctuations in the longitudinal field. However, there are many other physical situations for which this Hamiltonian is an approximate description, including quantum tunnelling in a double-well potential [LCD+87].

Since the frequency Δ can be small, even zero, we cannot make a RWA in this model. Nevertheless, we can follow the procedure in Section 3.2, where $\hat{H}_0$ comprises the first two terms in Eq. (3.42). We assume the bath to be in a thermal equilibrium state of temperature $1/(k_B\beta)$ with respect to its Hamiltonian. Then, replacing $\rho(t_1)$ by $\rho(t)$ in Eq. (3.10) yields the master equation with time-dependent coefficients [PZ01]

$$\dot{\rho}(t) = -\int_0^t dt_1 \left(\nu(t_1)[\hat{\sigma}_z(t), [\hat{\sigma}_z(t-t_1), \rho(t)]] \right.$$
$$\left. - i\eta(t_1)[\hat{\sigma}_z(t), \{\hat{\sigma}_z(t-t_1), \rho(t)\}] \right), \tag{3.43}$$

where $\{\hat{A}, \hat{B}\} = \hat{A}\hat{B} + \hat{B}\hat{A}$ is known as an *anticommutator*, and the kernels are given by

$$\nu(t_1) = \frac{1}{2} \sum_k g_k^2 \langle \{\hat{q}_k(t), \hat{q}_k(t-t_1)\} \rangle = \int_0^\infty d\omega \, J(\omega)\cos(\omega t_1)[1 + 2\bar{n}(\omega)],$$
$$\tag{3.44}$$

$$\eta(t_1) = \frac{i}{2} \sum_k g_k^2 \langle [\hat{q}_k(t), \hat{q}_k(t-t_1)] \rangle = \int_0^\infty d\omega \, J(\omega)\sin(\omega t_1). \tag{3.45}$$

Here the spectral density function is defined by

$$J(\omega) = \sum_k \frac{g_k^2 \delta(\omega - \omega_k)}{2m_k\omega_k}, \tag{3.46}$$

and $\bar{n}(\omega)$ is the mean occupation number of the environmental oscillator at frequency ω. It is given as usual by the Planck law $1 + 2\bar{n}(\omega) = \coth(\beta\hbar\omega/2)$ (where, in deference to Planck, we have restored his constant). The sinusoidal kernels in Eqs. (3.44) and (3.45) result from the oscillatory time dependence of $\hat{q}_k(t)$ from the bath Hamiltonian.

The time dependence of the operator $\hat{\sigma}_z(t)$, in the interaction frame with respect to $\hat{H}_0$, is given by

$$\hat{\sigma}_z(t) = \hat{\sigma}_z \cos(\Delta t) + \hat{\sigma}_y \sin(\Delta t). \tag{3.47}$$

Exercise 3.14 *Show this by finding and solving the Heisenberg equations of motion for $\hat{\sigma}_y$ and $\hat{\sigma}_z$, for the Hamiltonian $\hat{H}_0$.*

Substituting this into Eq.(3.43), and then moving out of the interaction frame, yields the Schrödinger-picture master equation[2]

$$\dot{\rho} = -i[\hat{H}_{nh}\rho - \rho\hat{H}_{nh}^\dagger] - \zeta^*(t)\hat{\sigma}_z\rho\hat{\sigma}_y - \zeta(t)\hat{\sigma}_y\rho\hat{\sigma}_z - D(t)[\hat{\sigma}_z, [\hat{\sigma}_z, \rho]]. \tag{3.48}$$

Here

$$\hat{H}_{nh} = \left(\frac{\Delta}{2} + \zeta(t) \right)\hat{\sigma}_x \tag{3.49}$$

---

[2] With minor corrections to the result in Ref. [PZ01].

is a non-Hermitian operator (the Hermitian part of which can be regarded as the Hamiltonian), while

$$\zeta(t) = \int_0^t dt_1(\nu(t_1) - i\eta(t_1))\sin(\Delta t_1), \tag{3.50}$$

$$D(t) = \int_0^t dt_1 \, \nu(t_1)\cos(\Delta t_1). \tag{3.51}$$

The environment thus shifts the free Hamiltonian for the system (via $\mathrm{Re}[\zeta]$) and introduces irreversible terms (via $\mathrm{Im}[\zeta]$ and D). Note that if $\Delta = 0$ only the final term in Eq. (3.48) survives.

To proceed further we need an explicit form of the spectral density function. The simplest case is known as *Ohmic dissipation*, in which the variation with frequency is linear at low frequencies. We take

$$J(\omega) = 2\eta\frac{\omega}{\pi}\frac{\Lambda^2}{\Lambda^2 + \omega^2}, \tag{3.52}$$

where Λ is a cut-off frequency, as required in order to account for the physically necessary fall-off of the coupling at sufficiently high frequencies, and η is a dimensionless parameter characterizing the strength of the coupling between the spin and the environment. After splitting $\zeta(t)$ into real and imaginary parts as $\zeta(t) = f(t) - i\gamma(t)$, we can easily do the integral to find the decay term $\gamma(t)$. It is given by

$$\gamma(t) = \gamma_\infty\left[1 - \left(\cos(\Delta t) + \frac{\Lambda}{\Delta}\sin(\Delta t)\right)e^{-\Lambda t}\right]. \tag{3.53}$$

This begins at zero and decays (at a rate determined by the high-frequency cut-off) to a constant $\gamma_\infty \propto \Lambda^2/(\Lambda^2 + \Delta^2)$. The other terms depend on the temperature of the environment and are not easy to evaluate analytically. The diffusion constant can be shown to approach the asymptotic value

$$D_\infty = \eta\Delta\frac{\Lambda^2}{\Lambda^2 + \Delta^2}\coth(\beta\Delta/2). \tag{3.54}$$

The function $f(t)$ also approaches (algebraically, not exponentially) a limiting value, which at high temperatures is typically much smaller than D_∞ (by a factor proportional to Λ).

In the limit that $\Delta \to 0$, we find

$$D_\infty \to 2\eta k_B T, \tag{3.55}$$

and, as mentioned previously, $\zeta(t)$ is zero in this limit. In this case the master equation takes the following simple form in the long-time limit:

$$\dot{\rho} = -2\eta k_B T[\hat{\sigma}_z, [\hat{\sigma}_z, \rho]]. \tag{3.56}$$

This describes dephasing of the spin in the x–y plane at rate $D_\infty/2$.

3.4.2 Quantum Brownian motion

Another important model for which the RWA cannot be used is quantum Brownian motion. In this case we have a single particle with mass M, with position and momentum operators $\hat{X}$ and $\hat{P}$. It may be moving in some potential, and is coupled to an environment of simple harmonic oscillators. This is described by the Hamiltonian

$$\hat{H} = \frac{\hat{P}^2}{2M} + V(\hat{X}) + \sum_k \left(\frac{\hat{p}_k^2}{2m_k} + \frac{m_k \omega_k^2 \hat{q}_k^2}{2} \right) + \hat{X} \sum_k g_k \hat{q}_k. \tag{3.57}$$

The derivation of the perturbative master equation proceeds as in the case of the spin–boson model [PZ01]. It is only for simple potentials, such as the harmonic $V(\hat{X}) = M\Omega^2 \hat{X}^2/2$, that the evolution generated by $\hat{H}_0$ can be solved analytically. The derivation is much as in the spin–boson case, but, for dimensional correctness, we must replace η by $M\gamma$, where γ is a rate. The result is

$$\dot{\rho} = -\mathrm{i}[\hat{P}^2/(2M) + M\tilde{\Omega}(t)^2 \hat{X}^2/2, \rho] - \mathrm{i}\gamma(t)[\hat{X}, \{\hat{P}, \rho\}]$$
$$- D(t)[\hat{X}, [\hat{X}, \rho]] - f(t)[\hat{X}, [\hat{P}, \rho]]. \tag{3.58}$$

Here $\tilde{\Omega}(t)$ is a shifted frequency and $\gamma(t)$ is a momentum-damping rate. $D(t)$ gives rise to diffusion in momentum and $f(t)$ to so-called anomalous diffusion.

If we again assume the Ohmic spectral density function (3.52) then we can evaluate these time-dependent coefficients. The coefficients all start at zero, and tend asymptotically to constants, with the same properties as in the spin–boson case. The shifted frequency $\tilde{\Omega}$ tends asymptotically to $\sqrt{\Omega - 2\gamma_\infty \Lambda}$, which is unphysical for Λ too large. In the high-temperature limit, $k_B T \gg \Omega$, with $\Lambda \gg \Omega$ one finds

$$D_\infty = M\gamma\Omega \frac{\Lambda^2}{\Lambda^2 + \Omega^2} \coth(\beta\Omega/2) \to 2\gamma_\infty k_B T M, \tag{3.59}$$

while $f(t)$ is negligible ($\propto \Lambda^{-1}$) compared with this.

Replacing the above time-dependent coefficients with their asymptotic values will be a bad approximation at short times, and indeed may well lead to nonsensical results (as will be discussed in Section 3.6). However, at long times it is reasonable to use the asymptotic values, giving the Markovian master equation

$$\dot{\rho} = -\mathrm{i}[\hat{P}^2/(2M) + M\tilde{\Omega}_\infty^2 \hat{X}^2/2, \rho] - \mathrm{i}\gamma_\infty[\hat{X}, \{\hat{P}, \rho\}] - 2\gamma_\infty k_B T M[\hat{X}, [\hat{X}, \rho]]. \tag{3.60}$$

Exercise 3.15 *Show that this is identical with the Markovian master equation of Eq. (3.12) for this case.*

The first irreversible term in Eq. (3.60) describes the loss, and the second the gain, of kinetic energy, as can be seen in the following exercise.

Exercise 3.16 *Derive the equations of motion for the means and variances of the position and momentum using the high-temperature Brownian-motion master equation, Eq. (3.60). Thus show that momentum is damped exponentially at rate $2\gamma_\infty$, but that momentum* *diffusion adds kinetic energy at rate* $2\gamma_\infty k_B T$. *Show that for* $\tilde{\Omega}_\infty = 0$ *the steady-state energy of the particle is* $k_B T/2$, *as expected from thermodynamics.*

3.5 Fermionic reservoirs

In the previous examples the environment was taken to be composed of a very large (essentially infinite) number of harmonic oscillators. Such an environment is called *bosonic*, because the energy quanta of these harmonic oscillators are analogous to bosonic particles, with the associated commutation relations for the annihilation and creation operators (3.14). There are some very important physical situations in which the environment of a local system is in fact *fermionic*. An example is a local quantum dot (which acts something like a cavity for a single electron) coupled via tunnelling to the many electron states of a resistor. The annihilation and creation operators for fermionic particles, such as electrons, obey *anticommutation relations*

$$\{\hat{a}_k, \hat{a}_l^\dagger\} = \delta_{kl}, \tag{3.61}$$

$$\{\hat{a}_k, \hat{a}_l\} = 0. \tag{3.62}$$

The study of such systems is the concern of the rapidly developing field of *mesoscopic electronics* [Dat95, Imr97]. Unfortunately, perturbative master equations might not be appropriate in many situations when charged fermions are involved, since such systems are strongly interacting. However, there are some experiments for which a perturbative master equation is a good approximation. We now consider one of these special cases to illustrate some of the essential differences between bosonic and fermionic environments.

The concept of a mesoscopic electronic system emerged in the 1980s as experiments on small, almost defect-free, conductors and semiconductors revealed unexpected departures from classical current–voltage characteristics at low temperatures. The earliest of these results indicated quantized conductance. The classical description of conductance makes reference to random scattering of carriers due to inelastic collisions. However, in mesoscopic electronic systems, the mean free path for inelastic scattering may be longer than the length of the device. Such systems are dominated by *ballistic* behaviour in which conduction is due to the transport of single carriers, propagating in empty electron states above a filled Fermi sea, with only elastic scattering from confining potentials and interactions with magnetic fields. As Landauer [Lan88, Lan92] and Büttiker [Büt88] first made clear, conductance in such devices is determined not by inelastic scattering, but by the quantum-mechanical transmission probability, T, across device inhomogeneities. If a single ballistic channel supports a single transverse Fermi mode (which comprises two modes when spin is included), the transmission probability is $T \approx 1$. The resulting conductance of that channel is the reciprocal of the *quantum of resistance*. This is given by the Landauer–Büttiker theory as [Dat95]

$$R_Q = \frac{\pi\hbar}{e^2} \approx 12.9\,\mathrm{k}\Omega. \tag{3.63}$$

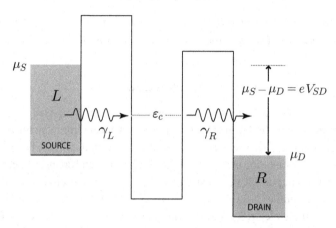

Fig. 3.1 A schematic representation of a quantum dot in the conduction band. Position runs from left to right and energy runs vertically. The quasibound state in the dot is labelled c. The grey regions labelled L and R represent metallic electronic states filled up to the local Fermi level. The difference in the Fermi levels between left and right is determined by the source–drain bias voltage as eV_{SD}.

A quantum dot is a three-dimensional confining potential for electrons or holes in a semiconductor, and can be fabricated in a number of ways [Tur95]. We will consider a very simple model in which the dot has only a single bound state for one electron. This is not as artificial as it may sound. Although a single quantum dot in fact contains a very large number of electrons, at low temperatures this system of electrons is close to the Fermi ground state. In a semiconductor dot the ground state is typically a filled valence band and an unoccupied conduction band. In a metallic dot (or grain as it sometimes called) the ground state is the conduction band filled up to the Fermi energy. At low temperatures and weak bias the current is carried by a few electrons near the Fermi energy and we are typically concerned only with *additional* electrons injected onto the dot. Because electrons are charged, large energy gaps can appear in the spectrum of multi-electron quantum dots, in addition to the quantization of energy levels due to confinement. This phenomenon is called *Coulomb blockade* [Kas93]. The Coulomb blockade energy required to add a single electron to a quantum dot is $e^2/(2C)$, where the capacitance C can be very small (less than 10^{-16} F) due to the small size of these systems. If the charging energy is large enough, compared with thermal energy, we can assume that only a single bound state for an additional electron is accessible in the quantum dot. Typically this would require a temperature below 1 K.

We also assume that the dot is coupled via tunnel junctions to two fermionic reservoirs; see Fig. 3.1. A tunnel junction is a region in the material from which charge carriers classically would be excluded by energy conservation. While propagating solutions of the Schrödinger equation cannot be found in such a region, exponentially decaying amplitudes can exist. We will assume that the region is not so extensive that all amplitudes decay to zero, but small enough for the coupling, due to the overlap of amplitudes inside and outside the region, to be small. In that case the coupling between propagating solutions on either side of the region can be treated perturbatively.

We assume that the reservoirs remain in the normal (Ohmic) conducting state. The total system is not in thermal equilibrium due to the bias voltage V_{SD} across the dot. However, the two reservoirs are held very close to thermal equilibrium at temperature T, but at *different* chemical potentials through contact to an external circuit via an *Ohmic contact*. We refer to the fermionic reservoir with the higher chemical potential as the *source* (also called the emitter) and the one with the lower chemical potential as the *drain* (also called the collector). The difference in chemical potentials is given by $\mu_S - \mu_D = eV_{SD}$. In this circumstance, charge may flow through the dot, and an external current will flow. The necessity to define a chemical potential is the first major difference between fermionic systems and the bosonic environments of quantum optics.

A perturbative master-equation approach to this problem is valid only if the resistance of the tunnel junction, R, is large compared with the quantum of resistance R_Q. The physical meaning of this condition is as follows. If for simplicity we denote the bias voltage of the junction as V, then the average current through the junction is V/R, so the tunnelling rate is $\Gamma = V/(eR)$. Thus the typical time between tunnelling events is $\Gamma^{-1} = eR/V$. Now, if the lifetime of the quasibound state is τ, then, by virtue of the time–energy uncertainty relation discussed in Section 3.3.1, there is an uncertainty in the energy level of order $\hbar/\tau$. If the external potential is to control the tunnelling then this energy uncertainty must remain less than eV. Thus the lifetime must be at least of order $\hbar/(eV)$. If we demand that the lifetime be much less than the time between tunnelling events, so that the events do not overlap in time, we thus require $\hbar/(eV) \ll eR/V$. This gives the above relation between R and the quantum of resistance.

The total Hamiltonian of a system composed of the two Fermi reservoirs, connected by two tunnel barriers to a single Fermi bound state, is (with $\hbar = 1$)

$$\hat{H}_{\text{QD+leads}} = \sum_k \varepsilon_k^S \hat{a}_k^\dagger \hat{a}_k + \varepsilon_c \hat{c}^\dagger \hat{c} + \sum_p \varepsilon_p^D \hat{b}_p^\dagger \hat{b}_p$$

$$+ \sum_k (T_k^S \hat{c}^\dagger \hat{a}_k + T_k^{S*} \hat{a}_k^\dagger \hat{c}) + \sum_p (T_p^D \hat{b}_p^\dagger \hat{c} + T_p^{D*} \hat{c}^\dagger \hat{b}_p). \qquad (3.64)$$

Here $a_k(a_k^\dagger)$, $c(c^\dagger)$ and $b_p(b_p^\dagger)$ are the fermion annihilation (creation) operators of electrons in the source (S) reservoir, in the central quantum dot and in the drain (D) reservoir, respectively. Because of the fermion anticommutation relations, the dot is described by just two states.

Exercise 3.17 *Show from Eqs. (3.61) and (3.62) that the eigenvalues for the fermion number operator $\hat{a}_l^\dagger \hat{a}_l$ are 0 and 1, and that, if the eigenstates are $|0\rangle$ and $|1\rangle$, respectively, then $\hat{a}_l^\dagger |0\rangle = |1\rangle$.*

The first three terms in Eq. (3.64) comprise $\hat{H}_0$. The energy of the bound state without bias is ε_0, which under bias becomes $\varepsilon_c = \varepsilon_0 - \alpha eV$, where α is a structure-dependent coefficient. The single-particle energies in the source and drain are, respectively, $\varepsilon_k^S = k^2/(2m)$ and $\varepsilon_p^D = p^2/(2m) - eV$. The energy reference is at the bottom of the conduction band of the source reservoir. Here, and below, we are assuming spin-polarized electrons so that we do not have to sum over the spin degree of freedom.

The fourth and fifth terms in the Hamiltonian describe the coupling between the quasi-bound electrons in the dot and the electrons in the reservoir. The tunnelling coefficients T_k^S and T_p^D depend upon the profile of the potential barrier between the dot and the reservoirs, and upon the bias voltage. We will assume that at all times the two reservoirs remain in their equlibrium states despite the tunnelling of electrons. This is a defining characteristic of a reservoir, and comes from assuming that the dynamics of the reservoirs are much faster than those of the quasibound quantum state in the dot.

In the interaction frame the Hamiltonian may be written as

$$\hat{V}(t) = \sum_{j=1}^{2} \hat{c}^\dagger \hat{\Upsilon}_j(t)e^{i\varepsilon_c t} + \hat{c}\hat{\Upsilon}_j^\dagger(t)e^{-i\varepsilon_c t}, \tag{3.65}$$

where the reservoir operators are given by

$$\hat{\Upsilon}_1(t) = \sum_k T_k^S \hat{a}_k e^{-i\varepsilon_k^S t}, \qquad \hat{\Upsilon}_2(t) = \sum_p T_p^D \hat{b}_p e^{-i\varepsilon_p^D t}. \tag{3.66}$$

We now obtain an equation of motion for the state matrix ρ of the bound state in the dot by following the standard method in Section 3.2. The only non-zero reservoir correlation functions we need to compute are

$$I_{jN}(t) = \int_0^t dt_1 \langle \Upsilon_j^\dagger(t)\Upsilon_j(t_1)\rangle e^{-i\varepsilon_c(t-t_1)}, \tag{3.67}$$

$$I_{jA}(t) = \int_0^t dt_1 \langle \Upsilon_j(t_1)\Upsilon_j^\dagger(t)\rangle e^{-i\varepsilon_c(t-t_1)}. \tag{3.68}$$

Here N and A stand for normal (annihilation operators after creation operators) and antinormal (vice versa) ordering of operators – see Section A.5. In order to illustrate the important differences between the fermionic case and the bosonic case discussed previously, we will now explicitly evaluate the first of these correlation functions, $I_{1N}(t)$.

Using the definition of the reservoir operators and the assumed thermal Fermi distribution of the electrons in the source, we find

$$I_{1N}(t) = \sum_k \bar{n}_k^S |T_k^S|^2 \int_0^t dt_1 \exp[i(\varepsilon_k^S - \varepsilon_c)(t - t_1)]. \tag{3.69}$$

Since the reservoir is a large system, we can introduce a density of states $\rho(\omega)$ as usual and replace the sum over k by an integral to obtain

$$I_{1N}(t) = \int_0^\infty d\omega\, \rho(\omega)\bar{n}^S(\omega)|T^S(\omega)|^2 \int_{-t}^0 d\tau\, e^{-i(\omega-\varepsilon_c)\tau}, \tag{3.70}$$

where we have also changed the variable of time integration. The dominant term in the frequency integration will come from frequencies near ε_c because the time integration is significant at that point. For fermionic reservoirs, the expression for the thermal occupation number is [Dat95]

$$\bar{n}^S(\omega) = [1 + e^{(\omega - \omega_f)/k_B T}]^{-1}, \tag{3.71}$$

where ω_f is the Fermi energy (recall that $\hbar = 1$). We assume that the bias is such that the quasibound state of the dot is below the Fermi level in the source. This implies that near $\omega = \varepsilon_c$, and at low temperatures, the average occupation of the reservoir state is very close to unity [Dat95].

Now we make the Markov approximation to derive an autonomous master equation as in Section 3.2. On extending the limits of integration from $-t$ to $-\infty$ in Eq. (3.70) as explained before, I_{1N} may be approximated by the constant

$$I_{1N}(t) \approx \pi \rho(\varepsilon_c) |T_S(\varepsilon_c)|^2 \equiv \gamma_L/2. \tag{3.72}$$

This defines the effective rate γ_L of injection of electrons from the source (the left reservoir in Fig. 3.1) into the quasibound state of the dot. This rate will have a complicated dependence on the bias voltage through both ε_c and the coupling coefficients $|T_S(\omega)|$, which can be determined by a self-consistent band calculation. We do not address this issue; we simply seek the noise properties as a function of the rate constants.

By evaluating all the other correlation functions under similar assumptions, we find that the quantum master equation for the state matrix representing the dot state in the interaction frame is given by

$$\frac{d\rho}{dt} = \frac{\gamma_L}{2}(2\hat{c}^\dagger \rho \hat{c} - \hat{c}\hat{c}^\dagger \rho - \rho \hat{c}\hat{c}^\dagger) + \frac{\gamma_R}{2}(2\hat{c}\rho\hat{c}^\dagger - \hat{c}^\dagger\hat{c}\rho - \rho\hat{c}^\dagger\hat{c}), \tag{3.73}$$

where γ_L and γ_R are constants determining the rate of injection of electrons from the source into the dot and from the dot into the drain, respectively.

From this master equation it is easy to derive the following equation for the mean occupation number $\langle n(t) \rangle = \mathrm{Tr}[\hat{c}^\dagger \hat{c}\rho(t)]$:

$$\frac{d\langle n \rangle}{dt} = \gamma_L(1 - \langle n \rangle) - \gamma_R\langle n \rangle. \tag{3.74}$$

Exercise 3.18 *Show this, and show that the steady-state occupancy of the dot is* $\langle n \rangle_{ss} = \gamma_L/(\gamma_L + \gamma_R)$.

The effect of Fermi statistics is evident in Eq. (3.74). If there is an electron on the dot, $\langle n \rangle = 1$, and the occupation of the dot can decrease only by emission of an electron into the drain at rate γ_R.

It is at this point that we need to make contact with measurable quantities. In the case of electron transport, the measurable quantities reduce to current $I(t)$ and voltage $V(t)$. The measurement results are a time series of currents and voltages, which exhibit both systematic and stochastic components. Thus $I(t)$ and $V(t)$ are classical conditional stochastic processes, driven by the underlying quantum dynamics of the quasibound state on the dot.

The reservoirs in the Ohmic contacts play a key role in defining the measured quantities and ensuring that they are ultimately classical stochastic processes. Transport through the dot results in charge fluctuations in either the left or the right channel. These fluctuations decay extremely rapidly, ensuring that the channels remain in thermal equilibrium with the respective Ohmic contacts. For this to be possible, charge must be able to flow into and out of the reservoirs from an external circuit.

If a single electron tunnels out of the dot into the drain between time t and $t + dt$, its energy is momentarily above the Fermi energy. This electron scatters very strongly from the electrons in the drain and propagates into the right Ohmic contact, where it is perfectly absorbed. The nett effect is a small current pulse in the external circuit of total charge $e_L = eC_R/(C_L + C_R)$. Here $C_{L/R}$ is the capacitance between the dot and the L/R reservoir, and we have ignored any parasitic capacitance between source and drain. This is completely analogous to perfect photodetection: a photon emitted from a cavity will be detected with certainty by a detector that is a perfect absorber. Likewise, when an electron in the right channel tunnels onto the dot, there is a rapid relaxation of this unfilled state back to thermal equilibrium as an electron is emitted from the right Ohmic contact into the depleted state of the source. This again results in a current pulse carrying charge $e_R = e - e_L$ in the circuit connected to the Ohmic contacts.

The energy gained when one electron is emitted from the left reservoir is, by definition, the chemical potential of that reservoir, μ_L, while the energy lost when one electron is absorbed into the right reservoir is μ_R. The nett energy transferred between reservoirs is $\mu_L - \mu_R$. This energy is supplied by the external voltage, V, and thus $\mu_L - \mu_R = eV$. On average, in the steady state, the same current flows in the source and drain:

$$J_{ss} \equiv e_L\gamma_L(1 - \langle n\rangle_{ss}) + e_R\gamma_R\langle n\rangle_{ss} \tag{3.75}$$

$$= e\gamma_L(1 - \langle n\rangle_{ss}) = e\gamma_R\langle n\rangle_{ss} \tag{3.76}$$

$$= e\frac{\gamma_L\gamma_R}{\gamma_L + \gamma_R}. \tag{3.77}$$

Exercise 3.19 *Verify the identity of these expressions.*

From this we see that the average tunnelling rate of the device, Γ as previously defined, is given by

$$\Gamma = \left(\frac{1}{\gamma_L} + \frac{1}{\gamma_R}\right)^{-1}. \tag{3.78}$$

Typical values for the tunnelling rates achievable in these devices are indicated by results from an experiment by Yacoby *et al.* [YHMS95] in which single-electron transmission through a quantum dot was measured. The quantum dot was defined by surface gates on a GaAs/AlGaAs two-dimensional electron gas. The quantum dot was 0.4 μm wide and 0.5 μm long and had an electron temperature of 100 mK. They measured a tunnelling rate of order 0.3 GHz.

Box 3.2 Quantum dynamical semigroups

Formally solving the master equation for the state matrix defines a map from the state matrix at time 0 to a state matrix at *later* times t by $\mathcal{N}_t$: $\rho(0) \rightarrow \rho(t) = \mathcal{N}_t \rho(0)$ for all times $t \geq 0$. This dynamical map must be completely positive (see Box 1.3). More formally, we require a *quantum dynamical semigroup* [AL87], which is a family of completely positive maps $\mathcal{N}_t$ for $t \geq 0$ such that

- $\mathcal{N}_t \mathcal{N}_s = \mathcal{N}_{t+s}$
- $\text{Tr}[(\mathcal{N}_t \rho)\hat{A}]$ is a continuous function of t for any state matrix ρ and Hermitian operator $\hat{A}$.

The family forms a semigroup rather than a group because there is not necessarily any inverse. That is, $\mathcal{N}_t$ is not necessarily defined for $t < 0$.

These conditions formally capture the idea of Markovian dynamics of a quantum system. (Note that there is no implication that all open-system dynamics must be Markovian.) From these conditions it can be shown that there exists a superoperator $\mathcal{L}$ such that

$$\frac{d\rho(t)}{dt} = \mathcal{L}\rho(t), \tag{3.79}$$

where $\mathcal{L}$ is called the generator of the map $\mathcal{N}_t$. That is,

$$\rho(t) = \mathcal{N}_t \rho(0) = e^{\mathcal{L}t}\rho(0). \tag{3.80}$$

Moreover, this $\mathcal{L}$ must have the Lindblad form.

3.6 The Lindblad form and positivity

We have seen a number of examples in which the dynamics of an open quantum system can be described by an automonous differential equation (a time-independent master equation) for the state matrix of the system. What is the most general form that such an equation can take such that the solution is always a valid state matrix? This is a dynamical version of the question answered in Box 1.3 of Chapter 1, which was as follows: what are the physically allowed operations on a state matrix? In fact, the question can be formulated in a way that generalizes the notion of operations to a *quantum dynamical semigroup* – see Box. 3.2

It was shown by Lindblad in 1976 [Lin76] that, for a Markovian master equation $\dot{\rho} = \mathcal{L}\rho$, the generator of the quantum dynamics must be of the form

$$\mathcal{L}\rho = -i[\hat{H}, \rho] + \sum_{k=1}^{K} \mathcal{D}[\hat{L}_k]\rho, \tag{3.81}$$

for $\hat{H}$ Hermitian and $\{\hat{L}_j\}$ arbitrary operators. Here $\mathcal{D}$ is the superoperator defined earlier in Eq. (3.29). For mathematical rigour [Lin76], it is also required that $\sum_{k=1}^{K} \hat{L}_k^\dagger \hat{L}_k$ be a bounded operator, but that is often not satisfied by the operators we use, so this requirement is usually ignored. This form is known as the *Lindblad form*, and the operators $\{\hat{L}_k\}$

are called Lindblad operators. The superoperator $\mathcal{L}$ is sometimes called the *Liouvillian* superoperator, by analogy with the operator which generates the evolution of a classical probability distribution on phase space, and the term Lindbladian is also used.

Each term in the sum in Eq. (3.81) can be regarded as an *irreversible channel*. It is important to note, however, that the decomposition of the generator into the Lindblad form is not unique. We can reduce the ambiguity by requiring that the operators $\hat{1}, \hat{L}_1, \hat{L}_2, \ldots, \hat{L}_k$ be linearly independent. We are still left with the possibility of redefining the Lindblad operators by an arbitrary $K \times K$ unitary matrix T_{kl}:

$$\hat{L}_k \rightarrow \sum_{l=1}^{K} T_{kl} \hat{L}_l. \tag{3.82}$$

In addition, $\mathcal{L}$ is invariant under c-number shifts of the Lindblad operators, accompanied by a new term in the Hamiltonian:

$$\hat{L}_k \rightarrow \hat{L}_k + \chi_k, \qquad \hat{H} \rightarrow \hat{H} - \frac{i}{2} \sum_{k=1}^{K} (\chi_k^* \hat{L}_k - \text{H.c.}). \tag{3.83}$$

Exercise 3.20 *Verify the invariance of the master equation under (3.82) and (3.83).*

In the case of a single irreversible channel, it is relatively simple to evaluate the completely positive map $\mathcal{N}_t = \exp(\mathcal{L}t)$ formally as

$$\mathcal{N}_t = \sum_{m=0}^{\infty} \mathcal{N}_t^{(m)}, \tag{3.84}$$

where the operations $\mathcal{N}^{(m)}(t)$ are defined by

$$\mathcal{N}_t^{(m)} = \int_0^t dt_m \int_0^{t_m} dt_{m-1} \cdots \int_0^{t_2} dt_1 \, \mathcal{S}(t - t_m) \mathcal{X}$$
$$\times \mathcal{S}(t_m - t_{m-1}) \mathcal{X} \cdots \mathcal{S}(t_2 - t_1) \mathcal{X} \mathcal{S}(t_1), \tag{3.85}$$

with $\mathcal{N}_t^{(0)} = \mathcal{S}(t)$. Here the superoperators $\mathcal{S}$ and $\mathcal{X}$ are defined by

$$\mathcal{S}(\tau) = \mathcal{J}\left[e^{-\tau(i\hat{H} + \hat{L}^\dagger \hat{L}/2)} \right], \tag{3.86}$$

$$\mathcal{X} = \mathcal{J}[\hat{L}], \tag{3.87}$$

where the superoperator $\mathcal{J}$ is as defined in Eq. (1.80).

Exercise 3.21 *Verify the above expression for $\mathcal{N}_t$ by calculating $\mathcal{N}_0$ and $\dot{\rho}(t)$, where $\rho(t) = \mathcal{N}_t \rho(0)$. Also verify that $\mathcal{N}_t$ is a completely positive map, as defined in Chapter 1.*

As we will see in Chapter 4, Eq. (3.85) can be naturally interpreted in terms of a stochastic evolution consisting of periods of smooth evolution, described by $\mathcal{S}(\tau)$, interspersed with jumps, described by $\mathcal{X}$.

Most of the examples of open quantum systems that we have considered above led, under various approximations, to a Markov master equation of the Lindblad form. However, as the example of Brownian motion (Section 3.4.2) showed, this is not always the case. It turns out that the time-dependent Brownian-motion master equation (3.58) does preserve positivity. It is only when making the approximations leading to the time-independent, but non-Lindblad, equation (3.60) that one loses positivity. Care must be taken in using master equations such as this, which are not of the Lindblad form, because there are *necessarily* initial states yielding time-evolved states that are non-positive (i.e. are not quantum states at all). Thus autonomous non-Lindblad master equations *must* be regarded as approximations, but, on the other hand, the fact that one has derived a Lindblad-form master equation does not mean that one has an exact solution. The approximations leading to the high-temperature spin–boson master equation (3.56) may be no more valid than those leading to the high-temperature Brownian-motion master equation (3.60), for example. Whether or not a given open system is well approximated by Markovian dynamics can be determined only by a detailed study of the physics.

3.7 Decoherence and the pointer basis

3.7.1 Einselection

We are now in a position to state, and address, one of the key problems of quantum measurement theory: what defines the measured observable? Recall the binary system and binary apparatus introduced in Section 1.2.4. For an arbitrary initial system (S) state, and appropriate initial apparatus (A) state, the final combined state after the measurement interaction is

$$|\Psi'\rangle = \sum_{x=0}^{1} s_x |x\rangle |y := x\rangle, \tag{3.88}$$

where $|x\rangle$ and $|y\rangle$ denote the system and apparatus in the *measurement* basis. A measurement of the apparatus in this basis will yield $Y = x$ with probability $|s_x|^2$, that is, with exactly the probability that a direct projective measurement of a physical quantity of the form $\hat{C} = \sum_x c(x)|x\rangle_S \langle x|$ on the system would have given. On the other hand, as discussed in Section 1.2.6, one could make a measurement of the apparatus in some other basis. For example, measurement in a complementary basis $|p\rangle_A$ yields no information about the system preparation at all.

In general one could read out the apparatus in the arbitrary orthonormal basis

$$|\phi_0\rangle = \alpha^*|0\rangle + \beta^*|1\rangle, \tag{3.89}$$

$$|\phi_1\rangle = \beta|0\rangle - \alpha|1\rangle, \tag{3.90}$$

where $|\alpha|^2 + |\beta|^2 = 1$. The state after the interaction between the system and the apparatus can now equally well be written as

$$|\Psi'\rangle = d_0|\psi_0\rangle_S \otimes |\phi_0\rangle_A + d_1|\psi_1\rangle_S \otimes |\phi_1\rangle_A, \tag{3.91}$$

where $d_0|\psi_0\rangle_S = \alpha s_0|0\rangle + \beta s_1|1\rangle$ and $d_1|\psi_1\rangle_S = \beta^* s_0|0\rangle - \alpha^* s_1|1\rangle$. Note that $|\psi_0\rangle$ and $|\psi_1\rangle$ are *not orthogonal* if $|\phi_0\rangle$ and $|\phi_1\rangle$ are different from $|0\rangle$ and $|1\rangle$.

Exercise 3.22 *Show that this is true except for the special case in which* $|s_0| = |s_1|$.

It is apparent from the above that there is only one basis (the measurement basis) in which one should measure the apparatus in order to make an effective measurement of the system observable $\hat{C}$. Nevertheless, measuring in other bases is equally permitted by the formalism, and yields different sorts of information. This does not seem to accord with our intuition that a particular measurement apparatus is constructed, often at great effort, to measure a particular system quantity. The flaw in the argument, however, is that it is often not possible on physical grounds to read out the apparatus in an arbitrary basis. Instead, there is a preferred apparatus basis, which is determined by the nature of the apparatus and its environment. This has been called the *pointer basis* [Zur81]. For a well-constructed apparatus, the pointer basis will correspond to the measurement basis as defined above.

The pointer basis of an apparatus is determined by how it is built, without reference to any intended measured system to which it may be coupled. One expects the measurement basis of the apparatus, $|0\rangle$, $|1\rangle$, to correspond to two macroscopic *classically distinguishable* states of a particular degree of freedom of the apparatus. This degree of freedom could, for example, be the position of a pointer, whence the name 'pointer basis'. An apparatus for which the pointer could be in a superposition of two distinct macroscopic states does not correspond to our intuitive idea of a pointer. Thus we expect that the apparatus can never enter a superposition of two distinct pointer states as Eq. (3.89) would require.

This is a kind of selection rule, called *einselection* (environmentally induced selection) by Zurek [Zur82]. In essence it is justified by an apparatus–environment interaction that very rapidly couples the pointer states to orthogonal environment (E) states:

$$|y\rangle|z := 0\rangle \rightarrow |y\rangle|z := y\rangle_E, \tag{3.92}$$

where here $|z\rangle$ denotes an environment state. This is identical in form to the original system–apparatus interaction. However, the crucial point is that now the total state is

$$|\Psi''\rangle = \sum_{x=0}^{1} s_x|x\rangle|y := x\rangle|z := x\rangle. \tag{3.93}$$

If we consider using a different basis $\{|\phi_0\rangle, |\phi_1\rangle\}$ for the apparatus, we find that it is not possible to write the total state in the form of Eq. (3.93). That is,

$$|\Psi''\rangle \neq \sum_{x=0}^{1} d_x|\psi_x\rangle_S|\phi_x\rangle_A|\theta_x\rangle_E, \tag{3.94}$$

for any coefficients d_x and states for the system and environment.

Exercise 3.23 *Show that this is true except for the special case in which* $|s_0| = |s_1|$.

Note that einselection does not solve the quantum measurement problem in that it does not explain how just one of the elements of the superposition in Eq. (3.93) appears to become real, with probability $|s_x|^2$, while the others disappear. The solutions to that problem are outside the scope of this book. What the approach of Zurek and co-workers achieves is to explain why, for macroscopic objects like pointers, some states are preferred over others in that they are (relatively) unaffected by decoherence. Moreover, they have argued plausibly that these states have classical-like properties, such as being localized in phase space. These states are not necessarily orthogonal states, as in the example above, but they are practically orthogonal if they correspond to distinct measurement outcomes [ZHP93].

3.7.2 A more realistic model

The above example is idealized in that we considered only two possible environment states. In reality the pointer may be described by continuous variables such as position. In this case, it is easy to see how physical interactions lead to an approximate process of einselection in the position basis. Most interactions depend upon the position of an object, and the position of a macroscopic object such as a pointer will almost instantaneously become correlated with many degrees of freedom in the environment, such as thermal photons, dust particles and so on. This process of decoherence rapidly destroys any coherence between states of macroscopically different position, but these states of relatively well-defined position are themselves little affected by the decoherence process (as expressed ideally in Eq. (3.92)).

Decoherence in this pointer basis can be reasonably modelled using the Brownian-motion master equation introduced in Section 3.4.2. In this situation, the dominant term in the master equation is the last one (momentum diffusion), so we describe the evolution of the apparatus state by

$$\dot{\rho} = -\gamma \lambda_T^{-2}[\hat{X}, [\hat{X}, \rho]]. \tag{3.95}$$

Here we have used γ for γ_∞, and λ_T is the thermal de Broglie wavelength, $(2Mk_BT)^{-1/2}$. It is called this because the thermal equilibrium state matrix for a free particle, in the position basis

$$\rho(x, x') = \langle x|\rho|x'\rangle, \tag{3.96}$$

has the form $\rho(x, x') \propto \exp[-(x - x')^2/(4\lambda_T^2)]$. That is, the characteristic coherence length of the quantum 'waves' representing the particle (first introduced by de Broglie) is λ_T. In this position basis the above master equation is easy to solve:

$$\rho(x, x'; t) = \exp[-\gamma t(x - x')^2/\lambda_T^2]\rho(x, x'; 0). \tag{3.97}$$

Exercise 3.24 *Show this. Note that this does not give the thermal equilibrium distribution in the long-time limit because the dissipation and free-evolution terms have been omitted.*

Let the initial state for the pointer be a superposition of two states, macroscopically different in position, corresponding to two different pointer readings. Let $2s$ be the separation of the states, and σ their width. For $s \gg \sigma$, the initial state matrix can be well approximated by

$$\rho(x, x'; 0) = (1/2)[\psi_-(x) + \psi_+(x)][\psi_-^*(x') + \psi_+^*(x')], \tag{3.98}$$

where

$$\psi_\pm(x) = (2\pi\sigma^2)^{-1/4} \exp[-(x \mp s)^2/(2\sigma^2)]. \tag{3.99}$$

That is, $\rho(x, x'; 0)$ is a sum of four equally weighted bivariate Gaussians, centred in (x, x')-space at $(-s, -s)$, $(-s, s)$, $(s, -s)$ and (s, s). But the effect of the decoherence (3.95) on these four peaks is markedly different. The off-diagonal ones will decay rapidly, on a time-scale

$$\tau_{\text{dec}} = \gamma^{-1} \left(\frac{\lambda_T}{2s}\right)^2. \tag{3.100}$$

For $s \gg \lambda_T$, as will be the case in practice, this decoherence time is much smaller than the dissipation time,

$$\tau_{\text{diss}} = \gamma^{-1}. \tag{3.101}$$

The latter will also correspond to the time-scale on which the on-diagonal peaks in $\rho(x, x')$ change shape under Eq. (3.97), provided that $\sigma \sim \lambda_T$. This seems a reasonable assumption, since one would wish to prepare a well-localized apparatus (small σ), but if $\sigma \ll \lambda_T$ then it would have a kinetic energy much greater than the thermal energy $k_B T$ and so would dissipate energy at rate γ anyway.

The above analysis shows that, under reasonable approximations, the coherences (the off-diagonal terms) in the state matrix decay much more rapidly than the on-diagonal terms change. Thus the superposition is transformed on a time-scale t, such that $\tau_{\text{dec}} \ll t \ll \tau_{\text{diss}}$, into a mixture of pointer states:

$$\rho(x, x'; t) \approx (1/2)[\psi_-(x)\psi_-^*(x') + \psi_+(x)\psi_+^*(x')]. \tag{3.102}$$

Moreover, for macroscopic systems this time-scale is very short. For example, if $s = 1$ mm, $T = 300$ K, $M = 1$ g and $\gamma = 0.01$ s^{-1}, one finds (upon restoring $\hbar$ where necessary) $\tau_{\text{dec}} \sim 10^{-37}$ s, an extraordinarily short time. On such short time-scales, it could well be argued that the Brownian-motion master equation is not valid, and that a different treatment should be used (see for example Ref. [SHB03]). Nevertheless, this result can be taken as indicative of the fact that there is an enormous separation of time-scales between that on which the pointer is reduced to a mixture of classical states and the time-scale on which those classical states evolve.

3.8 Preferred ensembles

In the preceding section we argued that the interaction of a macroscopic apparatus with its environment preserves classical states and destroys superpositions of them. From the simple model of apparatus–environment entanglement in Eq. (3.92), and from the solution to the (cut-down) Brownian-motion master equation (3.97), it is seen that the state matrix becomes diagonal in this pointer basis. Moreover, from Eq. (3.92), the environment carries the information about which pointer state the system is in. Any additional evolution of the apparatus (such as that necessary for it to measure the system of interest) could cause transitions between pointer states, but again this information would also be carried in the environment so that at all times an observer could know where the apparatus is pointing, so to speak.

It would be tempting to conclude from the above examples that all one need do to find out the pointer basis for a given apparatus is to find the basis which diagonalizes its state once it has reached equilibrium with its environment. However, this is *not* the case, for two reasons. The first reason is that the states forming the diagonal basis are not necessarily states that are relatively unaffected by the decoherence process. Rather, as mentioned above, the latter states will in general be non-orthogonal. In that case the preferred representation of the equilibrium state matrix

$$\rho_{ss} = \sum_k \wp_k \hat{\pi}_k \tag{3.103}$$

will be in terms of an ensemble $E = \{\wp_k, \hat{\pi}_k\}$ of pure states, with positive weights $\wp_k$, represented by non-orthogonal projectors: $\hat{\pi}_j \hat{\pi}_k \neq \delta_{jk}\hat{\pi}_k$. The second reason, which is generally ignored in the literature on decoherence and the pointer basis, is that the mere fact that the state of a system becomes diagonal in some basis, through entanglement with its environment, does not mean that by observing the environment one can find the system to be always in one of those diagonal states. Once again, it may be that one has to consider non-orthogonal ensembles, as in Eq. (3.103), in order to find a set of states that allows a classical description of the system. By this we mean that the system can be always known to be in one of those states, but to make transitions between them.

The second point above is arguably the more fundamental one for the idea that decoherence explains the emergence of classical behaviour. That is, the basic idea of einselection is that there is a preferred ensemble for ρ_{ss} for which an ignorance interpretation holds. With this interpretation of Eq. (3.103) one would claim that the system 'really' is in one of the pure states $\hat{\pi}_k$, but that one happens to be ignorant of which $\hat{\pi}_k$ (i.e. which k) pertains. The weight $\wp_k$ would be interpreted as the probability that the system has state $\hat{\pi}_k$. For this to hold, it is necessary that in principle an experimenter could know *which* state $\hat{\pi}_k$ the system is in at all times by performing continual measurements on the environment with which the system interacts. The pertinent index k would change stochastically such that the proportion of time for which the system has state $\hat{\pi}_k$ is $\wp_k$. This idea was first identified in Ref. [WV98]. The first point in the preceding paragraph then says that the states in the preferred ensemble should also be robust in the face of decoherence. For example, if the decoherence is described by a Lindbladian $\mathcal{L}$ then one could use the criterion adopted in Ref. [WV98]. This is that the average fidelity

$$F(t) = \sum_k \wp_k \, \mathrm{Tr}[\hat{\pi}_k \exp(\mathcal{L}t)\, \hat{\pi}_k] \tag{3.104}$$

should have a characteristic decay time that is as long as possible (and, for macroscopic systems, one hopes that this is much longer than that of a randomly chosen ensemble).

In the remainder of this section we are concerned with elucidating when an ignorance interpretation of an ensemble representing ρ_{ss} is possible. As well as being important in understanding the role of decoherence, it is also relevant to quantum control, as will be discussed in Chapter 6. Let us restrict the discussion to Lindbladians having a unique stationary state defined by

$$\mathcal{L}\rho_{ss} = 0. \qquad (3.105)$$

Also, let us consider only *stationary ensembles* for ρ_{ss}. Clearly, once the system has reached steady state such a stationary ensemble will represent the system *for all times t*. Then, as claimed above, it can be proven that for some ensembles (and, in particular, often for the orthogonal ensemble) there is no way for an experimenter continually to measure the environment so as to find out which state the system is in. We say that such ensembles are not *physically realizable* (PR). However, there are other stationary ensembles that *are* PR.

3.8.1 Quantum steering

To appreciate physical realizability of ensembles, it is first necessary to understand a phenomenon discovered by Schrödinger [Sch35a] and described by him as 'steering'[3] (we will call it quantum steering). This phenomenon was rediscovered (and generalized) by Hughston, Jozsa and Wootters [HJW93]. Consider a system with state matrix ρ that is mixed solely due to its entanglement with a second system, the environment. That is, there is a pure state $|\Psi\rangle$ in a larger Hilbert space of system plus environment such that

$$\rho = \mathrm{Tr}_{\mathrm{env}}[|\Psi\rangle\langle\Psi|]. \qquad (3.106)$$

This purification always exists, as discussed in Section A.2.2. Then, for any ensemble $\{(\hat{\pi}_k, \wp_k)\}_k$ that represents ρ, it is possible to measure the environment such that the system state is collapsed into one of the pure states $\hat{\pi}_k$ with probability $\wp_k$. This is sometimes known as the Schrödinger–HJW theorem.

Quantum steering gives rigorous meaning to the ignorance interpretation of any particular ensemble. It says that there will be a way to perform a measurement on the environment, without disturbing the system state on average, to obtain exactly the information as to which state the system is 'really' in. Of course, the fact that one can do this for any ensemble means that no ensemble can be fundamentally preferred over any other one, as a representation of ρ *at some particular time t*. To say that an ensemble is PR, however, requires justifying the ignorance interpretation at *all times* (after the system has reached steady state). We now establish the conditions for an ensemble to be PR.

---

[3] Schrödinger introduced this as an evocative term for the Einstein–Podolsky–Rosen effect [EPR35] involving entangled states. For a completely general formulation of steering in quantum information terms, see Refs. [WJD07, JWD07].

3.8.2 Conditions for physical realizability

According to quantum steering, it is always possible to realize a given ensemble at some particular time t by measuring the environment. This may involve measuring parts of the environment that interacted with the system an arbitrarily long time ago, but there is nothing physically impossible in doing this. Now consider the future evolution of a particular system state $\hat{\pi}_k$ following this measurement. At time $t + \tau$, it will have evolved to $\rho_k(t + \tau) = \exp(\mathcal{L}\tau)\hat{\pi}_k$. This is a mixed state because the system has now become re-entangled with its environment.

The system state can be repurified by making another measurement on its environment. However, if the same ensemble is to remain as our representation of the system state then the pure system states obtained as a result of this measurement at time $t + \tau$ must be contained in the set $\{\hat{\pi}_j : j\}$. Because of quantum steering, this will be possible if and only if $\rho_k(t + \tau)$ can be represented as a mixture of these states. That is, for all k there must exist a probability distribution $\{w_{jk}(\tau) : j\}$ such that

$$\exp(\mathcal{L}\tau)\hat{\pi}_k = \sum_j w_{jk}(\tau)\hat{\pi}_j. \tag{3.107}$$

If $w_{jk}(\tau)$ exists then it is the probability that the measurement at time $t + \tau$ yields the state $\hat{\pi}_j$.

Equation (3.107) is a necessary but not sufficient criterion for the ensemble $\{(\hat{\pi}_j, \wp_j) : j\}$ to be PR. We also require that the weights be stationary. That is, for all j and all τ,

$$\wp_j = \sum_k \wp_k w_{jk}(\tau). \tag{3.108}$$

Multiplying both sides of Eq. (3.107) by $\wp_k$, and summing over k, then using Eq. (3.108) and Eq. (3.103) gives $e^{\mathcal{L}\tau}\rho_{ss} = \rho_{ss}$, as required from the definition of ρ_{ss}.

One can analyse these conditions further to obtain simple criteria that can be applied in many cases of interest [WV01]. In particular, we will return to them in Chapter 6. For the moment, it is sufficient to prove that there are some ensembles that are PR and some that are not. This is what was called in Ref. [WV01] the *preferred-ensemble fact* (the 'preferred' ensembles are those that are physically realizable). Moreover, for some systems the orthogonal ensemble is PR and for others it is not. The models we consider are chosen for their simplicity (they are two-level systems), and are not realistic models for the decoherence of a macroscopic apparatus.

3.8.3 Examples

First we consider an example in which the orthogonal ensemble is PR: the high temperature spin–boson model. In suitably scaled time units, the Lindbladian in Eq. (3.56) is $\mathcal{L} = \mathcal{D}[\hat{\sigma}_z]$. In this example, there is no unique stationary state, but all stationary states are of the form

$$\rho_{ss} = \wp_-|\sigma_z := -1\rangle\langle\sigma_z := -1| + \wp_+|\sigma_z := 1\rangle\langle\sigma_z := 1|. \tag{3.109}$$

Exercise 3.25 *Show this.*

The orthogonal ensemble thus consists of the $\hat{\sigma}_z$ eigenstates with weights $\wp_\pm$. To determine whether this ensemble is PR, we must consider the evolution of its members under the Lindbladian. It is trivial to show that $\mathcal{L}|\sigma_z := \pm 1\rangle\langle\sigma_z := \pm 1| = 0$ so that these states do not evolve at all. In other words, they are perfectly robust, with $w_{jk}(\tau) = \delta_{jk}$ in Eq. (3.107). It is easy to see that Eq. (3.108) is also satisfied, so that the orthogonal ensemble is PR.

Next we consider an example in which the orthogonal ensemble is not PR: the driven, damped two-level atom. In the zero-detuning and zero-temperature limit, the Lindbladian is defined by

$$\mathcal{L}\rho = -\mathrm{i}\frac{\Omega}{2}[\hat{\sigma}_x, \rho] + \gamma \mathcal{D}[\hat{\sigma}_-]\rho. \tag{3.110}$$

The general solution of the corresponding Bloch equations is

$$x(t) = u\mathrm{e}^{-(\gamma/2)t}, \tag{3.111}$$

$$y(t) = c_+\mathrm{e}^{\lambda_+ t} + c_-\mathrm{e}^{\lambda_- t} + y_{ss}, \tag{3.112}$$

$$z(t) = c_+\frac{\gamma - 4\mathrm{i}\tilde{\Omega}}{4\Omega}\mathrm{e}^{\lambda_+ t} + c_-\frac{\gamma + 4\mathrm{i}\tilde{\Omega}}{4\Omega}\mathrm{e}^{\lambda_- t} + z_{ss}, \tag{3.113}$$

with eigenvalues defined by

$$\lambda_\pm = -\frac{3}{4}\gamma \pm \mathrm{i}\tilde{\Omega} \tag{3.114}$$

and $c_\pm$ are constants given by

$$c_\pm = \frac{1}{8\mathrm{i}\tilde{\Omega}}\left[\mp 4\Omega(w - z_{ss}) \pm (\gamma \pm 4\mathrm{i}\tilde{\Omega})(v - y_{ss})\right], \tag{3.115}$$

where u, v and w are used to represent the initial conditions of x, y and z. A modified Rabi frequency has been introduced,

$$\tilde{\Omega} = \sqrt{\Omega^2 - (\gamma/4)^2}, \tag{3.116}$$

which is real for $\Omega > \gamma/4$ and imaginary for $\Omega < \gamma/4$. The steady-state solutions are $x_{ss} = 0$, $y_{ss} = 2\Omega\gamma/(\gamma^2 + 2\Omega^2)$ and $z_{ss} = -\gamma^2/(\gamma^2 + 2\Omega^2)$, as shown in Exercise 3.10.

Exercise 3.26 *Derive the above solution, using standard techniques for linear differential equations.*

In the Bloch representation, the diagonal states of ρ_{ss} are found by extending the stationary Bloch vector forwards and backwards to where it intersects the surface of the Bloch sphere. That is, the two pure diagonal states are

$$\begin{pmatrix} u \\ v \\ w \end{pmatrix}_\pm = \begin{pmatrix} 0 \\ \pm 2\Omega \\ \mp \gamma \end{pmatrix}(4\Omega^2 + \gamma^2)^{-1/2}. \tag{3.117}$$

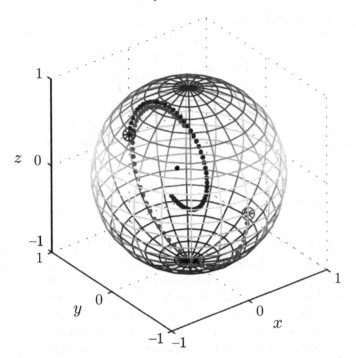

Fig. 3.2 Dynamics of two states on the Bloch sphere according to the master equation (3.110), for $\Omega = 10\gamma$, with points every $0.1\gamma^{-1}$. The initial states are those that diagonalize the stationary Bloch sphere, which are close to $y = \pm 1$. The stationary state is the dot close to the centre of the Bloch sphere.

Using these as initial conditions, it is easy to prove that, for general t,

$$\begin{pmatrix} x(t) \\ y(t) \\ z(t) \end{pmatrix}_{\pm} \neq w_{\pm+}(t) \begin{pmatrix} u \\ v \\ w \end{pmatrix}_{+} + w_{\pm-}(t) \begin{pmatrix} u \\ v \\ w \end{pmatrix}_{-} \tag{3.118}$$

for any weights $w_{\pm+}(t)$, $w_{\pm-}(t)$. That is, the diagonal states evolve into states that are not mixtures of the original diagonal states, so it is not possible for an observer to know at all times that the system is in a diagonal state. The orthogonal ensemble is not PR. This is illustrated in Fig. 3.2.

There are, however, *non-orthogonal* ensembles that *are* PR for this system. More-over, there is a PR ensemble with just two members, like the orthogonal ensemble. This is the ensemble $\{(\hat{\pi}_+, 1/2), (\hat{\pi}_-, 1/2)\}$, where this time the two states have the Bloch vectors

$$\begin{pmatrix} u \\ v \\ w \end{pmatrix}_{\pm} = \begin{pmatrix} \pm\sqrt{1 - y_{ss}^2 - z_{ss}^2} \\ y_{ss} \\ z_{ss} \end{pmatrix}. \tag{3.119}$$

Using these as initial conditions, we find

$$\begin{pmatrix} x(t) \\ y(t) \\ z(t) \end{pmatrix}_{\pm} = \begin{pmatrix} u_{\pm}e^{-\gamma t/2} \\ y_{ss} \\ z_{ss} \end{pmatrix}. \tag{3.120}$$

Obviously this *can* be written as a positively weighted sum of the two initial Bloch vectors, and averaging over the two initial states will give a sum that remains equal to the stationary Bloch vector. That is, the two conditions (3.107) and (3.108) are satisfied, and this ensemble is PR.

Exercise 3.27 *Prove the above by explicitly constucting the necessary weights $w_{\pm+}(t)$ and $w_{\pm-}(t)$.*

These results are most easily appreciated in the $\Omega \gg \gamma$ limit. Then the stationary solution is an almost maximally mixed state, displaced slightly from the centre of the Bloch sphere along the y axis. The diagonal states then are close to $\hat{\sigma}_y$ eigenstates, while the states in the PR ensemble are close to $\hat{\sigma}_x$ eigenstates. In this limit the master-equation evolution (3.110) is dominated by the Hamiltonian term, which causes the Bloch vector to rotate around the $\hat{\sigma}_x$ axis. Thus, the y eigenstates are rapidly rotated away from their original positions, so this ensemble is neither robust nor PR, but the x eigenstates are not rotated at all, and simply decay at rate $\gamma/2$ towards the steady state, along the line joining them. Thus this ensemble is PR. Moreover, it can be shown [WB00] that this is the most robust ensemble according to the fidelity measure Eq. (3.104), with a characteristic decay time (half-life) of $2\ln 2/\gamma$. These features are shown in Fig. 3.3.

The existence of a PR ensemble in this second case (where the simple picture of a diagonal pointer basis fails) is not happenstance. For any master equation there are in fact infinitely many PR ensembles. Some of these will be robust, and thus could be considered pointer bases, and some will not. A full understanding of how PR ensembles arise will be reached in Chapter 4, where we consider the conditional dynamics of a continuously observed open system.

3.9 Decoherence in a quantum optical system

3.9.1 Theoretical analysis

In recent years the effects of decoherence have been investigated experimentally, most notably in a quantum optical (microwave) cavity [BHD+96]. To appreciate this experiment, it is necessary to understand the effect of damping of the electromagnetic field in a cavity at zero temperature on a variety of initial states. This can be described by the interaction-frame master equation, Eq. (3.38):

$$\dot{\rho} = \gamma \mathcal{D}[\hat{a}]\rho. \tag{3.121}$$

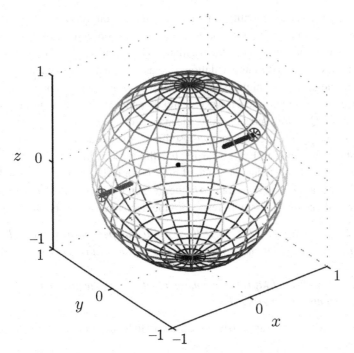

Fig. 3.3 Dynamics of two states on the Bloch sphere according to the master equation (3.110), for $\Omega = 10\gamma$, with points every $0.1\gamma^{-1}$. The initial states are the two non-orthogonal states $\hat{\pi}_\pm$ defined in Eq. (3.119), which are close to $x = \pm1$. The stationary state is the dot close to the centre of the Bloch sphere.

If we use the solution given in Eq. (3.85) we find that the general solution can be written as a Kraus sum,

$$\rho(t) = \sum_{m=0}^{\infty} \hat{M}_m(t)\rho(0)\hat{M}_m^\dagger(t), \tag{3.122}$$

where

$$\hat{M}_m(t) = \frac{(1 - e^{-\gamma t})^{m/2}}{\sqrt{m!}} e^{-\gamma t \hat{a}^\dagger \hat{a}/2} \hat{a}^m. \tag{3.123}$$

We can regard this as an expansion of the state matrix in terms of the number of photons *lost* from the cavity in time t.

Exercise 3.28 *Prove Eq. (3.122) by simplifying Eq. (3.85) using the property that $[\hat{a}^\dagger \hat{a}, \hat{a}^\dagger] = \hat{a}^\dagger$.*

Equation (3.122) can be solved most easily in the number-state basis. However, it is rather difficult to prepare a simple harmonic oscillator in a number eigenstate. We encounter simple harmonic oscillators regularly in classical physics; springs, pendula, cantilevers etc. What type of state describes the kinds of motional states in which such oscillators are typically found? We usually identify simple harmonic motion by observing oscillations. In the presence of friction oscillatory motion will decay. To observe a sustained oscillation, we need to provide a driving force to the oscillator. This combination of driving and friction reaches a steady state of coherent oscillation. In classical physics the resulting motion will have a definite energy, so one might expect that this would be a means of preparing a quantum oscillator in an energy eigenstate. However, a number state is actually time-independent, so this cannot be so. Quantum mechanically, the state produced by this mechanism does not have a definite energy. For weakly damped oscillators (as in quantum optics), it is actually a coherent state, as shown in Exercise 3.13. For the reasons just described, the coherent state is regarded as the most classical state of motion for a simple harmonic oscillator. It is often referred to as a *semiclassical* state. Another reason why coherent states are considered classical-like is their robustness with respect to the decoherence caused by damping.

Exercise 3.29 *From Eq. (3.122) show that, for a damped harmonic oscillator in the interaction frame, a coherent state simply decays exponentially. That is, if $|\psi(0)\rangle = |\alpha\rangle$ then $|\psi(t)\rangle = |\alpha e^{-\gamma t/2}\rangle$.*
Hint: *Consider the effect of $\hat{M}_m(t)$ on a coherent state, using the fact that $\hat{a}|\alpha\rangle = \alpha|\alpha\rangle$ and also using the number-state expansion for $|\alpha\rangle$.*

Suppose we somehow managed to prepare a cavity field in a superposition of two coherent states,

$$|\psi(0)\rangle = N(|\alpha\rangle + |\beta\rangle), \tag{3.124}$$

where the normalization constant is $N^{-1} = \sqrt{2 + 2\,\mathrm{Re}\langle\alpha|\beta\rangle}$. If $|\alpha - \beta| \gg 1$ then this corresponds to a superposition of macroscopically different fields. Such a superposition is often called a *Schrödinger-cat* state, after the thought experiment invented by Schrödinger which involves a superposition of a live cat and a dead cat [Sch35b].

We now show that such a superposition is very fragile with respect to even a very small amount of damping. Using the solution (3.122), the state will evolve to

$$\rho(t) \propto |\alpha(t)\rangle\langle\alpha(t)| + |\beta(t)\rangle\langle\beta(t)|$$

$$+ C(\alpha, \beta, t)|\alpha(t)\rangle\langle\beta(t)| + C^*(\alpha, \beta, t)|\beta(t)\rangle\langle\alpha(t)|, \tag{3.125}$$

where $\alpha(t) = \alpha e^{-\gamma t/2}$, $\beta(t) = \beta e^{-\gamma t/2}$ and

$$C(\alpha, \beta, t) = \exp\left\{-\frac{1}{2}\left[|\alpha(t)|^2 + |\beta(t)|^2 - 2\alpha(t)\beta^*(t)\right](1 - e^{\gamma t})\right\}. \tag{3.126}$$

Exercise 3.30 *Show this, by the same method as in Exercise 3.29.*

The state (3.125) is a superposition of two damped coherent states with amplitudes $\alpha(t)$ and $\beta(t)$ with a suppression of coherence between the states through the factor $C(\alpha, \beta, t)$. Suppose we now consider times much shorter than the inverse of the amplitude decay rate, $\gamma t \ll 1$. The coherence-suppression factor is then given by

$$C(\alpha, \beta, t) \approx \exp\left(-|\alpha - \beta|^2 \gamma t/2\right). \tag{3.127}$$

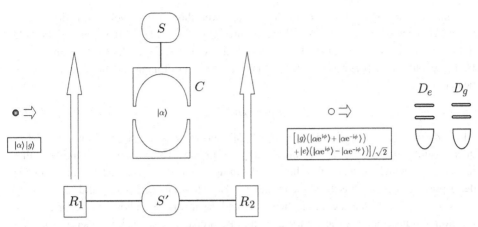

Fig. 3.4 A schematic diagram of the experiment performed by the Haroche group to investigate the decoherence of oscillator coherent states. The atom is prepared in an appropriate Rydberg state. The cavities R_1 and R_2 each apply a $\pi/2$ pulse. The interaction with the cavity field in C produces superpositions of coherent states. The final ionization detectors determine the atomic state of the atom. Figure 2 adapted with permission from M. Brune *et al.*, *Phys. Rev. Lett.* **77**, 4887, (1996). Copyrighted by the American Physical Society.

We thus see that at the very beginning the coherence does not simply decay at the same rate as the amplitudes, but rather at a decay rate that depends quadratically on the difference between the amplitudes of the initial superposed states. This is qualitatively the same as was seen for Brownian motion in Section 3.7.2. For macroscopically different states ($|\alpha - \beta| \gg 1$) the decoherence is very rapid. Once the coherence between the two states has become very small we can regard the state as a statistical mixture of the two coherent states with exponentially decaying coherent amplitudes. The quantum character of the initial superposition is rapidly lost and for all practical purposes we may as well regard the initial state as a classical statistical mixture of the two 'pointer states'. For this reason it is very hard to prepare an oscillator in a Schrödinger-cat state. However, the decoherence we have described has been observed experimentally for $|\alpha - \beta| \sim 1$.

3.9.2 Experimental observation

The experimental demonstration of the fast decay of coherence for two superposed coherent states was first performed by the Haroche group in Paris using the cavity QED system of Rydberg atoms in microwave cavities [BHD$^+$96]. The experiment is based on Ramsey fringe interferometry (see Box 1.4). A schematic diagram of the experiment is shown in Fig. 3.4.

A two-level atomic system with ground state $|g\rangle$ and excited state $|e\rangle$ interacts with a cavity field in C. This cavity field is well detuned from the atomic resonance. The ground and excited states of the atom correspond to Rydberg levels with principal quantum numbers 50 and 51. Such highly excited states have very large dipole moments and can thus interact very strongly with the cavity field even though it is well detuned from the cavity resonance. The effect of the detuned interaction is to change the phase of the field in the cavity. However, the sign of the phase shift is opposite for each of the atomic states. Using second-order perturbation theory, an effective Hamiltonian for this interaction can be derived:

$$\hat{H}_C = \chi \hat{a}^\dagger \hat{a} \hat{\sigma}_z, \tag{3.128}$$

where $\hat{\sigma}_z = |e\rangle\langle e| - |g\rangle\langle g|$, and $\chi = |\Omega|^2/(2\delta)$, where Ω is the single-photon Rabi frequency and $\delta = \omega_a - \omega_c$ is the atom–cavity detuning. Thus decreasing the detuning increases χ (which is desirable), but the detuning cannot be decreased too much or the description in terms of this effective interaction Hamiltonian becomes invalid.

Assume to begin that the cavity fields R_1 and R_2 in Fig. 3.4 above are resonant with the atomic transition. Say the cavity C is initially prepared in a weakly coherent state $|\alpha\rangle$ (in the experiment $|\alpha| = 3.1$) and the atom in the state $(|g\rangle + |e\rangle)/\sqrt{2}$, using a $\pi/2$ pulse in cavity R_1. Then in time τ the atom–cavity system will evolve under the Hamiltonian (3.128) to

$$|\psi(\tau)\rangle = \frac{1}{\sqrt{2}}\left(|g\rangle|\alpha e^{i\phi}\rangle + |e\rangle|\alpha e^{-i\phi}\rangle\right), \tag{3.129}$$

where $\phi = \chi\tau/2$.

Exercise 3.31 *Verify this.*

The state in Eq. (3.129) is an entangled state between a two-level system and an oscillator. Tracing over the atom yields a field state that is an equal mixture of two coherent states separated in phase by 2ϕ.

To obtain a state that correlates the atomic energy levels with coherent superpositions of coherent states, the atom is subjected to another $\pi/2$ pulse in cavity R_2. This creates the final state

$$|\psi\rangle_{\text{out}} = \frac{1}{\sqrt{2}}\left[|g\rangle(|\alpha e^{i\phi}\rangle + |\alpha e^{-i\phi}\rangle) + |e\rangle(|\alpha e^{i\phi}\rangle - |\alpha e^{-i\phi}\rangle)\right]. \tag{3.130}$$

If one now determines that the atom is in the state $|g\rangle$ at the final ionization detectors, the conditional state of the field is

$$|\psi^g\rangle_{\text{out}} = N_+(|\alpha e^{i\phi}\rangle + |\alpha e^{-i\phi}\rangle), \tag{3.131}$$

where N_+ is a normalization constant. Likewise, if the atom is detected in the excited state,

$$|\psi^e\rangle_{\text{out}} = N_-(|\alpha e^{i\phi}\rangle - |\alpha e^{-i\phi}\rangle). \tag{3.132}$$

These conditional states are superpositions of coherent states.

In the preceding discussion we ignored the cavity decay since this is small on the timescale of the interaction between a single atom and the cavity field. In order to see the effect of decoherence, one can use the previous method to prepare the field in a coherent superposition of coherent states and then let it evolve for a time T so that there is a significant probability that at least one photon is lost from the cavity. One then needs to probe the

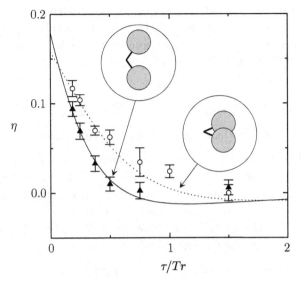

Fig. 3.5 A plot of the two-atom correlation η versus the delay time between successive atoms for two different values of the conditional phase shift. Figure 5(b) adapted with permission from M. Brune *et al.*, *Phys. Rev. Lett.* **77**, 4887, (1996). Copyrighted by the American Physical Society.

decohered field state. It is impossible to measure directly the state of a microwave cavity field at the quantum level because of the low energy of microwave photons compared with optical photons. Instead the Haroche team used a second atom as a probe for the field state. They then measured the state of the second atom, obtaining the conditional probabilities $p(e|g)$ and $p(e|e)$ (where the conditioning label refers to the result of the first atom measurement). Since the respective conditional field states after the first atom are different, these probabilities should be different. The extent of the difference is given by

$$\eta = p(e|e) - p(e|g). \tag{3.133}$$

From the result (3.125), after a time τ the two conditional states will have decohered to

$$\rho_{\text{out}}^{\overset{e}{g}}(\tau) \propto \left(|\alpha_\tau e^{i\phi}\rangle\langle\alpha_\tau e^{i\phi}| + |\alpha_\tau e^{-i\phi}\rangle\langle\alpha_\tau e^{-i\phi}|\right)$$
$$\pm \left(C(\tau)|\alpha_\tau e^{i\phi}\rangle\langle\alpha_\tau e^{-i\phi}| + C(\tau)^*|\alpha_\tau e^{-i\phi}\rangle\langle\alpha_\tau e^{i\phi}|\right), \tag{3.134}$$

where $\alpha_\tau < \alpha$ due to decay in the coherent amplitude and $|C(\tau)| < 1$ due to the decay in the coherences as before. In the limit $C(\tau) \to 0$, these two states are indistinguishable and so $\eta \to 0$. Thus, by repeating a sequence of double-atom experiments, the relevant conditional probabilities may be sampled and a value of η as a function of the delay time can be determined. (In the experiment, an extra averaging was performed to determine η, involving detuning the cavities R_1 and R_2 from atomic resonance ω_0 by a varying amount Δ.)

In Fig. 3.5 we reproduce the results of the experimental determination of η for two different values of the conditional phase shift, ϕ, as a function of the delay time τ in units of cavity relaxation lifetime $T_r = 1/\gamma$. As expected, the correlation signal decays to zero. Furthermore, it decays to zero more rapidly for larger conditional phase shifts. That is to say, it decays to zero more rapidly when the superposed states are further apart in phase-space. The agreement with the theoretical result is very good.

3.10 Other examples of decoherence

3.10.1 *Quantum electromechanical systems*

We now consider a simple model of a measured system in which the apparatus undergoes decoherence due to its environment. In the model the measured system is a two-level system (with basis states $|0\rangle$ and $|1\rangle$) while the apparatus is a simple harmonic oscillator, driven on resonance by a classical force. The coupling between the two-level system and the oscillator is assumed to change the frequency of the oscillator. The effective Hamiltonian for the system plus apparatus in the interaction frame is

$$\hat{H} = \epsilon(\hat{a}^\dagger + \hat{a}) + \chi\hat{a}^\dagger\hat{a}\hat{\sigma}_z, \tag{3.135}$$

where ϵ is the strength of the resonant driving force and χ is the strength of the coupling between the oscillator and the two-level system. The irreversible dynamics of the apparatus is modelled using the weak-damping, zero-temperature master equation of Eq. (3.121), giving the master equation

$$\dot{\rho} = -i\epsilon[\hat{a} + \hat{a}^\dagger, \rho] - i\chi[\hat{a}^\dagger\hat{a}\hat{\sigma}_z, \rho] + \gamma\mathcal{D}[\hat{a}]\rho. \tag{3.136}$$

There are numerous physical problems that could be described by this model. It could represent a two-level electric dipole system interacting with an electromagnetic cavity field that is far detuned, as can occur in cavity QED (see the preceding Section 3.9.2) and circuit QED (see the following Section 3.10.2). Another realization comes from the rapidly developing field of quantum electromechanical systems, as we now discuss.

Current progress in the fabrication of nano-electromechanical systems (NEMSs) will soon yield mechanical oscillators with resonance frequencies close to 1 GHz, and quality factors Q above 10^5 [SR05]. (The quality factor is defined as the ratio of the resonance frequency ω_0 to the damping rate γ.) At that scale, a NEMS oscillator becomes a quantum electromechanical system (QEMS). One way to define the quantum limit is for the thermal excitation energy to be less than the energy gap between adjacent oscillator energy eigenstates: $\hbar\omega_0 > k_B T$. This inequality would be satisfied by a factor of two or so with a device having resonance frequency $\omega_0 = 1 \times 2\pi$ GHz and temperature of $T_0 = 20$ mK.

In this realization, the two-level system or qubit could be a solid-state double-well structure with a single electron tunnelling between the wells (quantum dots). We will model this as an approximate two-state system. It is possible to couple the quantum-electromechanical oscillator to the charge state of the double dot via an external voltage gate. A possible device is shown in Fig. 3.6. The two wells are at different distances from the voltage gate and this distance is modulated as the oscillator moves. The electrostatic energy

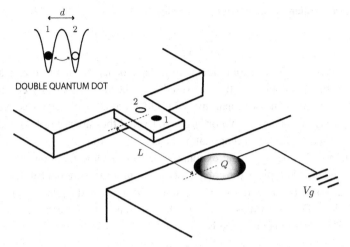

Fig. 3.6 A possible scheme for coupling a single-electron double-dot system to a nano-mechanical resonator. The double dot is idealized as a double-well potential for a single electron.

of the system depends on which well is occupied by the electron and on the square of the oscillator displacement. This leads to a shift in the frequency of the oscillator that depends on the location of the electron [CR98]. Currently such nano-mechanical electrometers are strongly dominated by thermal fluctuations and the irreversible dynamics are not well described by the decay term in Eq. (3.136). However, if quality factors and resonance frequencies continue to increase, these devices should enter a domain of operation where this description is acceptable.

At any time the state of the system plus apparatus may be written as

$$\rho(t) = \rho_{00} \otimes |0\rangle\langle 0| + \rho_{11} \otimes |1\rangle\langle 1| + \rho_{10} \otimes |1\rangle\langle 0| + \rho_{10}^{\dagger} \otimes |0\rangle\langle 1|, \qquad (3.137)$$

where ρ_{ij} is an operator that acts only in the oscillator Hilbert space. If we substitute this into Eq. (3.136), we find the following equations:

$$\dot{\rho}_{00} = -i\epsilon[\hat{a} + \hat{a}^{\dagger}, \rho_{00}] + i\chi[\hat{a}^{\dagger}\hat{a}, \rho_{00}] + \gamma \mathcal{D}[\hat{a}]\rho_{00}, \qquad (3.138)$$

$$\dot{\rho}_{11} = -i\epsilon[\hat{a} + \hat{a}^{\dagger}, \rho_{11}] - i\chi[\hat{a}^{\dagger}\hat{a}, \rho_{11}] + \gamma \mathcal{D}[\hat{a}]\rho_{11}, \qquad (3.139)$$

$$\dot{\rho}_{10} = -i\epsilon[\hat{a} + \hat{a}^{\dagger}, \rho_{10}] - i\chi\{\hat{a}^{\dagger}\hat{a}, \rho_{10}\} + \gamma \mathcal{D}[\hat{a}]\rho_{10}, \qquad (3.140)$$

where $\{\hat{A}, \hat{B}\} = \hat{A}\hat{B} + \hat{B}\hat{A}$ as usual. On solving these equations for the initial condition of an arbitary qubit state $c_0|0\rangle + c_1|1\rangle$ and the oscillator in the ground state, we find that the combined state of the system plus apparatus is

$$\rho(t) = |c_0|^2|\alpha_-(t)\rangle\langle\alpha_-(t)| \otimes |0\rangle\langle 0| + |c_1|^2|\alpha_+(t)\rangle\langle\alpha_+(t)| \otimes |1\rangle\langle 1|$$
$$+ \left[c_0 c_1^* C(t)|\alpha_+(t)\rangle\langle\alpha_-(t)| \otimes |1\rangle\langle 0| + \text{H.c.} \right], \qquad (3.141)$$

where $|\alpha_\pm(t)\rangle$ are coherent states with amplitudes

$$\alpha_\pm(t) = \frac{-i\epsilon}{\gamma/2 \pm i\chi}\left[1 - \mathrm{e}^{-(\gamma/2 \pm i\chi)t}\right]. \tag{3.142}$$

The coherence factor $C(t)$ has a complicated time dependence, but tends to zero as $t \to \infty$. Thus the two orthogonal states of the measured qubit become classically correlated with different coherent states of the apparatus. The latter are the pointer basis states of the apparatus, and may be approximately orthogonal. Even if they are not orthogonal, it can be seen that the qubit state becomes diagonal in the eigenbasis of $\hat{\sigma}_z$.

For short times $C(t)$ decays as an exponential of a quadratic function of time. Such a quadratic dependence is typical for coherence decay in a measurement model that relies upon an initial build up of correlations between the measured system and the pointer degree of freedom. For long times ($\gamma t \gg 1$), the coherence decays exponentially in time: $C(t) \sim \mathrm{e}^{-\Gamma t}$. The rate of decoherence is

$$\Gamma = \frac{2\epsilon^2\gamma\chi^2}{(\gamma^2/4 + \chi^2)^2}. \tag{3.143}$$

This qubit decoherence rate can be understood as follows. The long-time solution of Eq. (3.141) is

$$\rho_\infty = |c_0|^2|\alpha_-\rangle\langle\alpha_-| \otimes |0\rangle\langle 0| + |c_1|^2|\alpha_+\rangle\langle\alpha_+| \otimes |1\rangle\langle 1| \tag{3.144}$$

with $\alpha_\pm = -i\epsilon(\gamma/2 \pm i\chi)^{-1}$.

Exercise 3.32 *Verify by direct substitution that this is a steady-state solution of the master equation (3.136).*

The square separation $S = |\alpha_- - \alpha_+|^2$ between the two possible oscillator amplitudes in the steady state is given by

$$S = \frac{4\epsilon^2\chi^2}{(\gamma^2/4 + \chi^2)^2}. \tag{3.145}$$

Thus the long-time decoherence rate is $\Gamma = S\gamma/2$. This is essentially the rate at which information about which oscillator state is occupied (and hence which qubit state is occupied) is leaking into the oscillator's environment through the damping at rate γ. If $S \gg 1$, then the decoherence rate is much faster than the rate at which the oscillator is damped.

3.10.2 A superconducting box

The international effort to develop a quantum computer in a solid-state system is driving a great deal of fundamental research on the problem of decoherence. Recent experiments have begun to probe the mechanisms of decoherence in single solid-state quantum devices, particularly superconducting devices. In this section we will consider the physical mechanisms of decoherence in these devices and recent experiments. A superconducting box or *Cooper-pair box* (CPB) is essentially a small island of superconducting material separated

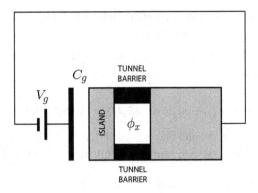

Fig. 3.7 A Cooper-pair box system. A superconducting metallic island is connected to a Cooper pair reservoir by a split tunnel junction, threaded by a magnetic flux ϕ_x. A DC bias gate with voltage V_g can make it energetically favourable for one or more Cooper-pairs to tunnel onto the island.

by a tunnel barrier from a reservoir of Cooper pairs. A Cooper pair (CP) is a pair of electrons bound together due to complex interactions with the lattice of the superconducting material [Coo56, BCS57]. Although electrons are fermions, a pair of electrons acts like a boson, and so can be described similarly to photons, using number states $|N\rangle$ for $N \in \mathbb{N}$.

A schematic representation of a CPB is shown in Fig. 3.7. The box consists of a small superconducting metallic island with oxide barrier tunnel junctions insulating it from the Cooper-pair reservoir. As the voltage V_g on the bias gate is changed, one or more Cooper pairs may tunnel onto the island. The tunnelling rate is determined by the Josephson energy E_J of the junction. This can be changed by adjusting the magnetic flux ϕ_x threading the loop: a so-called split-junction CPB.

In the experiment of Schuster *et al.* [SWB$^+$05] the CPB was placed inside a superconducting co-planar microwave LC-resonator. The resonator supports a quantized mode of the electromagnetic field, while the CPB acts like an atomic system. Thus the term 'circuit QED' (as opposed to the 'cavity QED' of Section 3.9.2) is used for these systems. The coupling between the CPB and the microwave field is given by

$$\hat{H} = \hbar \omega_r \hat{a}^\dagger \hat{a} - \frac{E_J}{2} \sum_N \left(|N\rangle\langle N+1| + |N+1\rangle\langle N| \right)$$

$$+ 4E_C \sum_N (N - \hat{n}_g)^2 |N\rangle\langle N|. \tag{3.146}$$

In the first term, $\omega_r = \sqrt{LC}$ is the frequency and $\hat{a}$ the annihilation operator for the microwave resonator field (note that in this section we are not setting $\hbar = 1$). The second term is the Josephson tunnelling term, with Josephson frequency $E_J/\hbar$. The third term is the coupling between the field and the CPB, in which $E_C = e^2/(2C_\Sigma)$ and $\hat{n}_g = C_g \hat{V}_g/(2e)$. Here C_Σ is the capacitance between the island and the rest of the circuit, C_g is the capacitance between the CPB island and the bias gate for the island, and $\hat{V}_g$ is the operator for the total voltage applied to the island by the bias gate. This voltage can be split as $\hat{V}_g = V_g^{(0)} + \hat{v}$, where $V_g^{(0)}$ is a DC field and $\hat{v}$ is the microwave field in the cavity, which is quantized. It is related to the cavity annihilation operator by

$$\hat{v} = (\hat{a} + \hat{a}^\dagger)\sqrt{\hbar\omega_r/(2C)}. \tag{3.147}$$

We can thus write

$$\hat{n}_g = n_g^{(0)} + \delta\hat{n}_g; \qquad \delta\hat{n}_g(t) = [C_g/(2e)]\hat{v}(t). \tag{3.148}$$

For $\delta\hat{n}_g$ small, we can choose the bias $n_g^{(0)}$ such that the CPB never has more than one CP on it at any time. That is, we can restrict the Hilbert space to the $N \in \{0, 1\}$ subspace, and write the Hamiltonian as

$$\hbar\omega_r\hat{a}^\dagger\hat{a} - 2E_C(1 - 2n_g^{(0)})\hat{\sigma}_z - \frac{E_J}{2}\hat{\sigma}_z - 4E_C\,\delta\hat{n}_g(t)(1 - 2n_g^{(0)} - \hat{\sigma}_x). \tag{3.149}$$

Here (differing from the usual convention) we have defined $\hat{\sigma}_x = |0\rangle\langle 0| - |1\rangle\langle 1|$ and $\hat{\sigma}_z = |1\rangle\langle 0| + |0\rangle\langle 1|$. If one chooses to operate at the charge-degeneracy point, $n_g^{(0)} = 1/2$, the Hamiltonian takes the form

$$\hat{H} = \hbar\omega_r\hat{a}^\dagger\hat{a} + \hbar\omega_a\hat{\sigma}_z/2 - \hbar g(\hat{a} + \hat{a}^\dagger)\hat{\sigma}_x, \tag{3.150}$$

where

$$g = e\frac{C_g}{C_\Sigma}\sqrt{\frac{\omega_r}{\hbar LC}}, \qquad \omega_a = \frac{E_J}{\hbar}. \tag{3.151}$$

Defining $\hat{H}_0 = \hbar\omega_a(\hat{a}^\dagger\hat{a} + \hat{\sigma}_z/2)$, we move to an interaction frame with respect to this Hamiltonian, and make a rotating-wave approximation as usual. The new Hamiltonian is

$$\hat{V} = \hbar\Delta\hat{a}^\dagger\hat{a} + \hbar g(\hat{a}\hat{\sigma}_+ + \hat{a}^\dagger\hat{\sigma}_-), \tag{3.152}$$

with $\Delta = \omega_r - \omega_a$ the detuning between the circuit frequency and the CPB tunnelling frequency. It is assumed to be small compared with ω_a. We can, however, still consider a detuning that is large compared with g. Treating the second term in Eq. (3.152) as a perturbation on the first term, it is possible to show using second-order perturbation theory that Eq. (3.152) may be approximated by the effective Hamiltonitan

$$\hat{V}_{\text{eff}} = \hbar\Delta\hat{a}^\dagger\hat{a} + \hbar\chi\hat{a}^\dagger\hat{a}\hat{\sigma}_z, \tag{3.153}$$

where $\chi = g^2/\Delta$. Moving frames again to the cavity resonance, and including a resonant microwave driving field ϵ, gives a Hamiltonian with the same form as Eq. (3.135).

Schuster *et al.* [SWB+05] recently implemented this system experimentally and measured the measurement-induced qubit dephasing rate given in Eq. (3.143). In their experiment, $\Delta/(2\pi) = 100$ MHz, $g/(2\pi) = 5.8$ MHz and the cavity decay rate was $\gamma/(2\pi) = 0.8$ MHz. Schoelkopf's team used a second probe microwave field tuned to the CPB resonance to induce coherence in the qubit basis. The measurement-induced decoherence time then appears as a broadening of the spectrum representing the response of the qubit to the probe. This spectrum is related to the norm squared of the Fourier transform of the coherence function in time. The results are found to be in good agreement with the theory presented here for small χ. Although the decay of coherence is exponential for long times with rate (3.143), for short times the decoherence is quadratic in time. This is manifested experimentally in the line shape of the probe absorption spectrum: the line shape deviates from the usual Lorentzian shape (corresponding to exponential decay) in its wings.

3.11 Heisenberg-picture dynamics

We saw in Section 1.3.2 that quantum dynamics, and even quantum measurement, can be formulated in the Heisenberg picture. It is thus not surprising that there is a Heisenberg-picture formulation for master equations. This formulation is sometimes called the quantum stochastic differential-equation technique, or the quantum Langevin approach [GC85] (after Paul Langevin, who developed the corresponding theory of classical stochastic differential equations early in the twentieth century [Lan08]). In this section we will develop the Heisenberg-picture description for a system coupled to a bosonic (harmonic-oscillator) bath. We will show how the Markovian limit can be elegantly formulated in the Heisenberg picture, and used to derive a Lindblad-form master equation.

Consider the interaction-frame coupling Hamiltonian in the rotating-wave approximation (3.24) derived in Section 3.3.1

$$\hat{V}_{\text{IF}}(t) = -i[\hat{b}(z := -t)\hat{c}^\dagger - \hat{b}^\dagger(z := -t)\hat{c}], \tag{3.154}$$

where

$$\hat{b}(z) = \gamma^{-1/2} \sum_k g_k \hat{b}_k e^{+i\delta_k z}, \tag{3.155}$$

where δ_k is the detuning of the bath mode k from the system, γ is the dissipation rate, and, as in the example of Section 3.3.1, $i\hat{c}$ is the system lowering operator multiplied by $\sqrt{\gamma}$.

We use z, rather than t, as the argument of $\hat{b}$ for two reasons. The first is that $\hat{b}(z := -t)$ is *not* at this stage a Heisenberg-picture operator. Rather, the interaction Hamiltonian is time-dependent because we are working in the interaction frame, and the operator $\hat{b}(z)$ is defined simply to make $\hat{V}_{\text{IF}}(t)$ simple in form. The second reason is that in some quantum-optical situations, such as the damping of a cavity mode at a single mirror, it is possible to consider the electromagnetic field modes which constitute the bath as being functions of one spatial direction only. On defining the speed of light *in vacuo* to be unity, and the origin as the location of the mirror, we have that, at time $t = 0$, $\hat{b}(z)$ relates to the bath at position $|z|$ away from the mirror. For $z < 0$ it represents a property of the incoming field, and for $z > 0$ it represents a property of the outgoing field. That is, $\hat{b}(z := -t)$ represents the field that will interact (for $t > 0$) or has interacted (for $t < 0$) with the system at time t. An explanation of this may be found in many textbooks, such as that of Gardiner and Zoller [GZ04].

Now, in the limit of a continuum bath as considered in Section 3.3.1, we find that

$$[\hat{b}(z), \hat{b}^\dagger(z')] = \gamma^{-1}\Gamma(z - z'). \tag{3.156}$$

In order to derive a Markovian master equation, it was necessary to assume that $\Gamma(\tau)$ was sharply peaked at $\tau = 0$. Ignoring the Lamb shift in Eq. (3.30), and taking the Markovian limit, we obtain

$$[\hat{b}(z), \hat{b}^\dagger(z')] = \delta(z - z').$$ (3.157)

Physically, this result cannot be exact because the bath modes all have positive frequency. Also, one must be careful using this result because of the singularity of the δ-function. Nevertheless, it is the result that must be used to obtain a strict correspondence with a Markovian master equation.

Before moving to the Heisenberg picture, it is useful to define the unitary operator for an infinitesimal evolution generated by Eq. (3.154):

$$\hat{U}(t + dt, t) = \exp\left[\hat{c}\, d\hat{B}^\dagger_{z:=-t} - \hat{c}^\dagger d\hat{B}_{z:=-t}\right].$$ (3.158)

Here we have defined a new infinitesimal operator,

$$d\hat{B}_z = \hat{b}(z)dt.$$ (3.159)

The point of defining this infinitesimal is that, although it appears to be of order dt, because of the singularity of the commutation relation (3.157), it is actually of order $\sqrt{dt}$. This can be seen by calculating its commutation relations

$$[d\hat{B}_z, d\hat{B}^\dagger_z] = dt,$$ (3.160)

where we have used the heuristic equation $\delta(0)dt = 1$. This can be understood by thinking of dt as the smallest unit into which time can be divided. Then the discrete approximation to a δ-function is a function which is zero everywhere except for an interval of size dt around zero, where it equals $\delta(0) = 1/dt$ (so that its area is unity). Because $d\hat{B}_z$ is of order $\sqrt{dt}$, it is necessary to expand $\hat{U}(t + dt, t)$ to second rather than first order in its argument. That is,

$$\hat{U}(t + dt, t) = \hat{1} + \left(\hat{c}\, d\hat{B}^\dagger_{z:=-t} - \hat{c}^\dagger\, d\hat{B}_{z:=-t}\right) - \tfrac{1}{2}\hat{c}^\dagger\hat{c}\, dt - \tfrac{1}{2}\{\hat{c}^\dagger, \hat{c}\}d\hat{B}^\dagger_{z:=-t}\, d\hat{B}_{z:=-t}$$
$$+ \tfrac{1}{2}\hat{c}^2\left(d\hat{B}^\dagger_{z:=-t}\right)^2 + \tfrac{1}{2}(\hat{c}^\dagger)^2\left(d\hat{B}_{z:=-t}\right)^2.$$ (3.161)

Exercise 3.33 *Show this, using Eq. (3.160).*

3.11.1 Quantum Langevin equations

To obtain the Heisenberg-picture dynamics of the system, one might think that all one need do is to write down the usual Heisenberg equations of motion (in the interaction frame) generated by the Hamiltonian (3.154). That is, for an arbitrary system operator $\hat{s}$,

$$\frac{d\hat{s}(t)}{dt} = [\hat{b}(z := -t, t)\hat{c}^\dagger(t) - \hat{b}^\dagger(z := -t, t)\hat{c}(t), \hat{s}(t)],$$ (3.162)

where now $\hat{b}(z, t)$ is also time-dependent through the evolution of $\hat{b}_k(t)$ in the Heisenberg picture. However, because of the singularity of the commutation relation (3.157), the situation is not so simple. The approach we will follow is different from that in most texts. It has the advantage of being simple to follow and of having a close relation to an analogous approach in classical Markovian stochastic differential equations. For a detailed discussion, see Appendix B.

In our approach, to find the correct Heisenberg equations one proceeds as follows (note that all unitaries are in the interaction frame as usual):

$$\hat{s}(t + dt) = \hat{U}^\dagger(t + dt, t_0)\hat{s}(t_0)\hat{U}(t + dt, t_0) \tag{3.163}$$

$$= \hat{U}^\dagger(t, t_0)\hat{U}^\dagger(t + dt, t)\hat{s}(t_0)\hat{U}(t + dt, t)\hat{U}(t, t_0) \tag{3.164}$$

$$= \hat{U}^\dagger_{\text{HP}}(t + dt, t)\hat{s}(t)\hat{U}_{\text{HP}}(t + dt, t), \tag{3.165}$$

where $\hat{s}(t) = \hat{U}^\dagger(t, t_0)\hat{s}(t_0)\hat{U}(t, t_0)$ and

$$\hat{U}_{\text{HP}}(t + dt, t) \equiv \hat{U}^\dagger(t, t_0)\hat{U}(t + dt, t)\hat{U}(t, t_0). \tag{3.166}$$

Here we are (just for the moment) using the subscript HP to denote that Eq. (3.166) is obtained by replacing the operators appearing in $\hat{U}(t + dt, t)$ by their Heisenberg-picture versions at time t. If we were to expand the exponential in $\hat{U}_{\text{HP}}(t + dt, t)$ to first order in its argument, we would simply reproduce Eq. (3.162). As motivated above, this will not work, and instead we must use the second-order expansion as in Eq. (3.161). First we define

$$d\hat{B}_{\text{in}}(t) \equiv d\hat{B}_{z:=-t}(t), \tag{3.167}$$

and $\hat{b}_{\text{in}}(t)$ similarly. These are known as *input field operators*. Note that as usual the t-argument on the right-hand side indicates that here $d\hat{B}_{z:=-t}$ is in the Heisenberg picture. Because of the bath commutation relation (3.157), this operator is unaffected by any evolution prior to time t, since $d\hat{B}_{\text{in}}(t)$ commutes with $d\hat{B}^\dagger_{\text{in}}(t)$ for non-equal times. Thus we could equally well have defined

$$d\hat{B}_{\text{in}}(t) \equiv d\hat{B}_{z:=-t}(t'), \quad \forall t' \leq t. \tag{3.168}$$

In particular, if $t' = t_0$, the initial time for the problem, then $d\hat{B}_{\text{in}}(t)$ is the same as the Schrödinger-picture operator $d\hat{B}_{z:=-t}$ appearing in $\hat{U}(t + dt, t)$ of Eq. (3.158).

If the bath is initially in the vacuum state, this leads to a significant simplification, as we will now explain. Ultimately we are interested in calculating the average of system (or bath) operators. In the Heisenberg picture, such an average is given by

$$\langle \hat{s}(t) \rangle = \text{Tr}[\hat{s}(t)\rho_{\text{S}} \otimes \rho_{\text{B}}] = \text{Tr}_{\text{S}}[\langle 0|\hat{s}(t)|0\rangle \rho_{\text{S}}], \tag{3.169}$$

where $|0\rangle$ is the vacuum bath state and ρ_{S} is the initial system state. Since $d\hat{B}_{\text{in}}(t)|0\rangle = 0$ for all t, any expression involving $d\hat{B}_{\text{in}}(t)$ and $d\hat{B}^\dagger_{\text{in}}(t)$ that is in normal order (see Section A.5) will contribute nothing to the average. Thus it is permissible to drop all normally ordered terms in Eq. (3.161) that are of second order in $d\hat{B}_{\text{in}}(t)$. That is to say, we can drop all second-order terms in $d\hat{B}_{in}(t)$ in Eq. (3.161) because we have already used

$$d\hat{B}_{in}(t)d\hat{B}_{in}^{\dagger}(t) = [d\hat{B}_{in}(t), d\hat{B}_{in}^{\dagger}(t)] + d\hat{B}_{in}^{\dagger}(t)d\hat{B}_{in}(t) = dt \qquad (3.170)$$

to obtain the non-zero second-order term $-\frac{1}{2}\hat{c}^{\dagger}\hat{c}\, dt$ in Eq. (3.161).

Although they do not contribute to $\langle d\hat{s}(t) \rangle$, first-order terms in the input field operator must be kept because they will in general contribute (via a non-normally ordered product) to the change in an operator product such as $\langle d(\hat{r}\hat{s}) \rangle$. That is because, not surprisingly, one must consider *second-order* corrections to the usual product rule:

$$d(\hat{r}\hat{s}) = (d\hat{r})\hat{s} + \hat{r}(d\hat{s}) + (d\hat{r})(d\hat{s}). \qquad (3.171)$$

One thus obtains from Eq. (3.165) the following Heisenberg equation of motion in the interaction frame:

$$d\hat{s} = dt\left(\hat{c}^{\dagger}\hat{s}\hat{c} - \frac{1}{2}\{\hat{c}^{\dagger}\hat{c}, \hat{s}\} + i[\hat{H}, \hat{s}]\right) - [d\hat{B}_{in}^{\dagger}(t)\hat{c} - \hat{c}^{\dagger}d\hat{B}_{in}(t), \hat{s}]. \qquad (3.172)$$

Here we have dropped the time arguments from all operators except the input bath operators. We have also included a system Hamiltonian $\hat{H}$, as could arise from having a non-zero $\hat{V}_S$, or a Lamb-shift term, as discussed in Section 3.3.1. Remember that we are still in the interaction frame – $\hat{H}$ here is not the same as the $\hat{H} = \hat{H}_0 + \hat{V}$ for the system plus environment with which we started the calculation.

We will refer to Eq. (3.172) as a quantum Langevin equation (QLE) for $\hat{s}$. The operator $\hat{s}$ may be a system operator or it may be a bath operator. Because $\hat{b}_{in}(t)$ is the bath operator *before* it interacts with the system, it is independent of the system operator $\hat{s}(t)$. Hence for system operators one can derive

$$\left\langle \frac{d\hat{s}}{dt} \right\rangle = \left\langle \left(\hat{c}^{\dagger}\hat{s}\hat{c} - \frac{1}{2}\hat{c}^{\dagger}\hat{c}\hat{s} - \frac{1}{2}\hat{s}\hat{c}^{\dagger}\hat{c}\right) + i[\hat{H}, \hat{s}]\right\rangle. \qquad (3.173)$$

Although the noise terms in (3.172) do not contribute to Eq. (3.173), they are necessary in order for Eq. (3.172) to be a valid Heisenberg equation of motion. If they are omitted then the operator algebra of the system will not be preserved.

Exercise 3.34 *Show this. For specificity, consider the case* $\hat{c} = \sqrt{\gamma}\hat{a}$, *where* $\hat{a}$ *is an annihilation operator, and show that, unless these terms are included,* $[\hat{a}(t), \hat{a}^{\dagger}(t)]$ *will not remain equal to unity.*

The master equation. Note that Eq. (3.173) is Markovian, depending only on the average of system operators at the same time. Therefore, it should be derivable from a Markovian evolution equation for the system in the Schrödinger picture. That is to say, there should exist a master equation for the system state matrix such that

$$\langle \hat{s}(t) \rangle = \text{Tr}[\hat{s}\hat{\rho}(t)]. \qquad (3.174)$$

Here, the placement of the time argument indicates the picture (Heisenberg or Schrödinger). By inspection of Eq. (3.173), the corresponding master equation is

$$\dot{\rho} = \mathcal{D}[\hat{c}]\rho - i[\hat{H}, \rho]. \tag{3.175}$$

As promised, this is of the Lindblad form.

3.11.2 Generalization for a non-vacuum bath

The above derivation relied upon the assumption that the bath was initially in the vacuum state $|0\rangle$. However, it turns out that there are other bath states for which it is possible to derive a Markovian QLE and hence a Markovian master equation. This generalization includes a bath with thermal noise and a bath with so-called broad-band squeezing. Instead of the equation $d\hat{B}_{in}(t)d\hat{B}_{in}^\dagger(t) = dt$, with all other second-order products ignorable and all first-order terms being zero on average, we have in the general case

$$d\hat{B}_{in}^\dagger(t)d\hat{B}_{in}(t) = N\,dt, \tag{3.176}$$

$$d\hat{B}_{in}(t)d\hat{B}_{in}^\dagger(t) = (N+1)dt, \tag{3.177}$$

$$d\hat{B}_{in}(t)d\hat{B}_{in}(t) = M\,dt, \tag{3.178}$$

$$\langle d\hat{B}_{in}(t)\rangle = \beta\,dt, \tag{3.179}$$

while Eq. (3.160) still holds. The parameter N is positive, while M and β are complex, with M constrained by

$$|M|^2 \leq N(N+1). \tag{3.180}$$

This type of input field is sometimes called a white-noise field, because the bath correlations are δ-correlated in time. That is, they are flat (like the spectrum of white light) in frequency space. A thermal bath is well approximated by a white-noise bath with $M = 0$ and $N = \{\exp[\hbar\omega_0/(k_B T)] - 1\}^{-1}$, where ω_0 is the frequency of the system's free oscillation. Only a pure squeezed (or vacuum) bath attains the equality in Eq. (3.180).

Using these rules in expanding the unitary operator in Eq. (3.165) gives the following general QLE for a white-noise bath:

$$d\hat{s} = i\,dt[\hat{H}, \hat{s}] + \frac{1}{2}\{(N+1)(2\hat{c}^\dagger\hat{s}\hat{c} - \hat{s}\hat{c}^\dagger\hat{c} - \hat{c}^\dagger\hat{c}\hat{s}) + N(2\hat{c}\hat{s}\hat{c}^\dagger - \hat{s}\hat{c}\hat{c}^\dagger - \hat{c}\hat{c}^\dagger\hat{s})$$

$$+ M[\hat{c}^\dagger, [\hat{c}^\dagger, \hat{s}]] + M^*[\hat{c}, [\hat{c}, \hat{s}]]\}dt - [d\hat{B}_{in}^\dagger\,\hat{c} - \hat{c}^\dagger\,d\hat{B}_{in}, \hat{s}]. \tag{3.181}$$

Here we have dropped time arguments but are still (obviously) working in the Heisenberg picture.

Exercise 3.35 *Derive Eq. (3.181).*

The corresponding master equation is evidently

$$\dot{\rho} = (N+1)\mathcal{D}[\hat{c}]\rho + N\mathcal{D}[\hat{c}^\dagger]\rho + \frac{M}{2}[\hat{c}^\dagger, [\hat{c}^\dagger, \rho]] + \frac{M^*}{2}[\hat{c}, [\hat{c}, \rho]]$$
$$- i[\hat{H} + i(\beta^*\hat{c} - \beta\hat{c}^\dagger), \rho]. \tag{3.182}$$

Note that the effect of the non-zero mean field (3.179) is simply to add a driving term to the existing system Hamiltonian $\hat{H}$. Although not obviously of the Lindblad form, Eq. (3.182) can be written in that form, with three irreversible terms, as long as Eq. (3.180) holds.

Exercise 3.36 *Show this.*
Hint: *Define N' such that $|M|^2 = N'(N'+1)$ and consider three Lindblad operators proportional to $\hat{c}$, $\hat{c}^\dagger$ and $[\hat{c}(N' + M^* + 1) - \hat{c}^\dagger(N' + M)]$.*

3.12 Further reading

There is a large and growing literature on describing the evolution of open quantum systems, both with and without the Markovian assumption. For a review, see the book by Breuer and Petruccione [BP02]. One of the interesting developments since that book was published is the derivation [PV05] of a Markovian master equation for Brownian motion starting from Einstein's original concept of Brownian motion. That is, instead of considering a particle coupled to a bath of harmonic oscillators, a massive particle is made to suffer collisions by being immersed in a bath of less massive particles in thermal equilibrium. Building on the work of Diósi [Dió93], Petruccione and Vacchini [PV05] have rigorously derived a *Lindblad*-form master equation that involves diffusion in position as well as momentum.

As we have discussed, the Lindblad form is the only form of a Markovian master equation that corresponds to a completely positive map for the state. The question of which *non-Markovian* master equations give rise to completely positive evolution has recently been addressed by Andersson, Cresser and Hall [ACH07]. They consider time-local non-Markovian master equations; that is, master equations with time-dependent coefficients such as those we discussed in Section 3.4. For finite-dimensional systems, they show how the state map for any time may be constructed from the master equation, and give a simple test for complete positivity. Conversely, they show that any continuous time-dependent map can be turned into a master equation.

In this chapter we have discussed master equations for systems that can exchange excitations with both fermionic and bosonic baths. However, when presenting the Heisenberg-picture dynamics (quantum Langevin equations) we considered only the case of a bosonic bath. The reason is that there is an important technical issue due to the anticommutation relations between the fermionic driving field and those system operators which can change the number of fermions within the system. This problem has been addressed in a recent paper by Gardiner [Gar04].

The decoherence 'programme' described briefly in Section 3.7 has been reviewed recently by Zurek, one of its chief proponents [Zur03]. For an excellent discussion of some of the conceptual issues surrounding decoherence and the quantum measurement problem, see the recent review by Schlosshauer [Sch04]. For an extensive investigation of physically realizable ensembles and robustness for various open quantum systems see Refs. [WV02a, WV02b, ABJW05]. Finally, we note that an improved version of the Schrödinger-cat decoherence experiment of Section 3.9.2 has been performed, also by the Haroche group. The new results [DDS$^+$08] allow reconstruction of the whole quantum state (specifically, its Wigner function – see Section A.5), showing the rapid vanishing of its nonclassical features under damping.

4

Quantum trajectories

4.1 Introduction

A very general concept of a quantum trajectory would be the path taken by the state of a quantum system over time. This state could be conditioned upon measurement results, as we considered in Chapter 1. This is the sort of quantum trajectory we are most interested in, and it is generally stochastic in nature. In ordinary use, the word trajectory usually implies a path that is also continuous in time. This idea is not always applicable to quantum systems, but we can maintain its essence by defining a quantum trajectory as the path taken by the conditional state of a quantum system for which the unconditioned system state evolves continuously. As explained in Chapter 1, the unconditioned state is that obtained by averaging over the random measurement results which condition the system.

With this motivation, we begin in Section 4.2 by deriving the simplest sort of quantum trajectory, which involves jumps (that is, *discontinuous* conditioned evolution). In the process we will reproduce Lindblad's general form for *continuous* Markovian quantum evolution as presented in Section 3.6. In Section 4.3 we relate these quantum jumps to photon-counting measurements on the bath for the model introduced in Section 3.11, and also derive correlation functions for these measurement records. In Section 4.4 we consider the addition of a coherent field (the 'local oscillator') to the output before detection. In the limit of a strong local oscillator this is called homodyne detection, and is described by a *continuous* (diffusive) quantum trajectory. In Section 4.5 we generalize this theory to describe heterodyne detection and even more general diffusive quantum trajectories. In Section 4.6 we illustrate the detection schemes discussed by examining the conditioned evolution of a simple system: a damped, driven two-level atom. In Section 4.7 we show that there is a complementary description of continuous measurement in the Heisenberg picture, and that this can also be used to derive correlation functions and other statistics of the measurement results. In Section 4.8 we show how quantum trajectory theory can be generalized to deal with imperfect detection, incorporating inefficiency, thermal and squeezed bath noise, dark noise and finite detector bandwidth. In Section 4.9 we turn from optical examples to mesoscopic electronics, including a discussion of imperfect detection. We conclude with further reading in Section 4.10.

4.2 Quantum jumps

4.2.1 Master equations and continuous measurements

The evolution of an isolated quantum system in the absence of measurement is Markovian:

$$|\psi(t + T)\rangle = \hat{U}(T)|\psi(t)\rangle = \exp(-i\hat{H}T)|\psi(t)\rangle, \tag{4.1}$$

where $\hat{H}$ is the Hamiltonian. This equation leads to a finite differential

$$\lim_{\tau \to 0} \frac{|\psi(t + \tau)\rangle - |\psi(t)\rangle}{\tau} = |\dot{\psi}(t)\rangle = -i\hat{H}(t)|\psi(t)\rangle = \text{finite} \tag{4.2}$$

and hence to continuous evolution. We now seek to generalize this unitary evolution by incorporating measurements. To consider the unconditioned state, averaged over the possible measurement results, we have to represent the system by a state matrix rather than a state vector. Then continuous evolution of ρ implies

$$\lim_{\tau \to 0} \frac{\rho(t + \tau) - \rho(t)}{\tau} = \dot{\rho}(t) = \text{finite}. \tag{4.3}$$

In order to obtain a differential equation for $\rho(t)$, we require the measurement time T to be infinitesimal. In this limit, we say that we are *monitoring* the system. Then, from Eq. (1.86), the state matrix at time $t + dt$, averaging over all possible results, is

$$\rho(t + dt) = \sum_r \mathcal{J}[\hat{M}_r(dt)]\rho(t). \tag{4.4}$$

If $\rho(t + dt)$ is to be infinitesimally different from $\rho(t)$, then a first reasonable guess at how to generalize Eq. (4.1) would be to consider just one r, say $r = 0$, and set

$$\hat{M}_0(dt) = \hat{1} - (\hat{R}/2 + i\hat{H})dt, \tag{4.5}$$

where $\hat{R}$ and $\hat{H}$ are Hermitian operators. However, we find that this single measurement operator does not satisfy the completeness condition (1.78), since, to order dt,

$$\hat{M}_0^\dagger(dt)\hat{M}_0(dt) = \hat{1} - \hat{R}\,dt \neq \hat{1}. \tag{4.6}$$

The above result reflects the fact that a measurement with only one possible result is not really a measurement at all and hence the 'measurement operator' (4.5) must be a unitary operator, as it is with $\hat{R} = 0$. If $\hat{R} \neq 0$ then we require at least one other possible result to enable $\sum_r \hat{M}_r(dt)^\dagger \hat{M}_r(dt) = \hat{1}$. The simplest suggestion is to consider two results 0 and 1. We let $\hat{M}_0(dt)$ be as above, and define

$$\hat{M}_1(dt) = \sqrt{dt}\,\hat{c}, \tag{4.7}$$

where $\hat{c}$ is an arbitrary operator obeying

$$\hat{c}^\dagger\hat{c} = \hat{R}, \tag{4.8}$$

which implies that we must have $\hat{R} \geq 0$, so that

$$\hat{M}_0^\dagger(dt)\hat{M}_0(dt) + \hat{M}_1^\dagger(dt)\hat{M}_1(dt) = \hat{1}. \tag{4.9}$$

This gives the non-selective evolution

$$\rho(t + \mathrm{d}t) = [\hat{1} - (\hat{c}^\dagger \hat{c}/2 + \mathrm{i}\hat{H})\mathrm{d}t]\rho(t)[\hat{1} - (\hat{c}^\dagger \hat{c}/2 - \mathrm{i}\hat{H})\mathrm{d}t] + \mathrm{d}t\, \hat{c}\rho(t)\hat{c}^\dagger, \qquad (4.10)$$

which has the differential form

$$\dot{\rho} = -\mathrm{i}[\hat{H}, \rho] + \mathcal{D}[\hat{c}]\rho \equiv \mathcal{L}\rho. \qquad (4.11)$$

Allowing for more than one irreversible term, we obtain exactly the master equation derived by Lindblad by more formal means [Lin76] (see Section 3.6).

4.2.2 Stochastic evolution

Let us consider the evolution implied by the two measurement operators above. The probability for the result $r = 1$ is

$$\wp_1(\mathrm{d}t) = \mathrm{Tr}[\mathcal{J}[\hat{M}_1(\mathrm{d}t)]\rho] = \mathrm{d}t\, \mathrm{Tr}\,[\hat{c}^\dagger \hat{c}\rho], \qquad (4.12)$$

which is infinitesimal (provided that $\hat{c}^\dagger \hat{c}$ is bounded as assumed by Lindblad [Lin76]). That is to say, for almost all infinitesimal time intervals, the measurement result is $r = 0$, because $\wp_0(\mathrm{d}t) = 1 - O(\mathrm{d}t)$. The result $r = 0$ is thus regarded as a null result. In the case of no result, the system state changes infinitesimally, but not unitarily, via the operator $\hat{M}_0(\mathrm{d}t)$. At random times, occurring at rate $\wp_1(\mathrm{d}t)/\mathrm{d}t$, there is a result $r = 1$, which we will call a *detection*. When this occurs, the system undergoes a finite evolution induced by the operator $\hat{M}_1(\mathrm{d}t)$. This change can validly be called a *quantum jump*. However, it must be remembered that it represents a sudden change in the observer's knowledge, not an objective physical event as in Bohr's original conception in the 1910s [Boh13].

Real measurements that correspond approximately to this ideal measurement model are made routinely in experimental quantum optics. If $\hat{c} = \sqrt{\gamma}\hat{a}$ then Eq. (4.11) is the damped-cavity master equation derived in Section 3.3.2, and this theory describes the system evolution in terms of photodetections of the cavity output. Note that we are ignoring the time delay between emission from the system and detection by the detector. Loosely, we can think of the conditioned state here as being the state the system *was* in at the time of emission. When it comes to considering feedback control of the system we will see that any time delay, whether between the emission and detection or between the detection and feedback action, must be taken into account.

Let us denote the number of photodetections up to time t by $N(t)$, and say for simplicity that the system state at time t is a pure state $|\psi(t)\rangle$. Then the stochastic increment $\mathrm{d}N(t)$ obeys

$$\mathrm{d}N(t)^2 = \mathrm{d}N(t), \qquad (4.13)$$

$$\mathrm{E}[\mathrm{d}N(t)] = \langle\hat{M}_1^\dagger(\mathrm{d}t)\hat{M}_1(\mathrm{d}t)\rangle = \mathrm{d}t\langle\psi(t)|\hat{c}^\dagger \hat{c}|\psi(t)\rangle, \qquad (4.14)$$

where a classical expectation value is denoted by E and the quantum expectation value by angle brackets. The first equation here simply says that $\mathrm{d}N$ is either zero or one, as it must be since it is the increment in N in an infinitesimal time. The second equation gives the mean of dN, which is identical with the probability of detecting a photon. This is an example of a point process – see Section B.6.

From the measurement operators (4.5) and (4.7) we see that, when $dN(t) = 1$, the state vector changes to

$$|\psi_1(t + dt)\rangle = \frac{\hat{M}_1(dt)|\psi(t)\rangle}{\sqrt{\langle \hat{M}_1^\dagger(dt)\hat{M}_1(dt)\rangle(t)}} = \frac{\hat{c}|\psi(t)\rangle}{\sqrt{\langle \hat{c}^\dagger \hat{c}\rangle(t)}}, \tag{4.15}$$

where the denominator gives the normalization. If there is no detection, $dN(t) = 0$ and

$$|\psi_0(t + dt)\rangle = \frac{\hat{M}_0(dt)|\psi(t)\rangle}{\sqrt{\langle \hat{M}_0^\dagger(dt)\hat{M}_0(dt)\rangle(t)}}$$

$$= \left\{\hat{1} - dt\left[i\hat{H} + \tfrac{1}{2}\hat{c}^\dagger \hat{c} - \tfrac{1}{2}\langle \hat{c}^\dagger \hat{c}\rangle(t)\right]\right\}|\psi(t)\rangle, \tag{4.16}$$

where the denominator has been expanded to first order in dt to yield the nonlinear term.

This stochastic evolution can be written explicitly as a nonlinear stochastic Schrödinger equation (SSE):

$$d|\psi(t)\rangle = \left[dN(t)\left(\frac{\hat{c}}{\sqrt{\langle \hat{c}^\dagger \hat{c}\rangle(t)}} - 1\right) + [1 - dN(t)]dt\left(\frac{\langle \hat{c}^\dagger \hat{c}\rangle(t)}{2} - \frac{\hat{c}^\dagger \hat{c}}{2} - i\hat{H}\right)\right]|\psi(t)\rangle.$$

$$\tag{4.17}$$

It is called a Schrödinger equation only because it preserves the purity of the state, like Eq. (4.1). We will call a solution to this equation a *quantum trajectory* for the system.

We can simplify the stochastic Schrödinger equation by using the rule (see Section B.6)

$$dN(t)dt = o(dt). \tag{4.18}$$

This notation means that the order of $dN(t)dt$ is smaller than that of dt and so the former is negligible compared with the latter. Then Eq. (4.17) becomes

$$d|\psi(t)\rangle = \left[dN(t)\left(\frac{\hat{c}}{\sqrt{\langle \hat{c}^\dagger \hat{c}\rangle(t)}} - \hat{1}\right) + dt\left(\frac{\langle \hat{c}^\dagger \hat{c}\rangle(t)}{2} - \frac{\hat{c}^\dagger \hat{c}}{2} - i\hat{H}\right)\right]|\psi(t)\rangle. \tag{4.19}$$

Exercise 4.1 *Verify that the only difference between the two equations is that the state vector after a jump is infinitesimally different. Since the total number of jumps in any finite time is finite, the difference between the two equations is negligible.*

From Eq. (4.19) it is simple to reconstruct the master equation using the rules (4.13) and (4.14). First define a projector

$$\hat{\pi}(t) = |\psi(t)\rangle\langle\psi(t)|, \tag{4.20}$$

and find (using the notation $|d\psi\rangle = d|\psi\rangle$)

$$d\hat{\pi}(t) = |d\psi(t)\rangle\langle\psi(t)| + |\psi(t)\rangle\langle d\psi(t)| + |d\psi(t)\rangle\langle d\psi(t)| \tag{4.21}$$

$$= \left\{dN(t)\mathcal{G}[\hat{c}] - dt\,\mathcal{H}\left[i\hat{H} + \tfrac{1}{2}\hat{c}^\dagger\hat{c}\right]\right\}\hat{\pi}(t). \tag{4.22}$$

Here, the nonlinear (in ρ) superoperators $\mathcal{G}$ and $\mathcal{H}$ are defined by

$$\mathcal{G}[\hat{r}]\rho = \frac{\hat{r}\rho\hat{r}^\dagger}{\text{Tr}[\hat{r}\rho\hat{r}^\dagger]} - \rho, \tag{4.23}$$

$$\mathcal{H}[\hat{r}]\rho = \hat{r}\rho + \rho\hat{r}^\dagger - \text{Tr}[\hat{r}\rho + \rho\hat{r}^\dagger]\rho. \tag{4.24}$$

Now define

$$\rho(t) = \text{E}[\hat{\pi}(t)], \tag{4.25}$$

that is, the state matrix is the expected value or ensemble average of the projector. From Eq. (B.54), the rule (4.14) generalizes to

$$\text{E}[dN(t)g(\hat{\pi}(t))] = dt\,\text{E}\left[\text{Tr}[\hat{\pi}(t)\hat{c}^\dagger\hat{c}]\,g(\hat{\pi}(t))\right], \tag{4.26}$$

for any function g. Using this yields finally

$$d\rho = -i\,dt[\hat{H}, \rho] + dt\,\mathcal{D}[\hat{c}]\rho, \tag{4.27}$$

as required.

Exercise 4.2 *Verify Eqs. (4.22) and (4.27) following the above steps.*

4.2.3 Quantum trajectories for simulations

For the most general master-equation evolution,

$$\dot{\rho} = -i[\hat{H}, \rho] + \sum_\mu \mathcal{D}[\hat{c}_\mu]\rho, \tag{4.28}$$

the above SSE can be generalized to

$$d|\psi(t)\rangle = \sum_\mu \left[dN_\mu(t)\left(\frac{\hat{c}_\mu}{\sqrt{\langle\hat{c}_\mu^\dagger\hat{c}_\mu\rangle(t)}} - \hat{1}\right) + dt\left(\frac{\langle\hat{c}_\mu^\dagger\hat{c}_\mu\rangle(t)}{2} - \frac{\hat{c}_\mu^\dagger\hat{c}_\mu}{2} - i\hat{H}\right)\right]|\psi(t)\rangle, \tag{4.29}$$

with

$$\text{E}[dN_\mu(t)] = \langle\psi(t)|\hat{c}_\mu^\dagger\hat{c}_\mu|\psi(t)\rangle dt, \tag{4.30}$$

$$dN_\mu(t)dN_\nu(t) = dN_\mu(t)\delta_{\mu\nu}. \tag{4.31}$$

Equations of this form have been used extensively since the mid 1990s in order to obtain numerical solutions of master equations [DZR92, MCD93]. The solution $\rho(t)$ is approximated by the ensemble average $\mathrm{E}[|\psi(t)\rangle\langle\psi(t)|]$ over a finite number $M \gg 1$ of numerical realizations of the stochastic evolution (4.29).

The advantage of doing this rather than solving the master equation (4.28) is that, if the system requires a Hilbert space of dimension N in order to be represented accurately, then in general storing the state matrix ρ requires of order N^2 real numbers, whereas storing the state vector $|\psi\rangle$ requires only of order N. For large N, the time taken to compute the evolution of the state matrix via the master equation scales as N^4, whereas the time taken to compute the ensemble of state vectors via the quantum trajectory scales as $N^2 M$, or just N^2 if parallel processors are available. Even though one requires $M \gg 1$, reasonable results may be obtainable with $M \ll N^2$. For extremely large N it may be impossible even to store the state matrix on most computers. In this case the quantum trajectory method may still be useful, if one wishes to calculate only certain system averages, rather than the entire state matrix, via

$$\mathrm{E}[\langle\psi(t)|\hat{A}|\psi(t)\rangle] = \mathrm{Tr}[\rho(t)\hat{A}]. \tag{4.32}$$

One area where this technique has been applied to good effect is the quantized motion of atoms undergoing spontaneuous emission [DZR92].

The simplest method of solution for Eq. (4.29) is to replace all differentials d by small but finite differences δ. That is, in a small interval of time δt, a random number $R(t)$ chosen uniformly from the unit interval is generated. If

$$R(t) < \wp_{\text{jump}} = \sum_{\mu} \langle\psi(t)|\hat{c}_{\mu}^{\dagger}\hat{c}_{\mu}|\psi(t)\rangle \delta t \tag{4.33}$$

then a jump happens. One of the possible jumps (μ) is chosen randomly using another (or the same) random number, with the weights $\langle\psi(t)|\hat{c}_{\mu}^{\dagger}\hat{c}_{\mu}|\psi(t)\rangle\delta t/\wp_{\text{jump}}$. The appropriate δN_{μ} is then set to 1, all others set to zero, and the increment (4.29) calculated.

In practice, this is not the most efficient method for simulation. Instead, the following method is generally used. Say the system starts at time 0. A random number R is generated as above. Then the unnormalized evolution

$$\frac{d}{dt}|\tilde{\psi}(t)\rangle = -\left(\sum_{\mu}\hat{c}_{\mu}^{\dagger}\hat{c}_{\mu}/2 + i\hat{H}\right)|\tilde{\psi}(t)\rangle \tag{4.34}$$

is solved for a time T such that $\langle\tilde{\psi}(T)|\tilde{\psi}(T)\rangle = R$. This time T will have to be found iteratively. However, since Eq. (4.34) is an ordinary linear differential equation, it can be solved efficiently using standard numerical techniques. The decay in the state-matrix norm $\langle\tilde{\psi}(t)|\tilde{\psi}(t)\rangle$ is because Eq. (4.34) keeps track only of the no-jump evolution, derived from the repeated action of $\hat{M}_0(dt)$. That is, the norm is equal to the probability of this series of results occurring (see Section 1.4). Thus this method generates T, the time at which the first jump occurs, with the correct statistics. Which jump (i.e. which μ) occurs at this time can be determined by the technique described above. The relevant collapse operator $\hat{c}_{\mu}$ is then applied to the system $|\tilde{\psi}(T)\rangle$, and the state normalized again. The simulation then repeats as if time $t = T$ were the initial time $t = 0$.

4.3 Photodetection

4.3.1 Photon emission and detection

We claimed above that jumpy quantum trajectories are realized routinely in quantum-optics laboratories by counting photons in the output beam from a cavity. We now show this explicitly for this sort of measurement which we call *direct detection*. Recall from Eq. (3.158) in Section 3.11 that the interaction between the system and the bath in the infinitesimal interval $[t, t + \mathrm{d}t)$ is described by the unitary operator

$$\hat{U}(t + \mathrm{d}t, t) = \exp[\hat{c}\,\mathrm{d}\hat{B}^\dagger - \hat{c}^\dagger\,\mathrm{d}\hat{B} - \mathrm{i}\hat{H}\,\mathrm{d}t]. \tag{4.35}$$

Here, for convenience, we are using the short-hand $\mathrm{d}\hat{B} = \mathrm{d}\hat{B}_{z:=-t}$, and we have included a system Hamiltonian $\hat{H}$, all in the interaction frame. If the system is in a pure state at time t (as it will be if the bath has been monitored up to that time), then we need consider only the bath mode on which $\mathrm{d}\hat{B}$ acts, which is initially in the vacuum state. That is, the entangled system–bath state at the end of the interval is $\hat{U}(t + \mathrm{d}t, t)|0\rangle|\psi(t)\rangle$.

Keeping only the non-normally ordered second-order terms $\mathrm{d}\hat{B}\,\mathrm{d}\hat{B}^\dagger = \mathrm{d}t$ (see Section 3.11.1), and given the fact that $\mathrm{d}\hat{B}|0\rangle = 0$, one finds

$$\hat{U}(t + \mathrm{d}t, t)|0\rangle|\psi(t)\rangle = [\hat{1} - \mathrm{d}t\,\hat{c}^\dagger\hat{c}/2 - \mathrm{i}\hat{H}\,\mathrm{d}t]|0\rangle|\psi(t)\rangle + \mathrm{d}\hat{B}^\dagger|0\rangle\hat{c}|\psi(t)\rangle. \tag{4.36}$$

Clearly $\mathrm{d}\hat{B}^\dagger|0\rangle$ is a bath state containing one photon. But it is not a normalized one-photon state; rather, it has a norm of

$$\langle 0|\mathrm{d}\hat{B}\,\mathrm{d}\hat{B}^\dagger|0\rangle = \mathrm{d}t. \tag{4.37}$$

Thus, the probability of finding one photon in the bath is

$$\langle 0|\mathrm{d}\hat{B}\,\mathrm{d}\hat{B}^\dagger|0\rangle\langle\psi(t)|\hat{c}^\dagger\hat{c}|\psi(t)\rangle = \wp_1(\mathrm{d}t), \tag{4.38}$$

where $\wp_1(\mathrm{d}t)$ is defined in Eq. (4.12). Moreover, from Eq. (4.36) it is apparent that the system state conditioned on the bath containing a photon is exactly as given in Eq. (4.15). The probability of finding no photons in the bath is $\wp_0(\mathrm{d}t) = 1 - \wp_1(\mathrm{d}t)$, and the conditioned system state is again as given previously in Eq. (4.16).

In the above, we have not specified whether the measurement on the bath is projective (which would leave the number of photons unchanged) or non-projective. In reality, photon detection, at least at optical frequencies, is done by absorption, so the field state at the end of the measurement is the vacuum state. However, it should be emphasized that this is in no way essential to the theory. The field state at the beginning of the next interval $[t + \mathrm{d}t, t + 2\,\mathrm{d}t)$ *is* a vacuum state, but not because we have assumed the photons to have been absorbed. Rather, it is a vacuum state for a *new* field operator, which pertains to the part of the field which has 'moved in' to interact with the system while the previous part (which has now become the emitted field) 'moves out' to be detected – see Section 3.11.

4.3.2 Output correlation functions

In experimental quantum optics, it is more usual to consider a photocurrent than a photocount. Any multiplier in the definition of a photocurrent is either purely conventional, or has meaning only for a particular detector, so we simply define the photocurrent to be

$$I(t) = dN(t)/dt. \tag{4.39}$$

Note that $dN(t)$, since its mean depends on the quantum state at time t, is conditioned on the record $dN(s)$ for $s < t$. That is, I is what is known as a self-exciting point process [Haw71]. We write the quantum state at time t (which in general may be mixed) as $\rho_I(t)$. Here the subscript I emphasizes that it is conditioned on the photocurrent. To consider mixed states, it is necessary to reformulate the stochastic Schrödinger equation (4.19) as a *stochastic master equation* (SME). This simply means replacing the projector $\hat{\pi}(t)$ in Eq. (4.22) by $\rho_I(t)$ to get

$$d\rho_I(t) = \left\{ dN(t)\mathcal{G}[\hat{c}] - dt\, \mathcal{H}\left[i\hat{H} + \tfrac{1}{2}\hat{c}^\dagger\hat{c}\right]\right\}\rho_I(t), \tag{4.40}$$

where the jump probability is

$$E[dN(t)] = dt\, \mathrm{Tr}\left[\hat{c}^\dagger\hat{c}\rho_I(t)\right]. \tag{4.41}$$

The photocurrent (4.39) is a singular quantity, consisting of a series of Dirac δ-functions at the times of photodetections. The most common ways to investigate its statistical properties are to find its mean and its autocorrelation function. The first of these statistics is simply

$$E[I(t)] = \mathrm{Tr}[\rho(t)\hat{c}^\dagger\hat{c}]. \tag{4.42}$$

The second is defined as

$$F^{(2)}(t, t + \tau) = E[I(t + \tau)I(t)]. \tag{4.43}$$

We will now show how this can be evaluated using Eq. (4.40). We take the state at time t to be a given $\rho(t)$.

First consider τ finite. Now $dN(t)$ is either zero or one. If it is zero, then the function is automatically zero. Hence,

$$F^{(2)}(t, t + \tau)(dt)^2 = \Pr[dN(t) = 1] \times E[dN(t + \tau)|dN(t) = 1], \tag{4.44}$$

where $E[A|B]$ means the expectation value of the variable A given the event B. This is equal to

$$F^{(2)}(t, t + \tau)(dt)^2 = \mathrm{Tr}[\hat{c}^\dagger\hat{c}\rho(t)]dt \times \mathrm{Tr}[\hat{c}^\dagger\hat{c}\, dt\, E[\rho_I(t + \tau)|dN(t) = 1]]. \tag{4.45}$$

Note that the ensemble average of $\rho_I(t + \tau)$ conditioned on $dN(t) = 1$ appears because we have no knowledge about photodetector clicks in the interval $[t + dt, t + \tau)$, and hence we must average over any such jumps. Now, if $dN(t) = 1$, then from Eq. (4.40) we obtain (to leading order in dt)

$$\rho_I(t + dt) = \rho(t) + \mathcal{G}[\hat{c}]\rho(t) = \hat{c}\rho(t)\hat{c}^\dagger/\mathrm{Tr}[\hat{c}\rho(t)\hat{c}^\dagger]. \tag{4.46}$$

By virtue of the linearity of the ensemble-average evolution (4.27),

$$E[\rho_I(t+\tau)|dN(t) = 1] = \exp(\mathcal{L}\tau)\hat{c}\rho(t)\hat{c}^\dagger/\text{Tr}[\hat{c}\rho(t)\hat{c}^\dagger]. \qquad (4.47)$$

Here the superoperator $e^{\mathcal{L}\tau}$ acts on the product of all operators to its right. Thus, the final expression for τ finite is

$$F^{(2)}(t, t+\tau) = \text{Tr}[\hat{c}^\dagger\hat{c}e^{\mathcal{L}\tau}\hat{c}\rho(t)\hat{c}^\dagger]. \qquad (4.48)$$

For $\hat{c}$ an annihilation operator for a cavity mode, this is equal to Glauber's second-order coherence function, $G^{(2)}(t, t+\tau)$ [Gla63].

If $\tau = 0$, then the expression (4.43) diverges, because $dN(t)^2 = dN(t)$. Naively,

$$F^{(2)}(t, t) = \text{Tr}[\hat{c}^\dagger\hat{c}\rho(t)]/dt. \qquad (4.49)$$

However, since we are effectively discretizing time in bins of size dt, the expression $1/dt$ at $\tau = 0$ is properly interpreted as the Dirac δ-function $\delta(\tau)$ – see the discussion following Eq. (3.160). Since this is infinitely larger than any finite term at $\tau = 0$, we can write the total expression as

$$F^{(2)}(t, t+\tau) = \text{Tr}[\hat{c}^\dagger\hat{c}e^{\mathcal{L}\tau}\hat{c}\rho(t)\hat{c}^\dagger] + \text{Tr}[\hat{c}^\dagger\hat{c}\rho(t)]\delta(\tau). \qquad (4.50)$$

Often we are interested in the *stationary* or *steady-state* statistics of the current, in which case the time argument t disappears and $\rho(t)$ is replaced by ρ_{ss}, the (assumed unique) stationary solution of the master equation:

$$\mathcal{L}\rho_{ss} = 0. \qquad (4.51)$$

4.3.3 Coherent field input

In the preceding section we have assumed a bath in the vacuum state. In Section 3.11.2 we showed how to derive the quantum Langevin equations and master equations for the case of a white-noise (thermal or squeezed) bath. If one tried to consider photodetection in such a case, one would run into the problem that a white-noise bath has a theoretically infinite photon flux. It is not possible to count the photons in such a beam; no matter how small the time interval, one would still expect a finite number of counts. In practice, such noise is not strictly white, and in any case the bandwidth of the detector will keep the count rate finite.[1] Nevertheless, in the limit where the white-noise approximation is a good one, the photon flux due to the bath will be much greater than that due to the system. Hence, direct detection will yield negligible information about the system state, so the quantum trajectory will simply be the unconditioned master equation (4.27).

However, the vacuum input considered so far can still be generalized by adding a coherent field, as explained in Section 3.11.2. Given that $d\hat{B}\,d\hat{B}^\dagger = dt$, to obtain a state with $\langle d\hat{B}\rangle = \beta\,dt$ as in Eq. (3.179), we can assume an initial state of the form $[\hat{1} + \beta\,d\hat{B}^\dagger]|0\rangle$,

---

[1] Or the detector may burn out!

which is normalized to leading order. Now recall from Section 4.3.1 that a normalized one-photon state is $d\hat{B}^\dagger|0\rangle/\sqrt{dt}$. Therefore, the jump measurement operator is

$$\hat{M}_1(dt) = (dt)^{-1/2}\langle 0|d\hat{B}\,\hat{U}(t+dt,t)[\hat{1}+\beta\,d\hat{B}^\dagger]|0\rangle, \qquad (4.52)$$

which to leading order evaluates to

$$\hat{M}_1(dt) = (dt)^{1/2}(\hat{c}+\beta). \qquad (4.53)$$

Similarly, the no-jump measurement operator

$$\hat{M}_0(dt) = \langle 0|\hat{U}(t+dt,t)[\hat{1}+\beta\,d\hat{B}^\dagger]|0\rangle \qquad (4.54)$$

evaluates to

$$\hat{M}_0(dt) = \hat{1} - \left[i\hat{H} + \tfrac{1}{2}\hat{c}^\dagger\hat{c} + \hat{c}^\dagger\beta\right]dt. \qquad (4.55)$$

Exercise 4.3 *Show these.*

These measurement operators define the following SSE:

$$d|\psi_I(t)\rangle = \left[dN(t)\left(\frac{\hat{c}+\beta}{\sqrt{\langle(\hat{c}^\dagger+\beta^*)(\hat{c}+\beta)\rangle_I(t)}}-1\right)+dt\right.$$
$$\left.\times\left(\frac{\langle\hat{c}^\dagger\hat{c}\rangle_I(t)}{2}-\frac{\hat{c}^\dagger\hat{c}}{2}+\frac{\langle\hat{c}^\dagger\beta+\beta^*\hat{c}\rangle_I(t)}{2}-\hat{c}^\dagger\beta-i\hat{H}\right)\right]|\psi_I(t)\rangle.$$
$$(4.56)$$

The ensemble-average evolution is the master equation

$$\dot\rho = \mathcal{D}[\hat{c}]\rho - i[\hat{H}+i\beta^*\hat{c}-i\hat{c}^\dagger\beta,\rho], \qquad (4.57)$$

which is as expected from Eq. (3.182) with $N=M=0$. To 'unravel' this master equation for purposes of numerical calculation, one could choose the SSE as for the vacuum input (4.19), merely changing the Hamiltonian as indicated in the master equation (4.57). However, this would be a mistake if the trajectories were meant to represent the actual conditional evolution of the system, which is given by Eq. (4.56).

4.4 Homodyne detection

4.4.1 Adding a local oscillator

As noted in Section 3.6, the master equation

$$d\rho = -i\,dt[\hat{H},\rho] + dt\,\mathcal{D}[\hat{c}]\rho \qquad (4.58)$$

is invariant under the transformation

$$\hat{c}\to\hat{c}+\gamma; \qquad \hat{H}\to\hat{H}-i\tfrac{1}{2}(\gamma^*\hat{c}-\gamma\hat{c}^\dagger), \qquad (4.59)$$

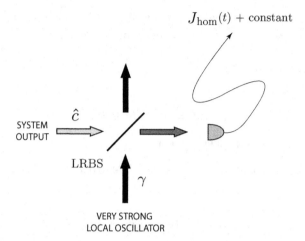

Fig. 4.1 A scheme for simple homodyne detection. A low-reflectivity beam-splitter (LRBS) transmits almost all of the system output, and adds only a small amount of the local oscillator through reflection. Nevertheless, the local oscillator is so strong that this reflected field dominates the intensity at the single photoreceiver. This is a detector that does not resolve single photons but rather produces a photocurrent proportional to $J_{\text{hom}}(t)$ plus a constant.

where γ is an arbitrary complex number (in particular, it is not related to the system damping rate for which we sometimes use γ). Under this transformation, the measurement operators transform to

$$\hat{M}_1(\mathrm{d}t) = \sqrt{\mathrm{d}t}(\hat{c} + \gamma), \tag{4.60}$$

$$\hat{M}_0(\mathrm{d}t) = \hat{1} - \mathrm{d}t\left[\mathrm{i}\hat{H} + \tfrac{1}{2}(\hat{c}\gamma^* - \hat{c}^\dagger\gamma) + \tfrac{1}{2}(\hat{c}^\dagger + \gamma^*)(\hat{c} + \gamma)\right]. \tag{4.61}$$

This shows that the *unravelling* of the deterministic master-equation evolution into a set of stochastic quantum trajectories is not unique.

Physically, the above transformation can be achieved by homodyne detection. In the simplest configuration (see Fig. 4.1), the output field of the cavity is sent through a beam-splitter of transmittance η. The transformation of a field operator $\hat{b}$ entering one port of a beam-splitter can be taken to be

$$\hat{b} \rightarrow \sqrt{\eta}\hat{b} + \sqrt{1 - \eta}\hat{o}, \tag{4.62}$$

where $\hat{o}$ is the operator for the field incident on the other port of the beam-splitter, which is reflected into the path of the transmitted beam. This other field transforms on transmission as $\hat{o} \rightarrow \sqrt{\eta}\hat{o} - \sqrt{1 - \eta}\hat{b}$. (Note the minus sign.) In the case of homodyne detection, this other field is a very strong coherent field. It has the same frequency as the system dipole, and is known as the local oscillator. It can be modelled as $\hat{o} = \gamma/\sqrt{1 - \eta} + \hat{v}$. The first part represents the coherent amplitude of the local oscillator, with $|\gamma|^2/(1 - \eta)$ being its photon flux. The second part represents the 'vacuum fluctuations'; $\hat{v}$ is a continuum field that satisfies $[\hat{v}(t), \hat{v}^\dagger(t')] = \delta(t - t')$ and can be assumed to act on the vacuum state. For η very close to unity, as is desired here, the transformation (4.62) reduces to

$$\hat{b} \to \hat{b} + \gamma, \qquad (4.63)$$

which is called a displacement of the field (see Section A.4). A perfect measurement of the photon number of the displaced field leads to the above measurement operators.

Exercise 4.4 *Convince yourself of this.*

Let the coherent field γ be real, so that the homodyne detection leads to a measurement of the x *quadrature* of the system dipole. This can be seen from the rate of photodetections at the detector:

$$E[dN(t)/dt] = \mathrm{Tr}[(\gamma^2 + \gamma\hat{x} + \hat{c}^\dagger\hat{c})\rho_I(t)]. \qquad (4.64)$$

Here we are defining the two system quadratures by[2]

$$\hat{x} = \hat{c} + \hat{c}^\dagger; \qquad \hat{y} = -\mathrm{i}(\hat{c} - \hat{c}^\dagger). \qquad (4.65)$$

In the limit that γ is much larger than $\langle\hat{c}^\dagger\hat{c}\rangle$, the rate (4.64) consists of a large constant term plus a term proportional to $\hat{x}$, plus a small term. From the measurement operators (4.60) and (4.61), the stochastic master equation for the conditioned state matrix is

$$d\rho_I(t) = \left\{ dN(t)\mathcal{G}[\hat{c} + \gamma] + dt\,\mathcal{H}\left[-\mathrm{i}\hat{H} - \gamma\hat{c} - \tfrac{1}{2}\hat{c}^\dagger\hat{c}\right] \right\}\rho_I(t). \qquad (4.66)$$

This SME can be equivalently written as the SSE

$$
d|\psi_I(t)\rangle = \left[dN(t)\left(\frac{\hat{c} + \gamma}{\sqrt{\langle(\hat{c}^\dagger + \gamma)(\hat{c} + \gamma)\rangle_I(t)}} - 1 \right) + dt \right.
$$
$$
\left. \times \left(\frac{\langle\hat{c}^\dagger\hat{c}\rangle_I(t)}{2} - \frac{\hat{c}^\dagger\hat{c}}{2} + \frac{\langle\hat{c}^\dagger\gamma + \gamma\hat{c}\rangle_I(t)}{2} - \gamma\hat{c} - \mathrm{i}\hat{H} \right) \right]|\psi_I(t)\rangle.
$$

$$(4.67)$$

This shows how the master equation (4.58) can be unravelled in a completely different manner from the usual quantum trajectory (4.19). Note the minor difference from the coherently driven SSE (4.56), which makes the latter simulate a different master equation (4.57).

4.4.2 The continuum limit

The ideal limit of homodyne detection is when the local oscillator amplitude goes to infinity. In this limit, the rate of photodetections goes to infinity, but the effect of each detection on the system goes to zero, because the field being detected is almost entirely due to the local

---

[2] Be aware of the following possibility for confusion: for a two-level atom we typically have $\hat{c} = \hat{\sigma}_-$, in which case $\hat{x} = \hat{\sigma}_x$ but $\hat{y} = -\hat{\sigma}_y$!

oscillator. Thus, it should be possible to approximate the photocurrent by a continuous function of time, and also to derive a smooth evolution equation for the system. This was done first by Carmichael [Car93]; the following is a somewhat more rigorous working of the derivation he sketched.

Let the system operators be of order unity, and let the local oscillator amplitude γ be an arbitrarily large real parameter. Consider a time interval $[t, t + \delta t)$, where $\delta t = O(\gamma^{-3/2})$. This scaling is chosen so that, within this time interval, the number of detections $\delta N \sim \gamma^2 \delta t = O(\gamma^{1/2})$ is very large, but the systematic change in the system, of $O(\delta t) = O(\gamma^{-3/2})$, is very small. Taking the latter change into account, the mean number of detections in this time will be

$$\mu = \text{Tr}\big[\big(\gamma^2 + \gamma\hat{x} + \hat{c}^\dagger\hat{c}\big)\big\{\rho_I(t) + O\big(\gamma^{-3/2}\big)\big\}\big]\delta t$$
$$= \big[\gamma^2 + \gamma\langle\hat{x}\rangle_I(t) + O\big(\gamma^{1/2}\big)\big]\delta t. \tag{4.68}$$

The error in μ (due to the change in the system over the interval) is larger than the contribution from $\hat{c}^\dagger\hat{c}$. The variance in δN will be dominated by the Poissonian number statistics of the coherent local oscillator (see Section A.4.2). Because the number of counts is very large, these Poissonian statistics will be approximately Gaussian. Specifically, it can be shown [WM93b] that the statistics of δN are consistent with those of a Gaussian random variable of mean (4.68) and variance

$$\sigma^2 = \big[\gamma^2 + O\big(\gamma^{3/2}\big)\big]\delta t. \tag{4.69}$$

The error in σ^2 is necessarily as large as expressed here in order for the statistics of δN to be consistent with Gaussian statistics. Thus, δN can be written as

$$\delta N = \gamma^2 \delta t[1 + \langle\hat{x}\rangle_I(t)/\gamma] + \gamma\,\delta W, \tag{4.70}$$

where the accuracy in both terms is only as great as the highest order expression in $\gamma^{-1/2}$. Here δW is a Wiener increment satisfying $E[\delta W] = 0$ and $E[(\delta W)^2] = \delta t$ (see Appendix B).

Now, insofar as the system is concerned, the time δt is still very small. Expanding Eq. (4.66) in powers of γ^{-1} gives

$$\delta\rho_I(t) = \delta N(t)\bigg(\frac{\mathcal{H}[\hat{c}]}{\gamma} + \frac{\langle\hat{c}^\dagger\hat{c}\rangle_I(t)\mathcal{G}[\hat{c}] - \langle\hat{x}\rangle_I(t)\mathcal{H}[\hat{c}]}{\gamma^2} + O\big(\gamma^{-3}\big)\bigg)\rho_I(t)$$
$$+ \delta t\,\mathcal{H}\big[-i\hat{H} - \gamma\hat{c} - \tfrac{1}{2}\hat{c}^\dagger\hat{c}\big]\rho_I(t), \tag{4.71}$$

where $\mathcal{G}$ and $\mathcal{H}$ are as defined previously in Eqs. (4.23) and (4.24).

Exercise 4.5 *Show this.*

Although Eq. (4.66) requires that $dN(t)$ be a point process, it is possible simply to substitute the expression obtained above for δN as a Gaussian random variable into Eq. (4.71). This is because each jump is infinitesimal, so the effect of many jumps is approximately equal to the effect of one jump scaled by the number of jumps. This can be justified by considering an expression for the system state given precisely δN detections, and then taking the large-δN limit [WM93b]. The simple procedure adopted here gives the correct answer more rapidly.

Keeping only the lowest-order terms in $\gamma^{-1/2}$ and letting $\delta t \to dt$ yields the SME

$$d\rho_J(t) = -i[\hat{H}, \rho_J(t)]dt + dt\,\mathcal{D}[\hat{c}]\rho_J(t) + dW(t)\mathcal{H}[\hat{c}]\rho_J(t), \tag{4.72}$$

where the subscript J is explained below, under Eq. (4.75). Here $dW(t)$ is an infinitesimal Wiener increment satisfying

$$dW(t)^2 = dt, \tag{4.73}$$

$$E[dW(t)] = 0. \tag{4.74}$$

That is, the jump evolution of Eq. (4.66) has been replaced by diffusive evolution.

Exercise 4.6 *Derive Eq. (4.72).*

By its derivation, Eq. (4.72) is an Itô stochastic differential equation, which we indicate by our use of the explicit increment (rather than the time derivative). It is trivial to see that the ensemble-average evolution reproduces the non-selective master equation by using Eq. (4.74) to eliminate the noise term. Readers unfamiliar with stochastic calculus, or unfamiliar with our conventions regarding the Itô and Stratonovich versions, are referred to Appendix B.

Just as $\gamma \to \infty$ leads to continuous evolution for the state, so does it change the point-process photocount into a continuous photocurrent with white noise. Removing the constant local oscillator contribution gives

$$J_{\mathrm{hom}}(t) \equiv \lim_{\gamma \to \infty} \frac{\delta N(t) - \gamma^2\,\delta t}{\gamma\,\delta t} = \langle \hat{x} \rangle_J(t) + \xi(t), \tag{4.75}$$

where $\xi(t) = dW(t)/dt$. This is why the subscript I has been replaced by J in Eq. (4.72).

Finally, it is worth noting that these equations can all be derived from balanced homodyne detection, in which the beam-splitter transmittance is one half, rather than close to unity (see Fig. 4.2). In that case, one photodetector is used for each output beam, and the signal photocurrent is the difference between the two currents. This configuration has the advantage of needing smaller local oscillator powers to achieve the same ratio of system amplitude to local oscillator amplitude, because all of the local oscillator beam is detected. Also, if the local oscillator has classical intensity fluctuations then these cancel out when the photocurrent difference is taken; with simple homodyne detection, these fluctuations are indistinguishable from (and may even swamp) the signal. Thus, in practice, balanced homodyne detection has many advantages over simple homodyne detection. But, in theory, the ideal limit is the same for both, which is why we have considered only simple homodyne detection. An analysis for balanced homodyne detection can be found in Ref. [WM93b].

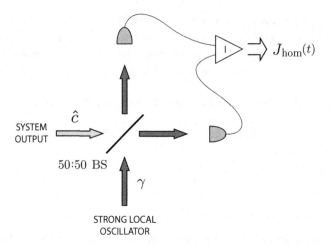

Fig. 4.2 A scheme for balanced homodyne detection. A 50 : 50 beam-splitter equally mixes the system output field and the local oscillator in both output ports. The local oscillator here can be weaker, but still dominates the intensity at the two photoreceivers. The difference between the two photocurrents is proportional to $J_{\text{hom}}(t)$.

4.4.3 Linear quantum trajectories

For pure initial states the homodyne SME (4.72) is equivalent to the SSE

$$d|\psi_J(t)\rangle = \left\{-i\hat{H}\,dt - \tfrac{1}{2}\big[\hat{c}^\dagger\hat{c} - 2\langle\hat{x}/2\rangle_J(t)\hat{c} + \langle\hat{x}/2\rangle_J^2(t)\big]dt \right.$$
$$\left. + [\hat{c} - \langle\hat{x}/2\rangle_J(t)]dW(t)\right\}|\psi_J(t)\rangle. \tag{4.76}$$

If one ignores the normalization of the state vector, then one gets the simpler equation

$$d|\bar{\psi}_J(t)\rangle = dt\big[-i\hat{H} - \tfrac{1}{2}\hat{c}^\dagger\hat{c} + J_{\text{hom}}(t)\hat{c}\big]|\bar{\psi}_J(t)\rangle. \tag{4.77}$$

This SSE (which is the form Carmichael originally derived [Car93]) very elegantly shows how the state is conditioned on the measured photocurrent. We have used a bar rather than a tilde to denote the unnormalized state because its norm alone does not tell us the probability for a measurement result, unlike in the case of the unnormalized states introduced in Section 1.4. Nevertheless, the linearity of this equation suggests that it should be possible to derive it simply using quantum measurement theory, rather than in the complicated way we derived it above. This is indeed the case, as we will show below. In fact, a derivation for quantum diffusion equations like Eq. (4.76) was first given by Belavkin [Bel88] along the same lines as below, but more rigorously.

Consider the infinitesimally entangled bath–system state introduced in Eq. (4.36):

$$|\Psi(t+dt)\rangle = [\hat{1} - i\hat{H}\,dt - \tfrac{1}{2}\hat{c}^\dagger\hat{c}\,dt + d\hat{B}^\dagger\hat{c}]|0\rangle|\psi(t)\rangle. \tag{4.78}$$

Now because $d\hat{B}|0\rangle = 0$, it is possible to replace $d\hat{B}^\dagger$ in Eq. (4.78) by $d\hat{B}^\dagger + d\hat{B}$, giving

$$|\Psi(t + dt)\rangle = \{\hat{1} - i\hat{H}\, dt - \tfrac{1}{2}\hat{c}^\dagger\hat{c}\, dt + \hat{c}[d\hat{B}^\dagger + d\hat{B}]\}|\psi(t)\rangle|0\rangle. \qquad (4.79)$$

This is useful since we wish to consider measuring the x quadrature of the bath after it has interacted with the system. This measurement is modelled by projecting the field onto the eigenstates $|J\rangle$, where

$$[\hat{b} + \hat{b}^\dagger]|J\rangle = J|J\rangle. \qquad (4.80)$$

The unnormalized system state conditioned on the measurement is thus

$$|\tilde{\psi}_J(t + dt)\rangle = \langle J|\Psi(t + dt)\rangle$$
$$= [\hat{1} - i\hat{H}\, dt - \tfrac{1}{2}\hat{c}^\dagger\hat{c}\, dt + \hat{c}J\, dt]|\psi(t)\rangle\sqrt{\wp_{\rm ost}(J)}, \qquad (4.81)$$

where $\wp_{\rm ost}(J)$ is the vacuum probability distribution for J,

$$\wp_{\rm ost}(J) = |\langle J|0\rangle|^2 = \sqrt{\frac{dt}{2\pi}}\exp\left[-\tfrac{1}{2}\, dt\, J^2\right]. \qquad (4.82)$$

Exercise 4.7 *Derive Eq. (4.82) from the wavefunction for the vacuum state (see Section A.4), remembering that $\sqrt{dt}\,\hat{b}$, not $\hat{b}$, acts like an annihilation operator.*

We will call $\wp_{\rm ost}(J)$ the *ostensible probability distribution* for J for reasons that will become obvious. The state-matrix norm of the state $|\tilde{\psi}_J(t + dt)\rangle$ gives the probability for the result J.

Taking the trace over the bath is the same as averaging over the measurement result, yielding

$$\rho(t + dt) = \int_{-\infty}^{\infty} dJ|\tilde{\psi}_J(t + dt)\rangle\langle\tilde{\psi}_J(t + dt)|. \qquad (4.83)$$

Exercise 4.8 *Show that this gives $d\rho(t) = dt\, \mathcal{L}\rho$, as in Eq. (4.11).*

Note, however, that, since J is a continuous variable, the choice of integration measure in this integral is not unique. That is, we can remove the factor $\sqrt{\wp_{\rm ost}(J)}$ in the state vector by changing the integration measure for the measurement result:

$$\rho(t + dt) = \int d\mu(J)|\bar{\psi}_J(t + dt)\rangle\langle\bar{\psi}_J(t + dt)|, \qquad (4.84)$$

where

$$|\bar{\psi}_J(t + dt)\rangle = [\hat{1} - i\hat{H}\, dt - \tfrac{1}{2}\hat{c}^\dagger\hat{c}\, dt + \hat{c}J\, dt]|\psi(t)\rangle \qquad (4.85)$$

and $d\mu(J) = \wp_{\rm ost}(J)dJ$. That is, we have a linear differential equation for a non-normalized state that nevertheless averages to the correct ρ, if $J(t)$ is chosen not according to its actual distribution, but according to its ostensible distribution, $\wp_{\rm ost}(J)$. This equation (4.85) is known as a *linear quantum trajectory*.

Now Eq. (4.85) is the same as Eq. (4.77), where here we see that the homodyne photocurrent J_{hom} as we have defined it in Eq. (4.75) is simply a measurement of $\hat{b}^\dagger + \hat{b}$. All that remains is to show that the above theory correctly predicts the statistics for J. The true probability distribution for J is

$$\wp(J)\mathrm{d}J = \langle\tilde{\psi}_J(t+\mathrm{d}t)|\tilde{\psi}_J(t+\mathrm{d}t)\rangle\mathrm{d}J = \langle\bar{\psi}_J(t+\mathrm{d}t)|\bar{\psi}_J(t+\mathrm{d}t)\rangle\mathrm{d}\mu(J). \qquad (4.86)$$

That is, the *actual* probability for the result J equals the *ostensible* probability multiplied by the state-matrix norm of $|\bar{\psi}_J(t+\mathrm{d}t)\rangle$. From Eq. (4.85), this evaluates to

$$\wp(J)\mathrm{d}J = \wp_{\text{ost}}(J)\{1 + J\langle\hat{x}\rangle\,\mathrm{d}t - \langle\hat{c}^\dagger\hat{c}\rangle[\mathrm{d}t - (J\,\mathrm{d}t)^2]\}\mathrm{d}J. \qquad (4.87)$$

We can clarify the orders of the terms here by defining a new variable $S = J\sqrt{\mathrm{d}t}$ which is of order unity. Then

$$\wp(S)\mathrm{d}S = (2\pi)^{-1/2}\exp(-S^2/2)[1 + S\langle\hat{x}\rangle\sqrt{\mathrm{d}t} + O(\mathrm{d}t)]\mathrm{d}S, \qquad (4.88)$$

or, to the same order in $\mathrm{d}t$,

$$\wp(S)\mathrm{d}S = (2\pi)^{-1/2}\exp[-(S - \langle\hat{x}\rangle\sqrt{\mathrm{d}t})^2/2]\mathrm{d}S. \qquad (4.89)$$

That is, the true distribution for the measured current J is

$$\wp(J)\mathrm{d}J = \wp_{\text{ost}}(J - \langle\hat{x}\rangle)\mathrm{d}J, \qquad (4.90)$$

which is precisely the statistics generated by Eq. (4.75).

Exercise 4.9 *Convince yourself of this.*

It is also useful to consider the state-matrix version of Eq. (4.85):

$$\mathrm{d}\bar{\rho}_J = \mathrm{d}t\,\mathcal{L}\bar{\rho}_J + J\,\mathrm{d}t(\hat{c}\bar{\rho}_J + \bar{\rho}_J\hat{c}^\dagger), \qquad (4.91)$$

where we have used $(J\,\mathrm{d}t)^2 = \mathrm{d}t$, which is true in the statistical sense under the ostensible distribution – see Section B.2 (the same holds for the actual distribution to leading order). From this form it is easy to see that the ensemble-average evolution (with J chosen according to its ostensible distribution) is the master equation $\dot{\rho} = \mathcal{L}\rho$, because the ostensible mean of J is zero. The actual distribution of J is again the ostensible distribution multiplied by the norm of $\bar{\rho}_J$.

It is not a peculiarity of homodyne detection that we can reformulate a *nonlinear* equation for a *normalized* state in which $\mathrm{d}W(t) = J(t)\mathrm{d}t - \langle J(t)\rangle\mathrm{d}t$ is white noise as a *linear* equation for a *non-normalized* state in which $J(t)$ has some other (ostensible) statistics. Rather, it is a completely general aspect of quantum or classical measurement theory. It is useful primarily in those cases in which the ostensible distribution for the measurement result $J(t)$ can be chosen so as to yield a particularly simple linear equation. That was the case above, where we chose $J(t)$ to have the ostensible statistics of white noise. Another convenient choice might be for the ostensible statistics of $J(t)$ to be Gaussian with a variance of $1/\mathrm{d}t$ (as in white noise) but a mean of μ. In that case Eq. (4.91) becomes

$$\mathrm{d}\bar{\rho}_J = \mathrm{d}t\,\mathcal{L}\bar{\rho}_J + (J - \mu)\mathrm{d}t(\hat{c}\bar{\rho}_J + \bar{\rho}_J\hat{c}^\dagger - \mu\bar{\rho}_J). \qquad (4.92)$$

Exercise 4.10 *Show that this does give the correct average ρ, and the correct actual distribution for $J(t)$.*

From this, we see that the *nonlinear* quantum trajectory (4.72) is just a special case in which μ is chosen to be equal to the *actual* mean of $J(t)$, since then

$$J(t) - \mu = J(t) - \langle J(t)\rangle_J = J(t) - \mathrm{Tr}[(\hat{c} + \hat{c}^\dagger)\rho_J(t)] = dW(t)/dt. \tag{4.93}$$

4.4.4 Output field correlation functions

As for direct detection, the dynamics of the system can be conveniently quantified by calculating the mean and autocorrelation function of the homodyne photocurrent. The mean is simply

$$E[J_{\mathrm{hom}}(t)] = \mathrm{Tr}[\rho(t)\hat{x}], \tag{4.94}$$

where $\hat{x} = \hat{c} + \hat{c}^\dagger$, as usual, and $\rho(t)$ is assumed given (it could be ρ_{ss}). The autocorrelation function is defined as

$$F_{\mathrm{hom}}^{(1)}(t, t + \tau) = E[J_{\mathrm{hom}}(t + \tau)J_{\mathrm{hom}}(t)]. \tag{4.95}$$

We use a superscript (1), rather than (2), because this function is related to Glauber's *first*-order coherence function [Gla63], as will be shown in Section 4.5.1.

From Eq. (4.75) and the fact that $\xi(t + \tau)$ is independent of the system at the past times t, this expression can be split into three terms,

$$F_{\mathrm{hom}}^{(1)}(t, t + \tau) = E[\langle\hat{x}\rangle_J(t + \tau)\xi(t)] + E[\xi(t + \tau)\xi(t)] + E[\langle\hat{x}\rangle_J(t + \tau)]\langle\hat{x}\rangle(t), \tag{4.96}$$

where the factorization of the third term is due to the fact that $\rho(t)$ is given. The second term here is equal to $\delta(\tau)$. The first term is non-zero because the conditioned state of the system at time $t + \tau$ depends on the noise in the photocurrent at time t. That noise enters by the conditioning equation (4.72), so

$$\rho_J(t + dt) = \rho(t) + O(dt) + dW(t)\mathcal{H}[\hat{c}]\rho(t). \tag{4.97}$$

The subsequent stochastic evolution of the system will be independent of the noise $\xi(t) = dW(t)/dt$ and hence may be averaged, giving

$$E[\langle\hat{x}\rangle_J(t + \tau)\xi(t)] = \mathrm{Tr}\Big[\hat{x}e^{\mathcal{L}\tau}E[\{1 + dW(t)\mathcal{H}[\hat{c}]\}\rho_J(t)dW(t)/dt]\Big]. \tag{4.98}$$

Using the Itô rules for $dW(t)$ and expanding the superoperator $\mathcal{H}$ yields

$$E[\langle\hat{x}\rangle_J(t + \tau)\xi(t)] = \mathrm{Tr}\Big[\hat{x}e^{\mathcal{L}\tau}\big(\hat{c}\rho(t) + \rho(t)\hat{c}^\dagger\big)\Big] - \mathrm{Tr}\Big[\hat{x}e^{\mathcal{L}\tau}\rho(t)\Big]\mathrm{Tr}[\hat{x}\rho(t)]. \tag{4.99}$$

The second term here cancels out the third term in Eq. (4.96), to give the final expression

$$F_{\mathrm{hom}}^{(1)}(t, t + \tau) = \mathrm{Tr}\Big[\hat{x}e^{\mathcal{L}\tau}\big(\hat{c}\rho(t) + \rho(t)\hat{c}^\dagger\big)\Big] + \delta(\tau). \tag{4.100}$$

Experimentally, it is more common to represent the information in the correlation function by its Fourier transform. At steady state, this is known as the *spectrum* of the homodyne photocurrent,

$$S(\omega) = \lim_{t \to \infty} \int_{-\infty}^{\infty} d\tau \, F_{\text{hom}}^{(1)}(t, t + \tau) e^{-i\omega\tau} \tag{4.101}$$

$$= 1 + \int_{-\infty}^{\infty} d\tau \, e^{-i\omega\tau} \, \text{Tr}\big[\hat{x} e^{\mathcal{L}\tau} (\hat{c}\rho_{\text{ss}} + \rho_{\text{ss}}\hat{c}^{\dagger})\big]. \tag{4.102}$$

The unit contribution is known as the local oscillator shot noise or vacuum noise because it is present even when there is no light from the system.

4.5 Heterodyne detection and beyond

4.5.1 Heterodyne detection

Homodyne detection has the advantage over direct detection that it can detect phase-dependent properties of the system. By choosing the phase of the local oscillator, any given quadrature of the system can be measured. However, only one quadrature can be measured at a time. It would be possible to obtain information about two orthogonal quadratures simultaneously by splitting the system output beam into two by use of a beam-splitter, and then homodyning each beam with the same local oscillator apart from a $\pi/2$ phase shift. An alternative way to achieve this double measurement is to detune the local oscillator from the system dipole frequency by an amount Δ much larger than any other system frequency. The photocurrent will then oscillate rapidly at frequency Δ, and the two Fourier components (cos and sin) of this oscillation will correspond to two orthogonal quadratures of the output field. This is known as heterodyne detection, which is the subject of this section

Begin with the homodyne SME (4.72), which assumed a constant local oscillator amplitude and phase. Detuning the local oscillator to a frequency Δ above that of the system will affect only the final (stochastic) term in Eq. (4.72). Its effect will be simply to replace $\hat{c}$ by $\hat{c} \exp(i\Delta t)$, giving

$$d\rho_J(t) = -i[\hat{H}, \rho_J(t)]dt + \mathcal{D}[\hat{c}]\rho_J(t)dt$$
$$+ dW(t)\big\{ e^{i\Delta t} [\hat{c}\rho_J(t) - \langle\hat{c}\rangle_J(t)\rho_J(t)] + \text{H.c.}\big\}. \tag{4.103}$$

Consider a time δt, small on a characteristic time-scale of the system, but large compared with Δ^{-1} so that there are many cycles due to the detuning. One might think that averaging the rotating exponentials over this time would eliminate the terms in which they appear. However, this is not the case because these terms are stochastic, and, since the noise is white by assumption, it will vary even faster than the rotation at frequency Δ. Define two new Gaussian random variables

$$\delta W_x(t) = \int_t^{t+\delta t} \sqrt{2} \cos(\Delta s) dW(s), \tag{4.104}$$

$$\delta W_y(t) = -\int_t^{t+\delta t} \sqrt{2} \sin(\Delta s) dW(s). \tag{4.105}$$

It is easy to show that, to zeroth order in Δ^{-1}, these obey

$$E[\delta W_q(t)\delta W_{q'}(t')] = \delta_{q,q'}(\delta t - |t - t'|)\Theta(\delta t - |t - t'|), \qquad (4.106)$$

where q and q' stand for x or y, and Θ is the Heaviside function, which is zero when its argument is negative and one when its argument is positive.

On the system's time-scale δt is infinitesimal. Thus the $\delta W_q(t)$ can be replaced by infinitesimal Wiener increments $dW_q(t)$ obeying

$$\langle \xi_q(t)\xi_{q'}(t')\rangle = \delta_{q,q'}\delta(t - t'), \qquad (4.107)$$

where $\xi_q(t) = dW_q(t)/dt$. Taking the average over many detuning cycles therefore transforms Eq. (4.103) into

$$d\rho_J(t) = -i[\hat{H}, \rho_J(t)]dt + \mathcal{D}[\hat{c}]\rho_J(t)dt$$
$$+ \sqrt{1/2}\big(dW_x(t)\mathcal{H}[\hat{c}] + dW_y(t)\mathcal{H}[-i\hat{c}]\big)\rho_J(t). \qquad (4.108)$$

Exercise 4.11 *Verify Eq. (4.106) and hence convince yourself of Eqs. (4.107) and (4.108).*
Hint: *Show that the right-hand side of Eq. (4.106) is zero when $|t - t'| > \delta t$, and that integrating over t or t' yields $(\delta t)^2$.*

This is equivalent to homodyne detection of the two quadratures simultaneously, each with efficiency $1/2$. (Non-unit efficiency will be discussed in Section 4.8.1). On defining a normalized complex Wiener process by

$$dZ = (dW_x + i\,dW_y)/\sqrt{2}, \qquad (4.109)$$

which satisfies $dZ^* dZ = dt$ but $dZ^2 = 0$, we can write Eq. (4.108) more elegantly as

$$d\rho_J(t) = -i[\hat{H}, \rho_J(t)]dt + \mathcal{D}[\hat{c}]\rho_J(t)dt + \mathcal{H}[dZ^*(t)\hat{c}]\rho_J(t). \qquad (4.110)$$

In order to record the measurements of the two quadratures, it is necessary to find the Fourier components of the photocurrent. These are defined by

$$J_x(t) = (\delta t)^{-1} \int_t^{t+\delta t} 2\cos(\Delta s)J_{\text{hom}}(s)ds, \qquad (4.111)$$

$$J_y(t) = -(\delta t)^{-1} \int_t^{t+\delta t} 2\sin(\Delta s)J_{\text{hom}}(s)ds. \qquad (4.112)$$

To zeroth order in Δ^{-1}, these are

$$J_x(t) = \langle\hat{x}\rangle_J(t) + \sqrt{2}\xi_x(t), \qquad (4.113)$$

$$J_y(t) = \langle\hat{y}\rangle_J(t) + \sqrt{2}\xi_y(t). \qquad (4.114)$$

(Recall that $\hat{x}$ and $\hat{y}$ are the quadratures defined in Eq. (4.65).) Again, these are proportional to the homodyne photocurrents that are expected for an efficiency of $1/2$ (see Section 4.8.1).

We can combine these photocurrents to make a complex heterodyne photocurrent

$$J_{\text{het}}(t) = (\delta t)^{-1} \int_t^{t+\delta t} \exp(-i\Delta s) J_{\text{hom}}(s) ds \tag{4.115}$$

$$= \tfrac{1}{2}[J_x(t) + iJ_y(t)] \tag{4.116}$$

$$= \langle \hat{c} \rangle_J(t) + dZ(t)/dt, \tag{4.117}$$

where dZ is as defined in Eq. (4.109). In terms of this current, one can derive an unnormalized SSE analogous to Eq. (4.77):

$$d|\bar{\psi}_J(t)\rangle = dt\big[-i\hat{H} - \tfrac{1}{2}\hat{c}^\dagger\hat{c} + J_{\text{het}}(t)^*\hat{c}\big]|\bar{\psi}_J(t)\rangle. \tag{4.118}$$

Equation (4.118), with the expression (4.117) in place of J_{het}, was introduced by Gisin and Percival in 1992 [GP92b] as 'quantum state diffusion'. However, they considered it to describe the objective evolution of a single open quantum system, rather than the conditional evolution under a particular detection scheme as we are interpreting it.

Using the same techniques as in Section 4.4.4, it is simple to show that the average complex heterodyne photocurrent is

$$E[J_{\text{het}}(t)] = \text{Tr}[\rho(t)\hat{c}], \tag{4.119}$$

and the autocorrelation function (note the use of the complex conjugate) is

$$E\big[J_{\text{het}}(t+\tau)^* J_{\text{het}}(t)\big] = \text{Tr}[\hat{c}^\dagger e^{\mathcal{L}\tau}\hat{c}\rho(t)] + \delta(\tau). \tag{4.120}$$

Ignoring the second (δ-function) term in this autocorrelation function, the remainder is simply Glauber's first-order coherence function $G^{(1)}(t, t+\tau)$ [Gla63]. In steady state this is related to the so-called *power spectrum* of the system by

$$P(\omega) = \frac{1}{2\pi} \int_{-\infty}^{\infty} d\tau\, e^{-i\omega\tau} \text{Tr}[\hat{c}^\dagger e^{\mathcal{L}\tau}\hat{c}\rho_{\text{ss}}]. \tag{4.121}$$

This can be interpreted as the photon flux in the system output per unit frequency (a dimensionless quantity).

Exercise 4.12 *Show that $\int_{-\infty}^{\infty} d\omega\, P(\omega) = \text{Tr}[\hat{c}^\dagger\hat{c}\rho_{\text{ss}}]$ and that this is consistent with the above interpretation.*

In practice it is this second interpretation that is usually used to measure $P(\omega)$. That is, the power spectrum is usually measured by using a spectrometer to determine the output intensity as a function of frequency, rather than by autocorrelating the heterodyne photocurrent.

4.5.2 Completely general dyne detection

Heterodyne detection, like homodyne detection, leads to an unravelling that is continuous in time. For convenience, we will call unravellings with this property *dyne* unravellings.

In this section, we give a complete classification of all such unravellings, for the general Lindblad master equation

$$\dot\rho = \mathcal{L}\rho \equiv -i[\hat{H}, \rho] + \hat{c}_k\rho\hat{c}_k^\dagger - \tfrac{1}{2}\left\{\hat{c}_k^\dagger\hat{c}_k, \rho\right\}. \tag{4.122}$$

Here, and in related sections, we are using the Einstein summation convention because this simplifies many of the formulae. That is, there is an implicit sum for repeated indices, which for k is from 1 to K. Using this convention, the most general SME was shown in Ref. [WD01] to be

$$d\rho_{\vec{J}} = dt\,\mathcal{L}\rho_{\vec{J}} + \left[(\hat{c}_k - \langle\hat{c}_k\rangle)\rho_{\vec{J}}\,dZ_k^* + \text{H.c.}\right]. \tag{4.123}$$

Here the dZ_k are complex Wiener increments satisfying

$$dZ_j(t)dZ_k^*(t) = dt\,\delta_{jk}, \tag{4.124}$$

$$dZ_j(t)dZ_k(t) = dt\,\Upsilon_{jk}. \tag{4.125}$$

The $\Upsilon_{jk} = \Upsilon_{kj}$ are arbitrary complex numbers subject only to the condition that the cross-correlations for Z are consistent with the self-correlations. This is the case iff (if and only if) the $2K \times 2K$ correlation matrix of the vector $\left(\text{Re}[d\vec{Z}], \text{Im}[d\vec{Z}]\right)$ is positive semi-definite.[3] That is,

$$\frac{dt}{2}\begin{pmatrix} I + \text{Re}[\Upsilon] & \text{Im}[\Upsilon] \\ \text{Im}[\Upsilon] & I - \text{Re}[\Upsilon] \end{pmatrix} \geq 0. \tag{4.126}$$

Here the real part of a matrix A is defined as $\text{Re}[A] = (A + A^*)/2$, and similarly $\text{Im}[A] = -i(A - A^*)/2$. Equation (4.126) is satisfied in turn iff the spectral norm of Υ is bounded from above by unity. That is,

$$\|\Upsilon\|^2 \equiv \lambda_{\max}(\Upsilon^\dagger\Upsilon) \leq 1, \tag{4.127}$$

where $\lambda_{\max}(A)$ denotes the maximum of the real parts of the eigenvalues of A. In the present context, the eigenvalues of A are real, of course, since $\Upsilon^\dagger\Upsilon$ is Hermitian.

Exercise 4.13 *Show that Eq. (4.126) is satisfied iff Eq. (4.127) is satisfied.*
Hint: *Consider the real symmetric matrix*

$$X = \begin{pmatrix} \text{Re}[\Upsilon] & \text{Im}[\Upsilon] \\ \text{Im}[\Upsilon] & -\text{Re}[\Upsilon] \end{pmatrix}. \tag{4.128}$$

Show that the eigenvalues of X are symmetrically placed around the origin. Thus we will have $I + X \geq 0$ iff $\|X\| \leq 1$. This in turn will be the case iff X^{2n} converges as $n \to \infty$. Show that

$$X^{2n} = \begin{pmatrix} \text{Re}[A^n] & \text{Im}[A^n] \\ -\text{Im}[A^n] & \text{Re}[A^n] \end{pmatrix}, \tag{4.129}$$

where $A = \Upsilon^\dagger\Upsilon$, and hence show the desired result.

---

[3] A positive semi-definite matrix A is an Hermitian matrix with no negative eigenvalues, indicated by $A \geq 0$.

Box 4.1 Gauge-invariance of SMEs

Consider the following transformation of a state vector:

$$|\psi(t)\rangle \rightarrow |\phi\rangle = \exp[i\chi(t)]|\psi(t)\rangle, \qquad (4.130)$$

where $\chi(t)$ is an arbitrary real function of time. This is known as a *gauge transformation*. It has no effect on any physical properties of the system. However, it can radically change the appearance of a stochastic Schrödinger equation, since $\chi(t)$ may be stochastic. Consider for example the simple SSE

$$d|\psi\rangle = \left[-i\hat{H} - \tfrac{1}{2}(\hat{x} - \langle\hat{x}\rangle)^2\right]dt|\psi\rangle + (\hat{x} - \langle\hat{x}\rangle)dZ^*|\psi\rangle, \qquad (4.131)$$

where $dZ\,dZ^* = dt$ and $dZ^2 = \upsilon\,dt$. Let the global phase χ obey the equation

$$d\chi = f\,dZ^* + f^*\,dZ, \qquad (4.132)$$

where $f(t)$ is an arbitrary smooth function of time that may even be a function of $|\psi\rangle$ itself. Then

$$|\phi\rangle + d|\phi\rangle = \left(1 + i\,d\chi - \tfrac{1}{2}\,d\chi\,d\chi\right)e^{i\chi(t)}(|\psi\rangle + d|\psi\rangle). \qquad (4.133)$$

The resultant equation for $|\phi\rangle$ is

$$\begin{aligned}
d|\phi\rangle = &\left[-i\hat{H} - \mathrm{Re}\left(f^2\upsilon^* + |f|^2\right)\right]dt|\phi\rangle \\
&- \tfrac{1}{2}(\hat{x} - \langle\hat{x}\rangle)\left(\hat{x} - \langle\hat{x}\rangle + if\upsilon^* + if^*\right)dt|\phi\rangle \\
&+ \left[(\hat{x} - \langle\hat{x}\rangle + if)\,dZ^* + if^*\,dZ\right]|\phi\rangle,
\end{aligned} \qquad (4.134)$$

which appears quite different from Eq. (4.131) (think of the case $f = -i\langle\hat{x}\rangle$, for example). By contrast, the SME is invariant under global phase changes:

$$d\rho = -i\,dt[\hat{H}, \rho] + dt\,\mathcal{D}[\hat{x}]\rho + \mathcal{H}[dZ^*\,\hat{x}]\rho. \qquad (4.135)$$

The above formulae apply for *efficient detection* (see Section 4.8 for a discussion of inefficient detection and Section 6.5.2 for the required generalization). Thus we could write the unravelling as a SSE rather than the SME (4.123). However, there are good reasons to prefer the SME form, even for efficient detection. First, it is more general in that it can describe the purification of an initially mixed state. Second, it is easier to see the relation between the quantum trajectories and the master equation which the system still obeys on average. Third, it is invariant under gauge transformations (see Box 4.1).

As expected from Section 4.4.3, the SME (4.123) can be derived directly from quantum measurement theory. We describe the measurement result in the infinitesimal time interval $[t, t + dt)$ by a vector of complex numbers $\vec{J}(t) = \{J_k(t)\}_{k=1}^K$. As functions of time, these are continuous but not differentiable, and, following the examples of homodyne and heterodyne detection, we will call them currents. Explicitly, they are given by

$$J_k \, dt = dt \left\langle \Upsilon_{kj} \hat{c}_j^\dagger + \hat{c}_k \right\rangle + dZ_k. \tag{4.136}$$

We can prove this relation between the noise in the quantum trajectory and the noise in the measurement record by using the measurement operators

$$\hat{M}_{\vec{J}} = \hat{1} - i\hat{H} \, dt - \tfrac{1}{2}\hat{c}_k^\dagger \hat{c}_k \, dt + J_k^* \hat{c}_k \, dt. \tag{4.137}$$

These obey the completeness relation

$$\int d\mu\left(\vec{J}\right) \hat{M}_{\vec{J}}^\dagger \hat{M}_{\vec{J}} = \hat{1} \tag{4.138}$$

if we choose $d\mu\left(\vec{J}\right)$ to be the measure yielding the ostensible moments

$$\int d\mu\left(\vec{J}\right)(J_k \, dt) = 0, \tag{4.139}$$

$$\int d\mu\left(\vec{J}\right)(J_j^* \, dt)(J_k \, dt) = \delta_{jk} \, dt, \tag{4.140}$$

$$\int d\mu\left(\vec{J}\right)(J_j \, dt)(J_k \, dt) = \Upsilon_{jk} \, dt. \tag{4.141}$$

Exercise 4.14 *Show this.*

These moments are the same as those of the Wiener increment $d\vec{Z}$ as defined above.

With this assignment of measurement operators $\hat{M}_{\vec{J}}$ and measure $d\mu\left(\vec{J}\right)$ we can easily show that the expected value of the result $\vec{J}$ is

$$E[J_k] = \int d\mu\left(\vec{J}\right) \text{Tr}\left[\hat{M}_{\vec{J}} \rho \hat{M}_{\vec{J}}^\dagger\right] J_k = \left\langle \Upsilon_{kj} \hat{c}_j^\dagger + \hat{c}_k \right\rangle. \tag{4.142}$$

This is consistent with the previous definition in Eq. (4.136). Furthermore, as in Section 4.4.3, we can show that the second moments of $\vec{J} \, dt$ are (to leading order in dt) independent of the system state and can be calculated using $d\mu$. In other words, they are identical to the statistics of $d\vec{Z}$ as defined above. This completes the proof that Eq. (4.136) gives the correct probability for the result $\vec{J}$.

The next step is to derive the conditioned state of the system after the measurement. This is given by

$$\rho + d\rho_{\vec{J}} = \frac{\hat{M}_{\vec{J}} \rho \hat{M}_{\vec{J}}^\dagger}{\text{Tr}\left[\hat{M}_{\vec{J}} \rho \hat{M}_{\vec{J}}^\dagger\right]}. \tag{4.143}$$

Expanding to order dt gives

$$d\rho_{\vec{J}} = -\tfrac{1}{2}\left\{\hat{c}_k^\dagger \hat{c}_k, \rho_{\vec{J}}\right\} dt + J_k^* \, dt \, \hat{c}_k \, \rho_{\vec{J}} \, \hat{c}_l^\dagger J_l \, dt + (J_k^* \, dt \, J_k - 1)\left\langle \hat{c}_j^\dagger \hat{c}_j \right\rangle \rho_{\vec{J}} \, dt$$

$$- i[\hat{H}, \rho_{\vec{J}}]dt + \left[J_k^*(\hat{c}_k - \langle \hat{c}_k \rangle)\rho_{\vec{J}} \, dt + \rho_{\vec{J}}\left(\hat{c}_k^\dagger - \left\langle \hat{c}_k^\dagger \right\rangle\right)J_k \, dt\right]$$

$$\times \left(1 - J_l^* \langle \hat{c}_l \rangle dt - J_l\left\langle \hat{c}_l^\dagger \right\rangle dt\right). \tag{4.144}$$

Exercise 4.15 *Verify Eqs. (4.142) and (4.144).*

Substituting into the above result (4.136) for $\vec{J}$ yields the required equation (4.123). From this it is again obvious that on average the system obeys the master equation (4.11).

Exercise 4.16 *Verify that Eq. (4.144) gives Eq. (4.123). Also find the linear version of Eq. (4.144) for the ostensible distribution used above.*

4.6 Illustration on the Bloch sphere

Before proceeding further with the theory, we will pause to illustrate the different sorts of quantum trajectories introduced above using a simple example. Perhaps the simplest nontrivial open quantum system is the driven and damped two-level atom as introduced in Section 3.3.1. Taking the bath temperature to be zero ($\bar{n} = 0$), the resonance fluorescence master equation (3.36) is

$$\dot{\rho} = -\mathrm{i}\left[\frac{\Omega}{2}\hat{\sigma}_x + \frac{\Delta}{2}\hat{\sigma}_z, \rho\right] + \gamma\mathcal{D}[\hat{\sigma}_-]\rho. \tag{4.145}$$

The solutions to this equation were the subject of Exercise 3.26. We now consider five different unravellings of this master equation.

4.6.1 Direct photodetection

The state of a classically driven two-level atom conditioned on direct photodetection of its resonance fluorescence was one of the inspirations for the development of the quantum trajectory theory in optics [CSVR89, DCM92, DZR92]. The treatment here is restricted to formulating the stochastic evolution in the manner of Section 4.2.2 and giving a closed-form expression for the stationary probability distribution of states on the Bloch sphere [WM93a].

Consider a two-level atom situated in an experimental apparatus such that the light it emits is all collected and enters a detector. (In principle this could be achieved by placing the atom at the focus of a large parabolic mirror, as shown in Fig. 4.3.) Then the direct detection theory of Section 4.3.1 can be applied, with $\hat{c} = \sqrt{\gamma}\hat{\sigma}_-$. The state vector of the atom conditioned on the photodetector count obeys the following SSE:

$$\mathrm{d}|\psi_I(t)\rangle = \left[\mathrm{d}N(t)\left(\frac{\hat{\sigma}_-}{\sqrt{\langle\hat{\sigma}_-^\dagger\hat{\sigma}_-\rangle_I(t)}} - 1\right)\right.$$

$$\left. - \mathrm{d}t\left(\frac{\gamma}{2}[\hat{\sigma}_-^\dagger\hat{\sigma}_- - \langle\hat{\sigma}_-^\dagger\hat{\sigma}_-\rangle_I(t)] + \mathrm{i}\hat{H}\right)\right]|\psi_I(t)\rangle, \tag{4.146}$$

where $\hat{H} = \frac{1}{2}(\Omega\hat{\sigma}_x + \Delta\hat{\sigma}_z)$ and the photocount increment $\mathrm{d}N(t)$ satisfies $\mathrm{E}[\mathrm{d}N(t)] = \gamma\langle\hat{\sigma}_-^\dagger\hat{\sigma}_-\rangle_I(t)\mathrm{d}t$.

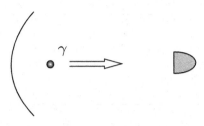

Fig. 4.3 A scheme for direct detection of an atom. The atom is placed at the focus of a parabolic mirror so that all the fluorescence emitted by the atom is detected by the photodetector. Figure 1 adapted with permission from J. Gambetta *et al.*, *Phys. Rev. A* **64**, 042105, (2001). Copyrighted by the American Physical Society.

With the conditioned subscript understood, one can write the conditioned state in terms of the Euler angles (ϕ, θ) parameterizing the surface of the Bloch sphere. Since we are assuming a pure state (see Box. 3.1),

$$|\psi(t)\rangle = c_g|g\rangle + c_e|e\rangle, \tag{4.147}$$

these angles are given by

$$\phi(t) = \arg[c_g(t)c_e^*(t)], \tag{4.148}$$

$$\theta(t) = 2\arctan[|c_g(t)/c_e(t)|]. \tag{4.149}$$

Exercise 4.17 *Show this, using the usual relation between* $(\phi, \theta, r := 1)$ *and* (x, y, z), *with* $|0\rangle = |g\rangle$ *and* $|1\rangle = |e\rangle$.

A typical stochastic trajectory is shown in Fig. 4.4. From an ensemble of these one could obtain the stationary distribution $\wp_{\text{ss}}(\phi, \theta)$ for the states on the Bloch sphere under direct detection.

In practice, it is easier to find the steady-state solution by returning to the SSE (4.146) and ignoring normalization terms. Consider the evolution of the system following a photodetection at time $t = 0$ so that $|\psi(0)\rangle = |g\rangle$. Assuming that no further photodetections take place, and omitting the normalization terms in Eq. (4.146), the state evolves via

$$\frac{d}{dt}|\tilde{\psi}_0(t)\rangle = -\left(\frac{\gamma}{2}\hat{\sigma}_-^\dagger\hat{\sigma}_- + i\hat{H}\right)|\tilde{\psi}_0(t)\rangle. \tag{4.150}$$

Here, the state vector has a state-matrix norm equal to the probability of it remaining in this no-jump state, as discussed in Section 4.2.3.

On writing the unnormalized conditioned state vector as

$$|\tilde{\psi}_0(t)\rangle = \tilde{c}_g(t)|g\rangle + \tilde{c}_e(t)|e\rangle, \tag{4.151}$$

the solution satisfying $\tilde{c}_j(0) = \delta_{j,g}$ is easily found to be

$$\tilde{c}_g(t) = \cos(\alpha t) + \frac{\gamma/2 + i\Delta}{2\alpha}\sin(\alpha t), \tag{4.152}$$

$$\tilde{c}_e(t) = -i\frac{\Omega}{2\alpha}\sin(\alpha t), \tag{4.153}$$

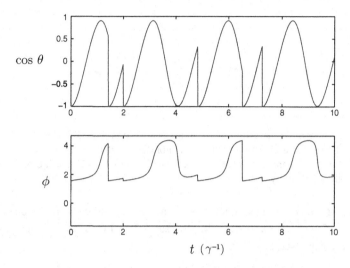

$$t \ (\gamma^{-1})$$

Fig. 4.4 A typical trajectory for the conditioned state of an atom under direct detection in terms of the Bloch angles ϕ and $\cos\theta$. The driving and detuning are $\Omega = 3$ and $\Delta = 0.5$, in units of the decay rate γ.

where

$$2\alpha = \left[(\Delta - i\gamma/2)^2 + \Omega^2 \right]^{1/2} \tag{4.154}$$

is a complex number that reduces to the detuned Rabi frequency as $\gamma \to 0$. One can still use the definitions (4.148) and (4.149) with c_j replaced by $\tilde{c}_j$, since they are insensitive to normalization. The significance of the normalization is that

$$S(t) = \langle \tilde{\psi}_0(t) | \tilde{\psi}_0(t) \rangle = |\tilde{c}_g(t)|^2 + |\tilde{c}_e(t)|^2 \tag{4.155}$$

is the probability of there being no detections from time 0 up to time t.

Exercise 4.18 *Show that for the case $\Delta = 0$, $\Omega > \gamma/2$, the no-jump quantum trajectory traverses the $x = 0$ great circle of the Bloch sphere.*

Whenever a photodetection does occur, the system returns to its state at $t = 0$. It then takes a finite time until the excited-state population becomes finite and the atom can re-emit. This gives the obvious interpretation for the photon *antibunching* predicted [CW76] and observed [KDM77] in the resonance fluorescence of a two-level atom. This is the phenomenon that, for some time following one detection, another detection is *less likely*. The term was defined to contrast with photon *bunching* (where following one detection another detection is *more likely*), which was the only correlation that had hitherto been observed. The re-excitation of the atom is always identical, so the stationary probability distribution on the Bloch sphere $\wp_{ss}(\phi, \theta)$ is confined to the curve parameterized by $(\phi(t), \theta(t))$, with each point weighted by the survival probability $S(t)$. (For $\Delta = 0$ this curve wraps around on itself, so that each point obtains multiple contributions to its weight.)

4.6.2 Adaptive detection with a local oscillator

Recall that in Section 3.8.2 we discussed the conditions under which a probability distribution of pure states $\{(\hat{\pi}_k, \wp_k)\}_k$ could be *physically realizable* (PR). An ensemble is PR if there is a continuous measurement scheme such that an experimenter implementing it would be able to know, at some time far in the future from the initial preparation, that the system is in one of the states $\hat{\pi}_k$, and the probability for each state would be $\wp_k$.

From the preceding section it is evident that the stationary distribution $\wp_{ss}(\phi, \theta)$ on the surface of the Bloch sphere introduced in the preceding subsection is just such a PR ensemble. The continuous measurement scheme which realizes the ensemble is, of course, direct detection. As discussed above, in the case $\Delta = 0$, $\wp_{ss}(\phi, \theta)$ is confined to the $x = 0$ great circle of the Bloch sphere.

In Section 3.8.3 we presented an example of a PR ensemble for the resonance fluorescence master equation (4.145) (for $\Delta = 0$) that was quite different from that in the preceding section. Specifically, the ensemble contained just two pure states $\hat{\pi}_{\pm}$, equally weighted, defined by the Bloch vectors

$$\begin{pmatrix} u \\ v \\ w \end{pmatrix}_{\pm} = \begin{pmatrix} \pm\sqrt{1 - y_{ss}^2 - z_{ss}^2} \\ y_{ss} \\ z_{ss} \end{pmatrix}. \tag{4.156}$$

It turns out that this can be generalized for $\Delta \neq 0$, but it is a lot more complicated, so in this section we retain $\Delta = 0$. For large Ω, these points on the Bloch sphere approach the antipodal pair at $x = \pm 1$.

Exercise 4.19 *Show this, and show that in the same limit the direct detection ensemble becomes equally spread over the $x = 0$ great circle.*

Thus, these two PR ensembles are as different as they possibly can be.

In Section 3.8.3 we did not identify the measurement scheme that realizes this ensemble. Since the elements of the ensemble are discrete, the unravelling must involve jumps. Since it is not the direct detection unravelling of the preceding section, it must involve a local oscillator, as introduced in Section 4.4.1. Since here we are using γ for the atomic decay rate, in this section we use $\sqrt{\gamma}\mu$ for the local oscillator amplitude. The no-jump and jump measurement operators are then

$$\hat{M}_0(dt) = \hat{1} - \left(i\frac{\Omega}{2}\hat{\sigma}_x + \frac{\gamma}{2}\hat{\sigma}_-^\dagger\hat{\sigma}_- + \mu^*\gamma\hat{\sigma}_- + \frac{\gamma|\mu|^2}{2}\right)dt, \tag{4.157}$$

$$\hat{M}_1(dt) = \sqrt{\gamma\,dt}\,(\hat{\sigma}_- + \mu). \tag{4.158}$$

Direct detection is recovered by setting $\mu = 0$.

If the atom radiates into a beam as considered previously, the above measurement can be achieved by mixing it with a resonant local oscillator at a beam-splitter, as shown in Fig. 4.5. The transmittance of the beam-splitter must be close to unity. The phase of μ is of course defined relative to the field driving the atom, parameterized by Ω.

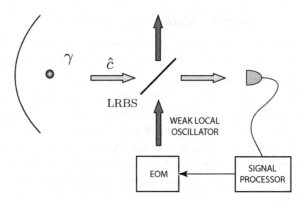

Fig. 4.5 A scheme for adaptive detection. The fluorescence emitted by the atom is coherently mixed with a weak local oscillator (LO) via a low-reflectivity beam-splitter (LRBS). The electro-optic modulator (EOM) reverses the amplitude of the LO every time the photodetector fires. Figure 5 adapted with permission from J. Gambetta *et al.*, *Phys. Rev. A* **64**, 042105, (2001). Copyrighted by the American Physical Society.

Since our aim is for the atom to remain in one of two fixed pure states, except when it jumps, we must examine the fixed points (i.e. eigenstates) of the operator $\hat{M}_0(\mathrm{d}t)$. It turns out that it has two fixed states, such that, if $\mathrm{Re}[\mu] \neq 0$, one is stable and one unstable. For $\mathrm{Re}[\mu] > 0$, the stable fixed state is

$$|\tilde{\psi}_{\mathrm{s}}^{\mu}\rangle = \left(\sqrt{\Omega^2 - 2\mathrm{i}\Omega\gamma\mu^* - \frac{\gamma^2}{4}} + \frac{\mathrm{i}\gamma}{2}\right)|g\rangle + \Omega|e\rangle. \tag{4.159}$$

Here the tilde denotes an unnormalized state. The corresponding eigenvalue is

$$\lambda_{\mathrm{s}}^{\mu} = -\gamma\frac{1 + 2|\mu|^2}{4} - \frac{\mathrm{i}}{2}\sqrt{\Omega^2 - 2\mathrm{i}\Omega\gamma\mu^* - \frac{\gamma^2}{4}}. \tag{4.160}$$

The unstable state and eigenvalue are found by replacing the square root by its negative. It is unstable in the sense that its eigenvalue is *more* negative, indicating that its norm will decay faster than that of the stable eigenstate. Thus, an initial superposition of these two (linearly independent) states will, when normalized, evolve towards the stable eigenstate.

Let us say $\mu = \mu_+$, with $\mathrm{Re}[\mu_+] > 0$, and assume the system is in the appropriate stable state $|\psi_{\mathrm{s}}^+\rangle$. When a jump occurs the new state of the system is proportional to

$$\hat{M}_1^+|\psi_{\mathrm{s}}^+\rangle \propto (\hat{\sigma}_- + \mu_+)|\psi_{\mathrm{s}}^+\rangle. \tag{4.161}$$

The new state will obviously be different from $|\psi_{\mathrm{s}}^+\rangle$ and so will not remain fixed. This is in contrast to what we are seeking, namely a system that will remain fixed between jumps. However, let us imagine that, immediately following the detection, the value of the local oscillator amplitude μ is changed to some new value, μ_-. This is an example of an *adaptive* measurement scheme as discussed in Section 2.5, in that the parameters defining the measurement depend upon the past measurement record. We want this new μ_- to be chosen such that the state $(\hat{\sigma}_- + \mu_+)|\psi_s^+\rangle$ is a stable fixed point of the new $\hat{M}_0^-(dt)$. The conditions for this to be so will be examined below. If they are satisfied then the state will remain fixed until another jump occurs. This time the new state will be proportional to

$$(\hat{\sigma}_- + \mu_-)(\hat{\sigma}_- + \mu_+)|\psi_s^+\rangle = [\mu_-\mu_+ + (\mu_- + \mu_+)\hat{\sigma}_-]|\psi_s^+\rangle. \qquad (4.162)$$

If we want jumps between just two states then we require this to be proportional to $|\psi_s^+\rangle$. Clearly this will be so if and only if

$$\mu_- = -\mu_+. \qquad (4.163)$$

Writing $\mu_+ = \mu$, we now return to the condition that $(\hat{\sigma}_- + \mu_+)|\psi_s^+\rangle$ be the stable fixed state of $\hat{M}_0^-(dt)$. From Eq. (4.159), and using Eq. (4.163), this gives the relation

$$\sqrt{\Omega^2 + 2i\Omega\gamma\mu^* - \frac{\gamma^2}{4}} = \sqrt{\Omega^2 - 2i\Omega\gamma\mu^* - \frac{\gamma^2}{4}} - \frac{\Omega}{\mu}. \qquad (4.164)$$

This has just two solutions,

$$\mu_\pm = \pm\frac{1}{2}, \qquad (4.165)$$

which, remarkably, are independent of the ratio γ/Ω. The stable and unstable fixed states for this choice are

$$|\psi_s^\pm\rangle = \frac{\pm\Omega - i\gamma}{\sqrt{2\Omega^2 + \gamma^2}}|g\rangle - \frac{\Omega}{\sqrt{2\Omega^2 + \gamma^2}}|e\rangle, \qquad (4.166)$$

$$|\psi_u^\pm\rangle = \frac{1}{\sqrt{2}}|g\rangle \pm \frac{1}{\sqrt{2}}|e\rangle, \qquad (4.167)$$

and the corresponding eigenvalues are

$$\lambda_s^\pm = -\frac{\gamma}{8} \pm \frac{i\Omega}{2}, \qquad (4.168)$$

$$\lambda_u^\pm = -\frac{5\gamma}{8} \mp \frac{i\Omega}{2}. \qquad (4.169)$$

Exercise 4.20 *Show that the stable eigenstates correspond to the two states $\hat{\pi}_\pm$ defined by the Bloch vectors in Eq. (4.156).*

We have thus constructed the measurement scheme that realizes the two-state PR ensemble for the two-level atom. Ignoring problems of collection and detector efficiency, it may seem that this adaptive measurement scheme would not be much harder to implement experimentally than homodyne detection; it requires simply an amplitude inversion of the local oscillator after each detection. In fact, this is very challenging, since the feedback delay must be very small compared with the characteristic time-scale of the system. For a typical atom the decay time ($\gamma^{-1} \sim 10^{-8}$ s) is shorter than currently feasible times for electronic feedback. Any experimental realization would have to use an atom with a very long lifetime, or some other equivalent effective two-level system with radiative transitions in the optical frequency range.

Another difference from homodyne detection is that the adaptive detection has a very small local oscillator intensity at the detector: it corresponds to half the photon flux of the atom's fluorescence if the atom were saturated. In either stable fixed state, the actual photon flux entering the detector in this scheme is

$$\langle \psi_s^{\pm} | (\mu_{\pm} + \hat{\sigma}_-^{\dagger})(\mu_{\pm} + \hat{\sigma}_-) | \psi_s^{\pm} \rangle = \frac{\gamma}{4}, \qquad (4.170)$$

which is again independent of Ω/γ. This rate is also, of course, the rate for the system to jump to the other stable fixed state, so that the two are equally occupied in steady state.

There are many similarities between the stochastic evolution under this adaptive unravelling and the conditioned evolution of the atom under spectral detection, as investigated in Ref. [WT99]. Spectral detection uses optical filters to resolve the different frequencies of the photons emitted by the atom. As a consequence it is not possible to formulate a trajectory for the quantum state of the atom alone. For details, see Ref. [WT99]. In the case of a strongly driven atom ($\Omega \gg \gamma$), photons are emitted with frequencies approximately equal to the atomic resonance frequency, ω_a, and to the 'sideband' frequencies $\omega_a \pm \Omega$. (This is the characteristic Mollow power spectrum of resonance fluorescence [Mol69].) In the interaction frame these frequencies are 0 and $\pm \Omega$, and can be seen in the imaginary parts of (respectively) the eigenvalues $-\gamma/2$ and $\lambda_{\pm}$ appearing in the solution to the resonance fluorescence master equation in Section 3.8.3. In this high-driving limit, the conditioned atomic state can be made approximately pure, and it jumps between states close to the $\hat{\sigma}_x$ eigenstates, just as in the adaptive detection discussed above. In this case the total detection rate is approximately $\gamma/2$, as expected for a strongly driven (saturated) atom. Of these detections, half are in the peak of the power spectrum near resonance (which do not give rise to state-changing jumps), while half are detections in the sidebands (which do). Thus the rate of state-changing jumps is approximately $\gamma/4$, just as in the case of adaptive detection.

4.6.3 Homodyne detection

We now turn to homodyne detection of the light emitted from the atom. Say the local oscillator has phase Φ. Then, from Eq. (4.77), the system obeys the following SSE:

$$d|\bar{\psi}_J(t)\rangle = \left\{ -\left(\frac{\gamma}{2} \hat{\sigma}_-^{\dagger} \hat{\sigma}_- + i\hat{H} \right) dt \right.$$
$$\left. + \left[\gamma \, dt \langle e^{-i\Phi} \hat{\sigma}_- + e^{i\Phi} \hat{\sigma}_-^{\dagger} \rangle_J(t) + \sqrt{\gamma} \, dW(t) \right] e^{-i\Phi} \hat{\sigma}_- \right\} |\bar{\psi}_J(t)\rangle, \quad (4.171)$$

where $dW(t)$ is a real infinitesimal Wiener increment.

In Fig. 4.6, we plot a typical trajectory on the Bloch sphere for two values of Φ, namely 0 and $\pi/2$, corresponding to measuring the quadrature of the spontaneously emitted light in phase and in quadrature with the driving field, respectively. In both plots $\Omega = 3\gamma$ and

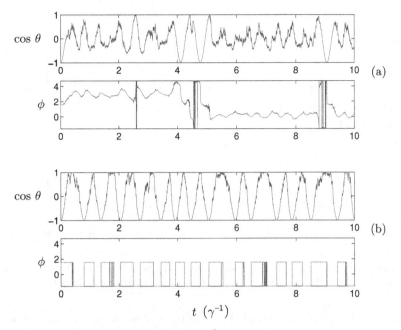

Fig. 4.6 A segment of a trajectory of duration $10\gamma^{-1}$ of an atomic state on the Bloch sphere under homodyne detection. The phase Φ of the local oscillator relative to the driving field is 0 in (a) and $\pi/2$ in (b). The driving and detuning are $\Omega = 3$ and $\Delta = 0$.

$\Delta = 0$, so the true distributions on the Bloch sphere are symmetric under reflection in the y–z plane. The effect of the choice of measurement is dramatic and readily understandable. The homodyne photocurrent from Eq. (4.75) is

$$J_{\text{hom}}(t) = \gamma(\langle\hat{\sigma}_x\rangle\cos\Phi - \langle\hat{\sigma}_y\rangle\sin\Phi) + \sqrt{\gamma}\xi(t). \tag{4.172}$$

When the local oscillator is in phase ($\Phi = 0$), the deterministic part of the photocurrent is proportional to $x(t)$. Under this measurement, the atom tends towards states with well-defined $\hat{\sigma}_x$. The eigenstates of $\hat{\sigma}_x$ are stationary states of the driving Hamiltonian, so this leads to trajectories that stay near these eigenstates for a relatively long time. This is seen in Fig. 4.6(a), where ϕ tends to stay around 0 or π. In contrast, measuring the $\Phi = \pi/2$ quadrature tries to force the system into an eigenstate of $\hat{\sigma}_y$. However, such an eigenstate will be rapidly spun around the Bloch sphere by the driving Hamiltonian. This effect is clearly seen in Fig. 4.6(b), where the trajectory is confined to the $\phi = \pm\pi/2$ great circle (like that for direct detection).

The above explanation for the nature of the quantum trajectories is also useful for understanding the noise spectra of the quadrature photocurrents in Eq. (4.172). The power spectrum (see Section 4.5.1) of resonance fluorescence of a strongly driven two-level atom has three peaks, as discussed in the preceding section. The central one is peaked at the atomic frequency, and the two sidebands (each of half the area) are displaced by the Rabi

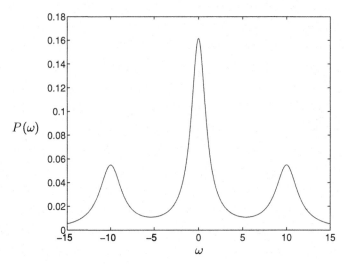

$P(\omega)$

Fig. 4.7 A power spectrum for resonance fluorescence with $\gamma = 1$, $\Delta = 0$ and $\Omega = 10$. Note that this Rabi frequency is larger than that in Fig. 4.6, in order to show clearly the Mollow triplet.

frequency [Mol69], as shown in Fig. 4.7. It turns out that the spectrum of the in-phase homodyne photocurrent (see Section 4.4.4) gives the central peak, while the quadrature photocurrent gives the two sidebands [CWZ84]. This is readily explained qualitatively from the evolution of the atomic state under homodyne measurements. When $\hat{\sigma}_x$ is being measured, it varies slowly, remaining near one eigenvalue on a time-scale like γ^{-1}. This gives rise to an exponentially decaying autocorrelation function for the photocurrent (4.172), or a Lorentzian with width scaling as γ in the frequency domain. When $\hat{\sigma}_y$ is measured, it undergoes rapid sinusoidal variation at frequency Ω, with noise added at a rate γ. This explains the side peaks.

4.6.4 Heterodyne detection

If the atomic fluorescence enters a perfect heterodyne detection device, then, from Eq. (4.118), the system evolves via the SSE

$$|\bar{\psi}_J(t + dt)\rangle = \left\{ \hat{1} - \left(\frac{\gamma}{2}\hat{\sigma}_-^\dagger \hat{\sigma}_- + i\hat{H} \right) dt \right.$$

$$\left. + \left[\gamma \, dt \langle \hat{\sigma}_-^\dagger \rangle_J(t) + \sqrt{\gamma} \, dZ(t) \right] \hat{\sigma}_- \right\} |\bar{\psi}_J(t)\rangle, \qquad (4.173)$$

where here $dZ(t)$ is the complex infinitesimal Wiener increment introduced in Eq. (4.109).

As for the previous case, the steady-state probability distribution can be represented by an ensemble of points on the Bloch sphere drawn (at many different times) from a single long-time solution of the SSE. In this case, the stationary probability distribution is spread fairly well over the entire Bloch sphere. This can be understood as the result of the two competing measurements ($\hat{\sigma}_x$ and $\hat{\sigma}_y$) combined with the driving Hamiltonian causing rotation around the x axis. The complex photocurrent as defined in Eq. (4.117) is

$$J_{\text{het}}(t) = \gamma \langle \hat{\sigma}_- \rangle + \sqrt{\gamma} \zeta(t), \tag{4.174}$$

where $\zeta(t) = dZ(t)/dt$. The spectrum of this photocurrent (the Fourier transform of the two-time correlation function (4.206)) gives the complete Mollow triplet, since y is rotated at frequency Ω with noise, while the dynamics of x is only noise.

4.7 Monitoring in the Heisenberg picture

4.7.1 Inputs and outputs

In this chapter so far we have described the monitoring of a quantum system in the Schrödinger picture, treating the measurement results as a classical record. However, we showed in Section 1.3.2 of Chapter 1 that it is possible to treat measurement results as operators, enabling a description in the Heisenberg picture. Moreover, we showed in Section 3.11 of Chapter 3 that the dynamics of open quantum systems can also be treated in this picture. Thus it is not surprising that we can describe monitoring in the Heisenberg picture. The crucial change from the earlier parts of this chapter is that now the measurement results are represented by bath operators.

Recall from Eq. (3.172) that, in the Heisenberg picture, the increment in an arbitrary system operator $\hat{s}$ is

$$d\hat{s} = dt \left(\hat{c}^\dagger \hat{s} \hat{c} - \tfrac{1}{2} \{ \hat{c}^\dagger \hat{c}, \hat{s} \} + i[\hat{H}, \hat{s}] \right) - [d\hat{B}_{\text{in}}^\dagger(t)\hat{c} - \hat{c}^\dagger \, d\hat{B}_{\text{in}}(t), \hat{s}]. \tag{4.175}$$

Now there was nothing in the derivation of this equation that required $\hat{s}$ to be specifically a system operator rather than a bath operator. Thus we could choose instead $\hat{s} = \hat{b}_{\text{in}}(t)$, the bath field operator interacting with the system at time t (see Section 3.11). Then, using Eq. (4.175) and Eq. (3.160), we find

$$d\hat{b}_{\text{in}}(t) = O(dt) + \hat{c}. \tag{4.176}$$

That is, the singularity of the bath commutation relations leads to a finite change in the bath field operator $\hat{b}_{\text{in}}$ in an infinitesimal time.

Because of this finite change, it is necessary to distinguish between the bath operator $\hat{b}_{\text{in}}$ interacting with the system at the beginning of the time interval $[t, t + dt)$ and that at the end, which we will denote $\hat{b}_{\text{out}}(t)$. Ignoring infinitesimal terms, Eq. (4.176) implies that

$$\hat{b}_{\text{out}} = \hat{b}_{\text{in}} + \hat{c}. \tag{4.177}$$

This is sometimes called the input–output relation [GC85]. To those unfamiliar with the Heisenberg picture, it may appear odd that a system operator appears in the expression for a bath operator, but this is just what one would expect with classical equations for dynamical variables. As explained in Section 1.3.2, Heisenberg equations such as this are the necessary counterpart to entanglement in the Schrödinger picture. If the system is an

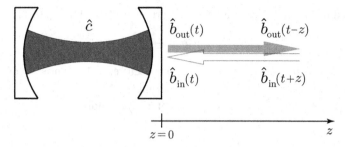

Fig. 4.8 A schematic diagram showing the relation between the input and output fields for a one-sided cavity with one relevant mode described by the annihilation operator $\hat{c}$. The distance from the cavity, z, appears as a time difference in the free field operators because we are using units such that the speed of light is unity. The commutation relations say that (for $z > 0$) $\hat{b}_{in}(t + z)$ and $\hat{b}_{out}(t - z)$ commute with any system operator at time t. This is readily understood from the figure, since these field operators apply to parts of the field that also exist (at a point in space removed from the system) at time t.

optical cavity, then this operator represents the field immediately after it has bounced off the cavity mirror. This is shown in Fig. 4.8.

Just as $\hat{b}_{in}(t)$ commutes with an arbitrary system operator $\hat{s}(t')$ at an earlier time $t' < t$, it can simply be shown that $\hat{b}_{out}(t)$ commutes with the system operators at a later time $t' > t$. As a consequence of this, the output field obeys the same commutation relations as the input field,

$$[\hat{b}_{out}(t), \hat{b}_{out}^{\dagger}(t')] = \delta(t - t'), \tag{4.178}$$

as required because it is a free field also.

Unlike $\hat{b}_{in}(t)$ for earlier times, $\hat{b}_{out}(t)$ for later times does depend on the state of the system. In fact, a measurement of the output field will yield information about the system. Monitoring of the system (as discussed above in the Schrödinger picture) corresponds to measuring, in every interval $[t, t + dt)$, a bath observable that is a function of $\hat{b}_{out}(t)$. Because $\hat{b}_{out}(t)$ commutes with $\hat{s}(t')$ for $t' > t$, the operator for the measurement record (which relates to the system in the past) will commute with all system operators in the present. That is, it is a c-number record insofar as the system is concerned. This fact is crucial to the Heisenberg-picture approach to feedback to be developed in Chapter 5. We will now examine how various monitoring schemes correspond to different operators for the measurement record.

4.7.2 Direct photodetection

From Section 4.3.1, it is apparent that the observable for photon flux is

$$\hat{I}_{out}(t) = \hat{b}_{out}^{\dagger}(t)\hat{b}_{out}(t). \tag{4.179}$$

It is easy to show that this is statistically identical to the classical photocurrent $I(t)$ used in Section 4.3.2. To see this, consider the Heisenberg operator $\mathrm{d}\hat{N}_{\text{out}}(t) = \hat{b}^\dagger_{\text{out}}(t)\hat{b}_{\text{out}}(t)\mathrm{d}t$ for the output field. This is found from Eq. (4.177) to be

$$\mathrm{d}\hat{N}_{\text{out}}(t) = [\hat{c}^\dagger(t) + \hat{v}^\dagger(t)][\hat{c}(t) + \hat{v}(t)]\mathrm{d}t. \tag{4.180}$$

Here we are using $\hat{v}(t)$ for $\hat{b}_{\text{in}}(t)$ to emphasize that the input field is in the vacuum state and so satisfies $\langle \hat{v}(t)\hat{v}^\dagger(t')\rangle = \delta(t - t')$, with all other second-order moments vanishing. It is then easy to show that

$$\langle \mathrm{d}\hat{N}_{\text{out}}(t)\rangle = \text{Tr}[\hat{c}^\dagger(t)\hat{c}(t)\rho], \tag{4.181}$$

$$\mathrm{d}\hat{N}_{\text{out}}(t)^2 = \mathrm{d}\hat{N}_{\text{out}}(t). \tag{4.182}$$

These moments are identical to those of the photocount increment, Eqs. (4.41) and (4.13), with a change from Schrödinger to Heisenberg picture.

The identity between the statistics of the output photon-flux operator $\hat{I}_{\text{out}}(t)$ and the photocurrent $I(t)$ does not stop at the single-time moments (4.181) and (4.182). As discussed in Section 4.3.2, the most commonly calculated higher-order moment is the autocorrelation function. In the Heisenberg picture this is defined as

$$F^{(2)}(t, t + \tau) = \langle \hat{I}_{\text{out}}(t + \tau)\hat{I}_{\text{out}}(t)\rangle. \tag{4.183}$$

From the expression (4.180), this is found to be

$$F^{(2)}(t, t + \tau) = \langle \hat{c}^\dagger(t)\hat{c}^\dagger(t + \tau)\hat{c}(t + \tau)\hat{c}(t)\rangle + \delta(\tau)\langle \hat{c}^\dagger(t)\hat{c}(t)\rangle. \tag{4.184}$$

Exercise 4.21 *Show this using the commutation relations (4.178) to put the field operators in normal order. (Doing this eliminates the input field operators, because they act directly on the vacuum, giving a null result.)*

The autocorrelation function can be rewritten in the Schrödinger picture as follows. First, recall that the average of a system operator at time t is

$$\langle \hat{s}(t)\rangle = \text{Tr}_{\text{S}}[\rho(t)\hat{s}] = \text{Tr}_{\text{S}}[\text{Tr}_{\text{B}}[W(t)]\hat{s}] = \text{Tr}[W(t)\hat{s}], \tag{4.185}$$

where $\rho(t)$ is the system (S) state matrix and $W(t)$ is the state matrix for the system plus bath (B). Now $W(t) = \hat{U}(t)W(0)\hat{U}^\dagger(t)$, where in the Markovian approximation the unitary evolution is such that

$$\text{Tr}_{\text{B}}[\hat{U}(t)\tilde{W}(0)\hat{U}^\dagger(t)] = \exp(\mathcal{L}t)\tilde{\rho}(0), \tag{4.186}$$

where $\tilde{W}(0) = \tilde{\rho}(0) \otimes \rho_{\text{B}}(0)$ and $\tilde{\rho}(0)$ is arbitrary. This result can be generalized to a two-time correlation function for system operators $\hat{s}$ and $\hat{z}$:

$$\langle \hat{s}(t)\hat{z}(0)\rangle = \text{Tr}[W(0)\hat{s}(t)\hat{z}(0)] \tag{4.187}$$

$$= \text{Tr}[W(0)\hat{U}^\dagger(t)\hat{s}\hat{U}(t)\hat{z}(0)] \tag{4.188}$$

$$= \text{Tr}[\hat{U}(t)\hat{z}(0)W(0)\hat{U}^\dagger(t)\hat{s}] \tag{4.189}$$

$$= \text{Tr}_{\text{S}}\big[\text{Tr}_{\text{B}}[\hat{U}(t)\hat{z}(0)W(0)\hat{U}^\dagger(t)]\hat{s}\big]. \tag{4.190}$$

Using the Markov assumption with $\tilde{\rho}(0) = \hat{z}(0)\rho(0)$,

$$\langle \hat{s}(t)\hat{z}(0)\rangle = \text{Tr}_S[\hat{s}\exp(\mathcal{L}t)\hat{z}\rho(0)] = \text{Tr}_S[\hat{s}\tilde{\rho}(t)], \tag{4.191}$$

where $\tilde{\rho}(t)$ is the solution of the master equation with the given initial conditions. This is sometimes known as the *quantum regression theorem*.

Exercise 4.22 *Generalize the above result to two-time correlation functions of the form* $\langle \hat{x}(0)\hat{s}(t)\hat{z}(0)\rangle$.

Applying the result of this exercise to the above autocorrelation function (4.184) gives

$$F^{(2)}(t, t+\tau) = \text{Tr}[\hat{c}^\dagger \hat{c} e^{\mathcal{L}\tau}\hat{c}\rho(t)\hat{c}^\dagger] + \text{Tr}[\hat{c}^\dagger \hat{c}\rho(t)]\delta(\tau). \tag{4.192}$$

This is identical to Eq. (4.50) obtained using the conditional evolution in Section 4.3.2. In fact, any statistical comparison between the two will agree because $\hat{I}_{\text{out}}(t)$ and $I(t)$ are merely different representations of the same physical quantity. Conceptually, the two representations are quite different. In the Heisenberg-operator derivation, the shot-noise term (the delta function) in the autocorrelation function arises from the commutation relations of the electromagnetic field. By contrast, the quantum trajectory model produces shot noise because photodetections are discrete events, which is a far more intuitive explanation. As we will see, some results may be more obvious using one method, others more obvious with the other, so it is good to be familiar with both.

4.7.3 Homodyne detection

Adding a local oscillator to the output field transforms it to

$$\hat{b}'_{\text{out}}(t) = \hat{b}_{\text{out}}(t) + \gamma = \hat{c}(t) + \hat{v}(t) + \gamma. \tag{4.193}$$

On dropping the time arguments, the photon-flux operator for this field is

$$\hat{I}'_{\text{out}} = \gamma^2 + \gamma(\hat{c} + \hat{c}^\dagger + \hat{v} + \hat{v}^\dagger) + (\hat{c}^\dagger + \hat{v}^\dagger)(\hat{c} + \hat{v}). \tag{4.194}$$

In the limit that $\gamma \to \infty$, the last term can be ignored for the homodyne photocurrent operator,

$$\hat{J}^{\text{hom}}_{\text{out}}(t) \equiv \lim_{\gamma\to\infty} \frac{\hat{I}'_{\text{out}}(t) - \gamma^2}{\gamma} = \hat{x}(t) + \hat{\xi}(t). \tag{4.195}$$

Here, $\hat{x}$ is the quadrature operator defined in Eq. (4.65) and $\hat{\xi}(t)$ is the vacuum input operator

$$\hat{\xi}(t) = \hat{v}(t) + \hat{v}^\dagger(t), \tag{4.196}$$

which has statistics identical to those of the normalized Gaussian white noise for which the same symbol is used.

Exercise 4.23 *Convince yourself of this.*

The operator nature of $\hat{\xi}(t)$ is evident only from its commutation relations with its conjugate variable $\hat{\upsilon}(t) = -i\hat{\upsilon}(t) + i\hat{\upsilon}^\dagger(t)$:

$$[\hat{\xi}(t), \hat{\upsilon}(t')] = 2i\delta(t - t'). \tag{4.197}$$

The output quadrature operator (4.195) evidently has the same single-time statistics as the homodyne photocurrent operator (4.75), with a mean equal to the mean of $\hat{x}$, and a white-noise term. The two-time correlation function of the operator $\hat{J}_{\text{out}}^{\text{hom}}(t)$ is defined by

$$F_{\text{hom}}^{(1)}(t, t + \tau) = \langle \hat{J}_{\text{out}}^{\text{hom}}(t + \tau)\hat{J}_{\text{out}}^{\text{hom}}(t)\rangle. \tag{4.198}$$

Using the commutation relations for the output field (4.178), this can be written as

$$F_{\text{hom}}^{(1)}(t, t + \tau) = \langle: \hat{x}(t + \tau)\hat{x}(t) :\rangle + \delta(\tau), \tag{4.199}$$

where the annihilation of the vacuum has been used as before. Here the colons denote time and normal ordering of the operators $\hat{c}$ and $\hat{c}^\dagger$. The meaning of this can be seen in the Schrödinger picture:

$$F_{\text{hom}}^{(1)}(t, t + \tau) = \text{Tr}\big[\hat{x}e^{\mathcal{L}\tau}\big(\hat{c}\rho(t) + \rho(t)\hat{c}^\dagger\big)\big] + \delta(\tau), \tag{4.200}$$

where $\mathcal{L}$ is as before and $\rho(t)$ is the state of the system at time t, which is assumed known.

Again, this is in exact agreement with that calculated from the quantum trajectory method in Section 4.4.4. The operator quantity $\hat{J}_{\text{out}}^{\text{hom}}(t)$ has the same statistics as those of the classical photocurrent $J_{\text{hom}}(t)$. Again the different conceptual basis is reflected in the origin of the delta function in the autocorrelation function: operator commutation relations in the Heisenberg picture versus local oscillator shot noise from the quantum trajectories.

4.7.4 Heterodyne detection

The equivalence between the Heisenberg picture and quantum trajectory calculations of the output field correlation functions applies for heterodyne detection in much the same way as for homodyne detection. For variation, we construct the Heisenberg operator for the 'heterodyne photocurrent' from two homodyne measurements, rather than from the Fourier components of the heterodyne signal. To make two homodyne measurements, it is necessary to use a 50 : 50 beam-splitter to divide the cavity output before the local oscillator is added. This gives two output beams,

$$\hat{b}_{\text{out}}^\pm = \sqrt{1/2}(\hat{\upsilon} + \hat{c} \pm \hat{\mu}). \tag{4.201}$$

Here we have introduced an ancilla vacuum annihilation operator $\hat{\mu}$ that enters at the free port of the beam-splitter.

Exercise 4.24 *Show that*

$$\hat{b}_{\text{out}}^\dagger \hat{b}_{\text{out}} = \hat{b}_{\text{out}}^{+\dagger}\hat{b}_{\text{out}}^+ + \hat{b}_{\text{out}}^{-\dagger}\hat{b}_{\text{out}}^-, \tag{4.202}$$

so that photon flux is preserved.

Let the $\hat{b}_{\text{out}}^{+}$ beam enter a homodyne apparatus to measure the x quadrature, and the $\hat{b}_{\text{out}}^{-}$ beam one to measure the y quadrature. Then the operators for the two photocurrents, normalized with respect to the system signal, are

$$\hat{J}_{\text{out}}^{x} = \hat{x} + [\hat{v} + \hat{v}^{\dagger} + \hat{\mu} + \hat{\mu}^{\dagger}], \tag{4.203}$$

$$\hat{J}_{\text{out}}^{y} = \hat{y} - \mathrm{i}[\hat{v} - \hat{v}^{\dagger} - \hat{\mu} + \hat{\mu}^{\dagger}], \tag{4.204}$$

where $\hat{y}$ is defined in Eq. (4.65). Defining the complex heterodyne photocurrent as in Eq. (4.116) gives

$$\hat{J}_{\text{out}}^{\text{het}} = \hat{c} + \hat{v} + \hat{\mu}^{\dagger}. \tag{4.205}$$

Exercise 4.25 *Convince yourself that $\hat{\zeta}(t) = \hat{v}(t) + \hat{\mu}^{\dagger}(t)$ has the same statistics as the complex Gaussian noise $\zeta(t)$ defined in Section 4.6.4.*

Thus, the operator (4.205) has the same statistics as the photocurrent (4.117). The autocorrelation function

$$F_{\text{het}}^{(1)}(t, t + \tau) = \langle \hat{J}_{\text{out}}^{\text{het}\,\dagger}(t + \tau) \hat{J}_{\text{out}}^{\text{het}}(t) \rangle, \tag{4.206}$$

evaluates to

$$F_{\text{het}}^{(1)}(t, t + \tau) = \langle \hat{c}^{\dagger}(t + \tau)\hat{c}(t) \rangle + \delta(\tau). \tag{4.207}$$

Again, this is equal to the Schrödinger-picture expression (4.120).

4.7.5 Completely general dyne detection

In the Heisenberg picture, the completely general Lindblad evolution (4.122) becomes

$$d\hat{s} = dt \left(\hat{c}_{k}^{\dagger}\hat{s}\hat{c}_{k} - \tfrac{1}{2}\left\{ \hat{c}_{k}^{\dagger}\hat{c}_{k}, \hat{s} \right\} + \mathrm{i}[\hat{H}, \hat{s}] \right)$$
$$- [d\hat{B}_{k;\text{in}}^{\dagger}(t)\hat{c}_{k} - \hat{c}_{k}^{\dagger}\, d\hat{B}_{k;\text{in}}(t), \hat{s}], \tag{4.208}$$

where we are using the Einstein summation convention as before and $d\hat{B}_{k;\text{in}} = \hat{b}_{k;\text{in}}\, dt$ where the $\hat{b}_{k;\text{in}}$ are independent vacuum field operators. The output field operators are

$$\hat{b}_{k;\text{out}} = \hat{b}_{k;\text{in}} + \hat{c}_{k}. \tag{4.209}$$

Recall that for a completely general dyne unravelling the measurement result was a vector of complex currents $J_{k}(t)$ given by (4.136), where the noise correlations are defined by a complex symmetric matrix Υ. In the Heisenberg picture the operators for these currents are

$$\hat{J}_{k} = \hat{b}_{k;\text{out}} + \Upsilon_{kj}\hat{b}_{j;\text{out}}^{\dagger} + \mathrm{T}_{kj}\hat{a}_{j}^{\dagger}. \tag{4.210}$$

Here $\hat{a}_{k}$ are ancillary annihilation operators, which are also assumed to act on a vacuum state, and obey the usual continuum-field commutation relations,

$$[\hat{a}_{j}(t), \hat{a}_{k}^{\dagger}(t')]dt = \delta_{jk}\delta(t - t'). \tag{4.211}$$

These ancillary operators ensure that all of the components $\hat{J}_k$ commute with one another. This is necessary since this vector operator represents an observable quantity. Assuming T (a capital τ) to be a symmetric matrix like Υ, we find

$$[\hat{J}_j, \hat{J}_k]dt = \Upsilon_{jk} - \Upsilon_{kj} = 0, \tag{4.212}$$

$$[\hat{J}_j, \hat{J}_k^\dagger]dt = \delta_{jk} - \Upsilon_{jl}\Upsilon_{lk}^* - T_{jl}T_{lk}^*. \tag{4.213}$$

Thus we require the matrix T to satisfy

$$T^*T = I - \Upsilon^*\Upsilon. \tag{4.214}$$

The right-hand side of this equation is always positive (see Section 4.5.2), so it is always possible to find a suitable T.

From these definitions one can show that

$$\hat{J}_k = \hat{c}_k + \Upsilon_{kj}\hat{c}_j^\dagger + \delta\hat{J}_k, \tag{4.215}$$

where

$$\delta\hat{J}_k = \hat{b}_{k;\text{in}} + \Upsilon_{kj}\hat{b}_{j;\text{in}}^\dagger + T_{kj}\hat{a}_j^\dagger \tag{4.216}$$

has a zero mean. Thus the mean of $\hat{J}_k$ is the same as that of the classical current in Section 4.5.2. Also, one finds that

$$(\delta\hat{J}_j\, dt)(\delta\hat{J}_k\, dt)^\dagger = \delta_{jk}\, dt, \tag{4.217}$$

$$(\delta\hat{J}_j\, dt)(\delta\hat{J}_k\, dt) = \Upsilon_{jk}\, dt, \tag{4.218}$$

the same correlations as for the noise in Section 4.5.2.

4.7.6 Relation to distribution functions

Obviously direct, homodyne and heterodyne detection are related to measurements of, respectively, the intensity, one quadrature and the complex amplitude of the radiating dipole of the system. Equally obviously, the measurements described are far removed from simple measurements of these quantities, such as by the projective measurements. Nevertheless, there must be some elementary relation between the two types of measurements for at least some cases. For example, counting the number of photons in a cavity should give the same results (statistically) irrespective of whether this is done by allowing the photons to escape gradually through an end mirror into a photodetector or whether the measurement is a projection of the cavity mode into photon-number eigenstates, provided that the system Hamiltonian commutes with photon number. The purpose of this section is to establish a relationship between quantum monitoring and simpler descriptions of measurements, for the three schemes discussed. In all cases we must wait until $t = \infty$ for the measurement to be complete.

For simplicity, we consider the case of a freely decaying optical cavity with unit linewidth, so that $\hat{c} = \hat{a}$ with $[\hat{a}, \hat{a}^\dagger] = 1$. We will see that the Heisenberg-picture formalism is a powerful framework for establishing the relations we seek. In this picture, the dynamics of the system is given by the quantum Langevin equation

$$d\hat{a}(t) = -\tfrac{1}{2}\hat{a}(t)dt + \hat{v}(t)dt,$$ (4.219)

which has the solution

$$\hat{a}(t) = \hat{a}(0)e^{-t/2} - \int_0^t e^{(s-t)/2}\hat{v}(s)ds.$$ (4.220)

Exercise 4.26 *Verify these equations.*

The output field is of course

$$\hat{b}_{\text{out}}(t) = \hat{a}(t) + \hat{v}(t).$$ (4.221)

Photon-number distribution. We begin with photon counting. The most obvious difference between a projective measurement of photon number and an external counting of escaped photons from a freely decaying cavity is that the final state of the cavity mode is the appropriate photon-number eigenstate in the first case and the vacuum in the second. The latter result comes about because the counting time must be infinite to allow all photons to escape. Although extra-cavity detection is not equivalent to projective detection, it should give the same statistics in the infinite time limit.

The output photon flux is

$$\hat{I}(t) = [\hat{a}^\dagger(t) + \hat{v}^\dagger(t)][\hat{a}(t) + \hat{v}(t)],$$ (4.222)

which using Eq. (4.220) evaluates to

$$\hat{I}(t) = \left[\hat{a}^\dagger(0)e^{-t/2} - \int_0^t e^{(s-t)/2}\hat{v}^\dagger(s)ds + \hat{v}^\dagger(t)\right]$$
$$\times \left[\hat{a}(0)e^{-t/2} - \int_0^t e^{(s-t)/2}\hat{v}(s)ds + \hat{v}(t)\right].$$ (4.223)

The operator for the total photocount is then

$$\hat{N} \equiv \int_0^\infty \hat{I}(t)dt.$$ (4.224)

On using integration by parts (we show this explicitly for the simpler cases of homodyne detection below), it is possible to evaluate this as

$$\hat{N} = \hat{a}^\dagger(0)\hat{a}(0) + \hat{n}.$$ (4.225)

Here $\hat{n}$ contains only bath operators, and annihilates on the vacuum. Hence it commutes with the first term and contributes nothing to the expectation value of any function of $\hat{N}$, provided that the bath is in a vacuum state. This confirms that the integral of the photocurrent does indeed measure the operator $\hat{a}^\dagger\hat{a}$ for the initial cavity state.

Quadrature distribution. Now consider homodyne detection. Unlike direct detection, one should not simply integrate the photocurrent from zero to infinity because even when all light has escaped the cavity the homodyne measurement continues to give a non-zero current (vacuum noise). Thus, for long times the additional current is merely adding noise to the record. This can be circumvented by properly mode-matching the local oscillator to the system. That is to say, by matching the decay rate, as well as the frequency, of the local oscillator amplitude to that of the signal. Equivalently, the current could be electronically multiplied by the appropriate factor and then integrated.

For the cavity decay in Eq. (4.220) the appropriately scaled homodyne current operator is

$$\hat{J}_{\text{out}}^{\text{hom}}(t) = e^{-t/2}[\hat{x}(t) + \hat{\xi}(t)]. \qquad (4.226)$$

From Eq. (4.220), this is equal to

$$\hat{J}_{\text{out}}^{\text{hom}}(t) = e^{-t/2}\left[\hat{x}(0)e^{-t/2} - \int_0^t e^{(s-t)/2}\hat{\xi}(s)ds + \hat{\xi}(t)\right], \qquad (4.227)$$

and so

$$\hat{X} \equiv \int_0^\infty \hat{J}_{\text{out}}^{\text{hom}}(t)dt \qquad (4.228)$$

$$= \hat{x}(0) + \int_0^\infty e^{-t/2}\hat{\xi}(t)dt - \int_0^\infty dt\, e^{-t} \int_0^t e^{s/2}\hat{\xi}(s)ds. \qquad (4.229)$$

Using integration by parts on the last term, it is easy to show that it cancels out the penultimate term, so the operator of the integrated photocurrent is simply

$$\hat{X} = \hat{x}(0). \qquad (4.230)$$

Thus it is possible to measure a quadrature of the field by homodyne detection. Unlike the case of direct detection, this derivation requires no assumptions about the statistics of the bath field.

Husimi distribution. Heterodyne detection is different from direct and homodyne detection in that it does not measure an Hermitian system operator. That is because it measures both quadratures simultaneously. However, it does measure a *normal operator* (see Box 1.1). The (normal) operator for the heterodyne photocurrent (4.205) is, with appropriate photocurrent scaling factor,

$$\hat{J}_{\text{out}}^{\text{het}}(t) = e^{-t/2}[\hat{a}(t) + \hat{\zeta}(t)], \qquad (4.231)$$

where $\hat{\zeta} = \hat{v} + \hat{\mu}^\dagger$. Proceeding as above, the operator for the measurement result is

$$\hat{A} \equiv \int_0^\infty \hat{J}_{\text{out}}^{\text{het}}(t)dt = \hat{a}(0) + \hat{e}^\dagger, \qquad (4.232)$$

where $\hat{e}^\dagger = \int_0^\infty e^{-t/2}\hat{\mu}^\dagger(t)dt$.

Exercise 4.27 *Show that $[\hat{e}^\dagger, \hat{e}] = -1$ and hence show that $\hat{A}$ commutes with its Hermitian conjugate.*

From this exercise it follows that $\hat{A}$ is a normal operator, and hence there is no ambiguity in the calculation of its moments. That is, all orderings are equivalent. The most convenient operator ordering is one of normal ordering with respect to $\hat{e}$. (Recall that normal ordering, defined in Section A.5, has nothing to do with normal operators.) For a vacuum-state bath, the expectation value of any expression normally ordered in $\hat{e}$ will have zero contribution from all terms involving $\hat{e}$. Now normal ordering with respect to $\hat{e}$ is *antinormal* ordering with respect to $\hat{a}$. Thus the statistics of $\hat{A}$ are the antinormally ordered statistics of $\hat{a}$. As shown in Section A.5, these statistics are those found from the so-called Husimi or Q function,

$$Q(\alpha)\mathrm{d}^2\alpha = \frac{\mathrm{d}^2\alpha}{\pi}\langle\alpha|\rho|\alpha\rangle, \qquad (4.233)$$

where $|\alpha\rangle$ is a coherent state as usual.

Adaptive measurements. One should not conclude from the above analyses that the Heisenberg picture is necessarily always more powerful than the Schrödinger picture. Take, for example, adaptive measurements, as considered in Sections 2.5 and 4.6.2. In these cases, the measured currents are fed back to alter the future conditions of the measurement. This leads to intractable nonlinear Heisenberg equations, even for a system as simple as a decaying cavity [Wis95]. However, the problem can be tackled using *linear* quantum trajectory theory as introduced in Section 4.4.3. By this method it is possible to obtain an exact analytical treatment of the adaptive phase-estimation algorithm [Wis95, WK97, WK98] implemented in the experiment [AAS⁺02] described in Section 2.6. This will be discussed in detail in Section 7.9.

4.8 Imperfect detection

In all of the above we have considered detection under perfect conditions. That is, the detectors were *efficient*, detecting all of the output field; the input field was in a pure state; the detectors added no electronic noise to the measured signal; and the detectors did not filter the measured signal. In reality some or all of these assumptions will be invalid, and this means that an observer will not have access to perfect information about the system. Since the quantum state of the system *is* the observer's knowledge about the system, imperfect knowledge means a different (more mixed) quantum state. In this section we consider each of these imperfections in turn and derive the appropriate quantum trajectory to describe it.

4.8.1 Inefficient detection

The simplest sort of imperfection to describe is inefficiency. A photodetector of efficiency η (with $0 \leq \eta \leq 1$) can be modelled as a perfect detector detecting only a proportion η of the output beam. Thus, we can split the general master equation Eq. (4.11) into

$$\dot{\rho} = -\mathrm{i}[\hat{H}, \rho] + (1 - \eta)\mathcal{D}[\hat{c}]\rho + \mathcal{D}[\sqrt{\eta}\,\hat{c}]\rho \qquad (4.234)$$

and unravel only the last term.

For direct detection, Eq. (4.40) is replaced by

$$\mathrm{d}\rho_I(t) = \left\{\mathrm{d}N(t)\mathcal{G}[\sqrt{\eta}\,\hat{c}] + \mathrm{d}t\,\mathcal{H}[-\mathrm{i}\hat{H} - \eta\tfrac{1}{2}\hat{c}^\dagger\hat{c}] + \mathrm{d}t(1 - \eta)\mathcal{D}[\hat{c}]\right\}\rho_I(t), \qquad (4.235)$$

where now Eq. (4.41) becomes

$$E[dN(t)] = \eta \operatorname{Tr}[\hat{c}\rho(t)\hat{c}^\dagger]dt. \tag{4.236}$$

In this case a SSE does not exist because the conditioned state will not remain pure even if it begins pure. Note that in the limit $\eta \to 0$ one obtains the unconditioned master equation.

For homodyne detection, the homodyne photocurrent is obtained simply by replacing $\hat{c}$ by $\sqrt{\eta}\,\hat{c}$ to give

$$J_{\text{hom}}(t) = \sqrt{\eta}\langle\hat{x}\rangle_J(t) + \xi(t). \tag{4.237}$$

Note that here the shot noise (the final term) remains normalized so as to give a unit spectrum; other choices of normalization are also used. The SME (4.72) is modified to

$$d\rho_J = -i[\hat{H}, \rho_J(t)]dt + \mathcal{D}[\hat{c}]\rho_J(t)dt + \sqrt{\eta}\,dW(t)\mathcal{H}[\hat{c}]\rho_J(t). \tag{4.238}$$

Again, there is no SSE. The generalization of the heterodyne SME (4.110) is left as an exercise for the reader.

4.8.2 Detection with a white-noise input field

So far, we have considered optical measurements with the input in a vacuum, or coherent, state. As explained in Section 4.3.3, the photon flux is infinite for an input bath in a more general state, such as with thermal or squeezed white noise. This indicates that direct detection is impossible (or at least useless, as explained in Section 4.3.3). However, the output field quadrature operators are well defined even with white noise, because they are only linear in the noise. Thus, field operators for homodyne- and heterodyne-detection photocurrents can be defined without difficulty, simply by replacing the vacuum operator $\hat{v}$ by the more general input bath operator $\hat{b}_0$. This indicates that it should be possible to treat homodyne and heterodyne detection in such situations. In this section, we develop the quantum trajectory theory for detection in the presence of white noise.

The homodyne-detection theory of Section 4.4 began with a finite local oscillator amplitude γ, so that the quantum trajectories were jump-like. Diffusive trajectories were obtained when the $\gamma \to \infty$ limit was taken. This approach would fail if the input field were contaminated by white noise, for the same reason as that rendering direct detection impossible with white noise: the infinite photon flux due to the noise would swamp the signal. However, as noted in Section 4.3.3, a physical noise source would not be truly white, and in any case the response of the detector would give a cut-off to the flux. If the local oscillator were made sufficiently intense, then the signal due to this would overcome that due to the noise. Thus, in order to treat detection with white noise, it is necessary to begin with an infinitely large local oscillator. Then one can assume that the homodyne detection effects an ideal measurement of the instantaneous quadrature of the output field, without worrying about individual jumps.

Rather than doing a completely general derivation, we consider first the case of a white-noise bath in a pure, but non-vacuum, state. That is, as in Section 3.11.2,

$$d\hat{B}^\dagger(t)d\hat{B}(t) = N\,dt, \tag{4.239}$$

$$d\hat{B}(t)d\hat{B}(t) = M\,dt, \tag{4.240}$$

$$[d\hat{B}(t), d\hat{B}^\dagger(t)] = dt, \tag{4.241}$$

but the inequality (3.180) is replaced by the equality

$$|M|^2 = N(N+1). \tag{4.242}$$

In this case the pure state of the bath, which we denote $|M\rangle$, obeys

$$[(N + M^* + 1)\hat{b}(t) - (N + M)\hat{b}^\dagger(t)]|M\rangle = 0. \tag{4.243}$$

Exercise 4.28 *From Eqs. (4.241)–(4.243), derive Eqs. (4.239) and (4.240).*

Now replacing the vacuum bath state $|0\rangle$ by $|M\rangle$ in Eq. (4.36) and expanding the unitary operator $\hat{U}(dt)$ to second order yields

$$|\Psi(t + dt)\rangle = \left\{ 1 - \tfrac{1}{2}dt\left[(N+1)\hat{c}^\dagger\hat{c} + N\hat{c}\hat{c}^\dagger - M\hat{c}^{\dagger 2} - M^*\hat{c}^2 \right] \right.$$
$$\left. + \hat{c}\,d\hat{B}^\dagger(t) - \hat{c}^\dagger\,d\hat{B}(t) \right\}|\psi(t)\rangle|M\rangle. \tag{4.244}$$

Consider homodyne detection on the output. Any multiple of the operator in Eq. (4.243) can be added to Eq. (4.244) without affecting it. Thus it is possible to replace $d\hat{B}^\dagger$ by

$$d\hat{B}^\dagger + \frac{N + M^* + 1}{L}d\hat{B} - \frac{N+M}{L}d\hat{B}^\dagger = \frac{N + M^* + 1}{L}[d\hat{B}^\dagger + d\hat{B}] \tag{4.245}$$

and $d\hat{B}$ by

$$d\hat{B} - \frac{N + M^* + 1}{L}d\hat{B} + \frac{N+M}{L}d\hat{B}^\dagger = \frac{N+M}{L}[d\hat{B}^\dagger + d\hat{B}], \tag{4.246}$$

where

$$L = 2N + M^* + M + 1. \tag{4.247}$$

This yields

$$|\Psi(t + dt)\rangle = \left\{ 1 - \tfrac{1}{2}dt\left[(N+1)\hat{c}^\dagger\hat{c} + N\hat{c}\hat{c}^\dagger - M\hat{c}^{\dagger 2} - M^*\hat{c}^2 \right] \right.$$
$$\left. + \left[\hat{c}(N + M^* + 1)/L - \hat{c}^\dagger(N+M)/L \right] \right\}$$
$$\times [d\hat{B}^\dagger(t) + d\hat{B}(t)]\right\}|\psi(t)\rangle|M\rangle. \tag{4.248}$$

Projecting onto eigenstates $|J\rangle$ of the output quadrature $\hat{b} + \hat{b}^\dagger$ then gives the unnormalized conditioned state

$$|\tilde{\psi}_J(t + dt)\rangle = \left\{ 1 - \tfrac{1}{2}dt\left[(N+1)\hat{c}^\dagger\hat{c} + N\hat{c}\hat{c}^\dagger - M\hat{c}^{\dagger 2} - M^*\hat{c}^2 \right] \right.$$
$$\left. + J\,dt\left[\hat{c}(N + M^* + 1)/L - \hat{c}^\dagger(N+M)/L \right] \right\}$$
$$\times |\psi(t)\rangle\sqrt{\wp_{\mathrm{ost}}^M(J)}, \tag{4.249}$$

where the state-matrix norm $\wp_{\text{ost}}^M(J)$ gives the probability of obtaining the result J and the ostensible probability distribution for J is

$$\wp_{\text{ost}}^M(J) = |\langle J|M\rangle|^2 = \sqrt{\frac{dt}{2\pi L}} \exp(-\tfrac{1}{2}J^2\,dt/L). \tag{4.250}$$

Note that from this ostensible distribution the variance of J^2 is L/dt, which, depending on the modulus and argument of M, may be larger or smaller than its vacuum value of $1/dt$.

Now, by following the method in Section 4.4.3, one obtains the following SSE for the unnormalized state vector:

$$d|\bar{\psi}_J(t)\rangle = \left\{ -\frac{dt}{2}\left[(N+1)\hat{c}^\dagger\hat{c} + N\hat{c}\hat{c}^\dagger - M\hat{c}^{\dagger 2} - M^*\hat{c}^2\right] \right.$$

$$\left. + J(t)dt\left(\hat{c}\frac{N+M^*+1}{L} - \hat{c}^\dagger\frac{N+M}{L}\right)\right\}|\bar{\psi}_J(t)\rangle, \tag{4.251}$$

where the *actual* statistics of the homodyne photocurrent J are given by

$$J_{\text{hom}}(t) = \langle\hat{x}(t)\rangle + \sqrt{L}\xi(t), \tag{4.252}$$

where $\xi(t) = dW(t)/dt$ is white noise as usual. Note that the high-frequency spectrum of the photocurrent is no longer unity, but L.

Turning Eq. (4.251) into an equation for $\bar{\rho}_J = |\bar{\psi}_J\rangle\langle\bar{\psi}_J|$ and then normalizing it yields

$$d\rho_J(t) = \left(dt\,\mathcal{L} + \frac{1}{\sqrt{L}}dW(t)\mathcal{H}\left[(N+M^*+1)\hat{c} - (N+M)\hat{c}^\dagger\right]\right)\rho_J(t), \tag{4.253}$$

where the unconditional evolution is

$$\mathcal{L}\rho = (N+1)\mathcal{D}[\hat{c}]\rho + N\mathcal{D}[\hat{c}^\dagger]\rho + \frac{M}{2}[\hat{c}^\dagger,[\hat{c}^\dagger,\rho]] + \frac{M^*}{2}[\hat{c},[\hat{c},\rho]] - i[\hat{H},\rho]. \tag{4.254}$$

Exercise 4.29 *Verify this.*

The general case of an impure bath is now obtained simply by relaxing the equality in Eq. (4.242) back to the inequality in Eq. (3.180).

Note that, even for $M = 0$, the quantum trajectory described by Eq. (4.253) is not simply the homodyne quantum trajectory for a vacuum bath (4.72) with the addition of the thermal terms

$$N\big(\mathcal{D}[\hat{c}] + \mathcal{D}[\hat{c}^\dagger]\big)\rho \tag{4.255}$$

in the non-selective evolution. Rather, the conditioning term depends upon N and involves both $\hat{c}$ and $\hat{c}^\dagger$. In quantum optics, N is typically negligible, but for quantum-electromechanical systems (see Section 3.10.1), N is not negligible. Thus, for continuous measurement of such devices by electro-mechanical means, it may be necessary to apply a SME of form similar to Eq. (4.253).

From this conditioning equation and the expressions for the photocurrent, it is easy to find the two-time correlation function for the output field using the method of Section 4.4.4.

The result is

$$F_{\text{hom}}^{(1)}(t, t+\tau) = \text{E}[J_{\text{hom}}(t+\tau)J_{\text{hom}}(t)] \tag{4.256}$$

$$= \text{Tr}\big[(\hat{c}+\hat{c}^\dagger)e^{\mathcal{L}\tau}\{(N+M^*+1)\hat{c}\rho(t) - (N+M)\hat{c}^\dagger\rho(t)$$

$$+ (N+M+1)\rho(t)\hat{c}^\dagger - (N+M^*)\rho(t)\hat{c}\}\big] + L\delta(\tau). \tag{4.257}$$

Note that, unlike in the case of a vacuum input, there is no simple relationship between this formula and the Glauber coherence functions.

This correlation function could be derived from the Heisenberg field operators. The relevant expression is

$$\langle[\hat{x}(t+\tau) + \hat{b}_{\text{in}}(t+\tau) + \hat{b}_{\text{in}}^\dagger(t+\tau)][\hat{x}(t) + \hat{b}_{\text{in}}(t) + \hat{b}_{\text{in}}^\dagger(t)]\rangle. \tag{4.258}$$

We will not attempt to prove that this evaluates to Eq. (4.257), because it is considerably more difficult than with a vacuum input $\hat{b}_{\text{in}} = \hat{v}$. The reason for this is that it is impossible to choose an operator ordering such that the contributions due to the bath input vanish. The necessary method would have to be more akin to that used in obtaining Eq. (4.257), where the stochastic equation analogous to the SME is the quantum Langevin equation (3.181).

4.8.3 Detection with dark noise

The next sort of imperfection we consider is that of dark noise in the detector. This terminology is used for electronic noise generated within the detector because it is present even when no field illuminates the detector (i.e. in the dark). For simplicity we will treat only the case of detection of the homodyne type. Once again, it is convenient to use the approach of Section 4.4.3. We use the linear SME (4.91)

$$d\bar{\rho} = dt\{\mathcal{L}\bar{\rho} + J_0(t)[\hat{c}\bar{\rho} + \text{H.c.}]\}, \tag{4.259}$$

where $\mathcal{L} = \mathcal{H}[-i\hat{H}] + \mathcal{D}[\hat{c}]$. This corresponds to detection by an ideal detector with current J_0 having an ostensible distribution corresponding to Gaussian white noise. We model the addition of dark noise by setting the output of the realistic detector to be the current

$$J(t)dt = [J_0(t)dt + \sqrt{N}\,dW_1]/\sqrt{1+N}, \tag{4.260}$$

where dW_1 is an independent Wiener increment and N is the dark-noise power relative to the shot noise. Note that we have included the normalization factor so that the ostensible distribution for $J(t)$ is also that of normalized Gaussian white noise.

The problem is to determine the quantum trajectory for the system state ρ_J conditioned on Eq. (4.260), rather than that conditioned on the ideal current J_0. To proceed, we rewrite J_0 as

$$J_0(t)dt = [J(t)dt + \sqrt{N}\,dW'(t)]/\sqrt{1+N}, \tag{4.261}$$

where

$$dW' = (\sqrt{N} J_0 dt - dW_1)/\sqrt{1 + N} \tag{4.262}$$

is ostensibly a Wiener increment and is independent of $J(t)$.

Exercise 4.30 *Verify that* $(J dt)^2 = (dW')^2 = dt$ *and that* $(J dt)dW' = 0$.

Substituting this into Eq. (4.259) yields

$$d\bar{\rho} = \mathcal{L}\bar{\rho}\, dt + (1 + N)^{-1/2}[J(t)dt + \sqrt{N}\, dW'][\hat{c}\bar{\rho} + \text{H.c.}]. \tag{4.263}$$

Now we can average over the unobserved noise to get

$$d\bar{\rho} = \mathcal{L}\bar{\rho}\, dt + (1 + N)^{-1/2} J(t)dt[\hat{c}\bar{\rho} + \text{H.c.}]. \tag{4.264}$$

Converting this into a nonlinear SME for the normalized state matrix gives

$$d\rho = \mathcal{L}\rho\, dt + \sqrt{\eta}\, dW(t)\mathcal{H}[\hat{c}]\rho, \tag{4.265}$$

where $\eta \equiv 1/(1 + N)$ and $dW(t)$ is the Gaussian white noise which appears in the actual photocurrent:

$$J(t) = \sqrt{\eta}\langle \hat{c} + \hat{c}^\dagger \rangle + dW(t)/dt. \tag{4.266}$$

In comparison with Section 4.8.1 we see that the addition of Gaussian white noise to the photocurrent before the experimenter has access to it is exactly equivalent to an inefficient homodyne detector.

For direct detection, dark noise is *not* the same as an inefficiency. Modelling it is actually more akin to the methods of the following subsection, explicitly involving the detector. The interested reader is referred to Refs. [WW03a, WW03b].

4.8.4 Detectors with a finite bandwidth

The final sort of detector imperfection we consider is a finite detector bandwidth. By contrast with dark noise, this produces something quite different from what we have seen before. Specifically, it leads to a non-Markovian quantum trajectory. The system still obeys a Markovian master equation on average, but, because the observer has access only to a filtered output from the system, the information the observer needs to update their system state $\rho(t)$ is not contained solely in $\rho(t)$ and the measurement result in the infinitesimal interval $[t, t + dt)$. We will see instead that the observer must keep track of a joint state of the quantum system and the classical detector, in which there are correlations between these two systems.

Actually, all of these remarks apply only if the realistic detector adds noise to the output of an ideal homodyne detector (as in the preceding section) as well as filtering it. For example, say the output of the ideal detector was $J(t)$, and the output of the realistic detector was put through a low-pass filter of bandwidth B:

$$Q(t) = \int_{-\infty}^t B e^{-B(t-s)} J(s) ds. \tag{4.267}$$

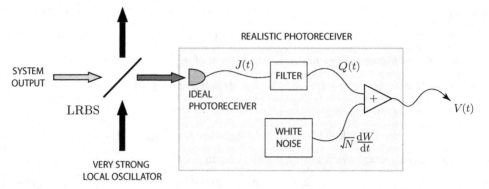

Fig. 4.9 A schematic diagram of the model for simple homodyne detection by a realistic photoreceiver. The realistic photoreceiver is modelled by a hypothetical ideal photoreciever, the output $J(t)$ of which is passed through a low-pass filter to give $Q(t)$. White noise $\sqrt{N}\, dW/dt$ is added to this to yield the observable output $V(t)$ of the realistic photoreceiver.

Then an observer could obtain J from Q simply as

$$J(t) = Q(t) + \dot{Q}(t)/B. \tag{4.268}$$

Thus, the quantum state conditioned on $Q(t)$ is the same as that conditioned on $J(t)$. However a nontrivial result is obtained if we say that the output is

$$V(t) = Q(t) + \sqrt{N}\, dW_1(t)/dt, \tag{4.269}$$

where dW_1 is an independent Wiener increment. This is shown in Fig. 4.9

An equation of the form of Eq. (4.269) can be derived by considering the output of a realistic homodyne detector, called a photoreceiver. This consists of a p–i–n photodiode (which acts as an ideal detector, apart from a small inefficiency), which produces a current $\propto J$ that is fed into an operational amplifier set up as a transimpedance amplifier. This has a low effective input impedance, so the diode acts as a current source, and J is converted into a charge $\propto Q$ on a capacitor of capacitance C in parallel with the feedback resistor of resistance R. The bandwidth B in Eq. (4.267) is equal to $1/(RC)$. The resistor, being at a finite temperature, introduces Johnson noise with power $\propto N$ into the amplifier output V, which is what the observer can see. For details, see Ref. [WW03a].

The general method we adopt is the following. First, we determine the SME for the state of the quantum system conditioned on J. Then we determine a stochastic Fokker–Planck equation (FPE) (see Section B.5) for the state of the detector. This is represented by $\wp(q)$, the probability distribution for $Q(t)$, and also depends upon J. Next we consider $\wp_V(q)$, the conditioned classical state based on our observation of V, which involves another stochastic process. Finally, we combine all of these equations together to derive an equation for the state of a *supersystem*

$$\rho_V(q) = \mathrm{E}_J[\rho^J \wp_V^J(q)]. \tag{4.270}$$

Here the J superscript represents dependence upon the unobservable microscopic current J. This is the variable which is averaged over to find the expectation value. From this equation we can obtain our state of knowledge of the system, conditioned on V, as

$$\rho_V = \int dq\, \rho_V(q). \tag{4.271}$$

Similarly, our state of knowledge of the detector is

$$\wp_V(q) = \text{Tr}[\rho_V(q)]. \tag{4.272}$$

Note, however, that $\rho(q)$ contains more information than do ρ and $\wp(q)$ combined, because of correlations between the quantum system and the classical detector.

The ideal stochastic master equation. As in the previous cases of imperfect detection, it is convenient to use the linear form of the SME as in Section 4.4.3. For a detector that is ideal apart from an efficiency η, this is

$$d\bar\rho^J = \mathcal{L}\bar\rho^J\, dt + \sqrt{\eta}[J(t)dt](\hat{c}\bar\rho^J + \bar\rho^J\hat{c}^\dagger), \tag{4.273}$$

where the ostensible distribution for the current $J(t)$ is that of Gaussian white noise.

The stochastic FPE for the detector. The detector state is given by $\wp(q)$, the probability distribution for Q. From Eq. (4.267), Q obeys the SDE

$$dQ = -B[Q - J(t)]dt, \tag{4.274}$$

where J as given above satisfies $(J\, dt)^2 = dt$. From Section B.5, the probability distribution $\wp(q)$ obeys the stochastic Fokker–Planck equation (FPE)

$$d\wp^J(q) = \left\{ B\frac{\partial}{\partial q}[q - J(t)]dt + \frac{1}{2}B^2\frac{\partial^2}{(\partial q)^2}dt \right\}\wp^J(q). \tag{4.275}$$

This assumes knowledge of J, as shown explicitly by the superscript. Note that this equation has the solution

$$\wp^J(q) = \delta(q - Q^J), \tag{4.276}$$

where Q^J is the solution of Eq. (4.274). It is only later, when we average over J, that we will obtain an equation with diffusion leading to a non-singular distribution. If we were to do this at this stage we would derive a FPE of the usual (deterministic) type, without the J-dependent term in Eq. (4.275).

The Zakai equation for the detector. Now we consider how $\wp(q)$ changes when the observer obtains the information in V. That is, we determine the conditioned state by Bayesian inference:

$$\wp_V(q) \equiv \wp(q|V) = \frac{\wp(V|q)\wp(q)}{\wp(V)}. \tag{4.277}$$

From Eq. (4.269), in any infinitesimal time interval, the noise in V is infinitely greater than the signal. Specifically, the root-mean-square noise in an interval of duration dt is

$\sqrt{N/dt}$, while the signal is Q. Thus, the amount of information about Q contained in $V(t)$ is infinitesimal, and hence the change from $\wp(q)$ to $\wp_V(q)$ is infinitesimal, of order $\sqrt{dt}$. This means that it is possible to derive a stochastic differential equation for $\wp_V(q)$. This is called a *Kushner–Stratonovich* equation (KSE). Such a KSE is exactly analogous to the stochastic master equation describing the update of an observer's knowledge of a quantum system.

Just as in the quantum case one can derive a linear version of the SME for an unnormalized state $\bar{\rho}$, one can derive a linear version of the KSE for an unnormalized probability distribution $\bar{\wp}_V(q)$. This involves choosing an ostensible distribution $\wp_{ost}(V)$, such as $\wp_{ost}(V) = \wp(V|q := 0)$. Then

$$\bar{\wp}_V(q) = \frac{\wp(V|q)\bar{\wp}(q)}{\wp_{ost}(V)} \tag{4.278}$$

has the interpretation that $\wp_{ost}(V) \int dq \, \bar{\wp}_V(q)$ is the actual probability distribution for V. Note the analogy with the quantum case in Section 4.4.3, where a trace rather than an integral is used (see also Table 1.1). The linear version of the KSE is called the *Zakai* equation and it will be convenient to use it in our derivation.

From Eq. (4.269), the distribution of V given that $Q = q$ is

$$\wp(V|q) = [dt/(2\pi N)]^{1/2} \exp[-(V - q)^2 \, dt/(2N)]. \tag{4.279}$$

Hence, with the above choice,

$$\wp_{ost}(V) = [dt/(2\pi N)]^{1/2} \exp[-V^2 \, dt/(2N)], \tag{4.280}$$

and the ratio multiplying $\bar{\wp}(q)$ in Eq. (4.278) is

$$\exp[(2Vq - q^2)dt/(2N)]. \tag{4.281}$$

Now the width of $\wp(q)$ will not depend upon dt (as we will see), so we can assume that $\wp(q)$ has support that is finite. Hence the q in Eq. (4.281) can be assumed finite. By contrast, V is of order $1/\sqrt{dt}$, as explained above. Thus, the leading term in the exponent of Eq. (4.281) is $Vq/(dt \, N)$, and this is of order $\sqrt{dt}$. Expanding the exponent to leading order gives

$$\bar{\wp}_V(q) = [1 + qV(t)dt/N]\bar{\wp}(q). \tag{4.282}$$

Expressing this in terms of differentials, we have the Zakai equation

$$d\bar{\wp}_V(q) = (1/\sqrt{N})[V(t)dt/\sqrt{N}]q\bar{\wp}(q), \tag{4.283}$$

where $V(t)dt/\sqrt{N}$ has the ostensible distribution of a Wiener process.

The joint stochastic equation. We now combine the three stochastic equations we have derived above (the SME for the system, the stochastic FPE for the detector and the Zakai equation for the detector) to obtain a joint stochastic equation. We define

$$\bar{\rho}_V^J(q) = \bar{\rho}^J \bar{\wp}_V^J(q). \tag{4.284}$$

On introducing a time-argument for ease of presentation, we have

$$\bar{\rho}_V^J(q; t + dt) = [1 + qV(t)dt/N][\wp^J(q; t) + d\wp^J(q; t)][\bar{\rho}^J(t) + d\bar{\rho}^J(t)], \quad (4.285)$$

where we have used Eq. (4.282), and $d\wp^J(q; t)$ is given by Eq. (4.275) and $d\bar{\rho}^J(t)$ by Eq. (4.273). By expanding these out we find

$$d\bar{\rho}_V^J(q) = dt \left\{ B \frac{\partial}{\partial q}[q - J(t)] + \frac{1}{2} B^2 \frac{\partial^2}{(\partial q)^2} + \mathcal{L} \right\} \bar{\rho}^J(q)$$

$$+ \sqrt{\eta}[J(t)dt] \left\{ 1 - [J(t)dt] B \frac{\partial}{\partial q} \right\} [\hat{c} \bar{\rho}^J(q) + \bar{\rho}^J(q) \hat{c}^\dagger]$$

$$+ (1/\sqrt{N})[V(t)dt/\sqrt{N}]q\bar{\rho}^J(q). \quad (4.286)$$

Averaging over unobserved processes. By construction, the joint stochastic equation in Eq. (4.286) will preserve the factorization of $\bar{\rho}^J(q)$ in the definition Eq. (4.284). This is because this equation assumes that J, the output of the ideal (apart from its inefficiency) detector, is known. In practice, the experimenter knows only V, the output of the realistic detector. Therefore we should average over J. Since we are using a linear SME, this means using the ostensible distribution for Jdt in which it has a mean of zero and a variance of dt. Thus we obtain

$$d\bar{\rho}_V(q) = dt \left\{ B \frac{\partial}{\partial q} q + \frac{1}{2} B^2 \frac{\partial^2}{(\partial q)^2} + \mathcal{L} \right\} \bar{\rho}(q)$$

$$- dt \sqrt{\eta} B \frac{\partial}{\partial q}[\hat{c} \bar{\rho}(q) + \bar{\rho}(q) \hat{c}^\dagger]$$

$$+ (1/\sqrt{N})[V(t)dt/\sqrt{N}]q\bar{\rho}(q). \quad (4.287)$$

We call this the superoperator Zakai equation.

The superoperator Kushner–Stratonovich equation. Let $\bar{\rho}(q)$ at the start of the interval be normalized. That is, let it equal $\rho(q)$, where

$$\int_{-\infty}^{\infty} dq \, \text{Tr}[\rho(q)] = 1. \quad (4.288)$$

Then the infinitesimally evolved unnormalized state determines the actual distribution for V according to

$$\wp(V) = \wp_{\text{ost}}(V) \int_{-\infty}^{\infty} dq \, \text{Tr}[\rho(q) + d\bar{\rho}_V(q)]. \quad (4.289)$$

Using the same arguments as in Section 4.4.3, we see that the actual statistics for V are

$$V dt = \langle Q \rangle dt + \sqrt{N} \, dW(t), \quad (4.290)$$

where

$$\langle Q \rangle = \int_{-\infty}^{\infty} dq \, \text{Tr}[\rho(q)] \, q. \quad (4.291)$$

Note that $dW(t)$ is *not* the same as $dW_1(t)$ in Eq. (4.269). Specifically,

$$\sqrt{N}\,dW = \sqrt{N}\,dW_1 + (Q - \langle Q \rangle). \qquad (4.292)$$

The realistic observer cannot find dW_1 from dW, since that observer does not know Q.

The final step is to obtain an equation for the normalized state, using the actual distribution for V. If we again assume that the state is normalized at the start of the interval, we find

$$d\rho_V(q) = \frac{\rho(q) + d\bar{\rho}_V(q)}{\int dq\,\mathrm{Tr}[\rho(q) + d\bar{\rho}_V(q)]} - \rho(q). \qquad (4.293)$$

Now taking the trace and integral of Eq. (4.287) gives zero for every term except for the last, so the denominator evaluates to

$$1 + (1/\sqrt{N})[V(t)dt/\sqrt{N}]\langle Q \rangle. \qquad (4.294)$$

By expanding the reciprocal of this to *second* order and using Eq. (4.290) to replace $[V(t)dt/\sqrt{N}]^2$ by dt, we find

$$d\rho_V(q) = dt \left\{ B\frac{\partial}{\partial q}q + \frac{1}{2}B^2\frac{\partial^2}{(\partial q)^2} + \mathcal{L} \right\}\rho_V(q)$$

$$- dt\sqrt{\eta}B\frac{\partial}{\partial q}[\hat{c}\rho_V(q) + \rho_V(q)\hat{c}^\dagger]$$

$$+ (1/\sqrt{N})dW(t)(q - \langle Q \rangle)\rho_V(q). \qquad (4.295)$$

Exercise 4.31 *Work this all through explicitly.*

We call Eq. (4.295) a superoperator Kushner–Stratonovich equation. Note that we have placed the V subscript on the state on the right hand side, because typically this will be conditioned on earlier measurements of the output V. The first line describes the evolution of the detector and quantum system separately. The second line describes the coupling of the system to the detector. The third line describes the acquisition of knowledge about the detector from its output Eq. (4.290). This term has the typical form of the information-gathering part of the classical Kushner–Stratonovich equation. Here this information also tells us about the quantum system $\rho_V = \int_{-\infty}^{\infty} dq\,\rho_V(q)$, because of the correlations between the system and the detector caused by their coupling.

If one ignores the output of the detector, then one obtains the unconditioned evolution equation

$$d\rho(q) = dt \left\{ B\frac{\partial}{\partial q}q + \frac{1}{2}B^2\frac{\partial^2}{(\partial q)^2} + \mathcal{L} \right\}\rho(q)$$

$$- dt\sqrt{\eta}B\frac{\partial}{\partial q}[\hat{c}\rho(q) + \rho(q)\hat{c}^\dagger]. \qquad (4.296)$$

The coupling term here will still generate correlations between the system and the detector. Note, however, that this coupling does not cause any back-action on the system, only forward-action on the detector. This can be verified by showing that the *unconditioned* system state $\rho = \int_{-\infty}^{\infty} dq \, \rho(q)$ obeys the original master equation

$$\dot{\rho} = \mathcal{L}\rho = -i[\hat{H}, \rho] + \mathcal{D}[\hat{c}]\rho. \tag{4.297}$$

Exercise 4.32 *Verify this, using the fact that $\wp(q)$, and hence $\rho(q)$, can be assumed to go smoothly to zero as $q \to \pm\infty$.*

The superoperator Kushner–Stratonovich equation is clearly considerably more complicated than the stochastic master equations used to describe other imperfections in detection. In general it is possible only to simulate it numerically. (One technique is given in
Ref. [WW03b], where it is applied to the homodyne detection of a two-level atom.) Nevertheless, it is possible to study some of its properties analytically [WW03a, WW03b].
From these, some general features of the quantum trajectories this equation generates can
be identified. First, in the limit $B \to \infty$, the detector simply adds dark noise and the appropriate SME (4.265) with effective efficiency $\eta/(1 + N)$) can be rederived. Second, for B
finite and $N \ll 1$, the detector has an *effective bandwidth* of

$$B_{\text{eff}} = B/\sqrt{N}, \tag{4.298}$$

which is much greater than B. That is, the detector is insensitive to changes in the system
on a time-scale less than $\sqrt{N}/B$. In the limit $N \to 0$ the effective bandwidth becomes
infinite and the noise becomes zero, so the detector is perfect (apart from η). That is, the
quantum trajectories reduce to those of the SME (4.238), as expected from the arguments
at the beginning of this subsection.

4.9 Continuous measurement in mesoscopic electronics

4.9.1 Monitoring a single quantum dot

In Section 3.5 we discussed the irreversible dynamics of a single-electron quantum dot
coupled to two fermionic reservoirs. Experiments using a two-dimensional electron gas,
confined to the interface between GaAs and AlGaAs and further confined using surface
gates, can be configured to enable real-time monitoring of tunnelling electrons [LJP+03,
BDD05, VES+04, SJG+07]. In this section we will consider a model of such an experiment.

Physical model. A quantum point contact or QPC is the simplest mesoscopic tunnelling
device and consists of two fermionic reservoirs connected by a single electrostatically
defined tunnelling barrier [BB00]. A QPC can be used as a sensitive electrometer if the
tunnelling barrier is modulated by the charge on a nearby quantum dot.

We will consider the device shown in Fig. 4.10. A quantum dot is connected to two Fermi
reservoirs biased so that electrons can tunnel onto the dot from the left reservoir and off the
dot onto the right reservoir. Close to the quantum dot is the QPC. As the electron moves
into the quantum dot it increases the height of the tunnel barrier for the nearby QPC. In

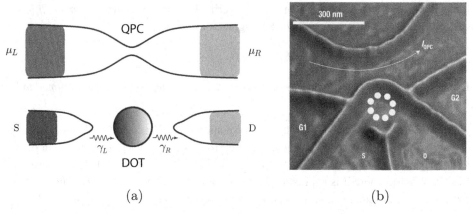

(a)   (b)

Fig. 4.10 (a) A quantum dot with a single-electron quasibound state is connected to two Fermi reservoirs, a source (S) and drain (D) by tunnel junctions. Tunnelling through the quantum dot modulates the tunnelling current through a quantum point contact. (b) An experimental realization using potentials (defined by surface gates) in a GaAs/AlGaAs two-dimensional electron-gas system. Part (b) is reprinted by permission from Macmillan Publishers Ltd, *Nature Physics*, E. V. Sukhorukov *et al.*, **3**, 243, Fig. 1(a), copyright 2007.

this way the modulated current through the QPC can be used continuously to monitor the occupation of the dot. We will follow the treatment given in Ref. [GM01].

The irreversible dynamics of the tunnelling though a single quasibound state on a quantum dot from source to drain is treated in Section 3.5. We assume here that the interaction of the dot with the QPC is weak, and hence does not significantly change the master equation derived there. Here we need a model to describe the tunnnelling current through the QPC and its interaction with the quantum dot. We use the following Hamiltonian (with $\hbar = 1$):

$$\hat{H} = \hat{H}_{\text{QD+leads}} + \hat{H}_{\text{QPC}} + \hat{H}_{\text{coup}}. \tag{4.299}$$

Here $\hat{H}_{\text{QD+leads}}$ is as in Eq. (3.64), and describes the tunnelling of electrons from the source to the dot and from the dot to the drain. This leads to the master equation (3.73) in which the tunnelling rate from source to dot is γ_L and that from dot to drain is γ_R. The new Hamiltonian terms in this chapter are

$$\hat{H}_{\text{QPC}} = \sum_k \left(\omega_k^L \hat{a}_{Lk}^\dagger \hat{a}_{Lk} + \omega_k^R \hat{a}_{Rk}^\dagger \hat{a}_{Rk} \right) + \sum_{k,q} \left(\tau_{kq} \hat{a}_{Lk}^\dagger \hat{a}_{Rq} + \tau_{qk}^* \hat{a}_{Rq}^\dagger \hat{a}_{Lk} \right), \tag{4.300}$$

$$\hat{H}_{\text{coup}} = \sum_{k,q} \hat{c}^\dagger \hat{c} \left(\chi_{kq} \hat{a}_{Lk}^\dagger \hat{a}_{Rq} + \chi_{qk}^* \hat{a}_{Rq}^\dagger \hat{a}_{Lk} \right). \tag{4.301}$$

Here, as in Section 3.5, $\hat{c}$ is the electron annihilation operator for the quantum dot. The Hamiltonian for the QPC detector is represented by $\hat{H}_{\text{QPC}}$, in which $\hat{a}_{Lk}$, $\hat{a}_{Rk}$ and ω_k^L, ω_k^R are, respectively, the electron (fermionic) annihilation operators and energies for the left and right reservoir modes for the QPC at wave number k. Also, there is tunnelling between these modes with amplitudes τ_{kq}. Finally, Eq. (4.301) describes the interaction between the detector and the dot: when the dot contains an electron, the effective tunnelling amplitudes of the QPC detector change from τ_{kq} to $\tau_{kq} + \chi_{kq}$.

In the interaction frame and Markovian approximation, the (unconditional) zero-temperature master equation of the reduced state matrix for the quantum-dot system is [Gur97, GMWS01]

$$\dot{\rho}(t) = \gamma_L \mathcal{D}[\hat{c}^\dagger]\rho + \gamma_R \mathcal{D}[\hat{c}]\rho + \mathcal{D}[\tilde{\tau} + \tilde{\chi}\hat{n}]\rho(t) \qquad (4.302)$$

$$\equiv \mathcal{L}\rho(t), \qquad (4.303)$$

where $\hat{n} = \hat{c}^\dagger \hat{c}$ is the occupation-number operator for the dot, $\tilde{\tau} = \alpha \tau_{00}$ and $\tilde{\chi} = \alpha \chi_{00}$. Here τ_{00} and χ_{00} are the tunnelling amplitudes for energies near the chemical potentials (μ_L and $\mu_R = \mu_L - eV$ in the left and right reservoirs, respectively), while $\alpha^2 = 2\pi(\mu_L - \mu_R)g_L g_R$, where g_L and g_R are the appropriate densities of states for the reservoirs.

Physically, the presence of the electron in the dot raises the effective tunnelling barrier of the QPC due to electrostatic repulsion. As a consequence, the effective tunnelling amplitude through the QPC becomes lower, i.e. $D' = |\tilde{\tau} + \tilde{\chi}|^2 < D = |\tilde{\tau}|^2$. This sets a condition on the relative phase θ between $\tilde{\chi}$ and $\tilde{\tau}$: $\cos\theta < -|\tilde{\chi}|/(2|\tilde{\tau}|)$.

The unconditional dynamics of the number of electrons on the dot is unchanged by the presence of the QPC, and so is given by Eq. (3.74), which we reproduce here:

$$\frac{d\langle \hat{n} \rangle}{dt} = \gamma_L (1 - \langle \hat{n} \rangle) - \gamma_R \langle \hat{n} \rangle. \qquad (4.304)$$

This is because the Hamiltonian describing the interaction between the dot and the QPC commutes with the number operator $\hat{n}$ – the measurement is a QND measurement of $\hat{n}$. However, if we ask for the *conditional* mean occupation of the dot *given* an observed current through the QPC, we do find a (stochastic) dependence on this current, as we will see later.

Exercise 4.33 *Show that the stationary solution to Eq. (4.302) is*

$$\rho_{ss} = \frac{\gamma_L}{\gamma_L + \gamma_R} |1\rangle\langle 1| + \frac{\gamma_R}{\gamma_L + \gamma_R} |0\rangle\langle 0|, \qquad (4.305)$$

and that this is consistent with the stationary solution of Eq. (4.304).

Currents and correlations. It is important to distinguish the two classical stochastic currents through this system: the current $I(t)$ through the QPC and the current $J(t)$ through the quantum dot. Equation (4.302) describes the evolution of the reduced state matrix of the quantum dot when these classical stochastic processes are averaged over. To study the stochastic evolution of the quantum-dot state, conditioned on a particular measurement realization, we need the conditional master equation. We first define the relevant point processes that are the source of the classically observed stochastic currents.

For the tunnelling onto and off the dot, we define two point processes:

$$[dM_c^b(t)]^2 = dM_c^b(t), \tag{4.306}$$

$$E[dM_c^L(t)] = \gamma_L \langle (1 - \hat{n}) \rangle_c(t) dt = \gamma_L \, \text{Tr}\big[\mathcal{J}[\hat{c}^\dagger]\rho_c(t)\big] \, dt, \tag{4.307}$$

$$E[dM_c^R(t)] = \gamma_R \langle \hat{n} \rangle_c(t) dt = \gamma_R \text{Tr}[\mathcal{J}[\hat{c}]\rho_c(t)] dt. \tag{4.308}$$

Here b takes the symbolic value L or R and we use the subscript c to emphasize that the quantities are conditioned upon previous observations (detection records) of the occurrences of electrons tunnelling through the quantum dot and also tunnelling through the QPC barrier (see below). The current through the dot is given by the classical stochastic process

$$J(t)dt = \big[e_L \, dM_c^L(t) + e_R \, dM_c^R(t)\big], \tag{4.309}$$

where e_L and e_R are as defined in Section 3.5 and sum to e.

Next, we define the point process for tunnelling through the QPC:

$$[dN_c(t)]^2 = dN_c(t), \tag{4.310}$$

$$E[dN_c(t)] = \text{Tr}[\mathcal{J}[\tilde{\tau} + \hat{n}\tilde{\chi}]\rho_c] = [D + (D' - D)\langle \hat{n} \rangle_c(t)]dt. \tag{4.311}$$

This is related to the current through the QPC simply by

$$I(t)dt = e \, dN_c(t). \tag{4.312}$$

The expected current is thus eD when the dot is empty and eD' when the dot is occupied.

Exercise 4.34 *Show that the steady-state currents through the quantum dot and QPC are, respectively,*

$$J_{ss} = \frac{e\gamma_L\gamma_R}{\gamma_L + \gamma_R}, \tag{4.313}$$

$$I_{ss} = eD\left(\frac{\gamma_R}{\gamma_L + \gamma_R}\right) + eD'\left(\frac{\gamma_L}{\gamma_L + \gamma_R}\right). \tag{4.314}$$

(Note that the expression for J_{ss} agrees with Eq. (3.77) from Section 3.5.)

The SME describing the quantum-dot state conditioned on the above three point processes is easily derived using techniques similar to those described in Section 4.3. The result is

$$d\rho_c = dM^L(t)\mathcal{G}[\sqrt{\gamma_L}\hat{c}^\dagger]\rho_c + dM^R(t)\mathcal{G}[\sqrt{\gamma_R}\hat{c}]\rho_c + dN(t)\mathcal{G}[\tilde{\tau} + \tilde{\chi}\hat{n}]\rho_c$$
$$- dt \tfrac{1}{2}\mathcal{H}\big[\gamma_L\hat{c}\hat{c}^\dagger + \gamma_R\hat{c}^\dagger\hat{c} + (D' - D)\hat{n}\big]\rho_c. \tag{4.315}$$

Equation (4.315) assumes that we can monitor the current through the quantum dot sufficiently quickly and accurately to distinguish the current pulses associated with the processes $dM^b(t)$. In this case, the dot occupation $\langle n \rangle_c(t)$ will jump between the values 0 and 1. It makes the transition $0 \to 1$ when an electron tunnels onto the dot, which occurs at rate γ_L. It makes the transition $1 \to 0$ when an electron leaves the quantum dot, which occurs at rate γ_R.

Exercise 4.35 *Convince yourself of this from Eq. (4.315).*

Next, we will also assume that D and D' are both much larger than γ_L or γ_R. This means that many electrons will tunnel through the QPC before the electron dot-occupation number changes. Thus the fluctuations in the QPC current $I_c(t)$ due to individual tunnelling events through the QPC (described by $dN(t)$) can be ignored, and it can be treated simply as a two-valued quantity:

$$\langle I \rangle_c (t) = eD + e(D' - D)\langle n \rangle_c(t). \tag{4.316}$$

The two values are eD and eD', depending on the value of $\langle n \rangle_c(t)$, and transitions between them are governed by the transition rates γ_L and γ_R. A process such as this is called a *random telegraph process*.

Using known results for a random telegraph process [Gar85], we can calculate the stationary two-time correlation function for the QPC current,

$$R(t - s) \equiv \mathrm{E}[I(t), I(s)]_{\mathrm{ss}}$$
$$= e^2(D - D')^2 \frac{\gamma_L \gamma_R}{(\gamma_l + \gamma_R)^2} e^{-(\gamma_R + \gamma_L)|t - s|}. \tag{4.317}$$

Here $\mathrm{E}[A, B] \equiv \mathrm{E}[AB] - \mathrm{E}[A]\,\mathrm{E}[B]$, and Eq. (4.317) is sometimes called the *reduced* correlation function because of the subtraction of the products of the means.

Exercise 4.36 *Derive Eq. (4.317) from the master equation (4.302) by identifying $I(t)$ with the observable $eD + e(D' - D)\hat{n}(t)$, and using the quantum regression theorem (4.191).*

As in the case of photocurrents, it is often convenient to characterize the dynamics by the *spectrum* of the current. We define it using the reduced correlation function:

$$S_{\mathrm{QPC}}(\omega) = \int_{-\infty}^{\infty} d\tau\, e^{-i\omega\tau} R(\tau). \tag{4.318}$$

Exercise 4.37 *Show that, for the situation considered above, the QPC spectrum is the Lorentzian*

$$S_{\mathrm{QPC}}(\omega) = e^2(D - D')^2 \frac{2\gamma_L \gamma_R}{(\gamma_L + \gamma_R)^2} \frac{\gamma_L + \gamma_R}{(\gamma_L + \gamma_R)^2 + \omega^2}. \tag{4.319}$$

Conditional dynamics. We now focus on the conditional dynamics of the quantum dot as the QPC current, $I(t)$, is monitored. That is, we average over the dot-tunnelling events described by $dM^b(t)$. Physically, this is reasonable because it would be very difficult to discern the charge pulses (less than one electron) associated with these jumps. It would be similarly difficult to discern the individual jumps $dN(t)$ which define the QPC current $I(t)$ according to Eq. (4.312). In the above, we avoided that issue by considering the limit in which the rate of tunnelling events through the QPC was so high that $I(t)$ could be treated as a random telegraph process, with randomness coming from the quantum-dot dynamics but no randomness associated with the QPC tunnelling itself. It is apparent from the results of the experiment (see Fig. 4.11) that this is a good, but not perfect, approximation. That

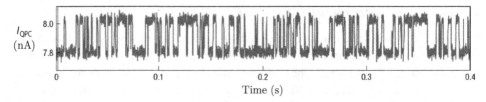

Fig. 4.11 A typical experimental record of the QPC tunnel current as it switches between two values depending on the occupation of the nearby quantum dot. Adapted by permission from Macmillan Publishers Ltd, *Nature Physics*, E. V. Sukhorukov *et al.*, **3**, 243, Fig. 1(c), copyright 2007.

is to say, it is evident that there is some noise on the current in addition to its switching between the values eD and eD'. This noise is due in part to the stochastic processes of electrons tunnelling through the QPC and in part to excess noise added in the amplification process. It is obvious that the individual jumps through the QPC are not resolved; rather, the noise appears to be that of a diffusion process.

We saw in Section 4.8.3 that, if the noise in an ideal current is a Wiener process and the excess noise is also a Wiener process, then the effect of the contaminating noise is simply to reduce the efficiency of the detection to $\zeta < 1$. Since this is the simplest model of imperfect detection, we will apply it in the present case. This requires finding a diffusive approximation to the quantum jump stochastic master equation for describing the conditional QPC current dynamics. As will be seen, this requires $|D' - D|/D \ll 1$. From Fig. 4.11 we see that in the experiment this ratio evaluated to approximately 0.03, so this approximation seems reasonable.

On averaging over the jump process dM^b, introducing an efficiency ζ as described above and assuming for simplicity that $\tilde{\tau}$ and $-\tilde{\chi}$ are real and positive (so that the relative phase $\theta = \pi$), Eq. (4.315) reduces to

$$d\rho_I = dN(t)\mathcal{G}[\sqrt{\zeta}(\tilde{\tau} + \tilde{\chi}\hat{n})]\rho_I + \zeta\tfrac{1}{2}\,dt\,\mathcal{H}\big[2\tilde{\tau}\tilde{\chi}\hat{n} - (\tilde{\chi}\hat{n})^2\big]\rho_I$$
$$+ dt\big\{\gamma_L\mathcal{D}[\hat{c}^\dagger] + \gamma_R\mathcal{D}[\hat{c}] + (1-\zeta)\mathcal{D}[\tilde{\tau} + \tilde{\chi}\hat{n}]\big\}\,\rho_I, \qquad (4.320)$$

where we have used a new subscript to denote conditioning on $I(t) = e\,dN/dt$, where

$$\mathrm{E}[dN(t)] = \mathrm{Tr}\Big[\mathcal{J}[\sqrt{\zeta}(\tilde{\tau} + \tilde{\chi}\hat{n})]\rho_c\Big] = \zeta[D + (D' - D)\langle n\rangle_c(t)]. \qquad (4.321)$$

This describes quantum jumps, in that every time there is a tunnelling event ($dN = 1$) the conditional state changes by a finite amount.

To derive a diffusive limit, we make the following identifications:

$$-\sqrt{\zeta}\,\tilde{\chi}\hat{n} \to \hat{c}; \qquad \sqrt{\zeta}\,\tilde{\tau} \to \gamma. \qquad (4.322)$$

Then, apart from the additional irreversible terms, Eq. (4.320) is identical to the conditional master equation for homodyne detection (4.66). Thus, if we assume that $\tilde{\tau} \gg -\tilde{\chi}$ (or, equivalently, $|D' - D|/D \ll 1$), we can follow the procedure of Section 4.4.2. We thus

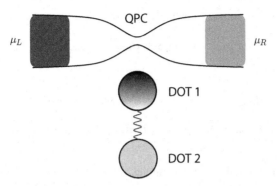

Fig. 4.12 A QPC is used to monitor the occupation of a quantum dot (1) coherently coupled by tunnelling to another quantum dot (2).

obtain the diffusive SME

$$d\rho_c(t) = dt\left(\gamma_L \mathcal{D}[\hat{c}^\dagger] + \gamma_R \mathcal{D}[\hat{c}] + \kappa \mathcal{D}[\hat{n}]\right)\rho_c(t) + \sqrt{\kappa\zeta}\ dW(t)\mathcal{H}[\hat{n}]\rho_c(t), \qquad (4.323)$$

where $\kappa = \tilde{\chi}^2$, and here $\kappa \ll D$. The current on which the state is conditioned is

$$I_c(t) = e\sqrt{\zeta D}\left[\sqrt{\zeta D} - \sqrt{\zeta\kappa}\langle\hat{n}\rangle_c(t) - \xi(t)\right], \qquad (4.324)$$

where $\xi(t) = dW/dt$ as usual. Note the negative signs in Eq. (4.324), because of the sign of $\tilde{\chi}$, and also the constant term which dominates over the term containing the conditional mean, since $\kappa \ll D$.

We can now find the dynamics of the dot occupation, conditioned on the observed QPC current, as

$$\frac{d\langle\hat{n}\rangle_c}{dt} = \gamma_L(1 - \langle\hat{n}\rangle_c) - \gamma_R\langle\hat{n}\rangle_c + 2\sqrt{\zeta\kappa}\langle\hat{n}\rangle_c(1 - \langle\hat{n}\rangle_c)\xi(t). \qquad (4.325)$$

Note that the noise 'turns off' when the dot is either occupied or empty ($\langle\hat{n}\rangle_c = 1$ or 0). This is necessary mathematically in order to prevent the occupation becoming less than 0 or greater than 1. Physically, if $\langle\hat{n}\rangle_c = 1$ (0), we are sure that there is (is not) an electron on the dot, so monitoring the dot cannot give us any more information about the state. Thus, there is no updating of the conditional mean occupation.

4.9.2 Monitoring a double quantum dot

In many ways the QPC coupled to a single quantum dot is not especially quantum; the state can be described by a single number, the occupation probability $\langle\hat{n}\rangle$, since quantum coherences are irrelevant. A modified model in which quantum coherence is important is the double quantum dot (DQD) depicted in Fig. 4.12. This was mentioned before briefly in Section 3.10.1. A system such as this can also be realized in a GaAs/AlGaAs structure [PJM+04, FHWH07]. Here the DQD is not connected to leads (it always contains exactly one electron), so the Hamiltonian for this system is

$$\hat{H} = \hat{H}_{DQD} + \hat{H}_{QPC} + \hat{H}_{coup}, \tag{4.326}$$

where

$$\hat{H}_{DQD} = \omega_1 \hat{c}_1^\dagger \hat{c}_1 + \omega_2 \hat{c}_2^\dagger \hat{c}_2 + \Omega(\hat{c}_1^\dagger \hat{c}_2 + \hat{c}_2^\dagger \hat{c}_1)/2, \tag{4.327}$$

$$\hat{H}_{coup} = \sum_{k,q} \hat{c}_1^\dagger \hat{c}_1 \left(\chi_{kq} \hat{a}_{Lk}^\dagger \hat{a}_{Rq} + \chi_{qk}^* \hat{a}_{Rq}^\dagger \hat{a}_{Lk} \right), \tag{4.328}$$

and $\hat{H}_{QPC}$ is the same as in the previous model.

A detailed analysis of this model is given in Ref. [GM01]. Following the method used above, the conditional evolution of the DQD is described by the SME

$$d\rho_I = dN(t)\mathcal{G}[\sqrt{\zeta}(\tilde{\tau} + \tilde{\chi}\hat{n}_1)]\rho_I + \zeta\tfrac{1}{2} dt\, \mathcal{H}[2\tilde{\tau}\tilde{\chi}\hat{n}_1 - (\tilde{\chi}\hat{n}_1)^2]\rho_I$$
$$- i\Big[(\omega_2 - \omega_1)(\hat{n}_1 - \hat{n}_2)/2 + \Omega(c_1^\dagger c_2 + c_2^\dagger c_1)/2, \rho_I\Big]dt, \tag{4.329}$$

where $\hat{n}_j = \hat{c}_j^\dagger \hat{c}_j$.

Averaging over the jump process gives the unconditional master equation

$$\dot{\rho} = \kappa\mathcal{D}[\hat{n}_1]\rho - i[\hat{V}, \rho]. \tag{4.330}$$

Here $\kappa = |\tilde{\chi}|^2$ as before, while $\hat{n}_j = \hat{c}_j^\dagger \hat{c}_j$ and the effective Hamiltonian is

$$\hat{V} = \frac{\Delta}{2}\hat{\sigma}_z + \frac{\Omega}{2}\hat{\sigma}_x \tag{4.331}$$

(*cf.* Eq. (3.31)). Here we have defined a Bloch representation by

$$\hat{\sigma}_x = \hat{c}_1^\dagger \hat{c}_2 + \hat{c}_2^\dagger \hat{c}_1, \tag{4.332}$$

$$\hat{\sigma}_y = i(\hat{c}_1^\dagger \hat{c}_2 - \hat{c}_2^\dagger \hat{c}_1), \tag{4.333}$$

$$\hat{\sigma}_z = \hat{c}_2^\dagger \hat{c}_2 - \hat{c}_1^\dagger \hat{c}_1, \tag{4.334}$$

so that $z(t) = 1$ and $z(t) = -1$ indicate that the electron is localized in dot 2 and dot 1, respectively.

Exercise 4.38 *Verify that the above operators $\hat{\sigma}_k$ are Pauli operators.*

The parameter Ω is the strength of the tunnelling from one dot to the other, while $\Delta = \omega_2 - \omega_1 + |\tilde{\tau}||\tilde{\chi}|\sin\theta$ is the difference in energy levels between the two dots, which is influenced by the coupling to the QPC unless $\sin\theta = 0$. (Recall that θ is the relative phase of the $\tilde{\chi}$ and $\tilde{\tau}$.) The coupling to the QPC also destroys coherence between the dots at the decoherence rate κ.

Exercise 4.39 *Derive the equations for the Bloch vector (x, y, z) from Eq. (4.330), and show that the steady state is $(0, 0, 0)$, a maximally mixed state.*

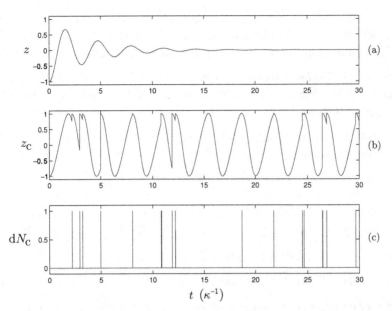

Fig. 4.13 Differences in behaviour between unconditional and conditional evolutions. The initial DQD state is $z = -1$ (dot 1). The parameters are $\zeta = 1$, $\Delta = 0$, $\theta = \pi$ and $D = \kappa = \Omega/2$, and time is in units of κ^{-1}. (Recall that $\kappa = |\tilde{\chi}|^2$ while $D = |\tilde{\tau}|^2$.) (a) Unconditional evolution of $z(t)$. (b) Conditional evolution of $z_c(t)$, interrupted by quantum jumps, corresponding to the stochastically generated QPC detection record shown in (c). Reprinted Figure 3 with permission from H-S. Goan and G. J. Milburn, *Phys. Rev.* B **64**, 235307, (2001). Copyright 2008 by the American Physical Society.

The unconditional evolution of $z(t)$ is shown in Fig. 4.13(a). It undergoes damped oscillations, tending towards zero, as expected for a maximally mixed steady state. The conditional time evolution is quite different. For simplicity we first consider the case, shown in Fig. 4.13(b), of $\theta = \pi$ and $D' = |\tilde{\tau} + \tilde{\chi}|^2 = 0$. That is, due to the electrostatic repulsion generated by the DQD electron, the QPC is blocked (no electron is transmitted) when dot 1 is occupied. As a consequence, whenever there is a detection of an electron tunnelling through the QPC barrier, the DQD state is collapsed into the state $z = 1$ (dot 2 occupied). Note that these jumps are more likely to happen when z is close to 1, so jumps tend to be small (contrast this with the conditioned evolution of a two-level atom under direct detection of its resonance fluorescence, Section 4.6.1). Between jumps, the evolution is smooth but non-unitary. Averaging over the many individual realizations shown in Fig. 4.13(b) leads to a closer and closer approximation of the ensemble average in Fig. 4.13(a).

The plot shown is for the case in which D is comparable to Ω. In the limit $D \gg \Omega$, the jumps become very frequent (with rate D) when the system is in (or close to) the state $z = 1$. Since each jump returns the system to state $z = 1$, as discussed above, the system tends to get stuck in state $z = 1$ (electron in dot 2). This is an example of the *quantum watched-pot effect*, or *quantum Zeno effect*.[4] In fact, the electron can still make a transition to dot 1 (the $z = -1$ state), but the rate of this is suppressed from $O(\Omega)$, in the no-measurement case, to $O(\Omega^2/D)$. In the limit $\Omega/D \to \infty$, the transition rate goes to zero.

The quantum diffusion limit. We saw in Section 4.9.1 that the quantum diffusion equations can be obtained from the quantum jump description under the assumption that $|\tilde{\tau}| \gg |\tilde{\chi}|$ (or equivalently $D \gg \kappa$).

Exercise 4.40 *Derive the diffusive SME for the double-dot case, and hence the diffusive stochastic Bloch equations:*

$$dx_c(t) = -[\Delta\, dt + \sqrt{\zeta\kappa}\, \sin\theta\, dW(t)]y_c(t) - (\kappa/2)x_c(t)dt$$
$$+ \sqrt{\zeta\kappa}\, \cos\theta\, z_c(t)x_c(t)dW(t), \tag{4.335}$$

$$dy_c(t) = [\Delta\, dt + \sqrt{\zeta\kappa}\, \sin\theta\, dW(t)]x_c(t) - \Omega z_c(t)dt - (\kappa/2)y_c(t)dt$$
$$+ \sqrt{\zeta\kappa}\, \cos\theta\, z_c(t)y_c(t)dW, \tag{4.336}$$

$$dz_c(t) = \Omega y_c(t)dt - \sqrt{\zeta\kappa}\, \cos\theta\big[1 - z_c^2(t)\big]dW. \tag{4.337}$$

The measured current gives information about which dot is occupied, as shown in the final term of Eq. (4.337), as expected. However, for $\sin\theta \neq 0$, it also gives information about the rotation around the z axis, as shown in the other two equations. That is, the effective detuning has a deterministic term Δ and a stochastic term proportional to the noise in the current.

In Figs. 4.14(a)–(d), we plot the conditional quantum-jump evolution of $z_c(t)$ and the corresponding detection record $dN_c(t)$, with various values of $(|\tilde{\tau}|/|\tilde{\chi}|)$. Each jump (discontinuity) in the $z_c(t)$ curves corresponds to the detection of an electron through the QPC barrier. One can clearly observe that, with increasing $(|\tilde{\tau}|/|\tilde{\chi}|)$, the rate of jumps increases, but the amplitude of the jumps decreases. When $D' = 0$ each jump collapses the DQD electron into dot 2 ($z = 1$), but as D' approaches D (from below) the jumps in z become smaller, although they are always positive. That is because, whenever there is a detection of an electron passing through QPC, dot 2 is more likely to be occupied than dot 1.

The case for quantum diffusion using Eqs. (4.335)–(4.337) is plotted in Fig. 4.14(e). In this case, infinitely small jumps occur infinitely frequently. We can see that the behaviour of $z_c(t)$ for $|\tilde{\tau}| = 5|\tilde{\chi}|$ in the quantum-jump case shown in Fig. 4.14(d) is already very close to that of quantum diffusion shown in Fig. 4.14(e). Note that for the case $\theta = \pi$ (which we have used in the simulations) the unconditional evolution does not depend on the parameter $|\tilde{\tau}|$. Thus all of these unravellings average to the same unconditioned evolution shown in Fig. 4.13(a).

The QPC current spectrum. We now calculate the stationary spectrum of the current fluctuations through the QPC measuring the coherently coupled DQD. This quantity is

---

[4] The former name alludes to the saying 'a watched pot never boils'; the latter alludes to the 'proof' by the Greek philosopher Zeno of Elea that motion is impossible. See Ref. [GWM93] for a review of this effect and an analysis using quantum trajectory theory.

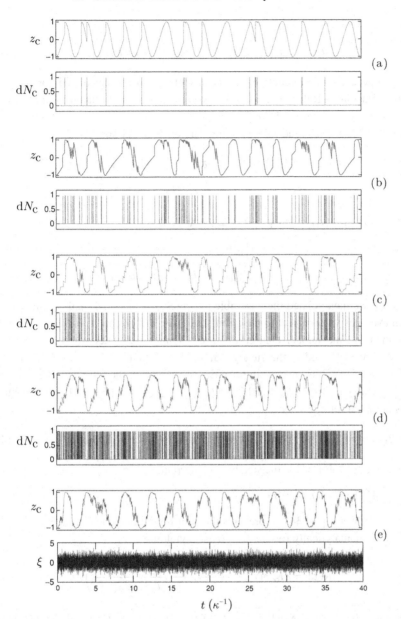

Fig. 4.14 Transition from quantum jumps to quantum diffusion. The parameters are $\zeta = 1$, $\Delta = 0$, $\theta = \pi$ and $\kappa = \Omega/2$, and time is in units of $\kappa^{-1} = |\tilde{\chi}|^{-2}$. In (a)–(d) are shown the quantum jump, conditional evolutions of $z_c(t)$ and corresponding detection moments with the following $|\tilde{\tau}|/|\tilde{\chi}|$ ratios: (a) 1, (b) 2, (c) 3 and (d) 5. In (e) the conditional evolutions of $z_c(t)$ in the quantum diffusive limit ($|\tilde{\tau}|/|\tilde{\chi}| \to \infty$) are shown. The variable $\xi(t) = dW/dt$, the noise in the QPC current in the quantum-diffusive limit, is scaled so as to have unit variance on the plot. Reprinted Figure 4 with permission from H-S. Goan and G. J. Milburn, *Phys. Rev.* B **64**, 235307, (2001). Copyright 2008 by the American Physical Society.

defined in Eq. (4.318), where

$$R(\tau) = E[I(t+\tau)I(t)] - E[I(t+\tau)]E[I(t)]. \tag{4.338}$$

This two-time correlation function for the current has been calculated for the case of quantum diffusion in Ref. [Kor01a]. Here we will present the quantum-jump case from Ref. [GM01], where $I(t) = e\,dN(t)/dt$.

Using the SME (4.329) and following the derivation in Section 4.3.2, we find that

$$E[I(t)] = eE[dN(t)]/dt = e\zeta\,\mathrm{Tr}\big[\{D + (D' - D)\hat{n}_1\}\rho(t)\big], \tag{4.339}$$

while, for $\tau \geq 0$,

$$\begin{aligned}
E[I(t+\tau)I(t)] &= e^2 E[dN_c(t+\tau)dN(t)]/(dt)^2 \\
&= e^2\zeta^2\mathrm{Tr}\big[\{D + (D' - D)\hat{n}_1\}e^{\mathcal{L}\tau}\{\mathcal{J}[\tilde{\tau} + \tilde{\chi}\hat{n}_1]\rho(t)\}\big] \\
&\quad + e^2\zeta\,\mathrm{Tr}\big[\{D + (D' - D)\hat{n}_1\}\rho(t)\big]\delta(\tau).
\end{aligned} \tag{4.340}$$

In this form, we have related the ensemble averages of a classical random variable to the quantum averages with respect to the qubit state matrix. The case $\tau < 0$ is covered by the fact that the two-time autocorrelation function is symmetric by definition.

Now we are interested in the steady-state case in which $t \to \infty$, so that $\rho(t) \to I/2$ (see Exercise 4.39.) Thus we can simplify Eq. (4.340) using the following identities for an arbitrary operator B: $\mathrm{Tr}[e^{\mathcal{L}\tau}B] = \mathrm{Tr}[B]$ and $\mathrm{Tr}[Be^{\mathcal{L}\tau}\rho_\infty] = \mathrm{Tr}[B]/2$, Hence we obtain the steady-state $R(\tau)$ for $\tau \geq 0$ as

$$R(\tau) = eI_\infty\delta(\tau) + e^2\zeta^2(D' - D)^2\big\{\mathrm{Tr}\big[\hat{n}_1 e^{\mathcal{L}\tau}(\hat{n}_1/2)\big] - \mathrm{Tr}[\hat{n}_1/2]^2\big\}, \tag{4.341}$$

where $I_\infty = e\zeta(D' + D)/2$ is the steady-state current.

Exercise 4.41 *Verify Eq. (4.341).*

The first term in Eq. (4.341) represents the shot-noise component. It is easy to evaluate the second term analytically for the $\Delta = 0$ case, yielding

$$R(\tau) = eI_\infty\delta(\tau) + \frac{(\delta I)^2}{4}\left(\frac{\mu_+ e^{\mu_-\tau} - \mu_- e^{\mu_+\tau}}{\mu_+ - \mu_-}\right), \tag{4.342}$$

where $\mu_\pm = -(\kappa/4) \pm \sqrt{(\kappa/4)^2 - \Omega^2}$, and $\delta I = e\zeta(D - D')$ is the difference between the two average currents.

Exercise 4.42 *Derive Eq. (4.342).*
Hint: *This can be done by solving the equations for the unconditioned Bloch vector – Eqs. (4.335)–(4.337) with $\Delta = \zeta = 0$ – with the appropriate initial conditions to represent the initial 'state' $\hat{n}_1/2$. This 'state' is not normalized, but the norm is unchanged by the evolution, so one can take out a factor of $1/2$ and use the normalized initial state $\hat{n}_1$.*

Fourier transforming this, as in Eq. (4.318), yields the spectrum of the current fluctuations as

$$S(\omega) = S_0 + \frac{\Omega^2(\delta I)^2 \kappa/4}{(\omega^2 - \Omega^2)^2 + (\kappa/2)^2 \omega^2}, \tag{4.343}$$

where $S_0 = eI_\infty$ represents the shot noise. Note that, from Eq. (4.343), the noise spectrum at $\omega = \Omega$ can be written as

$$\frac{S(\Omega) - S_0}{S_0} = \frac{(\delta I)^2}{(e\kappa)I_\infty}. \tag{4.344}$$

Exercise 4.43 *Show that this ratio cannot exceed 4ζ.*

For the case $\theta = \pi$ (real tunnelling amplitudes), this ratio can be written as

$$\frac{S(\Omega) - S_0}{S_0} = 2\zeta \frac{(\sqrt{D} + \sqrt{D'})^2}{(D + D')} \tag{4.345}$$

since $\kappa = (\sqrt{D} - \sqrt{D'})^2$ in this case. In the quantum-diffusive limit $(D + D') \gg (D - D')$, this ratio attains its upper bound, as was first shown in Ref. [Kor01a].

For $\kappa < 4\Omega$, the spectrum has a double-peak structure, indicating that coherent tunnelling is taking place between the two qubit states. This is illustrated in Figs. 4.15(a) and (b). Conditionally, the oscillations of $z_c(t)$ would be very nearly sinusoidal for $\kappa \ll 4\Omega$, but would become increasingly noisy and distorted as κ increases. When $\kappa \geq 4\Omega$, the measurement is strong enough substantially to destroy any coherence between the two dots, suppressing coherent oscillations. In this case, only a single peak, centred at $\omega = 0$, appears in the noise spectrum, as illustrated in Figure 4.15(c). In the limit $\kappa \gg 4\Omega$, the conditioned state of the DQD exhibits jump-like behaviour, with the electron being very likely to be in one well or the other at all times. That is, $z_c(t)$ can be modelled as a classical random telegraph process, with two states, $z_c = \pm 1$, just as in the single-dot case of Section 4.9.1.

4.9.3 Other theoretical approaches to monitoring in mesoscopic systems

In monitoring a solid-state quantum system, if one ignores or averages over the results, the only effect of the measurement is to decohere the system. The first step beyond this is to condition on a single property of the output (by which we mean the drain of the QPC, for example). The authors of Refs. [Gur97, SS98] considered conditioning on the number N of excess electrons that had tunnelled into the drain. This approach represents suboptimal conditioning of the system's state matrix because the information in the *times* of electron tunnelling events through the detector is unused. (In the case of a single-electron transistor (SET) as detector [SS98], further information is ignored – the times at which electrons tunnel onto the SET island from the source. See Ref. [Oxt07] for a discussion of this and other points in this section.)

A stochastic equation representing the conditional evolution of a solid-state system in real time was introduced into the phenomenological model by Korotkov [Kor99]. This was

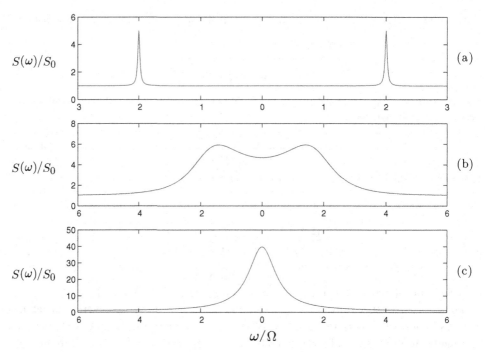

Fig. 4.15 A plot of the noise spectrum of the QPC current monitoring a double quantum dot. The spectrum is normalized by the shot-noise level, and is shown for various ratios of measurement strength κ to tunnelling strength Ω, with $\kappa/(4\Omega)$ equalling (a) 0.01, (b) 0.5 and (c) 2. For discussion, see the text. Reprinted Figure 5 with permission from H-S. Goan and G. J. Milburn, *Phys. Rev.* B **64**, 235307, (2001). Copyright 2008 by the American Physical Society.

later derived from a microscopic model in Ref. [Kor01b]. This model was restricted to a weakly responding detector (the diffusive limit discussed above), but has been extended to the case of a strongly responding detector (quantum jumps) [Kor03]. Korotkov's approach (which he called 'Bayesian') is completely equivalent to the quantum trajectory approach used above [GMWS01]. Quantum trajectories were first used in the solid-state context in Ref. [WUS+01], which included a derivation from a (rather simplistic) microscopic model.

From the beginning [Kor99, WUS+01], these theories of continuous monitoring in mesoscopic systems have allowed for non-ideal monitoring. That is, even if the initial state were pure, the conditional state would not in general remain pure; there is no description using a stochastic Schrödinger equation. The microscopic model in Ref. [WUS+01] is inherently non-ideal, while, in Ref. [Kor99], Korotkov introduced a phenomenological dephasing rate, which he later derived by introducing extra back-action noise from the detector [Kor01b]. Another sort of non-ideality was considered in Ref. [GM01], where the authors introduce 'inefficient' detection by a QPC, as used in Section 4.9.1 Here efficiency has the same sense as in quantum optics: some proportion $1 - \zeta$ of the detector output is 'lost'. This is of course equivalent to introducing extra decoherence as produced by an unmonitored detector. As shown generally in Section 4.8.3, in the diffusive limit, the same conditioning equation results if extra white noise ('dark noise') is added to the detector output before it is recorded.

The theory of quantum trajectories for mesoscopic systems has recently been extended to allow for the noisy filtering characteristic of amplifiers used in such experiments [OWW+05, OGW08]. This was done using the same theory as that presented in Section 4.8.4, but taking into account correlations between noise that disturbs the system and noise in the recorded current. Such 'realistic' quantum trajectories are essential for optimal feedback control, because the optimal algorithms are based upon the state conditioned on the measurement record, as will be discussed in Section 6.3.3. Note that the authors of Ref. [RK03] do consider the effect of a finite-bandwidth filter on a feedback algorithm, but that is a quite distinct idea. There, the feedback algorithm is not based on the state of the system conditioned on the filtered current, and indeed no such conditional state is calculated.

4.10 Further reading

There are several books that discuss continuous quantum measurement theory in the context of open quantum systems. *An Open Systems Approach to Quantum Optics* by Carmichael [Car93] was a very influential book in the quantum-optics community, and introduced many of the terms used here, such as 'quantum trajectory' and 'unravelling'. Much of this material is contained in volume 2 of Carmichael's recently published *Statistical Methods in Quantum Optics* [Car99, Car07]. *Quantum Noise* by Gardiner and Zoller [GZ04] is not confined to quantum optics. It has material on quantum trajectories and quantum stochastic differential equations, but also covers quantum noise in a more general sense. *The Theory of Open Quantum Systems* by Breuer and Pettrucione [BP02] emphasizes the application of stochastic Schrödinger equations for numerical simulations of open quantum systems, both Markovian and non-Markovian.

In this chapter we restricted our attention to systems obeying a Markovian master equation, and indeed a master equation of the Lindblad form. The reason is that it is only in this case that a measurement of the bath to which the system is coupled can be performed without altering the average evolution of the system [GW03, WG08]. In particular, the non-Markovian stochastic Schrödinger equations developed by Diósi, Strunz and Gisin [SDG99, Dió08] cannot be interpreted as evolution equations for a system conditioned upon a measurement record. This is despite the fact that in the Markovian limit they can reproduce the quantum trajectories for homodyne and heterodyne detection [GW02] and even for direct detection [GAW04]. In fact, the only physical interpretation for these equations is in terms of hidden-variable theories [GW03].

5

Quantum feedback control

5.1 Introduction

In the preceding chapter we introduced quantum trajectories: the evolution of the state of a quantum system conditioned on monitoring its outputs. As discussed in the preface, one of the chief motivations for modelling such evolution is for quantum feedback control. Quantum feedback control can be broadly defined as follows. Consider a detector continuously producing an output, which we will call a current. Feedback is any process whereby a physical mechanism makes the statistics of the present current at a later time depend upon the current at earlier times. Feedback control is feedback that has been engineered for a particular purpose, typically to improve the operation of some device. Quantum feedback control is feedback control that requires some knowledge of quantum mechanics to model. That is, there is some part of the feedback loop that must be treated (at some level of sophistication) as a quantum system. There is no implication that the whole apparatus must be treated quantum mechanically.

The structure of this chapter is as follows. The first quantum feedback experiments (or at least the first experiments specifically identified as such) were done in the mid 1980s by two groups [WJ85a, MY86]. They showed that the photon statistics of a beam of light could be altered by feedback. In Section 5.2 we review such phenomena and give a theoretical description using linearized operator equations. Section 5.3 considers the changes that arise when one allows the measurement to involve nonlinear optical processes. As well as explaining key results in quantum-optical feedback, these sections introduce important concepts for feedback in general, such as stability and robustness, and important applications such as noise reduction. These sections make considerable use of material from Ref. [Wis04].

From Section 5.4 onwards we turn from feedback on continuous fields to feedback on a localized system that is continuously monitored. We give a general description for feedback in such systems and show how, in the Markovian limit, the evolution including the feedback can be described by a new master equation. We formulate our description both in the Schrödinger picture and in the Heisenberg picture, and we discuss an elegant example for which the former description is most useful: protecting a Schrödinger-cat state from decoherence. In Section 5.5 we redevelop these results for the particular case of a measurement yielding a current with Gaussian white noise, such as homodyne detection. We include the effects of a white-noise (thermal or squeezed) bath. In Section 5.6 we apply this theory for homodyne-based feedback to a very simple family of quantum-optical systems with linear dynamics. We show that, although Markovian feedback can be described without reference to the conditional state, it is the conditional state that determines both the optimal feedback strength and how the feedback performs. In Section 5.7 we discuss a proposed application using (essentially) Markovian feedback to produce deterministic spin-squeezing in an atomic ensemble. Finally, in Section 5.8 we discuss other concepts and other applications of quantum feedback control.

5.2 Feedback with optical beams using linear optics

5.2.1 Linearized theory of photodetection

The history of feedback in quantum optics goes back to the observation of sub-shot-noise fluctuations in an in-loop photocurrent (defined below) in the mid 1980s by two groups [WJ85a, MY86]. A theory of this phenomenon was soon developed by Yamamoto and co-workers [HY86, YIM86] and by Shapiro *et al.* [SSH+87]. The central question they were addressing was whether this feedback was producing real squeezing (defined below), a question whose answer is not as straightforward as might be thought. These treatments were based in the Heisenberg picture. They used quantum Langevin equations where necessary to describe the evolution of source operators, but they were primarily interested in the properties of the beams entering the photodetectors, rather than their sources.

The Heisenberg picture is most convenient if (a) one is interested primarily in the properties of the beams and (b) an analytical solution is possible. To obtain analytical results, it is necessary to treat the quantum noise only within a linearized approximation. We begin therefore by giving the linearized theory for photodetection in the Heisenberg picture.

Using the theory from Section 4.7, the operator for the photon flux in a beam at longitudinal position z_1 is

$$\hat{I}(t) = \hat{b}_1^\dagger(t)\hat{b}_1(t), \tag{5.1}$$

where $\hat{b}_1(t) \equiv \hat{b}(z_1, t)$ as defined in Section 3.11. From that section, it should be apparent that it is not sensible to talk about a photodetector for photons of frequency ω_0 that has a response time comparable to or smaller than ω_0^{-1}, but, as long as we are not interested in times comparable to ω_0^{-1}, we can assume that the signal produced by an ideal photodetector at position z_1 is given by Eq. (5.1).

In experiments involving lasers, it is often the case (or at least it is sensible to assume [Møl97]) that $\hat{b}_1(t)$ has a mean amplitude β such that $\beta^2 = \langle \hat{b}_1^\dagger(t)\hat{b}_1(t) \rangle$. Here, without loss of generality, we have taken β to be real. Because point-process noise is often hard to treat analytically, it is common to linearize Eq. (5.1) by approximating it by

$$\hat{I}(t) = \beta^2 + \delta\hat{I}(t) = \beta^2 + \beta\hat{X}_1(t), \tag{5.2}$$

where the amplitude quadrature fluctuation operator

$$\hat{X}_1(t) = \hat{b}_1(t) + \hat{b}_1^\dagger(t) - 2\beta \tag{5.3}$$

is assumed to have zero mean. This approximation is essentially the same as that used in Section 4.4.2 to treat homodyne detection in the large-local-oscillator limit. It assumes that individual photon counts are unimportant, namely that the system fluctuations are evident only in large numbers of detections. This approximation will be valid if the correlations of interest in the system happen on a time-scale long compared with the inverse mean count rate and if the fluctuations are relatively small:

$$\langle \hat{X}_1(t + \tau)\hat{X}_1(t) \rangle \ll \beta^2 \text{ for } \tau \neq 0. \tag{5.4}$$

In all that follows we will consider only stationary stochastic processes, where the two-time correlation functions depend only on the time difference τ. We cannot consider $\tau = 0$ in Eq. (5.4), because the variance diverges due to vacuum fluctuations:[1]

$$\lim_{\tau \to 0}\langle \hat{X}_1(t + \tau)\hat{X}_1(t) \rangle = \lim_{\tau \to 0}\delta(\tau). \tag{5.5}$$

Note, however, that in this limit the linearized correlation function agrees with that from Eq. (5.1):

$$\lim_{\tau \to 0}\langle \hat{I}(t)\hat{I}(t + \tau) \rangle = \lim_{\tau \to 0}\langle \hat{I}(t) \rangle \delta(\tau) = \lim_{\tau \to 0}\beta^2\delta(\tau). \tag{5.6}$$

Just as we defined $\hat{x}$ and $\hat{y}$ quadratures for a system in Chapter 4, here it is also useful to define the phase quadrature fluctuation operator

$$\hat{Y}_1(t) = -i\hat{b}_1(t) + i\hat{b}_1^\dagger(t). \tag{5.7}$$

For free fields, where (taking the speed of light to be unity as usual)

$$\hat{b}(z, t + \tau) = \hat{b}(z - \tau, t), \tag{5.8}$$

the canonical commutation relation for the fields at different positions

$$[\hat{b}(z, t), \hat{b}^\dagger(z', t)] = \delta(z - z') \tag{5.9}$$

implies that the quadratures obey the temporal commutation relation

$$[\hat{X}_1(t), \hat{Y}_1(t')] = 2i\delta(t - t'). \tag{5.10}$$

We define the Fourier-transformed operator as follows:

$$\tilde{X}_1(\omega) = \int_{-\infty}^{\infty} dt\, \hat{X}_1(t)e^{-i\omega t}; \tag{5.11}$$

and similarly for $\tilde{Y}_1(\omega)$. Note that we use a tilde but drop the hat for notational convenience. Then it is simple to show that

$$[\tilde{X}_1(\omega), \tilde{Y}_1(\omega')] = 4\pi i\delta(\omega + \omega'). \tag{5.12}$$

---

[1] For thermal or squeezed white noise – see Section 4.8.2 – this δ-function singularity is multiplied by a non-unit constant.

For stationary statistics as we are considering, $\langle \hat{X}_1(t)\hat{X}_1(t')\rangle$ is a function of $t - t'$ only. From this it follows that

$$\langle \tilde{X}_1(\omega)\tilde{X}_1(\omega')\rangle \propto \delta(\omega + \omega'). \tag{5.13}$$

Exercise 5.1 *Verify Eqs. (5.12) and (5.13).*

Because of the singularities in Eqs. (5.12) and (5.13), to obtain a finite uncertainty relation it is more useful to consider the spectrum

$$S_1^X(\omega) = \frac{1}{2\pi} \int_{-\infty}^{\infty} \langle \tilde{X}_1(\omega)\tilde{X}_1(-\omega')\rangle d\omega' \tag{5.14}$$

$$= \int_{-\infty}^{\infty} e^{-i\omega t} \langle \hat{X}_1(t)\hat{X}_1(0)\rangle dt \; = \langle \tilde{X}_1(\omega)\hat{X}_1(0)\rangle. \tag{5.15}$$

Note that the final expression involves both $\hat{X}_1$ and $\tilde{X}_1$. Equation (5.15) is the same as the spectrum defined for a homodyne measurement of the x quadrature in Section 4.4.4 if we take the system quadrature x to have zero mean. In the present case, the spectrum can be experimentally determined as

$$S_1^X(\omega) = \langle \hat{I}(t)\rangle^{-1} \int_{-\infty}^{\infty} e^{-i\omega t} \langle \hat{I}(t), \hat{I}(0)\rangle dt, \tag{5.16}$$

where $\langle A, B\rangle \equiv \langle AB\rangle - \langle A\rangle\langle B\rangle$ as previously. In fact it is possible to determine $S_1^Q(\omega)$ for any quadrature Q in a similar way by adding a local oscillator of suitable amplitude and phase, as described in Section 4.4.

It can be shown [SSH+87] that, for a stationary free field, the commutation relations (5.10) imply that

$$S_1^X(\omega)S_1^Y(\omega) \geq 1. \tag{5.17}$$

This can be regarded as an uncertainty relation for continuum fields. A coherent continuum field where $\hat{b}_1|\beta\rangle = \beta|\beta\rangle$ has, for all ω,

$$S_1^Q(\omega) = 1, \tag{5.18}$$

where $Q = X$ or Y (or any intermediate quadrature). This is known as the *standard quantum limit* or *shot-noise limit*. A *squeezed* continuum field is one such that, for some ω and some Q,

$$S_1^Q(\omega) < 1. \tag{5.19}$$

This terminology is appropriate for the same reason as for single-mode squeezed states: the reduced noise in one quadrature gets 'squeezed' out into the other quadrature, because of Eq. (5.17).

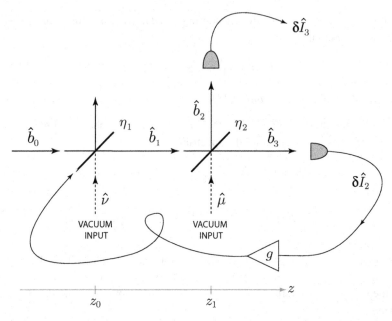

Fig. 5.1 A diagram for a travelling-wave feedback experiment. Travelling fields are denoted $\hat{b}$ and photocurrent fluctuations δI. The first beam-splitter transmittance, η_1, is variable, the second, η_2, fixed. The two vacuum field inputs are denoted $\hat{\nu}$ and $\hat{\mu}$. *Quantum Squeezing*, 2004, pp. 171–122, Chapter 6 'Squeezing and Feedback', H. M. Wiseman, Figure 6.1, Springer-Verlag, Berlin, Heidelberg. Redrawn and revised and adapted with kind permission of Springer Science+Business Media.

5.2.2 In-loop 'squeezing'

The simplest form of quantum optical feedback is shown in Fig. 5.1. This was the scheme considered by Shapiro *et al.* [SSH+87]. In our notation, we begin with a field $\hat{b}_0 = \hat{b}(z_0, t)$ as shown in Fig. 5.1. We will take this field to have stationary statistics with mean amplitude β and fluctuations

$$\tfrac{1}{2}[\hat{X}_0(t) + i\hat{Y}_0(t)] = \hat{b}_0(t) - \beta. \tag{5.20}$$

We take the amplitude and phase noises to be independent and characterized by arbitrary spectra $S_0^X(\omega)$ and $S_0^Y(\omega)$, respectively.

This field is then passed through a beam-splitter of transmittance $\eta_1(t)$. By unitarity, the diminution in the transmitted field by a factor $\sqrt{\eta_1(t)}$ must be accompanied by the addition of vacuum noise from the other port of the beam-splitter (see Section 4.4.1). The transmitted field is

$$\hat{b}_1(t) = \sqrt{\eta_1(t - \tau_1)}\, \hat{b}_0(t - \tau_1) + \sqrt{\bar{\eta}_1(t - \tau_1)}\, \hat{\nu}(t - \tau_1). \tag{5.21}$$

Here $\tau_1 = z_1 - z_0$ and we are using the notation

$$\bar{\eta}_j \equiv 1 - \eta_j. \tag{5.22}$$

The annihilation operator $\hat{v}(t)$ represents the vacuum fluctuations. The vacuum is a special case of a coherent continuum field of vanishing mean amplitude $\langle \hat{v}(t) \rangle = 0$, and so is completely characterized by its spectrum

$$S_v^Q(\omega) = 1. \tag{5.23}$$

Since the vacuum fluctuations are uncorrelated with any other field, and have stationary statistics, the time argument for $\hat{v}(t)$ is arbitrary (when it first appears).

The beam-splitter transmittance $\eta_1(t)$ in Eq. (5.21) is time-dependent. This time-dependence can be achieved experimentally by a number of means. For example, if the incoming beam is elliptically polarized then an electro-optic modulator (a device with a refractive index controlled by a current) will alter the orientation of the ellipse. A polarization-sensitive beam-splitter will then control the amount of the light which is transmitted, as done, for example, in [TWMB95]. As the reader will no doubt have anticipated, the current used to control the electro-optic modulator can be derived from a later detection of the light beam, giving rise to feedback. Writing $\eta_1(t) = \eta_1 + \delta\eta_1(t)$, and assuming that the modulation of the transmittance is small ($\delta\eta_1(t) \ll \eta_1, \bar{\eta}_1$), one can write

$$\sqrt{\eta_1(t)} = \sqrt{\eta_1} + (1/\sqrt{\eta_1})\tfrac{1}{2}\,\delta\eta_1(t). \tag{5.24}$$

Continuing to follow the path of the beam in Fig. 5.1, it now enters a second beam-splitter of constant transmittance η_2. The transmitted beam annihilation operator is

$$\hat{b}_2(t) = \sqrt{\eta_2}\,\hat{b}_1(t - \tau_2) + \sqrt{\bar{\eta}_2}\,\hat{\mu}(t - \tau_2), \tag{5.25}$$

where $\tau_2 = z_2 - z_1$ and $\hat{\mu}(t)$ represents vacuum fluctuations like $\hat{v}(t)$. The reflected beam operator is

$$\hat{b}_3(t) = \sqrt{\bar{\eta}_2}\,\hat{b}_1(t - \tau_2) - \sqrt{\eta_2}\,\hat{\mu}(t - \tau_2). \tag{5.26}$$

Using the approximation (5.24), the linearized quadrature fluctuation operators for $\hat{b}_2$ are

$$\hat{X}_2(t) = \sqrt{\eta_2\eta_1}\,\hat{X}_0(t - T_2) + \sqrt{\eta_2/\eta_1}\,\beta\,\delta\eta_1(t - T_2)$$
$$+ \sqrt{\eta_2\bar{\eta}_1}\,\hat{X}_v(t - T_2) + \sqrt{\bar{\eta}_2}\,\hat{X}_\mu(t - T_2), \tag{5.27}$$

$$\hat{Y}_2(t) = \sqrt{\eta_2\eta_1}\,\hat{Y}_0(t - T_2)$$
$$+ \sqrt{\eta_2\bar{\eta}_1}\,\hat{Y}_v(t - T_2) + \sqrt{\bar{\eta}_2}\,\hat{Y}_\mu(t - T_2), \tag{5.28}$$

where $T_2 = \tau_2 + \tau_1$. Here, for simplicity, we have shifted the time argument of the μ vacuum quadrature operators by τ_1. This is permissible because the vacuum noise is a stationary process (regardless of any other noise processes). Similarly, for $\hat{b}_3$ we have

$$\hat{X}_3(t) = \sqrt{\bar{\eta}_2\eta_1}\,\hat{X}_0(t - T_2) + \sqrt{\bar{\eta}_2/\eta_1}\,\beta\,\delta\eta_1(t - T_2)$$
$$+ \sqrt{\bar{\eta}_2\bar{\eta}_1}\,\hat{X}_v(t - T_2) - \sqrt{\eta_2}\,\hat{X}_\mu(t - T_2), \tag{5.29}$$

$$\hat{Y}_3(t) = \sqrt{\bar{\eta}_2\eta_1}\,\hat{Y}_0(t - T_2) + \sqrt{\bar{\eta}_2\bar{\eta}_1}\,\hat{Y}_v(t - T_2)$$
$$- \sqrt{\eta_2}\,\hat{Y}_\mu(t - T_2). \tag{5.30}$$

Exercise 5.2 *Derive Eqs. (5.27)–(5.30).*

The mean fields for $\hat{b}_2$ and $\hat{b}_3$ are $\sqrt{\eta_1 \eta_2}\,\beta$ and $\sqrt{\eta_1 \bar{\eta}_2}\,\beta$, respectively. Thus, if these fields are incident upon photodetectors, the respective linearized photocurrent fluctuations are, as explained in Section 5.2.1,

$$\delta \hat{I}_2(t) = \sqrt{\eta_1 \eta_2}\,\beta \hat{X}_2(t), \tag{5.31}$$

$$\delta \hat{I}_3(t) = \sqrt{\eta_1 \bar{\eta}_2}\,\beta \hat{X}_3(t). \tag{5.32}$$

Here we have assumed perfect-efficiency detectors. Inefficient detectors can be modelled by further beam-splitters (see Section 4.8), and the effect of this has been considered in detail in Refs. [MPV94, TWMB95].

Having obtained an expression for $\delta \hat{I}_2(t)$, we are now in a position to follow the next stage in Fig. 5.1 and complete the feedback loop. We set the modulation in the transmittance of the first beam-splitter to be

$$\delta \hat{\eta}_1(t) = \frac{g}{\eta_2 \beta^2} \int_0^\infty h(t')\delta \hat{I}_2(t - \tau_0 - t')\mathrm{d}t', \tag{5.33}$$

where g is a dimensionless parameter. It represents the low-frequency gain of the feedback, as will be seen. The response of the feedback loop, including the electro-optic elements, is assumed to be linear for small fluctuations and is characterized by the electronic delay time τ_0 and the response function $h(t')$, which satisfies $h(t) = 0$ for $t < 0$, $h(t) \geq 0$ for $t > 0$ and $\int_0^\infty h(t')\mathrm{d}t' = 1$.

The appearance of a hat on the beam-splitter transmittance η_1 in Eq. (5.33) may give the impression that one is giving a quantum-mechanical treatment of a macroscopic system. This is a false impression. The only features of the operators which are important are their stochastic nature and their correlations with the source. The macroscopic apparatus has not been quantized; it is simply correlated to the fluctuations in the observed photocurrent, which are represented by an operator, as explained in Sections 1.3 and 4.7. In the quantum trajectory description of feedback, which is considered in the later sections of this chapter, this false impression would never arise. The feedback apparatus is treated as a completely classical system, which is of course how experimentalists would naturally regard it.

5.2.3 Stability

Clearly the feedback can affect only the amplitude quadrature $\hat{X}$. Putting Eq. (5.33) into Eq. (5.27) yields

$$\hat{X}_2(t) = \sqrt{\eta_2 \eta_1}\,\hat{X}_0(t - T_2) + g \int_0^\infty h(t')\hat{X}_2(t - T - t')\mathrm{d}t'$$

$$+ \sqrt{\eta_2 \bar{\eta}_1}\,\hat{X}_\nu(t - T_2) + \sqrt{\bar{\eta}_2}\,\hat{X}_\mu(t - T_2), \tag{5.34}$$

where $T = \tau_0 + T_2 = \tau_0 + \tau_1 + \tau_2$. This is easy to solve in Fourier space, provided that $\hat{X}_2$ is a stationary stochastic process. This will be the case if and only if the feedback loop is stable.

The problem of stability in feedback loops is a difficult one in general. However, for a simple linear system such as we are considering here, there are criteria for testing whether a given loop is stable and techniques for designing loops that will satisfy these criteria. For the interested reader, an elementary introduction can be found in Ref. [SSW90]. Here we will not cover this theory in detail, but give only the bare essentials needed for this case. Our aim is to acquire some insight into the stability criteria, and especially the trade-off among the bandwidth of the loop, its gain and the time delay.

We begin by writing Eq. (5.34) in the form

$$\hat{X}_2(t) - g \int_0^\infty h(t')\hat{X}_2(t - T - t')dt' = \hat{f}(t), \tag{5.35}$$

where $\hat{f}(t)$ represents all of the (stationary) noise processes in Eq. (5.34). Now the solution to this equation can be found by taking the Laplace transform:

$$[1 - gh^{\mathrm{L}}(s)\exp(-sT)]\hat{X}_2^{\mathrm{L}}(s) = \hat{f}^{\mathrm{L}}(s), \tag{5.36}$$

where $h^{\mathrm{L}}(s) = \int_0^\infty dt\, e^{-st}h(t)$ etc.

Exercise 5.3 *Derive Eq. (5.36) and show that the low-frequency ($s = 0$) equation is*

$$\hat{X}_2^{\mathrm{L}}(0) = g\hat{X}_2^{\mathrm{L}}(0) + \hat{f}^{\mathrm{L}}(0), \tag{5.37}$$

which shows that g is indeed the low-frequency gain of the feedback.

A stable feedback loop means that, for a spectral-bounded noise input $\hat{f}(t)$, the solution $\hat{X}(t)$ will also be spectral-bounded for all times. (Here by 'spectral-bounded' we mean that the spectrum, as defined in Eq. (5.14), is bounded from above.) This will be the case if and only if

$$\mathrm{Re}[s] < 0, \tag{5.38}$$

where s is *any* solution of the characteristic equation

$$1 - gh^{\mathrm{L}}(s)\exp(-sT) = 0. \tag{5.39}$$

This is known as the Nyquist stability criterion [SSW90].

First we show that a *sufficient* condition for stability is $|g| < 1$. Looking for instability, assume that $\mathrm{Re}[s] > 0$. Then

$$\left|h^{\mathrm{L}}(s)e^{-sT}\right| = \left|\int_0^\infty dt\, e^{-s(t+T)}h(t)\right| < \int_0^\infty dt\, h(t) = 1. \tag{5.40}$$

Thus under this assumption the characteristic equation cannot be satisfied for $|g| < 1$. That is, the $|g| < 1$ regime will always be stable. On the other hand, if $g > 1$ then there is an s with a positive real part that will solve Eq. (5.39).

Box 5.1 The feedback gain–bandwidth relation

Consider feedback as described in the text with the simple smoothing function $h(t) = \Gamma e^{-\Gamma t}$ for $t > 0$. In this case the characteristic equation (5.39) becomes (for g negative)

$$s + \Gamma + |g|\Gamma e^{-sT} = 0. \tag{5.41}$$

On setting $s = T^{-1}(\lambda + i\omega)$ for λ and ω real, and defining $\gamma = \Gamma T$, this is equivalent to the following equations in real, dimensionless variables:

$$\lambda + \gamma = -|g|\gamma e^{-\lambda} \cos \omega, \tag{5.42}$$

$$\omega = +|g|\gamma e^{-\lambda} \sin \omega. \tag{5.43}$$

By combining these equations, one finds the following relation:

$$\lambda = -\gamma - \omega \cot \omega. \tag{5.44}$$

Thus, as long as $\omega \cot \omega$ is positive, the real part of s (that is, $T^{-1}\lambda$) will be negative, as is required for stability. Substituting Eq. (5.44) into Eq. (5.42) or Eq. (5.43) to eliminate λ yields

$$\frac{\omega}{\sin \omega} = |g|\gamma e^{\gamma} e^{\omega \cot \omega}. \tag{5.45}$$

This has no analytical solution, but as long as

$$|g|\gamma e^{\gamma} \leq \pi/2 \tag{5.46}$$

all solutions of this equation satisfy $n\pi \leq |\omega| \leq (n + 1/2)\pi$, for n an integer, and so guarantee that $\omega \cot \omega$ is positive. Moreover, for large gain $|g|$, if $|g|\gamma e^{\gamma}$ is significantly larger than $\pi/2$ then there exist solutions ω to Eq. (5.45) such that λ in Eq. (5.44) is positive.

Thus we have a stability condition (Eq. (5.46)) that is sufficient and, for large feedback gain, not too far from necessary. Since $|g| \gg 1$ requires $\gamma \ll 1$, in this regime the condition simplifies to $|g|\gamma \leq \pi/2$. Returning to the original variables, we thus have the following approximate gain–bandwidth relation:

$$\Gamma \lesssim \frac{\pi}{2T|g|} \ll T^{-1}. \tag{5.47}$$

That is, a finite delay time T and large negative feedback $-g \gg 1$ puts an upper bound on the bandwidth $B = 2\Gamma$ of the feedback (here defined as the full-width at half-maximum height of $|\tilde{h}(\omega)|^2$). Moreover, this upper bound implies that in the average signal delay in the feedback loop, $\int_T^{\infty} h(\tau)\tau \, d\tau = \Gamma^{-1} + T$, the response-function decay time Γ^{-1} dominates over the raw delay time T.

Exercise 5.4 *Prove this by considering the left-hand side of Eq. (5.39) as a function of s on the interval $[0, \infty)$ on the real line. In particular, consider its value at 0 and its value at ∞.*

Thus it is a *necessary* condition to have $g < 1$. If $g < -1$, the stability of the feedback depends on T and the shape of $h(t)$. However, it turns out that it is possible to have arbitrarily large negative low-frequency feedback (that is, $-g \gg 1$), for any feedback loop delay T, provided that $h(t)$ is broad enough. That is, the price to be paid for strong low-frequency negative feedback is a reduction in the bandwidth of the feedback, namely the width of $|\tilde{h}(\omega)|^2$. A simple example of this is considered in Box 5.1, to which the following exercise pertains.

Exercise 5.5 *Convince yourself of the statements following Eq. (5.46) by graphing both sides of Eq. (5.45) for different values of $|g|\gamma e^\gamma$.*

5.2.4 In-loop and out-of-loop spectra

Assuming, then, that the feedback is stable, we can solve Eq. (5.34) for $\hat{X}_2$ in the Fourier domain:

$$\tilde{X}_2(\omega) = \exp(-i\omega T_2)\frac{\sqrt{\eta_2\eta_1}\,\tilde{X}_0(\omega) + \sqrt{\eta_2\bar{\eta}_1}\,\tilde{X}_\nu(\omega) + \sqrt{\bar{\eta}_2}\,\tilde{X}_\mu(\omega)}{1 - g\tilde{h}(\omega)\exp(-i\omega T)}. \tag{5.48}$$

From this the in-loop amplitude quadrature spectrum is easily found from Eqs. (5.13) and (5.14) to be

$$S_2^X(\omega) = \frac{\eta_1\eta_2 S_0^X(\omega) + \eta_2\bar{\eta}_1 S_\nu^X(\omega) + \bar{\eta}_2 S_\mu^X(\omega)}{|1 - g\tilde{h}(\omega)\exp(-i\omega T)|^2}$$

$$= \frac{1 + \eta_1\eta_2[S_0^X(\omega) - 1]}{|1 - g\tilde{h}(\omega)\exp(-i\omega T)|^2}. \tag{5.49}$$

Exercise 5.6 *Derive these results.*

From these formulae the effect of feedback is obvious: it multiplies the amplitude quadrature spectrum at a given frequency by the factor $|1 - g\tilde{h}(\omega)\exp(-i\omega T)|^{-2}$. At low frequencies, this factor is simply $(1 - g)^{-2}$, which is why the feedback was classified on this basis into positive ($g > 0$) and negative ($g < 0$) feedback. The former will increase the noise at low frequency and the latter will decrease it. However, at higher frequencies, and in particular at integer multiples of π/T, the sign of the feedback will reverse and $g < 0$ will result in an increase in noise and vice versa. All of these results make perfect sense in the context of classical light signals, except that classically we would not worry about vacuum noise. Ignoring the vacuum noise is equivalent to assuming that the noise in the input beam is far above the shot-noise limit, so that one can replace $1 + \eta_1\eta_2[S_0^X(\omega) - 1]$ by $\eta_1\eta_2 S_0^X(\omega)$. This gives the result expected from classical signal processing: the signal is attenuated by the beam-splitters and either amplified or suppressed by the feedback.

The most dramatic effect is, of course, for large negative feedback. For sufficiently large $-g$ it is clear that one can make

$$S_2^X(\omega) < 1 \tag{5.50}$$

for some ω. This effect has been observed experimentally many times with different systems involving feedback; see for example Refs. [WJ85a, MY86, YIM86, MPV94, TWMB95]. Without a feedback loop this sub-shot-noise photocurrent would be seen as evidence for squeezing. However, there are several reasons to be very cautious about applying the description squeezing to this phenomenon relating to the in-loop field. Two of these reasons are theoretical, and are discussed in the following two subsections. The more practical reason relates to the out-of-loop beam $\hat{b}_3$.

From Eq. (5.29), the X quadrature of the beam $\hat{b}_3$ is, in the Fourier domain,

$$\tilde{X}_3(\omega) = \exp(-i\omega T_2)\left[\sqrt{\bar{\eta}_2\eta_1}\,\tilde{X}_0(\omega) + \sqrt{\bar{\eta}_2\bar{\eta}_1}\,\tilde{X}_\nu(\omega) - \sqrt{\eta_2}\,\tilde{X}_\mu(\omega)\right]$$
$$+ \sqrt{\bar{\eta}_2/\eta_2}\,g\tilde{h}(\omega)\exp(-i\omega T)\tilde{X}_2(\omega). \tag{5.51}$$

Here we have substituted for $\delta\eta_1$ in terms of $\hat{X}_2$. Now using the above expression (5.48) gives

$$\tilde{X}_3(\omega) = \exp(-i\omega T_2)\left\{\frac{\sqrt{\bar{\eta}_2\eta_1}\,\tilde{X}_0(\omega) + \sqrt{\bar{\eta}_2\bar{\eta}_1}\,\tilde{X}_\nu(\omega)}{1 - g\tilde{h}(\omega)\exp(-i\omega T)}\right.$$
$$\left. - \frac{\sqrt{\eta_2}\,[1 - g\tilde{h}(\omega)\exp(-i\omega T)/\eta_2]\tilde{X}_\mu(\omega)}{1 - g\tilde{h}(\omega)\exp(-i\omega T)}\right\}. \tag{5.52}$$

This yields the spectrum

$$S_3^X(\omega) = \frac{1 + \bar{\eta}_2\eta_1[S_0^X(\omega) - 1]}{|1 - g\tilde{h}(\omega)\exp(-i\omega T)|^2}$$
$$+ \frac{-2\,\mathrm{Re}[g\tilde{h}(\omega)\exp(-i\omega T)] + g^2|\tilde{h}(\omega)|^2/\eta_2}{|1 - g\tilde{h}(\omega)\exp(-i\omega T)|^2}. \tag{5.53}$$

The denominators are identical to those in the in-loop case, as is the numerator in the first line, but the additional term in the numerator of the second line indicates that there is extra noise in the out-of-loop signal.

The expression (5.53) can be rewritten as

$$S_3^X(\omega) = 1 + \frac{\bar{\eta}_2\eta_1[S_0^X(\omega) - 1] + g^2|\tilde{h}(\omega)|^2\bar{\eta}_2/\eta_2}{|1 - g\tilde{h}(\omega)\exp(-i\omega T)|^2}. \tag{5.54}$$

From this it is apparent that, unless the initial beam is amplitude-squeezed (that is, unless $S_0^X(\omega) < 1$ for some ω), the out-of-loop spectrum will always be greater than the shot-noise limit of unity. In other words, it is not possible to extract the apparent squeezing in the feedback loop by using a beam-splitter. In fact, in the limit of large negative feedback (which gives the greatest noise reduction in the in-loop signal), the low-frequency out-of-loop amplitude spectrum approaches a constant. That is,

$$\lim_{g \to -\infty} S_3^X(0) = \eta_2^{-1}. \tag{5.55}$$

Thus the more light one attempts to extract from the feedback loop, the higher above shot noise the spectrum becomes. Indeed, this holds for any frequency such that $\tilde{h}(\omega) \neq 0$, but recall that for large $|g|$ the bandwidth of $|\tilde{h}(\omega)|^2$ must go to zero (see Box. 5.1).

This result is counter to an intuition based on classical light signals, for which the effect of a beam-splitter is simply to split a beam in such way that both outputs would have the same statistics. The reason why this intuition fails is precisely because this is not all that a beam-splitter does; it also introduces vacuum noise, which is *anticorrelated* at the two output ports. The detector for beam $\hat{b}_2$ measures the amplitude fluctuations $\hat{X}_2$, which are a combination of the initial fluctuations $\hat{X}_0$ and the two vacuum fluctuations $\hat{X}_v$ and $\hat{X}_\mu$. The first two of these are common to the beam $\hat{b}_3$, but the last, $\hat{X}_\mu$, appears with opposite sign in $\hat{X}_3$. As the negative feedback is turned up, the first two components are successfully suppressed, but the last is actually amplified.

5.2.5 Commutation relations

Under normal circumstances (without a feedback loop) one would expect a sub-shot-noise amplitude spectrum to imply a super-shot-noise phase spectrum. However, that is not what is found from the theory presented here. Rather, the in-loop phase quadrature spectrum is unaffected by the feedback, being equal to

$$S_2^Y(\omega) = 1 + \eta_1 \eta_2 [S_0^Y(\omega) - 1]. \tag{5.56}$$

It is impossible to measure this spectrum without disturbing the feedback loop because all of the light in the $\hat{b}_2$ beam must be incident upon the photodetector in order to measure $\hat{X}_2$. However, it is possible to measure the phase-quadrature of the out-of-loop beam by homodyne detection. This was done in [TWMB95], which verified that this quadrature is also unaffected by the feedback, with

$$S_3^Y(\omega) = 1 + \eta_1 \bar{\eta}_2 [S_0^Y(\omega) - 1]. \tag{5.57}$$

For simplicity, consider the case in which the initial beam is coherent with $S_0^X(\omega) = S_0^Y(\omega) = 1$. Then $S_2^Y(\omega) = 1$ and

$$S_2^Y(\omega) S_2^X(\omega) = |1 - g\tilde{h}(\omega) \exp(-i\omega T)|^{-2}. \tag{5.58}$$

This can clearly be less than unity. This represents a violation of the uncertainty relation (5.17) which follows from the commutation relations (5.12). In fact it is easy to show (as done first by Shapiro *et al.* [SSH$^+$87]) from the solution (5.48) that the commutation relations (5.12) are false for the field $\hat{b}_2$ and must be replaced by

$$[\tilde{X}_2(\omega), \tilde{Y}_2(\omega')] = \frac{4\pi i\delta(\omega + \omega')}{1 - g\tilde{h}(\omega)\exp(-i\omega T)}, \qquad (5.59)$$

which explains how Eq. (5.58) is possible.

At first sight, this apparent violation of the canonical commutation relations would seem to be a major problem of this theory. In fact, there are no violations of the canonical commutation relations. As emphasized in Section 5.2.1, the canonical commutation relations (5.9) are between fields at different points in space, *at the same time*. It is only for free fields (travelling forwards in space for an indefinite time) that one can replace the space difference z by a time difference $t = z$ (remember that we are setting the speed of light $c = 1$). Field $\hat{b}_3$ is such a free field, since it can be detected an arbitrarily large distance away from the apparatus. Thus its quadratures at a particular point do obey the two-time commutation relations (5.10) and the corresponding Fourier-domain relations (5.12). But the field $\hat{b}_2$ cannot travel an indefinite distance before being detected. The time from the second beam-splitter to the detector τ_2 is a physical parameter in the feedback system.

For time differences shorter than the total feedback loop delay T it can be shown that the usual commutation relations hold:

$$[\hat{X}_2(t), \hat{Y}_2(t')] = 2i\delta(t - t') \quad \text{for} \quad |t - t'| < T. \qquad (5.60)$$

The field $\hat{b}_2$ is only in existence for a time τ_2 before it is detected. Because $\tau_2 < T$, this means that the two-time commutation relations between different parts of field $\hat{b}_2$ are actually preserved for any time such that those parts of the field are in existence, travelling through space towards the detector. It is only at times greater than the feedback loop delay time T that non-standard commutation relations hold. To summarize, the commutation relations between any of the fields at different spatial points always hold, but there is no reason to expect the time or frequency commutation relations to hold for an in-loop field. Without these relations, it is not clear how 'squeezing' should be defined. Indeed, it has been suggested [BGS+99] that 'squashing' would be a more appropriate term for in-loop 'squeezing' because the uncertainty has actually been squashed, rather than squeezed out of one quadrature and into another.

A second theoretical reason against the use of the word squeezing to describe the sub-shot-noise in-loop amplitude quadrature is that (provided that beam $\hat{b}_0$ is not squeezed), the entire apparatus can be described semiclassically. In a semiclassical description there is no noise except classical noise in the field amplitudes, and shot noise is a result of a quantum detector being driven by a classical beam of light. That such a description exists might seem surprising, given the importance of vacuum fluctuations in the explanation of the spectra in Section 5.2.4. However, the semiclassical explanation, as discussed for example in Refs. [SSH+87] and [TWMB95], is at least as simple. Nevertheless, this theory is less general than the quantum theory (it cannot treat an initial squeezed beam) so we do not develop it here.

5.2.6 QND measurements of in-loop beams

The existence of a semiclassical theory as just mentioned suggests that all of the calculations of noise spectra made above relate only to the noise of photocurrents and say nothing about the noise properties of the light beams themselves. However, this is not really true, as can be seen from considering quantum non-demolition (QND) measurements. In QND measurements, the measured observable (one quadrature in this case) is unchanged by the measurement, unlike in an ordinary absorptive measurement in optics. QND measurements cannot be described by semiclassical theory, which again shows that this theory is not complete.

A specific model for a QND device will be considered in Section 5.3.1. Here we simply assume that a perfect QND device can measure the amplitude quadrature X of a continuum field without disturbing it beyond the necessary back-action from the Heisenberg uncertainty principle. In other words, the QND device should give a readout at time t that can be represented by the operator $\hat{X}(t)$. The correlations of this readout will thus reproduce the correlations of $\hat{X}(t)$. For a perfect QND measurement of $\hat{X}_2$ and $\hat{X}_3$, the spectrum will reproduce those of the conventional (demolition) photodetectors which measure these beams. This confirms that these detectors (assumed perfect) are indeed recording the true quantum fluctuations of the light impinging upon them.

What is more interesting is to consider a QND measurement on $\hat{X}_1$. That is because the set-up in Fig. 5.1 is equivalent (as mentioned above) to a set-up without the second beam-splitter, but instead with an in-loop photodetector with efficiency η_2. In this version, the beams $\hat{b}_2$ and $\hat{b}_3$ do not physically exist. Rather, $\hat{b}_1$ is the in-loop beam and $\hat{X}_2$ is the operator for the noise in the photocurrent produced by the detector. As shown above, $\hat{X}_2$ can have vanishing noise at low frequencies for $g \to -\infty$. However, this is not reflected in the noise in the in-loop beam, as recorded by our hypothetical QND device. Following the methods of Section 5.2.4, the spectrum of $\hat{X}_1$ is

$$S_1^X(\omega) = \frac{1 + \eta_1[S_0(\omega) - 1] + g^2|\tilde{h}(\omega)|^2\bar{\eta}_2/\eta_2}{|1 - g\tilde{h}(\omega)\exp(-i\omega T)|^2}. \tag{5.61}$$

In the limit $g \to -\infty$, this becomes at low frequencies

$$S_1^X(0) \to \frac{1 - \eta_2}{\eta_2}, \tag{5.62}$$

which is not zero for any detection efficiency η_2 less than unity. Indeed, for $\eta_2 < 0.5$ it is above shot noise.

The reason why the in-loop amplitude quadrature spectrum is not reduced to zero for large negative feedback is that the feedback loop is feeding back noise $\hat{X}_\mu(t)$ in the photocurrent fluctuation operator $\hat{X}_2(t)$ that is independent of the fluctuations in the amplitude quadrature $\hat{X}_1(t)$ of the in-loop light. The smaller η_2, the larger the amount of extraneous noise in the photocurrent and the larger the noise introduced into the in-loop light. In order to minimize the low-frequency noise in the in-loop light, there is an optimal feedback gain.

In the case $S_0(\omega) = 1$ (a coherent input), this is given by

$$g_{\text{opt}} = -\frac{\eta_2}{1 - \eta_2},$$ (5.63)

giving a minimum in-loop low-frequency noise spectrum

$$S_1^X(0)|_{g=g_{\text{opt}}} = 1 - \eta_2.$$ (5.64)

The fact that the detection efficiency does matter in the attainable squashing (in-loop squeezing) shows that these are true quantum fluctuations.

5.2.7 Applications to noise reduction

Although the feedback device discussed in this section cannot produce a free squeezed beam, it is nevertheless useful for reducing classical noise in the output beam $\hat{b}_3$. It is easy to verify the result of [TWMB95] that, if one wishes to reduce classical noise $S_3^X(\omega) - 1$ at a particular frequency ω, then the optimal feedback is such that

$$g\tilde{h}(\omega)\exp(-i\omega T) = -\eta_1\eta_2[S_0^X(\omega) - 1].$$ (5.65)

This gives the lowest noise level in the amplitude of $\hat{b}_3$ at that frequency,

$$S_3^X(\omega)_{\text{opt}} = 1 + \frac{\bar{\eta}_2\eta_1[S_0^X(\omega) - 1]}{1 + \eta_2\eta_1[S_0^X(\omega) - 1]}.$$ (5.66)

For large classical noise we have feedback proportional to $S_0^X(\omega)$ and an optimal noise value of $1/\eta_2$, as this approaches the limit of Eq. (5.55). The interesting regime [TWMB95] is the opposite one, where $S_0^X(\omega) - 1$ is small, or even negative. The case of $S_0^X(\omega) - 1 < 0$ corresponds to a squeezed input beam. Putting squeezing through a beam-splitter reduces the squeezing in both output beams. In this case, with no feedback the residual squeezing in beam $\hat{b}_3$ would be

$$S_3^X(\omega)|_{g=0} = 1 + \bar{\eta}_2\eta_1[S_0^X(\omega) - 1],$$ (5.67)

which is closer to unity than $S_0^X(\omega)$. The optimal feedback (the purpose of which is to reduce noise) is, according to Eq. (5.65), positive. That is to say, destabilizing feedback actually puts back into beam $\hat{b}_3$ some of the squeezing lost through the beam-splitter. Since the required round-loop gain (5.65) is less than unity, the feedback loop remains stable (see Section 5.2.3).

This result highlights the nonclassical nature of squeezed fluctuations. When an amplitude-squeezed beam strikes a beam-splitter, the intensity at one output port is anticorrelated with that at the other, hence the need for positive feedback. Of course, the feedback can never put more squeezing into the beam than was present at the start. That is, $S_3^X(\omega)_{\text{opt}}$ always lies between $S_0^X(\omega)$ and $S_3^X(\omega)|_{g=0}$. However, if we take the limit $\eta_1 \to 1$ and $S_0^X(\omega) \to 0$ (perfect squeezing to begin with) then all of this squeezing can be recovered, for any η_2.

5.3 Feedback with optical beams using nonlinear optics

5.3.1 QND measurements

Section 5.2.2 showed that it was not possible to create squeezed light in the conventional sense using ordinary photodetection and linear feedback. Although the quantum theory appeared to show that the light which fell on the detector in the feedback loop was sub-shot noise, this could not be extracted because it was demolished by the detector. An obvious way around this would be to use a QND quadrature detector to control the feedback modulation. One way to achieve a QND measurement is for fields of different frequency to interact via a nonlinear refractive index. In order to obtain a large effect, large intensities are required. It is easiest to build up large intensities by using a resonant cavity. To describe this requires the quantum Langevin equations (QLEs) and input–output theory presented in Section 4.7.

Consider the apparatus shown in Fig. 5.2. The purpose of the detector in the feedback loop is to make a QND measurement of the quadrature $X^b_{\rm in}$ of the field $\hat{b}_{\rm in} \equiv \hat{b}_1$. This field drives a cavity mode with decay rate κ described by annihilation operator $\hat{a}$. This mode is coupled to a second mode with annihilation operator $\hat{c}$ and decay rate γ. We will assume an ideal QND coupling between the two modes

$$\hat{H} = \frac{\chi}{2}\hat{x}^a\hat{y}^c, \tag{5.68}$$

where

$$\hat{x}^a = \hat{a} + \hat{a}^\dagger; \qquad \hat{y}^c = -i\hat{c} + i\hat{c}^\dagger. \tag{5.69}$$

As described in [AMW88], this Hamiltonian could in principle be realized by two simultaneous processes, assuming that modes a and c have the same frequency. The first process would be simple linear mixing of the modes (e.g. by an intracavity beam-splitter). The second process would require an intracavity crystal with a $\chi^{(2)}$ nonlinearity, pumped by a classical field at twice the frequency of modes a and c. The Hamiltonian (5.68) commutes with the x^a quadrature of mode a, and causes this to drive the x^c quadrature of mode c. Thus measuring the $X^d_{\rm out}$ quadrature of the output field $\hat{d}_{\rm out}$ from mode c will give a QND measurement of $\hat{a} + \hat{a}^\dagger$, which is approximately a QND measurement of $\hat{X}^b_{\rm in}$.

The QLEs in the interaction frame for the quadrature operators are

$$\frac{d}{dt}\hat{x}^a = -\frac{\kappa}{2}\hat{x}^a - \sqrt{\kappa}\,\hat{X}^b_{\rm in}, \tag{5.70}$$

$$\frac{d}{dt}\hat{x}^c = -\frac{\gamma}{2}\hat{x}^c - \sqrt{\gamma}\,\hat{X}^d_{\rm in} + \chi\hat{x}^a. \tag{5.71}$$

Exercise 5.7 *Derive these.*

These can be solved in the frequency domain as

$$\tilde{X}^a(\omega) = -\frac{\sqrt{\kappa}\,\tilde{X}^b_{\rm in}(\omega)}{\kappa/2 + i\omega}, \tag{5.72}$$

$$\tilde{X}^c(\omega) = -\frac{\sqrt{\gamma}\,\tilde{X}^d_{\rm in}(\omega) + \sqrt{\kappa}\chi\tilde{X}^b_{\rm in}(\omega)/(\kappa/2 + i\omega)}{\gamma/2 + i\omega}. \tag{5.73}$$

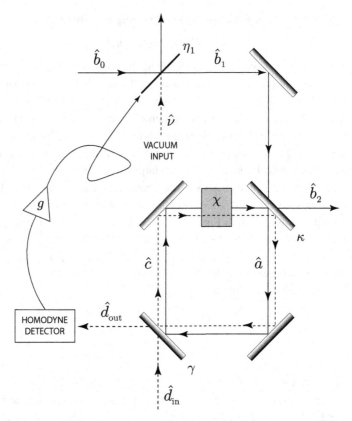

Fig. 5.2 A diagram for a travelling-wave feedback experiment based on a QND measurement. Travelling fields are denoted b and d. The first beam-splitter transmittance η_1 is variable. A cavity (drawn as a ring cavity for convenience) supports two modes, a (solid line) and c (dashed line). The decay rates for these two modes are κ and γ, respectively. They are coupled by a nonlinear optical process indicated by the crystal labelled χ. The perfect homodyne detection at the detector yields a photocurrent proportional to $\hat{X}_{\text{out}}^d = \hat{d}_{\text{out}} + \hat{d}_{\text{out}}^\dagger$. *Quantum Squeezing*, 2004, pp. 171–122, Chapter 6 'Squeezing and Feedback', H. M. Wiseman, Figure 6.2, Springer-Verlag, Berlin, Heidelberg. Redrawn and revised and adapted with kind permission of Springer Science+Business Media.

The quadrature of the output field $\hat{d}_{\text{out}} = \hat{d}_{\text{in}} + \sqrt{\gamma}\hat{c}$ is therefore

$$\tilde{X}_{\text{out}}^d(\omega) = -\frac{\gamma \kappa Q \tilde{X}_{\text{in}}^b(\omega)/(\kappa + 2\mathrm{i}\omega) + (\gamma - 2\mathrm{i}\omega)\tilde{X}_{\text{in}}^d(\omega)}{\gamma + 2\mathrm{i}\omega}, \tag{5.74}$$

where we have defined a quality factor for the measurement

$$Q = 4\chi / \sqrt{\gamma \kappa}. \tag{5.75}$$

In the limits $Q \gg 1$, and $\omega \ll \kappa, \gamma$, we have

$$\tilde{X}_{\text{out}}^d(\omega) \simeq -Q \tilde{X}_{\text{in}}^b(\omega), \tag{5.76}$$

which shows that a measurement (by homodyne detection) of the X quadrature of $\hat{d}_{\text{out}}$ can indeed effect a measurement of the low-frequency variation in $\hat{X}_{\text{in}}^{b}$.

To see that this measurement is a QND measurement, we have to calculate the statistics of the output field from mode a, that is $\hat{b}_{\text{out}}$. From the solution (5.72) we find

$$\tilde{X}_{\text{out}}^{b}(\omega) = -\frac{\kappa/2 - i\omega}{\kappa/2 + i\omega}\tilde{X}_{\text{in}}^{b}(\omega). \tag{5.77}$$

That is, for frequencies small compared with κ, the output field is identical to the input field, as required for a QND measurement. Of course, we cannot expect the other quadrature to remain unaffected, because of the uncertainty principle. Indeed, we find

$$\tilde{Y}_{\text{out}}^{b}(\omega) = -\frac{\kappa - 2i\omega}{\kappa + 2i\omega}\tilde{Y}_{\text{in}}^{b}(\omega) + \frac{Q\gamma\kappa\,\tilde{Y}_{\text{in}}^{d}(\omega)/(\gamma + 2i\omega)}{\kappa + 2i\omega}, \tag{5.78}$$

which shows that noise has been added to $\hat{Y}_{\text{in}}$. Indeed, in the good measurement limit which gave the result (5.76), we find the phase quadrature output to be dominated by noise:

$$\tilde{Y}_{\text{out}}^{b}(\omega) \simeq Q\tilde{Y}_{\text{in}}^{d}(\omega). \tag{5.79}$$

5.3.2 QND-based feedback

We now wish to show how a QND measurement, such as that just considered, can be used to produce squeezing via feedback. The physical details of how the feedback can be achieved are as outlined in Section 5.2.2 In particular, in the limit that the transmittance η_1 of the modulated beam-splitter goes to unity, the effect of the modulation is simply to add an arbitrary signal to the amplitude quadrature of the controlled beam. That is, the modulated beam can be taken to be

$$\hat{b}_1(t) = \hat{b}_0(t - \tau_1) + \beta\tfrac{1}{2}\,\delta\eta_1(t - \tau_1), \tag{5.80}$$

where $\hat{b}_0$ is the beam incident on the modulated beam-splitter, as in Section 5.2.2. In the present case, $\hat{b}_1(t)$ is then fed into the QND device, as shown in Fig. 5.2, so $\hat{b}_{\text{in}}(t) = \hat{b}_1(t)$ again, and the modulation is controlled by the photocurrent from an (assumed perfect) homodyne measurement of X_{out}^{d}:

$$\delta\hat{\eta}_1(t) = \frac{g}{-Q\beta}\int_0^\infty h(t')\hat{X}_{\text{out}}^{d}(t - \tau_0 - t')\mathrm{d}t'. \tag{5.81}$$

Here τ_0 is the delay time in the feedback loop, including the time of flight from the cavity for mode c to the homodyne detector, and $h(t)$ is as before.

On substituting Eqs. (5.80) and (5.81) into the results of the preceding subsection we find

$$\tilde{X}_{\text{out}}^{b}(\omega) = -\frac{\kappa - 2i\omega}{\kappa + 2i\omega}\frac{\tilde{X}_{\text{in}}^{b}(\omega) + \tilde{X}_{\text{in}}^{d}(\omega)g\tilde{h}(\omega)Q^{-1}(\gamma - 2i\omega)/(\gamma + 2i\omega)}{1 - g\tilde{p}(\omega)\tilde{h}(\omega)\exp(-i\omega T)}, \tag{5.82}$$

where $T = \tau_1 + \tau_0$ is the total round-trip delay time and

$$\tilde{p}(\omega) = \frac{\gamma \kappa}{(\kappa + 2i\omega)(\gamma + 2i\omega)} \tag{5.83}$$

represents the frequency response of the two cavity modes. If we assume that the field $\hat{d}_{\text{in}}$ is in the vacuum state then we can evaluate the spectrum of amplitude fluctuations in $\hat{X}_{\text{out}}^b$ to be

$$S_{\text{out}}^X(\omega) = \frac{S_{\text{in}}^X(\omega) + g^2 Q^{-2}}{|1 - g\,\tilde{p}(\omega)\tilde{h}(\omega)\exp(-i\omega T)|^2}. \tag{5.84}$$

Clearly, for a sufficiently high-quality measurement ($Q \to \infty$), the added noise term in the amplitude spectrum can be ignored. Then, for sufficiently large negative g, the feedback will produce a sub-shot-noise spectrum. Note the difference between this case and that of Section 5.2.2. Here the squeezed light is not part of the feedback loop; it is a free beam. The ultimate limit to how squeezed the beam can be is determined by the noise in the measurement, as will be discussed in Section 5.3.4.

Since $\hat{b}_{\text{out}}$ is a free field, not part of any feedback loop, it should obey the standard commutation relations. This is the case, as can be verified from the expression (5.78) for $\hat{Y}_{\text{out}}^b$ (which is unaffected by the feedback). Consequently, the spectrum for the phase quadrature

$$S_{\text{out}}^Y(\omega) = S_{\text{in}}^Y(\omega) + |Q\tilde{p}(\omega)|^2 \tag{5.85}$$

shows the expected increase in noise.

Exercise 5.8 *Derive Eqs. (5.84) and (5.85) and verify that the uncertainty product $S_{\text{out}}^Y(\omega)S_{\text{out}}^X(\omega)$ is always greater than or equal to unity, as required for free fields.*

5.3.3 Parametric down-conversion

The preceding section showed that feedback based on a perfect QND measurement can produce squeezing. This has never been done experimentally because of the difficulty of building a perfect QND measurement apparatus. However, it turns out that QND measurements are not the only way to produce squeezing via feedback. Any mechanism that produces correlations between the beam of interest and another beam that are more 'quantum' than the correlations between the two outputs of a linear beam-splitter can be the basis for producing squeezing via feedback. Such a mechanism must involve some sort of optical nonlinearity, hence the title of Section 5.3.

The production of amplitude squeezing by feeding back a quantum-correlated signal was predicted [JW85] and observed [WJ85b] by Walker and Jakeman in 1985, using the process of parametric down-conversion. An improved feedback scheme for the same system was later used by Tapster, Rarity and Satchell [TRS88] to obtain an inferred amplitude spectrum $S_1^X(\omega)_{\text{min}} = 0.72$ over a limited frequency range.

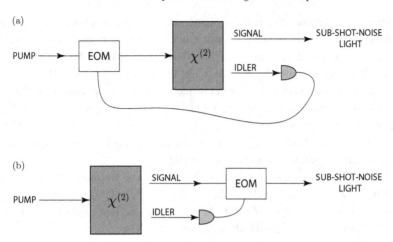

Fig. 5.3 A diagram showing two ways of producing sub-shot-noise light from parametric down-conversion. (a) Feedback, as first used by Walker and Jakeman [WJ85b]. (b) Feedforward, as first used by Mertz *et al.* [MHF+90]. Figure 1 adapted with permission from J. Mertz *et al.*, *Phys. Rev.* A **44**, 3229, (1991). Copyrighted by the American Physical Society.

The essential idea is as follows. Non-degenerate parametric down-conversion can be realized by pumping a crystal with a $\chi^{(2)}$ nonlinearity by a laser of frequency ω_1 and momentum $\mathbf{k}_1$. For particular choices of ω_1 the crystal can be aligned so that a pump photon can be transformed into a pair of photons with frequencies ω_2, ω_3 and momenta $\mathbf{k}_2$, $\mathbf{k}_3$ with $\omega_1 = \omega_2 + \omega_3$ and $\mathbf{k}_1 = \mathbf{k}_2 + \mathbf{k}_3$. A special case (taking place inside a pair of cavities, for two modes with $\omega_2 = \omega_3$) has been mentioned already in Section 5.3.1. A pair of down-converted photons will thus be correlated in time because they are produced from a single pump photon. On a more macroscopic level, this means that the amplitude quadrature fluctuations in the two down-converted beams are positively correlated, and the correlation coefficient can in principle be very close to unity. In this ideal limit, measuring the intensity of beam 2 (called the idler) should give a readout identical to that obtained from measuring beam 3 (the signal). In effect, the measurement of the idler is like a QND measurement of the signal. Thus, feeding back the photocurrent from the idler with a negative gain should be able to reduce the noise in the signal below the shot-noise limit, as shown in Fig. 5.3(a).

5.3.4 Feedback, feedforward and robustness

In Refs. [WJ85b, TRS88] the negative feedback was effected by controlling the power of the pump laser (which controls the rate at which photon pairs are produced). This maintains the symmetry of the experiment so that the photocurrent fed back from the idler has the same statistics as the photocurrent from the free beam, the signal. In other words, they will both be below the shot noise, whereas without feedback they are both above the shot noise. However, it is not necessary to preserve the symmetry in this way. The measured photocurrent from the idler can be fed *forwards* to control the amplitude fluctuations in the signal (for example by using an electro-optic modulator as described in Section 5.2.2). This feedforward was realized experimentally by Mertz *et al.* [MHF+90, MHF91], achieving similar results to that obtained by feedback. The two schemes are contrasted in Fig. 5.3.

Thus, unless one is concerned with light inside a feedback loop, there is no difference in theory between feedback and feedforward. Indeed, the squeezing produced by QND-based feedback discussed in Section 5.3.2 could equally well have been produced by QND-based feedforward. However, from a practical point of view, feedback has the advantage of being more *robust* with respect to parameter uncertainties.

Consider the QND-based feedback in Section 5.3.2, and for simplicity allow the QND cavity to be very heavily damped and the feedback to be very fast so that we may make the approximation $\tilde{h}(\omega)\exp(-i\omega T) = 1$. Also, say that the input light is coherent so that $S_{\text{in}}^{X}(\omega) = 1$. Then the output amplitude spectrum reduces to

$$S_{\text{out}}^{X} = \frac{1 + g^2 Q^{-2}}{(1 - g)^2}. \tag{5.86}$$

This has a minimum of

$$S_{\text{out;min}}^{X} = (1 + Q^2)^{-1} \tag{5.87}$$

at $g = -Q^2$. For large Q (high-quality QND measurement) this is much less than unity.

Let us say that the experimenter does not know Q precisely, or cannot control g precisely, so that in the experiment the actual feedback loop has

$$g = -Q^2(1 + \epsilon), \tag{5.88}$$

where ϵ is a small relative error. To second order in ϵ this gives a new squeezed noise level of

$$S_{\text{out}}^{X} = (1 + Q^2)^{-1}\left[1 + \frac{\epsilon^2}{1 + Q^2}\right]. \tag{5.89}$$

Exercise 5.9 *Show this.*

The relative size of the extra noise term decreases with increasing Q, and, as long as $|\epsilon| \ll Q$, the increase in the noise level is negligible.

Now consider feedforward. Under the above conditions, the measured current is represented by the operator

$$\hat{X}_{\text{out}}^{d} = Q\hat{X}_{\text{in}}^{b} + \hat{X}_{\text{in}}^{d}. \tag{5.90}$$

This is fed forwards to create a coherent field of amplitude $(g/Q)\hat{X}_{\text{out}}^{d}$, which is added to the output of the system. Here g is the feedforward gain and the new output of the system will be

$$\hat{X}_{\text{out}}^{b} = \hat{X}_{\text{in}}^{b} + g(\hat{X}_{\text{in}}^{b} + \hat{X}_{\text{in}}^{d}/Q). \tag{5.91}$$

This has a noise level of

$$S_{\text{out}}^X = (1 + g)^2 + g^2 Q^{-2}, \tag{5.92}$$

which has a minimum of

$$S_{\text{out;min}}^X = (1 + Q^2)^{-1}, \tag{5.93}$$

exactly the same as in the feedback case (as expected), but with an open-loop gain of $g = -Q^2/(1 + Q^2)$.

The difference between feedback and feedforward comes when we consider systematic errors. Again assuming a relative error in g of ϵ, so that

$$g = -\frac{Q^2(1 + \epsilon)}{1 + Q^2}, \tag{5.94}$$

the new noise level is exactly given by

$$S_{\text{out}}^X = (1 + Q^2)^{-1}(1 + Q^2\epsilon^2). \tag{5.95}$$

Now the relative size of the extra term actually *increases* as the quality of the measurement increases. In order for this term to be negligible, the systematic error must be extremely small: $|\epsilon| \ll Q^{-1}$. Thus, the feedforward approach is much less robust with respect to systematic errors due to uncertainties in the system parameters or inability to control the modulation exactly. This is a generic advantage of feedback over feedforward, and justifies our emphasis on the former in this book.

In the example above the distinction between feedback and feedforward is obvious. In the former case the measurement record (the current) used for control is affected by the controls applied at earlier times; in the latter it is not. This distinction will always apply for a continuous (in time) control protocol. However, discrete protocols may also be considered, and indeed one can consider the case of a single measurement result being used to control the system. In this case, one could argue that all control protocols are necessarily feedforward. However, the term feedback is often used in that case also, and we will follow that loose usage at times.

5.4 Feedback control of a monitored system

Having considered feedback on continuum fields, we now turn to a sort of feedback that would be more familiar in conception to a classical control engineer: the feedback control of a monitored system. We are concerned with quantum, rather than classical, systems, but many of the ideas from classical control theory carry over. For instance, it is often more convenient to describe the feedback using the conditioned state of the system (which classically would be a probability distribution as explained in Chapter 1). This is as opposed to describing the feedback using equations for the system operators (which classically would be equations for the system variables). Of course, the two methods are equally valid, and we will cover both in the remaining sections. However, the quantum trajectory (i.e. conditioned states) method will be seen in this chapter to be more useful in a number of applications, and often to have more explanatory power. These advantages are further developed in the next chapter. Hence we begin our treatment of feedback control of a quantum system by reconsidering quantum trajectories. In this section we consider jumpy trajectories (as arise from direct detection in quantum optics).

5.4.1 General feedback

Consider for simplicity an open quantum system with a single decoherence channel described by the Markovian master equation

$$\dot{\rho} = -\mathrm{i}[\hat{H}, \rho] + \mathcal{D}[\hat{c}]\rho. \tag{5.96}$$

As derived in Section 4.2, the simplest unravelling for this master equation is in terms of quantum jumps. In quantum optics, these correspond to δ-function spikes in the photocurrent $I(t)$ that are interpreted as the detection of a photon emitted by the system. We restate Eq. (4.40), the SME for the conditioned state $\rho_I(t)$:

$$\mathrm{d}\rho_I(t) = \left\{\mathrm{d}N(t)\mathcal{G}[\hat{c}] - \mathrm{d}t\,\mathcal{H}\left[\mathrm{i}\hat{H} + \tfrac{1}{2}\hat{c}^\dagger\hat{c}\right]\right\}\rho_I(t), \tag{5.97}$$

where the point process $\mathrm{d}N(t) = I(t)\mathrm{d}t$ is defined by

$$\mathrm{E}[\mathrm{d}N(t)] = \mathrm{Tr}[\hat{c}^\dagger\hat{c}\rho_I(t)], \tag{5.98}$$

$$\mathrm{d}N(t)^2 = \mathrm{d}N(t). \tag{5.99}$$

The current $I(t) = \mathrm{d}N/\mathrm{d}t$ could be used to alter the system dynamics in many different ways. Some examples from quantum optics are the following: modulating the pump rate of a laser, the amplitude of a driving field, the cavity length, or the cavity loss rate. The last three examples could be effected by using an electro-optic modulator (a device with a refractive index controlled by a current), possibly in combination with a polarization-dependent beam-splitter. The most general expression for the effect of the feedback would be

$$[\dot{\rho}_I(t)]_{\mathrm{fb}} = \mathcal{F}[t, \mathbf{I}_{[0,t)}]\rho_I(t). \tag{5.100}$$

Here $\mathbf{I}_{[0,t)}$ represents the complete photocurrent record from the beginning of the experiment up to the present time. Thus the superoperator $\mathcal{F}[t, \mathbf{I}_{[0,t)}]$ (which may be explicitly time-dependent) is a *functional* of the current for all past times. This functional dependence describes the response of the feedback loop, which may be nonlinear, and realistically must include some smoothing in time. The complete description of this feedback is given simply by adding Eq. (5.100) to Eq. (5.97). In general the resulting equation would have to be solved by numerical simulation.

To make progress towards understanding quantum feedback control, it is helpful to make simplifying assumptions. Firstly, let us consider a linear functional giving feedback evolution of the form

$$[\dot{\rho}_I(t)]_{\text{fb}} = \int_0^\infty h(s)I(t-s)\mathcal{K}\rho_I(t)ds, \tag{5.101}$$

where $\mathcal{K}$ is an arbitrary Liouville superoperator. Later in this section we will consider the Markovian limit in which the response function $h(s)$ goes to $\delta(s)$. To find this limit, it is first useful to consider the case $h(s) = \delta(s-\tau)$, where the feedback has a fixed delay τ. Then the feedback evolution is

$$[\dot{\rho}_I(t)]_{\text{fb}} = I(t-\tau)\mathcal{K}\rho_I(t). \tag{5.102}$$

Because there is no smoothing response function in Eq. (5.102), the right-hand side of the equation is a mathematically singular object, with $I(t)$ being a string of δ-functions. If it is meant to describe a physical feedback mechanism, then it is necessary to interpret the equation as an implicit stochastic differential equation, as explained in Section B.6. This is indicated already in the notation of using a fluxion on the left-hand side. The alternative interpretation as the explicit equation

$$[d\rho_I(t)]_{\text{fb}} = dN(t-\tau)\mathcal{K}\rho_I(t) \tag{5.103}$$

yields nonsense.

Exercise 5.10 *Show that Eq. (5.103) does not even preserve positivity.*

In order to combine Eq. (5.102) with Eq. (5.97), it is necessary to convert it from an implicit to an explicit equation. As explained in Section B.6, this is easy to accomplish because of the linearity (with respect to ρ) of Eq. (5.102). The result is

$$\rho_I(t) + [d\rho_I(t)]_{\text{fb}} = \exp[\mathcal{K}\, dN(t-\tau)]\rho_I(t). \tag{5.104}$$

Using the rule (5.99) and adding this evolution to that of the SME (5.97) gives the total conditioned evolution of the system

$$d\rho_I(t) = \left\{ dN(t)\mathcal{G}[\hat{c}] - dt\, \mathcal{H}\big[i\hat{H} + \tfrac{1}{2}\hat{c}^\dagger\hat{c}\big] + dN(t-\tau)\big(e^{\mathcal{K}} - 1\big) \right\}\rho_I(t). \tag{5.105}$$

It is not possible to turn this stochastic equation into a master equation by taking an ensemble average, as was possible with Eq. (5.97). This is because the feedback noise term (with argument $t - \tau$) is not independent of the state at time t. Physically, it is not possible to derive a master equation because the evolution including feedback (with a time delay) is not Markovian.

5.4.2 The feedback master equation

In order to make Eq. (5.105) more useful, it would be desirable to take the Markovian limit ($\tau \to 0$) in order to derive a non-selective master equation. Simply putting $\tau = 0$ in this equation and taking the ensemble average using Eq. (5.98) fails. The resultant evolution equation would be nonlinear in ρ and so could not be a valid master equation.

Exercise 5.11 *Show this.*

The reason why the nonlinearity is admissible in Eq. (5.105) is that ρ_I is a conditioned state for an individual system, not the unconditioned state $\rho = \mathrm{E}[\rho_I]$. Physically, putting $\tau = 0$ fails because the feedback must act after the measurement even in the Markovian limit. The correct limit can be achieved by the equation

$$\rho_I(t + dt) = \exp[dN(t - \tau)\mathcal{K}]\{1 + dN(t)\mathcal{G}[\hat{c}] + dt\,\mathcal{H}[-i\hat{H} - \tfrac{1}{2}\hat{c}^\dagger\hat{c}]\}\rho_I(t). \quad (5.106)$$

For τ finite, this reproduces Eq. (5.105). However, if $\tau = 0$, expanding the exponential gives

$$d\rho_I(t) = \{dN(t)[e^{\mathcal{K}}(\mathcal{G}[\hat{c}] + 1) - 1] + dt\,\mathcal{H}[-i\hat{H} - \tfrac{1}{2}\hat{c}^\dagger\hat{c}]\}\rho_I(t). \quad (5.107)$$

In this equation, it is possible to take the ensemble average because $dN(t)$ can simply be replaced by its expectation value (5.98), giving

$$\dot{\rho} = e^{\mathcal{K}}\mathcal{J}[\hat{c}]\rho - \mathcal{A}[\hat{c}]\rho - i[\hat{H}, \rho] = \mathcal{L}\rho. \quad (5.108)$$

Recall from Eq. (1.80) that $\mathcal{J}[\hat{A}]\hat{b} \equiv \hat{A}\hat{B}\hat{A}^\dagger$, while

$$\mathcal{A}[\hat{A}]\hat{B} = \tfrac{1}{2}\{\hat{A}^\dagger\hat{A}\hat{B} + \hat{B}\hat{A}^\dagger\hat{A}\}. \quad (5.109)$$

Exercise 5.12 *Derive Eqs. (5.107) and (5.108), and show that the latter is of the Lindblad form.*
Hint: *Remember that $e^{\mathcal{K}}$ is an operation.*

That is, we have a new master equation incorporating the effect of the feedback. This master equation could have been guessed from an understanding of quantum jumps. However, the derivation here has the advantage of making clear the relation of the superoperator $\mathcal{K}$ to experiment via Eq. (5.102). In the special case in which $\mathcal{K}\rho = -i[\hat{Z}, \rho]$, the conditioned SME with feedback can also be expressed as a SSE of the form of Eq. (4.19), with $\hat{c}$ replaced by $e^{-i\hat{Z}}\hat{c}$.

Producing nonclassical light. Just as feedback based on absorptive photodetection cannot create a free squeezed beam (as shown in Section 5.2) by linear optics, so feedback based on direct detection cannot create a nonclassical state of a cavity mode by linear optics. By linear optics we mean processes that take coherent states to coherent states: linear damping, driving and detuning. By a nonclassical state we mean one that cannot be expressed as a mixture of coherent states. This concept of nonclassicality is really just a statement of what sort of quantum states are easy to produce, like the concept of the standard quantum limit.

In the present case, we can understand this limitation on feedback by considering the quantum trajectories. Both the jump and the no-jump evolution for a freely decaying cavity take a coherent state to a coherent state, in the former case with no change and in the latter with exponential decay of its amplitude.

Exercise 5.13 *Show this.*

Hint: *First show that the measurement operators, for a decay rate γ, can be written as*

$$\hat{M}_1(dt) = \sqrt{\gamma\, dt}\, \hat{a}, \qquad \hat{M}_0(dt) = \exp(-\hat{a}^\dagger \hat{a} \gamma\, dt/2), \qquad (5.110)$$

and recall Exercise 3.29.

If the post-jump feedback evolution $e^{\mathcal{K}}$ also takes a coherent state to a coherent state (or to a mixture of coherent states), it is clear that a nonclassical state can never be produced.

However, just as in the case of beams, feedback can cause the in-loop photocurrent to have nonclassical statistics. For direct detection the simplest form of nonclassical statistics is sub-Poissonian statistics. That is, the number of photons detected in some time interval has a variance less than its mean. For a field in a coherent state, the statistics will be Poissonian, and for a process that produces a mixture of coherent states (of different intensities) the statistics will be super-Poissonian.

In the quantum trajectory formalism, the explanation for anomalous (e.g. sub-Poissonian) in-loop photocurrent statistics lies in the modification of the jump measurement operator by the feedback as in Eq. (5.107). That is, the in-loop photocurrent autocorrelation function (from which the statistics can be determined) is modified from Eq. (4.50) to

$$E[I(t')I(t)] = \text{Tr}\left[\gamma \hat{a}^\dagger \hat{a} e^{\mathcal{L}(t'-t)} e^{\mathcal{K}} \gamma \hat{a} \rho(t) \hat{a}^\dagger\right] + \text{Tr}\left[\gamma \hat{a}^\dagger \hat{a} \rho(t)\right]\delta(t'-t), \qquad (5.111)$$

where $\mathcal{L}$ is as defined in Eq. (5.108) with $\hat{c} = \sqrt{\gamma}\hat{a}$.

Exercise 5.14 *Show this using the same style of argument as in Section 4.3.2.*

It is the effect of the feedback specific to the in-loop current via $e^{\mathcal{K}}$, not the overall evolution including feedback via $e^{\mathcal{L}(t'-t)}$, that may cause the sub-Poissonian in-loop statistics even if only linear optics is involved.

5.4.3 Application: protecting Schrödinger's cat

We now give an application (not yet experimentally realized) for Markovian feedback based on direct detection that shows the usefulness of the quantum trajectory approach. This example is due to Horoshko and Kilin [HK97].

In Section 3.9.1 we showed how damping of an oscillator leads to the rapid destruction of quantum coherence terms in a macroscopic superposition of two coherent states (a Schrödinger-cat state). Consider the particular cat state defined by

$$|\alpha; \phi\rangle_{\text{cat}} = [2(1 + e^{-2|\alpha|^2} \cos \phi)]^{-1/2}\left(|\alpha\rangle + e^{i\phi}|-\alpha\rangle\right). \qquad (5.112)$$

Under the damping master equation

$$\dot{\rho} = \gamma \mathcal{D}[\hat{a}]\rho, \qquad (5.113)$$

the quantum coherence terms in ρ decay as $\exp(-2\gamma|\alpha|^2 t)$, while the coherent amplitudes themselves decay as $\exp(-\gamma t/2)$ (see Section 3.9.1).

If we consider a direct-detection unravelling of this master equation, then the no-jump evolution leads solely to the decay of the coherent amplitudes.

Exercise 5.15 *Show this.*

Thus it is the jumps that are responsible for the destruction of the superposition. This makes sense, since the rate of decay of the coherence terms scales as the rate of jumps. We can see explicitly how this happens from the following:

$$\hat{a}|\alpha;\phi\rangle_{\text{cat}} \propto \alpha\big[|\alpha\rangle - e^{i\phi}|-\alpha\rangle\big] \propto |\alpha;\phi-\pi\rangle_{\text{cat}}. \tag{5.114}$$

That is, upon a detection the phase of the superposition changes by π, which leads to the decoherence.

Exercise 5.16 *Show that, for $|\alpha| \gg 1$, an equal mixture of $|\alpha;\phi\rangle_{\text{cat}}$ and $|\alpha;\phi-\pi\rangle_{\text{cat}}$ has no coherence, since it is the same as an equal mixture of $|\alpha\rangle$ and $|-\alpha\rangle$.*

Of course, if one keeps track of the detections then one knows which cat state the system is in at all times. (A similar analysis of the case for homodyne detection is done in Ref. [CKT94].) It would be preferable, however, to have a deterministic cat state. If one had arbitrary control over the cavity state then this could always be done by feedback following each detection since any two pure states are always unitarily related. However, this observation is not particularly interesting unless the feedback can be implemented with a practically realizable interaction. As Horoshko and Kilin pointed out, for the state $|\alpha;\pi/2\rangle_{\text{cat}}$ this is the case, since a simple rotation of the state in phase-space has the following effect:

$$e^{-i\pi\hat{a}^\dagger\hat{a}}|\alpha;-\pi/2\rangle_{\text{cat}} = -i|\alpha;\pi/2\rangle_{\text{cat}}. \tag{5.115}$$

That is, if one uses the feedback Hamiltonian

$$\hat{H}_{\text{fb}}(t) = I(t)\pi\hat{a}^\dagger\hat{a}, \tag{5.116}$$

with $I(t)$ the photocurrent from direct detection of the cavity output, then, following each detection that causes ϕ to change from $\pi/2$ to $-\pi/2$, the feedback changes ϕ back to $\pi/2$. Thus the effect of the jumps is nullified, and the time-evolved state is simply $|\alpha e^{-\gamma t/2};\pi/2\rangle_{\text{cat}}$.

Exercise 5.17 *Verify Eq. (5.115) and also that $|\alpha e^{-\gamma t/2};\pi/2\rangle_{\text{cat}}$ is a solution of the feedback-modified master equation*

$$\dot{\rho} = \gamma\mathcal{D}[e^{-i\pi\hat{a}^\dagger\hat{a}}\hat{a}]\rho.$$

Practicalities of optical feedback. Physically, the Hamiltonian (5.116) requires the ability to control the frequency of the cavity mode. Provided that it is done slowly enough, this can be achieved simply by changing the physical properties of the cavity, such as its length, or the refractive index of some material inside it. Here 'slowly enough' means slow compared with the separation of the eigenstates of the Hamiltonian, so that the adiabatic theorem [BF28] applies. Assuming we can treat just a single mode in the cavity (as will be the case if it is small enough), this energy separation equals the resonance frequency ω_0. On the other hand, the δ-function in Eq. (5.116) implies a modulation that is fast enough for one to ignore any other evolution during its application. As we have seen, the fastest of the two evolution rates in the problem (without feedback) is $2\gamma|\alpha|^2$. Thus the time-scale T for the modulation of the cavity frequency must satisfy

$$\gamma|\alpha|^2 \ll T^{-1} \ll \omega_0. \tag{5.117}$$

Now $\gamma \ll \omega_0$ is necessary for the derivation of the master equation (5.113). Moreover, a typical ratio on these time-scales (the quality factor of the cavity) is of order 10^8. Thus, if both $\ll$ signs in Eq. (5.117) are assumed to be satisfied by ratios of 10^{-2}, the scheme could protect Schrödinger cats with $|\alpha| \sim 100$, which is arguably macroscopic. In practice, other physical limitations are going to be even more important.

First, realistic feedback will not be Markovian, but will have some time delay τ. For the Markovian approximation to be valid, this must be much less than the time-scale for photon loss: $\tau^{-1} \gg \gamma|\alpha|^2$. Even with very fast detectors and electronics it would be difficult to make the effective delay τ less than 10 ns [SRO+02]. Also, even very good optical cavities have γ at least of order 10^4 s^{-1}. Again assuming that the above inequality is satisfied by a factor of 10^2, the limit now becomes $|\alpha| \sim 10$.

Second, realistic detectors do not have unit efficiency, as discussed in Section 4.8.1. For photon counting, as required here, $\eta = 0.9$ is an exceptionally good figure at present. Taking into account inefficiency, the feedback-modified master equation is

$$\dot{\rho} = \gamma\eta\mathcal{D}[e^{-i\pi\hat{a}^\dagger\hat{a}}\hat{a}]\rho + \gamma(1-\eta)\mathcal{D}[\hat{a}]\rho. \tag{5.118}$$

Even with $\eta = 0.9$ the decay rate for the coherence terms, $2\gamma(1-\eta)|\alpha|^2$, will still be greater than the decay rate for the coherent amplitude, $\gamma/2$, unless $|\alpha| \sim 1.5$ or smaller. In other words, until ultra-high-efficiency photodetectors are manufactured, it is only Schrödinger 'kittens' that may be protected to any significant degree.

5.4.4 Feedback in the Heisenberg picture

The example in the preceding subsection shows how useful the quantum trajectory analysis is for designing feedback control of a quantum system, and how convenient the Schrödinger picture (in particular the feedback-modified master equation) is for determining the effect of the feedback. Nevertheless, it is possible to treat feedback control of a quantum system in the Heisenberg picture, as we used for the propagating fields in Sections 5.1 and 5.2. In this section, we present that theory.

Recall from Eq. (4.35) that the unitary operator generating the system and bath evolution for an infinitesimal interval is

$$\hat{U}_0(t + dt, t) = \exp\left[\hat{c}\, d\hat{B}_0(t)^\dagger - \hat{c}^\dagger\, d\hat{B}_0(t) - i\hat{H}\, dt\right]. \tag{5.119}$$

Here $d\hat{B}_0 = \hat{b}_0(t)dt$, where $\hat{b}_0$ is the annihilation operator for the input field which is in the vacuum state, so that $d\hat{B}_0\, d\hat{B}_0^\dagger = dt$ but all other second-order moments vanish. Using this, we obtain the QLE for an arbitrary system operator corresponding to the master equation (5.96):

$$d\hat{s} = \hat{U}_0^\dagger(t + dt, t)\hat{s}\hat{U}_0(t + dt, t) - \hat{s} \tag{5.120}$$

$$= i[\hat{H}, \hat{s}]dt + \left(\hat{c}^\dagger\hat{s}\hat{c} - \tfrac{1}{2}\hat{s}\hat{c}^\dagger\hat{c} - \tfrac{1}{2}\hat{c}^\dagger\hat{c}\hat{s}\right)dt - [d\hat{B}_0^\dagger\,\hat{c} - \hat{c}^\dagger\, d\hat{B}_0, \hat{s}]. \tag{5.121}$$

Similarly, the output field is the infinitesimally evolved input field:

$$\hat{b}_1(t) = \hat{U}_0^\dagger(t + dt, t)\hat{b}_0(t)\hat{U}_0(t + dt, t)$$

$$= \hat{b}_0(t) + \hat{c}(t). \tag{5.122}$$

The output photon-flux operator (equivalent to the photocurrent derived from a perfect detection of that field) is $\hat{I}_1(t) = \hat{b}_1^\dagger(t)\hat{b}_1(t)$. This suggests that the feedback considered in Section 5.4.1 could be treated in the Heisenberg picture by using the feedback Hamiltonian

$$\hat{H}_{\text{fb}}(t) = \hat{I}_1(t - \tau)\hat{Z}(t), \tag{5.123}$$

where each of these quantities is an operator. Here, the feedback superoperator $\mathcal{K}$ used in Section 5.4.1 would be defined by $\mathcal{K}\rho = -i[\hat{Z}, \rho]$. The generalization to arbitrary $\mathcal{K}$ would be possible by involving auxiliary systems.

It might be thought that there is an ambiguity of operator ordering in Eq. (5.123), because $\hat{I}_1$ contains system operators. In fact, the ordering is not important because $\hat{b}_1(t)$ commutes with all system operators at a later time as discussed in Section 4.7.1, so $\hat{I}_1(t)$ does also. Of course, $\hat{b}_1(t)$ will not commute with system operators for times after $t + \tau$ (when the feedback acts), but $\hat{I}_1(t)$ still will because it is not changed by the feedback interaction. (It commutes with the feedback Hamiltonian.) This fact would allow one to use the formalism developed here to treat feedback of a photocurrent smoothed by time-averaging. That is to say, there is still no operator-ordering ambiguity in the expression

$$\hat{H}_{\text{fb}}(t) = \hat{Z}(t) \int_0^\infty h(s)\hat{I}_1(t - s)ds, \tag{5.124}$$

or even for a general Hamiltonian functional of the current, as in Eq. (5.100). For a sufficiently broad response function $h(s)$, there is no need to use stochastic calculus for the feedback; the explicit equation of motion due to the feedback would simply be

$$d\hat{s}(t) = i[\hat{H}_{\text{fb}}(t), \hat{s}(t)]dt. \tag{5.125}$$

However, this approach makes the Markovian limit difficult to find. Thus, as in Section 5.4.1, the response function will be assumed to consist of a time delay only, as in Eq. (5.123).

In order to treat Eq. (5.123) it is necessary once again to use the stochastic calculus of Appendix B to find the explicit effect of the feedback. As in Section 3.11.1, the key is to expand the unitary operator for feedback

$$\hat{U}_{\text{fb}}(t+dt,t) = \exp[-i\hat{H}_{\text{fb}}(t)dt] \tag{5.126}$$

to as many orders as necessary. Since this evolution commutes with the no-feedback evolution (5.121), the feedback simply adds the following extra term to Eq. (5.121):

$$[d\hat{s}]_{\text{fb}} = \hat{U}_{\text{fb}}^{\dagger}(t+dt,t)\hat{s}(t)\hat{U}_{\text{fb}}(t+dt,t) - \hat{s}(t), \tag{5.127}$$

which evaluates to

$$[d\hat{s}]_{\text{fb}} = d\hat{N}_1(t-\tau)\left(e^{i\hat{Z}}\hat{s}e^{-i\hat{Z}} - \hat{s}\right). \tag{5.128}$$

Exercise 5.18 *Show this.*

Including the no-feedback evolution and expanding $d\hat{N}_1(t)$ using Eq. (5.122) gives

$$d\hat{s} = i[\hat{H},\hat{s}]dt + \left(\hat{c}^{\dagger}\hat{s}\hat{c} - \tfrac{1}{2}\hat{s}\hat{c}^{\dagger}\hat{c} - \tfrac{1}{2}\hat{c}^{\dagger}\hat{c}\hat{s}\right)dt - [d\hat{B}_0^{\dagger}\,\hat{c} - \hat{c}^{\dagger}\,d\hat{B}_0, \hat{s}]$$

$$+ [\hat{c}^{\dagger}(t-\tau) + \hat{b}_0^{\dagger}(t-\tau)]\left(e^{i\hat{Z}}\hat{s}e^{-i\hat{Z}} - \hat{s}\right)[\hat{c}(t-\tau) + \hat{b}_0(t-\tau)]dt. \tag{5.129}$$

Exercise 5.19 *Verify that this is a valid non-Markovian QLE. That is to say, that, for arbitrary system operators $\hat{s}_1$ and $\hat{s}_2$, $d(\hat{s}_1\hat{s}_2)$ is correctly given by $(d\hat{s}_1)\hat{s}_2 + \hat{s}_1(d\hat{s}_2) + (d\hat{s}_1)(d\hat{s}_2)$.*

Again, all time arguments are t unless otherwise indicated. This should be compared with Eq. (5.105). The obvious difference is that Eq. (5.105) explicitly describes direct photodetection, followed by feedback, whereas the irreversibility in Eq. (5.129) does not specify that the output has been detected. Indeed, the original Langevin equation (5.121) is unchanged if the output is subject to homodyne detection, rather than direct detection. This is the essential difference between the quantum fluctuations of Eq. (5.129) and the fluctuations due to information gathering in Eq. (5.105).

5.4.5 Markovian feedback in the Heisenberg picture

In Eq. (5.129), the vacuum field operators $\hat{b}_0(t)$ have deliberately been moved to the outside (using the fact that $\hat{b}_1(t-\tau)$ commutes with system operators at time t). This has been done for convenience, because, in this position, they disappear when the trace is taken over the bath density operator. Taking the total trace over system and bath density operators gives

$$\langle d\hat{s}\rangle = \left\langle i[\hat{H},\hat{s}] + \left(\hat{c}^{\dagger}\hat{s}\hat{c} - \tfrac{1}{2}\hat{s}\hat{c}^{\dagger}\hat{c} - \tfrac{1}{2}\hat{c}^{\dagger}\hat{c}\hat{s}\right)\right\rangle dt$$

$$+ \left\langle \hat{c}^{\dagger}(t-\tau)\left(e^{i\hat{Z}}\hat{s}e^{-i\hat{Z}} - \hat{s}\right)\hat{c}(t-\tau)\right\rangle dt. \tag{5.130}$$

In the limit $\tau \to 0$, so that $\hat{c}(t - \tau)$ differs negligibly from $\hat{c}(t)$, this gives

$$\langle \mathrm{d}\hat{s} \rangle = \left\langle \mathrm{i}[\hat{H}, \hat{s}] + \left(\hat{c}^\dagger \hat{s} \hat{c} - \tfrac{1}{2} \hat{s} \hat{c}^\dagger \hat{c} - \tfrac{1}{2} \hat{c}^\dagger \hat{c} \hat{s} \right) + \hat{c}^\dagger \left(e^{\mathrm{i}\hat{Z}} \hat{s} e^{-\mathrm{i}\hat{Z}} - \hat{s} \right) \hat{c} \right\rangle \mathrm{d}t. \tag{5.131}$$

In terms of the system density operator,

$$\langle \mathrm{d}\hat{s} \rangle = \mathrm{Tr}\left[\hat{s} \left(-\mathrm{i}[\hat{H}, \rho] + \mathcal{D}[e^{-\mathrm{i}\hat{Z}} \hat{c}] \rho \right) \mathrm{d}t \right]. \tag{5.132}$$

This is precisely what would have been obtained from the Markovian feedback master equation (5.108) for $\mathcal{K}\rho = -\mathrm{i}[\hat{Z}, \rho]$.

Moreover, it is possible to set $\tau = 0$ in Eq. (5.129) and still obtain a valid QLE:

$$\mathrm{d}\hat{s} = \mathrm{i}[\hat{H}, \hat{s}]\mathrm{d}t - [\hat{s}, \hat{c}^\dagger]\left(\tfrac{1}{2}\hat{c} + \hat{b}_0 \right)\mathrm{d}t + \left(\tfrac{1}{2}\hat{c}^\dagger + \hat{b}_0^\dagger \right)[\hat{s}, \hat{c}]\mathrm{d}t$$

$$+ (\hat{c}^\dagger + \hat{b}_0^\dagger)\left(e^{\mathrm{i}\hat{Z}} \hat{s} e^{-\mathrm{i}\hat{Z}} - \hat{s} \right)(\hat{c} + \hat{b}_0)\mathrm{d}t. \tag{5.133}$$

This equation is quite different from Eq. (5.129) because it is Markovian. This implies that, in this equation, it is no longer possible freely to move $\hat{b}_1 = (\hat{c} + \hat{b}_0)$, since it now has the same time argument as the other operators, rather than an earlier one.

Exercise 5.20 *Show that Eq. (5.133) is a valid QLE, bearing in mind that now it is $\hat{b}_0$ rather than $\hat{b}_1$ that commutes with all system operators.*

This trick with time arguments and commutation relations enables the correct QLE describing feedback to be derived without worrying about the method of dealing with the $\tau \to 0$ limit used in Section 5.4.2. There are subtleties involved in using this method in the Heisenberg picture, as will become apparent in Section 5.5.3.

5.5 Homodyne-mediated feedback control

Although homodyne detection can be considered the limit of a jump process (see Section 4.4), it is convenient to develop separately the theory of quantum feedback of currents with Gaussian white noise. In fact, it is necessary to do so in order to treat feedback based on homodyne detection in the presence of a broad-band non-vacuum bath, as we will do.

5.5.1 The homodyne feedback master equation

As shown in Section 4.8.1, the SME for homodyne detection of efficiency η is

$$\mathrm{d}\rho_J(t) = -\mathrm{i}[\hat{H}, \rho_J(t)]\mathrm{d}t + \mathrm{d}t\, \mathcal{D}[\hat{c}]\rho_J(t) + \sqrt{\eta}\, \mathrm{d}W(t)\mathcal{H}[\hat{c}]\rho_J(t). \tag{5.134}$$

The homodyne photocurrent, normalized so that the deterministic part does not depend on the efficiency, is

$$J_{\mathrm{hom}}(t) = \langle \hat{x} \rangle_J(t) + \xi(t)/\sqrt{\eta}, \tag{5.135}$$

where $\xi(t) = dW(t)/dt$ and $\hat{x} = \hat{c} + \hat{c}^\dagger$ as usual. Unlike the direct-detection photocurrent, this current may be negative because the constant local oscillator background has been subtracted. This means that, if one were to feed back this current as in Section 5.4.1, with

$$[\dot{\rho}_J(t)]_{\text{fb}} = J_{\text{hom}}(t - \tau)\mathcal{K}\rho_J(t), \tag{5.136}$$

then the superoperator $\mathcal{K}$ must be such as to give valid evolution irrespective of the sign of time. That is to say, it must give reversible evolution with

$$\mathcal{K}\rho \equiv -i[\hat{F}, \rho] \tag{5.137}$$

for some Hermitian operator $\hat{F}$.

To treat this feedback we use a similar analysis to that of Section 5.4. The only difference is that, because the stochasticity in the measurement (5.134) and the feedback (5.136) is Gaussian white noise, the feedback superoperator $\exp[\mathcal{K}J_{\text{hom}}(t - \tau)dt]$ need only be expanded to second order. The result for the total conditioned evolution of the system is

$$\rho_J(t + dt) = \Big\{1 + \mathcal{K}[\langle\hat{c} + \hat{c}^\dagger\rangle_J(t - \tau)dt + dW(t - \tau)/\sqrt{\eta}] + [1/(2\eta)]\mathcal{K}^2\,dt\Big\}$$
$$\times \Big\{1 + \mathcal{H}[-i\hat{H}]dt + \mathcal{D}[\hat{c}]dt + \sqrt{\eta}\,dW(t)\mathcal{H}[\hat{c}]\Big\}\rho_J(t). \tag{5.138}$$

For τ finite, this becomes

$$d\rho_J(t) = dt\Big\{\mathcal{H}[-i\hat{H}] + \mathcal{D}[\hat{c}] + \langle\hat{c} + \hat{c}^\dagger\rangle_J(t - \tau)\mathcal{K} + \frac{1}{2\eta}\mathcal{K}^2\Big\}\rho_J(t)$$
$$+ dW(t - \tau)\mathcal{K}\rho_J(t)/\sqrt{\eta} + \sqrt{\eta}\,dW(t)\mathcal{H}[\hat{c}]\rho_J(t). \tag{5.139}$$

On the other hand, putting $\tau = 0$ in Eq. (5.138) gives

$$d\rho_J(t) = dt\Big\{-i[\hat{H}, \rho_J(t)] + \mathcal{D}[\hat{c}]\rho_J(t) - i[\hat{F}, \hat{c}\rho_J(t) + \rho_J(t)\hat{c}^\dagger]\Big\}$$
$$+ dt\,\mathcal{D}[\hat{F}]\rho_J(t)/\eta + dW(t)\mathcal{H}[\sqrt{\eta}\hat{c} - i\hat{F}/\sqrt{\eta}]\rho_J(t). \tag{5.140}$$

For $\eta = 1$ and an initially pure state, this can be alternatively be expressed as a SSE. Ignoring normalization, this is simply

$$d|\bar{\psi}_J(t)\rangle = dt\Big[-i\hat{H} - \tfrac{1}{2}\big(\hat{c}^\dagger\hat{c} + 2i\hat{F}\hat{c} + \hat{F}^2\big) + J_{\text{hom}}(t)\big(\hat{c} - i\hat{F}\big)\Big]|\bar{\psi}_J(t)\rangle. \tag{5.141}$$

Exercise 5.21 *Verify this, by finding the SME for $\bar{\rho}_J \equiv |\bar{\psi}_J\rangle\langle\bar{\psi}_J|$ and then adding the terms necessary to preserve the norm.*

The non-selective evolution of the system is easier to find from the SME (5.140). This is a true Itô equation, so that taking the ensemble average simply removes the stochastic term. This gives the homodyne feedback master equation

$$\dot{\rho} = -i[\hat{H}, \rho] + \mathcal{D}[\hat{c}]\rho - i[\hat{F}, \hat{c}\rho + \rho\hat{c}^\dagger] + \frac{1}{\eta}\mathcal{D}[\hat{F}]\rho. \tag{5.142}$$

An equation of this form was derived by Caves and Milburn [CM87] for an idealized model for position measurement plus feedback, with $\hat{c}$ replaced by $\hat{x}$ and η set to 1. The first feedback term, linear in $\hat{F}$, is the desired effect of the feedback which would dominate in the classical regime. The second feedback term causes diffusion in the variable conjugate to $\hat{F}$. It can be attributed to the inevitable introduction of noise by the measurement step in the quantum-limited feedback loop. The lower the efficiency, the more noise introduced.

The homodyne feedback master equation can be rewritten in the Lindblad form (4.28) as

$$\dot{\rho} = -\mathrm{i}\big[\hat{H} + \tfrac{1}{2}(\hat{c}^{\dagger}\hat{F} + \hat{F}\hat{c}), \rho\big] + \mathcal{D}[\hat{c} - \mathrm{i}\hat{F}]\rho + \frac{1-\eta}{\eta}\mathcal{D}[\hat{F}]\rho \equiv \mathcal{L}\rho. \quad (5.143)$$

In this arrangement, the effect of the feedback is seen to replace $\hat{c}$ by $\hat{c} - \mathrm{i}\hat{F}$ and to add an extra term to the Hamiltonian, plus an extra diffusion term that vanishes for perfect detection. In what follows, η will be assumed to be unity unless stated otherwise, since the generalization is usually obvious from previous examples.

The two-time correlation function of the current can be found from Eq. (5.140) to be

$$\mathrm{E}[J_{\mathrm{hom}}(t')J_{\mathrm{hom}}(t)] = \mathrm{Tr}\Big\{(\hat{c} + \hat{c}^{\dagger})\mathrm{e}^{\mathcal{L}(t'-t)}[(\hat{c} - \mathrm{i}\hat{F})\rho(t) + \mathrm{H.c.}]\Big\} + \delta(\tau). \quad (5.144)$$

Exercise 5.22 *Verify this using the method of Section 4.4.4.*

Again, note that the feedback affects the term in square brackets, as well as the evolution by $\mathcal{L}$ for time $t' - t$. This means that the in-loop photocurrent may have a sub-shot-noise spectrum, even if the light in the cavity dynamics is classical. From the same reasoning as in Section 5.4.2, the feedback will not produce nonclassical dynamics for a damped harmonic oscillator ($\hat{c} \propto \hat{a}$) if $\hat{F}$ is a Hamiltonian corresponding to linear optical processes, that is, if $\hat{F}$ is linear in $\hat{a}$ or proportional to $\hat{a}^{\dagger}\hat{a}$.

5.5.2 Feedback with white noise

From one point of view, the results just obtained are simply a special case of those of Section 5.4. Consider the quantum jump SME for homodyne detection with finite local oscillator amplitude, as in Eq. (4.66). Now add feedback according to

$$[\dot{\rho}_J(t)]_{\mathrm{fb}} = -\mathrm{i}[\hat{F}, \rho_J(t)]\frac{\mathrm{d}N(t) - \gamma^2\,\mathrm{d}t}{\gamma\,\mathrm{d}t}, \quad (5.145)$$

where this is understood to be the $\tau \to 0$ limit. Using Section 5.4, the feedback master equation is

$$\dot{\rho} = -\mathrm{i}\big[\hat{H} + \mathrm{i}\tfrac{1}{2}(-\hat{c}\gamma + \hat{c}^{\dagger}\gamma) - \hat{F}\gamma, \rho\big] + \mathcal{D}\big[\mathrm{e}^{-\mathrm{i}\hat{F}/\gamma}(\hat{c} + \gamma)\big]\rho. \quad (5.146)$$

Expanding the exponential to second order in $1/\gamma$ and then taking the limit $\gamma \to \infty$ reproduces (5.143). The correlation functions follow similarly as a special case.

Exercise 5.23 *Show these results.*

However, in another sense, feedback based on homodyne detection is more general: it is possible to treat detection, and hence feedback, in the presence of thermal or squeezed white noise, as in Section 4.8.2.

The relevant conditioning equation for homodyne detection with a white-noise bath parameterized by N and M is Eq. (4.253), reproduced here:

$$d\rho_J(t) = \left\{ dt\, \mathcal{L} + \frac{1}{\sqrt{L}}\, dW(t)\mathcal{H}\left[(N + M^* + 1)\hat{c} - (N + M)\hat{c}^\dagger\right] \right\} \rho_c(t), \quad (5.147)$$

where $L = 2N + M + M^* + 1$, and the photocurrent is

$$J_{\text{hom}}(t) = \langle \hat{c} + \hat{c}^\dagger \rangle_J(t) + \sqrt{L}\xi(t). \quad (5.148)$$

Adding feedback as in Eq. (5.136), which is the same as introducing a feedback Hamiltonian

$$\hat{H}_{\text{fb}}(t) = \hat{F} J_{\text{hom}}(t), \quad (5.149)$$

and following the method above yields the master equation

$$\dot{\rho} = (N+1)\left\{\mathcal{D}[\hat{c}]\rho - i[\hat{F}, \hat{c}\rho + \rho\hat{c}^\dagger]\right\} + N\left\{\mathcal{D}[\hat{c}^\dagger]\rho + i[\hat{F}, \hat{c}^\dagger\rho + \rho\hat{c}]\right\}$$
$$+ M\left\{\tfrac{1}{2}[\hat{c}^\dagger, [\hat{c}^\dagger, \rho]] + i[\hat{F}, [\hat{c}^\dagger, \rho]]\right\} + M^*\left\{\tfrac{1}{2}[\hat{c}, [\hat{c}, \rho]] - i[\hat{F}, [\hat{c}, \rho]]\right\}$$
$$+ L\mathcal{D}[\hat{F}]\rho - i[\hat{H}, \rho]. \quad (5.150)$$

For the case $N = M = 0$, this reduces to Eq. (5.142) with $\eta = 1$.

5.5.3 Homodyne feedback in the Heisenberg picture

The quantum Langevin treatment of quadrature flux feedback (corresponding to homodyne detection) is relatively straightforward, because of the Gaussian nature of the noise. The homodyne photocurrent is identified with the quadrature of the outgoing field,

$$\hat{J}_{\text{hom}}(t) = \hat{b}_1(t) + \hat{b}_1^\dagger(t) = \hat{c}(t) + \hat{c}^\dagger(t) + \hat{b}_0(t) + \hat{b}_0^\dagger(t). \quad (5.151)$$

The feedback Hamiltonian is defined as

$$\hat{H}_{\text{fb}}(t) = \hat{F}(t)\hat{J}_{\text{hom}}(t - \tau). \quad (5.152)$$

The time delay τ ensures that the output quadrature operator $\hat{J}_{\text{hom}}(t - \tau)$ commutes with all system operators at time t. Thus, it will commute with $\hat{F}(t)$ and there is no ambiguity in the operator ordering in Eq. (5.152). Treating the equation of motion generated by this Hamiltonian as a Stratonovich (or implicit) equation, the Itô (or explicit) equation is

$$[d\hat{s}(t)]_{\text{fb}} = i\hat{J}_{\text{hom}}(t - \tau)[\hat{F}(t), \hat{s}(t)]dt - \tfrac{1}{2}[\hat{F}(t), [\hat{F}(t), \hat{s}(t)]]dt. \quad (5.153)$$

Adding in the non-feedback evolution gives the total explicit equation of motion

$$d\hat{s} = i[\hat{H}, \hat{s}]dt + i[\hat{c}^\dagger(t - \tau)dt + d\hat{B}_0^\dagger(t - \tau)][\hat{F}, \hat{s}]$$

$$+ i[\hat{F}, \hat{s}][\hat{c}(t - \tau)dt + d\hat{B}_0(t - \tau)] - \tfrac{1}{2}[\hat{F}, [\hat{F}, \hat{s}]]dt$$

$$+ \left(\hat{c}^\dagger \hat{s}\hat{c} - \tfrac{1}{2}\hat{s}\hat{c}^\dagger\hat{c} - \tfrac{1}{2}\hat{c}^\dagger\hat{c}\hat{s}\right)dt - [d\hat{B}_0^\dagger \hat{c} - \hat{c}^\dagger d\hat{B}_0, \hat{s}]. \tag{5.154}$$

Here, all time arguments are t unless indicated otherwise.

In Eq. (5.154), we have once again used the commutability of the output operators with system operators to place them suitably on the exterior of the feedback expression. This ensures that, when an expectation value is taken, the input noise operators annihilate the vacuum and hence give no contribution. This is the same trick as used in Section 5.4.4, and putting $\tau = 0$ in Eq. (5.154) also gives a valid Heisenberg equation of motion. That equation is the counterpart to the homodyne feedback master equation (5.143). However, this trick will not work if the input field is not in the vacuum state, but rather, for example, in a thermal state. For direct detection (without filtering), it is necessary to restrict to a vacuum bath, in which case the operator-ordering trick is perfectly legitimate. However, for quadrature-based feedback, as explained in Section 5.5.2, it is possible to treat white noise. Thus, it is necessary to have a method of treating the Markovian ($\tau \to 0$) limit that will work in this general case (although we will not present the general theory here). The necessary method is essentially the same as that used with the quantum trajectories, ensuring that the feedback acts later than the measurement. In applying it to Heisenberg equations of motion, it will be seen that one has to be quite careful with operator ordering.

In the $\tau = 0$ limit the feedback Hamiltonian (5.152) has an ordering ambiguity, because the bath operator $\hat{b}_1(t)$ does not commute with an arbitrary system operator $\hat{s}(t)$ at the same time. This problem would not occur if the feedback Hamiltonian were instead

$$\hat{H}_0^{\text{fb}}(t) = \hat{F}(t)[\hat{b}_0(t) + \hat{b}_0^\dagger(t)], \tag{5.155}$$

because $\hat{b}_0(t)$ does commute with $\hat{s}(t)$. At first sight it would not seem sensible to use Eq. (5.155) because $\hat{b}_0(t) + \hat{b}_0^\dagger(t)$ is the quadrature of the vacuum input, which is independent of the system and so (it would seem) cannot describe feedback. However, Eq. (5.155) is the correct Hamiltonian to use as long as we ensure that the feedback-coupling between the system and the bath occurs *after* the usual coupling between system and bath. That is, the total unitary operator evolving the system and bath at time t is

$$\hat{U}(t + dt, t) = e^{-i\hat{H}_0^{\text{fb}}(t)dt}\hat{U}_0(t + dt, t), \tag{5.156}$$

where $\hat{U}_0(t + dt, t)$ is defined in Eq. (5.119). In the Heisenberg picture, the system evolves via

$$\hat{s}(t + dt) = \hat{U}^\dagger(t + dt, t)\hat{s}(t)\hat{U}(t + dt, t) \tag{5.157}$$

$$= \hat{U}_0^\dagger(t + dt, t)e^{+i\hat{H}_0^{\text{fb}}(t)dt}\hat{s}(t)e^{-i\hat{H}_0^{\text{fb}}(t)dt}\hat{U}_0(t + dt, t). \tag{5.158}$$

Note that in Eq. (5.158) the feedback appears to act first because of the reversal of the order of unitary operators in the Heisenberg picture. If desired, one could rewrite Eq. (5.158) in a (perhaps) more intuitive order as

$$\hat{s}(t + dt) = e^{+i\hat{H}_1^{fb}(t)dt} \hat{U}_0^\dagger(t + dt, t)\hat{s}(t)\hat{U}_0(t + dt, t)e^{-i\hat{H}_1^{fb}(t)dt}. \qquad (5.159)$$

Here

$$\hat{H}_1^{fb}(t) = \hat{U}_0^\dagger(t + dt, t)\hat{H}_0^{fb}(t)\hat{U}_0(t + dt, t) \qquad (5.160)$$

$$= \hat{F}(t + dt)\hat{J}_{hom}(t). \qquad (5.161)$$

That is, we regain the output quadrature (or homodyne photocurrent operator), as well as replacing $\hat{F}(t)$ by $\hat{F}(t + dt)$. This ensures that, once again, there is no operator ambiguity in Eq. (5.161) because the $\hat{J}_{hom}$ represents the result of the homodyne measurement at a time t earlier (albeit infinitesimally) than the time argument for the system operator $\hat{F}$. Again, this makes sense physically because the feedback must act after the measurement.

Expanding the exponentials in Eq. (5.158) or Eq. (5.159), the quantum Itô rules give

$$d\hat{s} = i[\hat{H}, \hat{s}]dt - [\hat{s}, \hat{c}^\dagger]\left(\tfrac{1}{2}\hat{c}\,dt + d\hat{B}_0\right) + \left(\tfrac{1}{2}\hat{c}^\dagger dt + d\hat{B}_0^\dagger\right)[\hat{s}, \hat{c}]$$

$$+ i[\hat{c}^\dagger dt + d\hat{B}_0^\dagger][\hat{F}, \hat{s}] + i[\hat{F}, \hat{s}][\hat{c}\,dt + d\hat{B}_0] - \tfrac{1}{2}[\hat{F}, [\hat{F}, \hat{s}]]dt. \qquad (5.162)$$

Exercise 5.24 *Verify this, and show that this is a valid Markovian QLE that is equivalent to the homodyne feedback master equation (5.142).*

5.6 Markovian feedback in a linear system

In order to understand the nature of quantum-limited feedback, it is useful to consider a simple linear system that can be solved exactly. We use the example of Ref. [WM94c]: a single optical mode in a cavity, so that the two quadratures x and y, obeying $[\hat{x}, \hat{y}] = 2i$, form a complete set of observables. To obtain linear equations of motion for these observables it is necessary first to restrict the Hamiltonian $\hat{H}$ to be a quadratic function of $\hat{x}$ and $\hat{y}$. Second, the Lindblad operator $\hat{c}$ must be a linear function of $\hat{x}$ and $\hat{y}$, which it will be for cavity damping in which $\hat{c} = \hat{a} = (\hat{x} + i\hat{y})/2$. Third, the measured current must be a linear function of these variables, as in homodyne detection of the cavity output (which is what we will assume). Finally, the feedback Hamiltonian must be such that $\hat{F}$ is a linear function also, as in classical driving of a cavity. These basic ideas will be treated much more generally in Chapter 6.

5.6.1 The linear system

Since in this chapter we are seeking merely a simple example, we make the further assumption that we are interested only in $\hat{x}$ and that it obeys a linear QLE independent of $\hat{y}$. In this case, all Markovian linear dynamics (in the absence of feedback) can be composed of damping, driving and parametric driving. Damping will be assumed to be always present, since we will assume homodyne detection of the output field from our system. We there-fore take the damping rate to be unity. Constant linear driving simply shifts the stationary state away from $\langle \hat{x} \rangle = 0$, and will be ignored. Stochastic linear driving in the white-noise approximation causes diffusion in the x quadrature, at a rate l. Finally, if the strength of the parametric driving ($\hat{H} \propto \hat{x}\hat{y} + \hat{y}\hat{x}$) is χ (where $\chi = 1$ would represent a parametric oscillator at threshold), then the master equation for the system is

$$\dot{\rho} = \mathcal{D}[\hat{a}]\rho + \tfrac{1}{4}l\mathcal{D}[\hat{a}^{\dagger} - \hat{a}]\rho + \tfrac{1}{4}\chi[\hat{a}^2 - \hat{a}^{\dagger 2}, \rho] \equiv \mathcal{L}_0\rho, \qquad (5.163)$$

where $\hat{a}$ is the annihilation operator for the cavity mode as above.

Although parametric driving is an example of a nonlinear optical process (which can generate nonclassical optical states such as squeezed states), it gives linear dynamics in that the mean and variance of $\hat{x} = \hat{a} + \hat{a}^{\dagger}$ obey linear equations of the following form:

$$\frac{\mathrm{d}}{\mathrm{d}t}\langle \hat{x} \rangle = -k\langle \hat{x} \rangle, \qquad (5.164)$$

$$\frac{\mathrm{d}}{\mathrm{d}t}V = -2kV + D. \qquad (5.165)$$

Exercise 5.25 *Show that for the particular master equation above (the properties of which will be denoted by the subscript 0) these equations hold, with*

$$k_0 = \tfrac{1}{2}(1 + \chi), \qquad (5.166)$$

$$D_0 = 1 + l. \qquad (5.167)$$

Hint: *Remember that* $\mathrm{d}\langle \hat{x} \rangle/\mathrm{d}t = \mathrm{Tr}[\hat{x}\dot{\rho}]$ *and that* $\mathrm{d}\langle \hat{x}^2 \rangle/\mathrm{d}t = \mathrm{Tr}[\hat{x}^2\dot{\rho}]$.

For a *stable* system with $k > 0$, there is a steady state with $\langle x \rangle = 0$ and

$$V = \frac{D}{2k}. \qquad (5.168)$$

It turns out that the first two moments (the mean and variance) are actually sufficient to specify the stationary state of the system because it is a Gaussian state. That is, its Wigner function (see Section A.5) is Gaussian. The probability distribution for x (which is all we are interested in here) is just the marginal distribution for the Wigner function, so it is also Gaussian. Moreover, if the distribution is originally Gaussian (as for the vacuum, for example), then it will be Gaussian at all times. This can be seen by considering the equation of motion for the probability distribution for x,

$$\wp(x) = \langle x|\rho|x \rangle. \qquad (5.169)$$

This equation of motion can be derived from the master equation by considering the operator correspondences for the Wigner function (see Section A.5). Because here we have $[\hat{x}, \hat{y}] = 2\mathrm{i}$, if we identify $\hat{x}$ with $\hat{Q}$ then we must identify $\hat{y}$ with $2\hat{P}$. On doing this we find that

$\wp(x)$ obeys the following Fokker–Planck equation (see Section B.5)

$$\dot{\wp}(x) = \left(\frac{\partial}{\partial x}kx + \tfrac{1}{2}D\frac{\partial^2}{\partial x^2}\right)\wp(x). \tag{5.170}$$

This particular form, with linear drift kx and constant diffusion $D > 0$, is known as an Ornstein–Uhlenbeck equation (OUE).

Exercise 5.26 *Derive Eq. (5.170) from Eq. (5.163), and show by substitution that it has a Gaussian solution with mean and variance satisfying Eqs. (5.164) and (5.165), respectively.*

In the present case, $V_0 = (1 + l)/(1 + \chi)$. If this is less than unity, the system exhibits squeezing of the x quadrature. It is useful to characterize the squeezing by the normally ordered variance (see Section A.5)

$$U \equiv \langle\hat{a}^\dagger\hat{a}^\dagger + 2\hat{a}^\dagger\hat{a} + \hat{a}\hat{a}\rangle - \langle\hat{a}^\dagger + \hat{a}\rangle^2, \tag{5.171}$$

Exercise 5.27 *Show from this definition that $U = V - 1$.*

For this system, the normally ordered variance takes the value

$$U_0 = \frac{l - \chi}{1 + \chi}. \tag{5.172}$$

Now, if the system is to stay below threshold (so that the variance in the y quadrature does not become unbounded), then the maximum value for χ is unity.

Exercise 5.28 *Show this from the master equation (5.163)*

At this value, $U_0 = -1/2$ when the x-diffusion rate $l = 0$. Therefore the minimum value of squeezing which this linear system can attain as a stationary value is half of the theoretical minimum of $U_0 = -1$.

In quantum optics, the output light is often of more interest than the intracavity light. Therefore it is useful to compute the output noise statistics. For squeezed systems, the relevant quantity is the spectrum of the homodyne photocurrent, as introduced in Section 4.4.4,

$$S(\omega) = \lim_{t\to\infty}\int_{-\infty}^{\infty} d\tau\, E[J_{\text{hom}}(t + \tau)J_{\text{hom}}(t)]e^{-i\omega\tau}. \tag{5.173}$$

Given the drift and diffusion coefficients for the dynamics, the spectrum in the present case is

$$S(\omega) = 1 + \frac{D - 2k}{\omega^2 + k^2}. \tag{5.174}$$

Exercise 5.29 *Show this using the results from Section 4.4.4.*
Hint: *Remember that, for example, $\text{Tr}\big[\hat{x}e^{\mathcal{L}\tau}(\hat{a}\rho_{\text{ss}})\big]$ is just the expectation value of $\hat{x}$ at time τ using the 'state' with initial condition $\rho(0) = \hat{a}\rho_{\text{ss}}$. Thus, since the mean of $\hat{x}$ obeys the linear equation (5.164), it follows that this expression simplifies to $e^{-k\tau}\,\text{Tr}[\hat{x}\hat{a}\rho_{\text{ss}}]$.*

The spectrum consists of a constant term representing shot noise plus a Lorentzian, which will be negative for squeezed systems. The spectrum can be related to the intracavity squeezing by subtracting the vacuum noise:

$$\frac{1}{2\pi} \int_{-\infty}^{\infty} d\omega [S(\omega) - 1] = \frac{D - 2k}{2k} = U. \tag{5.175}$$

That is, the total squeezing integrated across all frequencies in the output is equal to the intracavity squeezing. However, the minimum squeezing, which for a simple linear system such as this will occur at zero frequency,[2] may be greater than or less than U. It is useful to define it by another parameter,

$$R = S(0) - 1 = 2U/k. \tag{5.176}$$

For the particular system considered above,

$$R_0 = \frac{l - \chi}{\frac{1}{4}(1 + \chi)^2}. \tag{5.177}$$

In the ideal limit ($\chi \to 1, l \to 0$), the zero-frequency squeezing approaches the minimum value of -1.

5.6.2 Adding linear feedback

Now consider adding feedback to try to reduce the fluctuations in x. Restricting the feedback to linear optical processes suggests the feedback operator

$$\hat{F} = -\lambda \hat{y}/2. \tag{5.178}$$

As a separate Hamiltonian, this translates a state in the negative x direction for λ positive. By controlling this Hamiltonian by the homodyne photocurrent, one thus has the ability to change the statistics for x and perhaps achieve better squeezing. Substituting Eq. (5.178) into the general homodyne feedback master equation (5.142) and adding the free dynamics (5.163) gives

$$\dot{\rho} = \mathcal{L}_0 \rho + \frac{\lambda}{2}[\hat{a} - \hat{a}^\dagger, \hat{a}\rho + \rho\hat{a}^\dagger] + \frac{\lambda^2}{4\eta}\mathcal{D}[\hat{a} - \hat{a}^\dagger]\rho. \tag{5.179}$$

Here η is the proportion of output light used in the feedback loop, multiplied by the efficiency of the detection. For the x distribution $\wp(x)$ one finds that it still obeys an OUE, but now with

$$k = k_0 + \lambda, \tag{5.180}$$

$$D = D_0 + 2\lambda + \lambda^2/\eta. \tag{5.181}$$

---

[2] In reality, the minimum noise is never found at zero frequency, because virtually all experiments are subject to non-white noise of technical origin, which can usually be made negligible at high frequencies, but whose spectrum grows without bound as $\omega \to 0$. Often, the spectrum scales at $1/\omega$, or $1/f$, where $f = \omega/(2\pi)$, in which case it is known as $1/f$ noise.

Exercise 5.30 *Show this.*

Provided that $\lambda + k_0 > 0$, there will exist a stable Gaussian solution to the master equation (5.179). The new intracavity squeezing parameter is

$$U_\lambda = (k_0 + \lambda)^{-1}\left(k_0 U_0 + \frac{\lambda^2}{2\eta}\right). \tag{5.182}$$

An immediate consequence of this expression is that U_λ can be negative only if U_0 is. That is to say, the feedback cannot produce squeezing, as explained in Section 5.5. On minimizing U_λ with respect to λ one finds

$$U_{\min} = \eta^{-1}\left(-k_0 + \sqrt{k_0^2 + 2\eta k_0 U_0}\right), \tag{5.183}$$

when

$$\lambda = -k_0 + \sqrt{k_0^2 + 2\eta k_0 U_0}. \tag{5.184}$$

Note that this λ has the same sign as U_0. That is to say, if the system produces squeezed light, then the best way to enhance the squeezing is to add a force that displaces the state in the direction of the difference between the measured photocurrent and the desired mean photocurrent. This positive feedback is the opposite of what would be expected classically, and can be attributed to the effect of homodyne measurement on squeezed states, as will be explained in Section 5.6.3. Obviously, the best intracavity squeezing will be when $\eta = 1$, in which case the intracavity squeezing can be simply expressed as

$$U_{\min} = k_0\left(-1 + \sqrt{1 + R_0}\right). \tag{5.185}$$

Although linear optical feedback cannot produce squeezing, this does not mean that it cannot reduce noise. In fact, it can be proven that $U_{\min} \leq U_0$, with equality only if $\eta = 0$ or $U_0 = 0$.

Exercise 5.31 *Show this for $\eta = 1$ using the result $\sqrt{1 + R_0} \leq 1 + R_0/2$ (since $R_0 \geq -1$). The result for any η follows by application of the mean-value theorem.*

This result implies that the intracavity variance in x can always be reduced by homodyne-mediated linear optical feedback, unless it is at the vacuum noise level. In particular, intracavity squeezing can always be enhanced. For the parametric oscillator defined originally in Eq. (5.163), with $l = 0$, $U_{\min} = -\chi/\eta$. For $\eta = 1$, the (symmetrically ordered) x variance is $V_{\min} = 1 - \chi$. The y variance, which is unaffected by feedback, is seen from Eq. (5.163) to be $(1 - \chi)^{-1}$. Thus, with perfect detection, it is possible to produce a minimum-uncertainty squeezed state with arbitrarily high squeezing as $\chi \to 1$. This is not unexpected, since parametric driving in an undamped cavity also produces minimum-uncertainty squeezed states (but there is no steady state). The feedback removes the noise that was added by the damping that enables the measurement used in the feedback.

Next, we turn to the calculation of the output squeezing. Here, it must be remembered that at least a fraction η of the output light is being used in the feedback loop. Thus, the fraction θ of cavity emission available as an output of the system is at best $1 - \eta$. Integrated over all frequencies, the total available output squeezing is thus θU_λ. It can be shown that

$$\theta U_{\min} \geq U_0 \text{ for } U_0 < 0. \tag{5.186}$$

Exercise 5.32 *Convince yourself that this is true.*

That is, dividing the cavity output and using some in a feedback loop produces worse squeezing in the remaining output than was present in the original, undivided output. Note, however, that, if the cavity output is inherently divided (which is often the case, with two output mirrors), then using one output in the feedback loop would enhance squeezing in the other output. This is because the squeezing in the system output of interest would have changed from θU_0 to $\theta U_{\min}$.

Rather than integrating over all frequencies, experimentalists are often more interested in the optimal noise reduction at some frequency, which is to say at zero frequency here. With no feedback, this is given by R_0. With feedback, it is given by

$$R_\lambda = \theta \frac{2U_\lambda}{k_0 + \lambda} = \theta \frac{2k_0 U_0 + \lambda^2/\eta}{(k_0 + \lambda)^2}. \tag{5.187}$$

In all cases, R_λ is minimized for a different value of λ from that which minimizes U_λ. One finds

$$R_{\min} = \frac{\theta R_0}{1 + R_0 \eta} \tag{5.188}$$

when

$$\lambda = 2\eta U_0. \tag{5.189}$$

Again, λ has the same sign as U_0. It follows immediately from Eq. (5.188) that, since $R_0 \geq -1$ and $\theta \leq 1 - \eta$,

$$R_{\min} \geq R_0 \text{ for } R_0 < 0. \tag{5.190}$$

That is to say, dividing the cavity output to add a homodyne-mediated classical feedback loop cannot produce better output squeezing at any frequency than would be available from an undivided output with no feedback. These 'no-go' results are analogous to those obtained for the feedback control of optical beams derived in Section 5.2.

5.6.3 Understanding feedback through conditioning

The preceding section gave the limits to noise reduction by classical feedback for a linear system, both intracavity and extracavity. In this section, we give an explanation for the intracavity results, in terms of the conditioning of the state by the measurement on which the feedback is based. To find this link between conditioning and feedback it is necessary to return to the selective stochastic master equation (5.138) for the conditioned state matrix $\rho_J(t)$,

$$d\rho_J(t) = dt\left(\mathcal{L}_0\rho_J(t) + \mathcal{K}[a\rho_J(t) + \rho_J(t)a^\dagger] + \frac{1}{2\eta}\mathcal{K}^2\rho_J(t)\right)$$

$$+ dW(t)\left(\sqrt{\eta}\mathcal{H}[\hat{a}] + \mathcal{K}/\sqrt{\eta}\right)\rho_J(t). \qquad (5.191)$$

Here, $\mathcal{L}_0$ is as defined in Eq. (5.163) and $\mathcal{K}\rho = -i[\hat{F}, \rho]$, where $\hat{F}$ is defined in Eq. (5.178). Changing this to a stochastic FPE for the conditioned marginal Wigner function gives

$$d\wp_J(x) = dt\left[\frac{\partial}{\partial x}(k_0 + \lambda)x + \frac{1}{2}\frac{\partial^2}{\partial x^2}(D_0 + 2\lambda + \lambda^2/\eta)\right]\wp_J(x)$$

$$+ dW(t)\left[\sqrt{\eta}\left(x - \bar{x}_J(t) + \frac{\partial}{\partial x}\right) + (\lambda/\sqrt{\eta})\frac{\partial}{\partial x}\right]\wp_J(x), \qquad (5.192)$$

where $\bar{x}_J(t)$ is the mean of the distribution $\wp_J(x)$.

Exercise 5.33 *Show this using the Wigner-function operator correspondences.*

This equation is obviously no longer a simple OUE. Nevertheless, it still has a Gaussian as an exact solution. Specifically, the mean $\bar{x}_J$ and variance V_J of the conditioned distribution obey the following equations (recall that $\xi(t) = dW/dt$):

$$\dot{\bar{x}}_J = -(k_0 + \lambda)\bar{x}_J + \xi(t)[\sqrt{\eta}(V_J - 1) - (\lambda/\sqrt{\eta})], \qquad (5.193)$$

$$\dot{V}_J = -2k_0V_J + D_0 - \eta(V_J - 1)^2. \qquad (5.194)$$

Exercise 5.34 *Show this by considering a Gaussian ansatz for Eq. (5.192).*
Hint: *Remember that, for any b, $1 + dW(t)b = \exp[dW(t)b - dt\, b^2/2]$.*

Two points about the evolution equation for V_J are worth noting: it is completely deterministic (no noise terms); and it is not influenced by the presence of feedback.

The equation for the conditioned variance is more simply written in terms of the conditioned normally ordered variance $U_J = V_J - 1$,

$$\dot{U}_J = -2k_0U_J - 2k_0 + D_0 - \eta U_J^2. \qquad (5.195)$$

On a time-scale as short as a cavity lifetime, U_J will approach its steady-state value of

$$U_J^{ss} = \eta^{-1}\left(-k_0 + \sqrt{k_0^2 + \eta(-2k_0 + D_0)}\right). \qquad (5.196)$$

Note that this is equal to the minimum *unconditioned* variance with feedback – the U_{min} of Eq. (5.183). The explanation for this will become evident shortly. Substituting the steady-state conditioned variance into Eq. (5.193) gives

$$\dot{\bar{x}}_J = -(k_0 + \lambda)\bar{x}_J + \xi(t)\frac{1}{\sqrt{\eta}}\left[-k_0 + \sqrt{k_0^2 + \eta(-2k_0 + D_0)} - \lambda\right]. \qquad (5.197)$$

If one were to choose $\lambda = -k_0 + \sqrt{k_0^2 + \eta(-2k_0 + D_0)}$ then there would be no noise at all in the conditioned mean and so one could set $\bar{x}_J = 0$ in steady state. This value of

λ is precisely that value derived as Eq. (5.184) to minimize the unconditioned variance under feedback. Now one can see why this minimum unconditioned variance is equal to the conditioned variance. The feedback works simply by suppressing the fluctuations in the conditioned mean.

In general, the unconditioned variance will consist of two terms, the conditioned quantum variance in x plus the classical (ensemble) average variance in the conditioned mean of x:

$$U_\lambda = U_J + E[\bar{x}_J^2]. \tag{5.198}$$

The latter term is found from Eq. (5.197) to be

$$E[\bar{x}_J^2] = \eta^{-1} \frac{1}{2(k_0 + \lambda)} \left[-(k_0 + \lambda) + \sqrt{k_0^2 + \eta(-2k_0 + D_0)} \right]^2. \tag{5.199}$$

Adding Eq. (5.196) gives

$$U_\lambda = \eta^{-1} \frac{1}{2(k_0 + \lambda)} [\lambda^2 + \eta(-2k_0 + D_0)]. \tag{5.200}$$

Exercise 5.35 *Verify that this is identical to the expression (5.182) derived in the preceding subsection using the unconditioned master equation.*

Using the conditioned equation, there is an obvious way to understand the feedback. The homodyne measurement reduces the conditioned variance (except when it is equal to the classical minimum of 1). The more efficient the measurement, the greater the reduction. Ordinarily, this reduced variance is not evident because the measurement gives a random shift to the conditional mean of x, with the randomness arising from the shot noise of the photocurrent. By appropriately feeding back this photocurrent, it is possible to counteract precisely this shift and thus observe the conditioned variance.

The sign of the feedback parameter λ is determined by the sign of the shift which the measurement gives to the conditioned mean $\bar{x}_J$. For classical statistics ($U \geq 0$), a higher than average photocurrent reading ($\xi(t) > 0$) leads to an increase in $\bar{x}_J$ (except if $U = 0$, in which case the measurement has no effect). However, for nonclassical states with $U < 0$, the classical intuition fails since a positive photocurrent fluctuation causes $\bar{x}_J$ to decrease. This explains the counter-intuitive negative value of λ required in squeezed systems, which naively would be thought to destabilize the system and increase fluctuations. However, the value of the positive feedback required, given by Eq. (5.184), is such that the overall decay rate $k_0 + \lambda$ is still positive.

It is worth remarking that the above conclusions are not limited to Markovian feedback, which is all that we have analysed. One could consider a feedback Hamiltonian proportional to an arbitrary (even nonlinear) function of the photocurrent $J(t)$, and the equation for the conditional variance (5.194) will remain exactly as it is. Only the equation for the mean will be changed. Although this equation might not be solvable, Eq. (5.198) guarantees that the unconditioned variance cannot be less than the conditioned variance. Moreover, if the feedback Hamiltonian is a linear functional of the photocurrent then the equation for the mean will be solvable in Fourier space, provided that the feedback is stable. That is, for linear systems one can solve for arbitrary feedback using the theory of quantum trajectories in exactly the same manner as we did for QLEs in Section 5.2. The interested reader is referred to Ref. [WM94c].

To conclude, one can state succinctly that conditioning can be made practical by feedback. The intracavity noise reduction produced by 'classical' feedback can never be better than that produced (conditionally) by the measurement. Of course, nonclassical feedback (such as using the photocurrent to influence nonlinear intracavity elements) may produce nonclassical states, but such elements can produce nonclassical states without feedback, so this is hardly surprising. In order to produce nonclassical states by feedback with linear optics, it would be necessary to have a nonclassical measurement scheme. That is to say, one that does not rely on measurement of the extracavity light to procure information about the intracavity state. For example, a QND measurement of one quadrature would produce a squeezed conditioned state and hence allow the production of unconditional intracavity (and extracavity) squeezing by feedback. Again, the interested reader is referred to Ref. [WM94c]. This is essentially the same conclusion as that which was reached for feedback on optical beams in Section 5.3. In the following section we consider feedback based on QND measurements in an atomic (rather than optical) system.

5.7 Deterministic spin-squeezing

Performing a QND measurement of an optical quadrature, as discussed in the preceding section and Section 5.3.1, is very difficult experimentally. However, for atomic systems it should be easier to achieve a QND measurement near the quantum limit. In this section we apply the theory developed in this chapter to the deterministic production of spin-squeezing, as proposed in Ref. [TMW02b]. Here the spin refers to the z component of angular momentum of atoms. A spin-squeezed state [KU93] is an entangled state of an ensemble of such atoms, such that the total z component of angular momentum has a smaller uncertainty than if all atoms were prepared identically without entanglement.

5.7.1 Spin-squeezing

Consider an atom with two relevant levels, with the population difference operator being the Pauli operator $\hat{\sigma}_z$. The collective properties of N such atoms prepared identically are conveniently described by a spin-J system for $J = N/2$. The collective angular-momentum operators, $\hat{\mathbf{J}}$, are $\hat{J}_\alpha = \sum_{k=1}^{N} \hat{j}_\alpha^{(k)}$, where $\alpha = x, y, z$ and where $\hat{j}_\alpha^{(k)} = \hat{\sigma}_\alpha^{(k)}/2$ is the angular-momentum operator for the kth atom. $\hat{\mathbf{J}}$ obey the cyclic commutation relations $[\hat{J}_x, \hat{J}_y] = i\epsilon_{xyz}\hat{J}_z$.

Exercise 5.36 *Verify this from the commutation relations for the Pauli matrices. See Box. 3.1.*

The corresponding uncertainty relations are

$$(\Delta J_y)^2(\Delta J_z)^2 \geq \tfrac{1}{4}|\langle J_x\rangle|^2, \tag{5.201}$$

plus cyclic permutations. The operator $\hat{J}_z$ represents half the total population difference and is a quantity that can be measured, for example by dispersive imaging techniques as will be discussed.

For a coherent spin state (CSS) of a spin-J system, the elementary spins all point in the same direction, with no correlations. An example of such a state is a $\hat{J}_x$ eigenstate of maximum eigenvalue $J = N/2$. Such a state achieves the minimum of the uncertainty relation (5.201), with the variance of the two components normal to the mean direction (in this case, J_z and J_y) equal to $J/2$. If quantum-mechanical correlations are introduced among the atoms it is possible to reduce the fluctuations in one direction at the expense of the other. This is the idea of a squeezed spin state (SSS) introduced by Kitagawa and Ueda [KU93]. That is, the spin system is squeezed when the variance of one spin component normal to the mean spin vector is smaller than the standard quantum limit (SQL) of $J/2$.

There are many ways to characterize the degree of spin-squeezing in a spin-J system. We will use the criteria of Sørensen and co-workers [SDCZ01] and Wang [Wan01], where the squeezing parameter is given by

$$\xi_{n_1}^2 = \frac{N(\Delta J_{n_1})^2}{\langle J_{n_2}\rangle^2 + \langle J_{n_3}\rangle^2}, \tag{5.202}$$

where $\hat{J}_n \equiv \mathbf{n}\cdot\hat{\mathbf{J}}$ and $\mathbf{n}_i$ for $i = 1, 2, 3$ are orthogonal unit vectors. Systems with $\xi_n^2 < 1$ are said to be spin-squeezed in the direction $\mathbf{n}$. It has also been shown that this indicates that the atoms are in an entangled state [SDCZ01]. This parameter also has the appealing property that, for a CSS, $\xi_n^2 = 1$ for all n [Wan01]. In all that follows, we consider spin-squeezing in the z direction and hence drop the subscript on ξ^2.

The ultimate limit to ξ^2 (set by the Heisenberg uncertainty relations) is of order $1/N$. Since N is typically very large experimentally (of order 10^{11}), the potential for noise reduction is enormous. However, so far, the degree of noise reduction achieved experimentally has been modest, with $\xi^2 \sim 10^{-1} \gg N^{-1}$. The amount of entanglement in such states is relatively small, so it is a good approximation to assume that the atoms are unentangled when evaluating the denominator of Eq. (5.202). That is, for example, if the mean spin is in the x direction, we can say that $\langle J_x\rangle = J$ and that $\langle J_y\rangle = \langle J_z\rangle = 0$. For squeezing in the z direction, the squeezing parameter is thus given by

$$\xi^2 \simeq \langle J_z^2\rangle/(J/2). \tag{5.203}$$

Spin-squeezed states have potential applications in fields such as quantum information [JKP01] and high-precision time-keeping [WBI$^+$92].

5.7.2 *Measurement and feedback*

Let the two relevant internal states of each atom be the magnetic sublevels of a $j = \frac{1}{2}$ state, such as the ground state of an alkali atom. In the absence of a magnetic field, these levels are degenerate. For such an atom, a probe beam that is suitably polarized and suitably detuned from a particular excited atomic level will not be absorbed by the atoms. Rather, it will suffer a phase shift of different size depending upon which state the atom is in. That is, the phase shift will be linear in the atomic operator $\hat{j}_z$. If all atoms are coupled equally to the probe beam then the total phase shift will depend only on the total spin operator $\hat{J}_z$. The phase shift can then be measured by homodyne detection.

A simple model for this that captures the most important physics of the measurement is the SME

$$d\rho_Y = M\mathcal{D}[\hat{J}_z]\rho_Y \, dt + \sqrt{M} \, dW(t)\mathcal{H}[\hat{J}_z]\rho_Y. \tag{5.204}$$

Here ρ_Y is the state of the spin system conditioned on the current

$$Y(t) = 2\sqrt{M}\langle J_z \rangle_Y + dW(t)/dt, \tag{5.205}$$

where unit detection efficiency has been assumed. We are using $Y(t)$, rather than $J(t)$ as has been our convention, in order to avoid confusion with the total spin J. For a probe beam in free space, the QND measurement strength (with dimensions of T^{-1}) is

$$M = P\hbar\omega[\gamma^2/(8A \, \Delta I_{\text{sat}})]^2, \tag{5.206}$$

where P is the probe power, $\omega = 2\pi c/\lambda$ is the probe frequency, A is the cross-sectional area of the probe, γ is the spontaneous emission rate from the excited state and I_{sat} is the saturation intensity for the transition, which equals $2\pi^2\hbar\omega\gamma/\lambda^2$ for a two-level atom [Ash70]. Note that Eq. (5.204) ignores spontaneous emission by the atoms – for a discussion see Ref. [TMW02a] and for more complete treatments see Refs. [BSSM07, NM08].

Using optical pumping, the atomic sample can be prepared such that all the atoms are in one of their internal states. A fast $\pi/2$-pulse can then be applied, coherently transferring all atoms into an equal superposition of the two internal states, which is an eigenstate of the $\hat{J}_x$ operator with eigenvalue J. As described earlier, the CSS is a minimum-uncertainty state, so in this case the variances of both J_z and J_y are $J/2$.

The dominant effects of the conditioned evolution (5.204) on the initial CSS are a decrease in the uncertainty of J_z (since we are measuring J_z) with corresponding noise increases in J_y and J_x, and a stochastic shift of the mean J_z away from its initial value of zero. If we were concerned only with J_z then we could think of this shift as arising from the measurement starting to reveal the 'true' initial value of J_z, somewhere within the probability distribution for J_z with variance $J/2$. However, we know that this picture is not really true, because if J_z had a predefined value then J_x and J_y would be completely undefined, which is not the case with our state. In any case, the shift can be calculated to be

$$d\langle J_z \rangle_Y = 2\sqrt{M} \, dW(t)(\Delta J_z)_Y^2$$

$$= 2\sqrt{M}Y(t)dt\langle J_z^2 \rangle_Y. \tag{5.207}$$

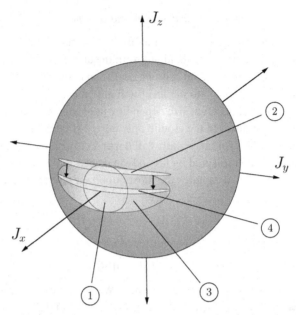

Fig. 5.4 Schematic quasiprobability distributions for the spin states, represented by ellipses on the Bloch sphere of radius J. The initial CSS, spin polarized in the x direction, is given by state 1. State 2 is one particular conditioned spin state after a measurement of J_z, while state 3 is the corresponding unconditioned state due to averaging over all possible conditioned states. The effect of the feedback is shown by state 4: a rotation about the y axis shifts the conditioned state 2 back to $\langle J_z \rangle_Y = 0$. The ensemble average of these conditioned states will then be similar to state 4. This is a reproduction of Fig. 2 of Ref. [TMW02a]. Based on Figure 2 from L. K. Thomsen *et al.*, Continuous Quantum Nondemolition Feedback and Unconditional Atomic Spin Squeezing, *Journal of Physics B: At. Mol. Opt. Phys.* **35**, 4937, (2002), IOP Publishing Ltd.

Exercise 5.37 *Show this.*
Hint: *Initially* $\langle J_z \rangle_Y = 0$ *so that* $Y(t) = dW/dt$.

Because the mean spin is initially in the x direction, a small shift in the mean J_z is equivalent to a small rotation of the mean spin about the y axis by an angle $d\phi \approx d\langle J_z \rangle_Y / J$. This is illustrated in Fig. 5.4, on a sphere. We call this the Bloch sphere, even though previously we have reserved this term for the $J = 1/2$ case.

Conditioning on the results of the measurement reduces the uncertainty in J_z below the SQL of $J/2$, while the deterministic term in the SME (5.204) increases the uncertainty in J_y. However, since the mean of J_z stochastically varies from zero (as shown in Eq. (5.207)), the atomic system conditioned on a particular measurement result is a squeezed spin state with just enough randomness in the direction of the mean spin to mask this spin-squeezing. The unconditioned system evolution $\dot{\rho} = M\mathcal{D}[\hat{J}_z]\rho$ is obtained by averaging over all the possible conditioned states, and this leads to a spin state with $(\Delta J_z)^2 = J/2$, the same as in a CSS. In other words, the squeezed character of individual conditioned system states is lost in the ensemble average. This is illustrated by the states 2 and 3 in Fig. 5.4.

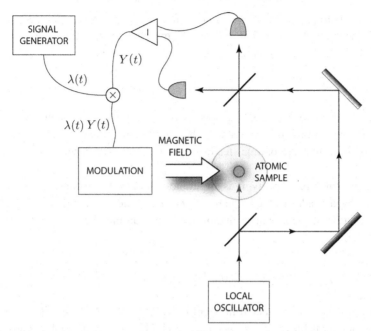

Fig. 5.5 A schematic diagram of an experimental apparatus that could be used for production of spin-squeezing via feedback. The laser probe passes through the ensemble of atoms and is detected by balanced homodyne detection. The current $Y(t)$ is fed back to control the magnetic field surrounding the atoms.

To retain the reduced fluctuations of J_z in the average evolution, we need a way of locking the conditioned mean spin direction. This can be achieved by feeding back the measurement results to drive the system continuously into the same squeezed state. The idea is to cancel out the stochastic shift of $\langle J_z \rangle_Y$ due to the measurement [TMW02b]. This simply requires a rotation of the mean spin about the y axis equal and opposite to that caused by Eq. (5.207). This is illustrated by state 4 in Fig. 5.4.

To make a rotation about the y axis proportional to the measured photocurrent $Y(t)$, we require a Hamiltonian of the form

$$\hat{H}_{\mathrm{fb}}(t) = \lambda(t)Y(t)\hat{J}_y/\sqrt{M} = \hat{F}(t)Y(t), \qquad (5.208)$$

where $\hat{F}(t) = \lambda(t)\hat{J}_y/\sqrt{M}$ and $\lambda(t)$ is the feedback strength. We have assumed instantaneous feedback because that is the form required to cancel out Eq. (5.207). Such a Hamiltonian can be effected by modulating the magnetic field in the region of the sample. This is shown in Fig. 5.5 from the experiment.

Assuming as above that $\langle J_z \rangle_Y = 0$, this feedback Hamiltonian leads to a shift in the mean J_z of

$$\mathrm{d}\langle J_z \rangle_{\mathrm{fb}} \approx -\lambda(t)Y(t)\mathrm{d}t\langle J_x \rangle_Y/\sqrt{M}. \qquad (5.209)$$

Since the idea is to produce $\langle J_z \rangle_Y = 0$ via the feedback, the approximations above and in Eq. (5.207) apply, and we can find a feedback strength such that Eq. (5.207) is cancelled out by Eq. (5.209). The required feedback strength for our scheme is thus

$$\lambda(t) = 2M\langle J_z^2 \rangle_Y / \langle J_x \rangle_Y. \tag{5.210}$$

This use of feedback to cancel out the noise in the conditional mean is the same technique as that which was found to be optimal in the linear system analysed in Section 5.6. The difference here is that the optimal feedback strength (5.210) depends upon conditional averages.

For experimental practicality it is desirable to replace the conditional averages in Eq. (5.210) by a predetermined function $\lambda(t)$ that could be stored in a signal generator. The evolution of the system including feedback can then be described by the master equation

$$\dot{\rho} = M\mathcal{D}[\hat{J}_z]\rho - i\lambda(t)[\hat{J}_y, \hat{J}_z\rho + \rho\hat{J}_z] + \frac{\lambda(t)^2}{M}\mathcal{D}[\hat{J}_y]\rho. \tag{5.211}$$

The choice of $\lambda(t)$ was considered in detail in Ref. [TMW02a], where it was shown that a suitable choice enables the master equation (5.211) to produce a Heisenberg-limited spin-squeezing ($\xi^2 \sim N^{-1}$) at a time $t \sim M^{-1}$.

5.7.3 *Experimental considerations*

For atoms in free space (as opposed to a cavity [NM08]), as discussed above and illustrated in Fig. 5.5, it is actually not possible to achieve Heisenberg-limited spin-squeezing by this method. This is because the decoherence time due to spontaneous emission is much shorter than the time $\sim M^{-1}$ required in order to attain the Heisenberg limit [TMW02a, NM08]. The relevant time-scale for a free-space configuration is in fact of order $(MN)^{-1}$, for which the degree of squeezing produced is moderate. In this limit it is possible to obtain a much simpler expression for the $\lambda(t)$ that should be used over the shorter time.

First note that the $\langle J_x \rangle_Y$ in Eq. (5.210) can be approximated by its initial value of J since the degree of squeezing is moderate. However, $\langle J_z^2 \rangle$ does change over this time, since this is precisely the squeezing we wish to produce. To find an expression for $\langle J_z^2 \rangle$, we assume that the atomic sample will approximately remain in a minimum-uncertainty state for J_z and J_y. This is equivalent to assuming that the feedback, apart from maintaining $\langle J_z \rangle = 0$, does not significantly alter the decreased variance of J_z that was produced by the measurement. This gives $\langle J_z^2 \rangle \approx J^2/(4\langle J_y^2 \rangle)$, where we have again used $\langle J_x \rangle \approx J$. This procedure is essentially a linear approximation represented by replacing J_x by J in the commutator $[\hat{J}_y, \hat{J}_z] = i\hat{J}_x$. From the above master equation we obtain (under the same approximation) $\langle J_y^2 \rangle \approx J^2 Mt + J/2$.

Exercise 5.38 *Verify this.*

By substituting the approximation for $\langle J_y^2 \rangle$ into the expressions for $\langle J_x^2 \rangle$ we obtain

$$\lambda(t) \approx M(1 + NMt)^{-1}. \qquad (5.212)$$

With this choice, a squeezing of

$$\xi^2(t) \approx (1 + NMt)^{-1} \qquad (5.213)$$

will be produced at time t, and this will be valid as long as $Mt \ll 1$. Spontaneous emission, and other imperfections, are of course still being ignored.

Exercise 5.39 *Show that, if λ is held fixed, rather than varied, the variance for the conditioned mean $\langle J_z \rangle_Y$ at time t is*

$$[e^{-4\lambda Jt} - (1 + 2JMt)^{-1}]\frac{J}{2} + (1 - e^{-4\lambda Jt})\frac{\lambda J}{4M}. \qquad (5.214)$$

Hint: *Remember that for linear systems* $\langle J_z^2 \rangle = \langle J_z^2 \rangle_Y + \langle J_z \rangle_Y^2$.

An experiment along the lines described above (with λ fixed) has been performed, with results that appeared to reflect moderate spin-squeezing [GSM04]. Unfortunately the published results were not reproducible and exhibited some critical calibration inconsistencies. The authors have since concluded that they must have been spurious and have retracted the paper [GSM08], saying that 'analyzing Faraday spectroscopy of alkali clouds at high optical depth in precise quantitative detail is surprisingly challenging'. High optical depth leads to significant difficulties with the accurate determination of effective atom number and degree of polarization (and thus of the CSS uncertainty level), while technical noise stemming from background magnetic fields and probe polarization or pointing fluctuations can easily mask the atomic spin projection signal. An additional complication relative to the simple theory given above is that the experiment was performed by probing cesium atoms on an optical transition with many coupled hyperfine-Zeeman states, rather than the two levels considered above. There is still a linear coupling of the probe field to the angular-momentum operator $\hat{j}_z$ defined on the entire hyperfine-Zeeman manifold, which can in principle be utilized to generate measurement-induced spin-squeezing. However, there is also a nonlinear atom–probe interaction that can corrupt the Faraday-rotation signals if it is not suppressed by very careful probe-polarization alignment. For more details see Ref. [Sto06]. Continuing research has led to the development of technical noise-suppression techniques and new modelling and calibration methods that enable accurate high-confidence determination of the CSS uncertainty level [MCSM], providing hope for improved experimental results regarding the measurement and control of spin-squeezing in the future.

5.8 Further reading

5.8.1 Coherent quantum feedback

It was emphasized in Section 5.2 that even when we use an operator to describe the feedback current, as is necessary in the Heisenberg picture, we do not mean to imply that the feedback apparatus is truly quantum mechanical. That is, the feedback Hamiltonians we use are model Hamiltonians that produce the correct evolution. They are not to be taken seriously as dynamical Hamiltonians for the feedback apparatus.

However, there are situations in which we might wish to consider a very small apparatus, and to treat it seriously as a quantum system, with no classical measurement device taking part in the dynamics. We could still consider this to be a form of quantum feedback if the Hamiltonian for the system of interest S and apparatus A were such that the following applied.

1. S and A evolve for time $t_{\rm m}$ under a joint Hamiltonian $\hat{H}_{\rm coup}$.
2. S and A may then evolve independently.
3. S and A then evolve for time $t_{\rm c}$ under another joint Hamiltonian $\hat{H}_{\rm fb} = \hat{F}_S \otimes \hat{J}_A$.

The form of $\hat{H}_{\rm fb}$ ensures that, insofar as S is concerned, the dynamics could have been implemented by replacing step 3 by the following two steps.

3. The apparatus obervable $\hat{J}_A$ is measured yielding result J.
4. S then evolves for time $t_{\rm c}$ under the Hamiltonian $\hat{H}_{\rm fb} = \hat{F}_S \times J$.

That is, the feedback could have been implemented classically.

Lloyd has called feedback of this sort (without measurement) *coherent quantum feedback* [Llo00], and it was demonstrated experimentally [NWCL00] using NMR quantum information-processing techniques. Previously, this concept was introduced in a quantum-optics context as *all-optical feedback* (as opposed to electro-optical feedback) [WM94b]. An important feature of coherent feedback is that, although on average the evolution of system S is the same as for measurement-based feedback, it would be possible to measure the apparatus A (after the action of $\hat{H}_{\rm fb}$) in a basis in which $\hat{H}_{\rm fb}$ is not diagonal. This would produce conditional states of system S incompatible with the conditional states of measurement-based feedback, in which $\hat{J}$ really was measured.

Exercise 5.40 *Convince yourself of this.*

We will consider an example of 'coherent' quantum feedback in Section 7.7.

An even more general concept of quantum feedback is to drop the above constraints on the system–apparatus coupling, but still to engineer the quantum apparatus so as to achieve some goal regarding the system state or dynamics. Exmples of this from all-optical feedback were considered in Ref. [WM94b], and there has recently been renewed interest in this area [Mab08]. This concept is so general that it encompasses any sort of engineered interaction between quantum systems. However, under some circumstances, ideas from engineering control theory naturally generalize to the fully quantum situation, in which case it is sensible to consider this to be an aspect of quantum control. For details see Ref. [Mab08] and references contained therein.

5.8.2 *Other applications of quantum feedback*

There are many other instances for the application of Markovian quantum feedback besides those mentioned in this chapter. Here are a few of them.

Back-action elimination. Recall from Section 1.4.2 that, for efficient measurements, any measurement can be considered as a minimally disturbing measurement followed by unitary evolution that depends on the result. This leads naturally to the idea, proposed in Ref. [Wis95], of using feedback to eliminate this unnecessary unitary back-action. A quantum-optical realization of a QND measurement of a quadrature using this technique was also proposed there. Courty, Heidman and Pinard [CHP03] have proposed using this principle to eliminate the radiation-pressure back-action in the interferometric measurement of mirror position. Their work has important implications for gravitational-wave detection.

Decoherence control. The Horoshko–Kilin scheme (see Section 5.4.3) for protecting a Schrödinger cat from decoherence due to photon loss works only if the lost photons are detected (and the information fed back). In the microwave regime lost photons cannot in practice be detected, so an alternative approach is necessary. Several feedback schemes have been suggested – see Ref. [ZVTR03] and references therein. In Ref. [ZVTR03], Zippilli *et al.* showed that the parity of the microwave cat state can be probed by passing an atom through the cavity. The same atom, containing the result of the 'measurement', can then be used to implement feedback on the mode during the latter part of its passage through the cavity. This is thus an example of the coherent feedback discussed above. They showed that the lifetime of the cat state can, in principle, be arbitrarily enhanced by this technique.

Engineering invariant attractive subsystems. The preparation of a two-level quantum system in a particular state by Markovian feedback was considered in Refs. [HHM98, WW01, WWM01] (see also Section 6.7.2). A much more general approach is discussed in Ref. [TV08], namely engineering attractive and invariant dynamics for a subsystem. Technically, a subsystem is a system with Hilbert space $\mathbb{H}_S$ such that the total Hilbert space can be written as $(\mathbb{H}_S \otimes \mathbb{H}_F) \oplus \mathbb{H}_R$. Attractive and invariant dynamics for the subsystem means that in the long-time limit the projection of the state onto $\mathbb{H}_R$ is zero, while the state on $\mathbb{H}_S \otimes \mathbb{H}_F$ has the form $\rho_S \otimes \rho_F$, for a particular ρ_S. Ticozzi and Viola discuss the conditions under which such dynamics can be engineered using Markovian feedback, for a given measurement interaction and Markovian decoherence. This is potentially useful for preparing systems for quantum information processing (see Chapter 7).

Linewidth narrowing of an atom laser. A continuous atom laser consists of a mode of bosonic atoms continuously damped, so as to form a beam, and continuously replenished. Such a device, which has yet to be realized, will almost certainly have a spectral linewidth dominated by the effect of the atomic interaction energy, which turns fluctuations in the condensate atom number into fluctuations in the condensate frequency. These correlated fluctuations mean that information about the atom number could be used to reduce the frequency fluctuations, by controlling a spatially uniform potential. Obtaining information about the atom number by a quantum-non-demolition measurement (similar to that discussed in Section 5.7) is a process that itself causes phase fluctuations, due to measurement back-action. Nevertheless, it has been shown that Markovian feedback based upon such a measurement could reduce the linewidth by many orders of magnitude [WT01, TW02].

Cooling of a trapped ion (theory and experiment). The motion of a single ion in a Paul trap [WPW99] can be treated as a harmonic oscillator. By using its internal electronic states and coupling to external lasers, it can be cooled using so-called Doppler cooling [WPW99]. However, the equilibrium thermal occupation number (the number of phonons of motion) is still large. It was shown by Steixner, Rabl and Zoller [SRZ05] that in this process some of the light emitted from the atom can be detected in a manner that allows a measurement of one quadrature of the ion's motion (similar to homodyne detection of an optical field). They then show, using Markovian feedback theory as presented here, that the measured current can be fed back to the trap electrodes to cool the motion of the ion. Moreover, this theory has since been verified experimentally by the group of Blatt [BRW+06], demonstrating cooling by more than 30% below the Doppler limit.

Linewidth narrowing of an atom by squashed light. It has been known for some time [Gar86] that a two-level atom strongly coupled to a beam of broad-band squeezed light will have the decay rate of one of its dipole quadratures changed by an amount proportional to the normally ordered spectrum of squeezing (i.e. the decay rate will be reduced). This could be seen by observing a narrow feature in the power spectrum of the emission of the atom into the non-squeezed modes to which it is coupled. It was shown in Ref. [Wis98] that the same phenomenon occurs for a squashed beam (see Section 5.2.5) as produced by a feedback loop. Note that an atom is a nonlinear optical element, so that a semiclassical theory of squashing cannot explain this effect, which has yet to be observed.

Applications of quantum feedback control in quantum information will be considered in Chapter 7.

6
State-based quantum feedback control

6.1 Introduction

In the preceding chapter we introduced quantum feedback control, devoting most space to the continuous feedback control of a localized quantum system. That is, we considered feeding back the current resulting from the monitoring of that system to control a parameter in the system Hamiltonian. We described feedback both in terms of Heisenberg-picture operator equations and in terms of the stochastic evolution of the conditional state. The former formulation was analytically solvable for linear systems. However, the latter could also be solved analytically for simple linear systems, and had the advantage of giving an explanation for how well the feedback could perform.

In this chapter we develop further the theory of quantum feedback control using the conditional state. The state can be used not only as a basis for understanding feedback, but also as the basis for the feedback itself. This is a simple but elegant idea. The conditional state is, by definition, the observer's knowledge about the system. In order to control the system optimally, the observer should use this knowledge. Of course a very similar idea was discussed in Section 2.5 in the context of adaptive measurements. There, one's joint knowledge of a quantum system and a classical parameter was used to choose future measurements so as to increase one's knowledge of the classical parameter. The distinction is that in this chapter we consider state-based feedback to control the quantum system itself.

This chapter is structured as follows. Section 6.2 introduces the idea of state-based feedback by discussing the first experimental implementation of a state-based feedback protocol to control a quantum state. This experiment, in a cavity QED system, was in the 'deep' quantum regime, for which there is no classical analogue. By contrast, the remainder of the chapter is oriented towards state-based control in linear quantum systems, for which there is a classical analogue. Hence we begin this part with an analysis of state-based feedback in general classical systems, in Section 6.3, and in linear classical systems, in Section 6.4. These sections introduce ideas that will apply in the quantum case also, such as optimal control, stability, detectability and stabilizability. We contrast Markovian feedback control with optimal feedback control and also analyse a classical Markovian feedback experiment. In Sections 6.5 and 6.6 we discuss state-based control in general quantum systems and in linear quantum systems, respectively. As discussed in the preface, these sections contain unpublished results obtained by one of us (H. M. W.) in collaboration with Andrew Doherty and (latterly) Andy Chia. Finally, we conclude as usual with a discussion of further reading.

6.2 Freezing a conditional state

This section is devoted to the first quantum feedback experiment [SRO+02] that implemented a state-based feedback protocol for controlling a quantum system. In calling the feedback protocol state-based we mean that it would have been infeasible for it to have been invented without explicitly modelling the quantum trajectories for the conditioned system state. The quantum feedback protocol discussed here has no classical analogue. This is unlike the experiments on linear systems discussed in the preceding chapter and unlike the adaptive phase measurements discussed in Chapter 2 (where the light source was a coherent state and hence could be treated semiclassically). This jusifies its claim to be the first continuous (in time) feedback experiment in the the 'deep' quantum regime.

The experiment was performed using a cavity QED system, as shown in Fig. 6.1. A small number of atoms (in this case, in a beam) is made to interact strongly with a cavity mode, as in the experiment discussed in Section 1.5. In this case, the cavity mode is very weakly driven (by a laser) and the atoms are initially in their ground state, so that in steady state the system as a whole (atoms plus field) is always close to the ground state. Moreover, to the extent that it is excited, it is a superposition (rather than a mixture) of ground and excited states. That is, the steady state of the system can be approximated by a pure state, $|\psi_{ss}\rangle$. We will show why this is the case later.

Quantum trajectories had previously been applied in this system to calculate and understand the correlation functions of the light emitted from the cavity [CCBFO00], which have also been measured experimentally [FMO00, FSRO02]. The link between correlation functions and conditioned states was explained in Section 4.3.2. Consider the $g^{(2)}(\tau)$ correlation function for the direct detection photocurrent. This is a normalized version of Glauber's second-order coherence function, $G^{(2)}(t, t+\tau)$:

$$g^{(2)}(\tau) = \frac{G^{(2)}_{ss}(t, t+\tau)}{G^{(2)}_{ss}(t, t+\infty)} = \frac{\langle I(t+\tau)I(t)\rangle_{ss}}{\langle I(t)\rangle^2_{ss}}. \tag{6.1}$$

For $\tau > 0$, we can use the approach of Section 4.3.2 to rewrite this in terms of conditional measurements as

$$g^{(2)}(\tau) = \langle I(\tau)\rangle_c / \langle I(0)\rangle_{ss}, \tag{6.2}$$

where here the subscript c means 'conditioned on a detection at time 0, when the system has reached its steady state'.

That is, $g^{(2)}(\tau)$ is the probability of getting a second detection a time τ after the first detection (which occurs when the system has reached steady state), divided by the *unconditioned* probability for the second detection. From Eq. (6.1), the function for $\tau < 0$ can be found by symmetry.

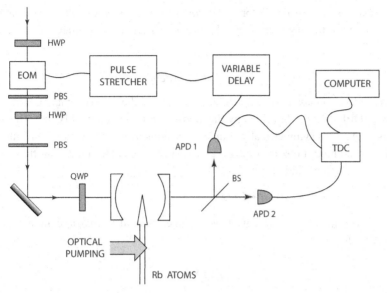

Fig. 6.1 A simplified diagram of the experimental set-up of Smith *et al.*, as depicted by Fig. 6 in Ref. [RSO⁺04]. Rubidium atoms in a beam are optically pumped into a ground state that couples to a cavity mode before entering the cavity. Two avalanche photo-diodes (APDs) measure the intensity of the light emitted by the cavity. The correlation between the detectors is processed using gating electronics, a time-to-digital converter (TDC) and a histogramming memory and computer. Photodetections at APD 1 trigger a change in the intensity injected into the cavity via an electro-optic modulator (EOM). The optics shown are relevant for control of the size of the intensity step and the polarization of the light injected into the cavity. HWP and QWP denote half- and quarter-wave plate, respectively, and PBS, polarization beam-splitter. Figure 6 adapted with permission from J. E. Reiner *et al.*, *Phys. Rev. A* **70**, 023819, (2004). Copyrighted by the American Physical Society.

Since the stationary system state is almost a pure state, we know from quantum trajectory theory that, immediately following the first detection, the conditional state is $|\psi_c(0)\rangle \propto \hat{a}|\psi_{ss}\rangle$, where $\hat{a}$ is the annihilation operator for the cavity mode. The correlation function (6.2) can thus be reformulated as

$$g^{(2)}(\tau) = \frac{\langle \psi_c(\tau)|\hat{a}^\dagger \hat{a}|\psi_c(\tau)\rangle}{\langle \psi_{ss}|\hat{a}^\dagger \hat{a}|\psi_{ss}\rangle}. \tag{6.3}$$

Here $|\psi_c(\tau)\rangle$ is the conditional state for $\tau > 0$, which relaxes back to $|\psi_{ss}\rangle$ as $\tau \to \infty$. In other words, measuring $g^{(2)}(\tau)$ for $\tau > 0$ is directly probing a property (the mean photon number) of the *conditional* state.

The next step taken in Ref. [SRO⁺02] was to *control* the conditional state (prepared by a photodetection), rather than simply *observing* it. That is, by altering the system dynamics subsequent to the first photodetection the conditional state could be altered, and hence $g^{(2)}(\tau)$ changed for $\tau > 0$. Specifically, it was shown that the dynamics of the conditional state could be *frozen* for an indefinite time, making $g^{(2)}(\tau)$ constant. The state could then be released to resume its (oscillatory) relaxation to $|\psi_{ss}\rangle$. This was done by changing the coherent driving of the cavity at a suitable time $\tau = T$ after the first detection.

6.2.1 The system

A simple model that captures the essential behaviour of the experimental system is the N-atom cavity QED system described by the Tavis–Cummings model [TC68]. This consists of N two-level atoms symmetrically coupled to a single mode of a cavity with annihilation operator $\hat{a}$. Assuming a resonant interaction between the cavity mode and the atoms, the Hamiltonian in the interaction frame is

$$\hat{H} = ig(\hat{a}^\dagger \hat{J} - \hat{a}\hat{J}^\dagger) + iE(\hat{a}^\dagger - \hat{a}) \tag{6.4}$$

(compare with Eq. (1.180)). Here we have assumed that all atoms are coupled with equal strength g, so that

$$\hat{J} = \sum_{k=1}^{N} \hat{\sigma}_k, \tag{6.5}$$

where $\hat{\sigma}_k$ is the lowering operator for atom k. We have also included coherent driving (E) of the cavity mode by a resonant laser.

Including damping of the atoms (primarily due to spontaneous emission through the sides of the cavity) and cavity (primarily due to transmission through the end mirrors), we can describe the system by the master equation (in the interaction frame)

$$\dot{\rho} = \left[E(\hat{a}^\dagger - \hat{a}) + g(\hat{a}^\dagger \hat{J} - \hat{a}\hat{J}^\dagger), \rho \right] + \kappa \mathcal{D}[\hat{a}]\rho(t) + \gamma \sum_k \mathcal{D}[\hat{\sigma}_k]\rho(t). \tag{6.6}$$

This describes a damped harmonic oscillator (the cavity) coupled to a damped anharmonic oscillator (the atoms). The anharmonicity is a result of the fact that $\hat{J}$ and $\hat{J}^\dagger$ obey different commutation relations from $\hat{a}$ and $\hat{a}^\dagger$. This is necessary since the maximum number of atomic excitations is N, which we are assuming is finite.

The evolution generated by Eq. (6.6) is very rich [Ber94]. Much simpler, but still interesting, dynamics results in the limit $E \ll \kappa \sim \gamma \sim g$ [CBR91]. In particular, in this weak-driving limit the steady state of the system is approximately a pure state. To understand why this is the case, consider a more general system consisting of damped and coupled oscillators, which could be harmonic or anharmonic. Let us denote the ground state by $|\psi_0\rangle$, and take the coupling rates and damping rates to be of order unity. For the system above, define a parameter

$$\lambda = \frac{E}{\kappa + 4\Omega^2/\gamma}, \tag{6.7}$$

where $\Omega = g\sqrt{N}$ is the N-atom single-photon Rabi frequency. For the case of weak driving, $\lambda \ll 1$. In this limit, λ is equal to the stationary value for $\langle \hat{a} \rangle$, as we will see. We now show that the steady state of the system ρ_{ss} is pure to order λ^2. That is, one can use the approximation

$$\rho_{ss} = |\psi_{ss}\rangle\langle\psi_{ss}| + O(\lambda^3), \tag{6.8}$$

where

$$|\psi_{ss}\rangle = |\psi_0\rangle + \lambda|\psi_1\rangle + \lambda^2|\psi_2\rangle + O(\lambda^3), \tag{6.9}$$

where $|\psi_1\rangle$ and $|\psi_2\rangle$ are states with norm of order unity having, respectively, one and two excitations (in the joint system of atom and cavity mode). Here and in the remainder of this section we are only bothering to normalize the states to lowest order in λ.

Consider unravelling the master equation of the system by unit-efficiency quantum jumps (corresponding to the emission of photons from the system). It is simple to verify that the no-jump evolution will take the system into a pure state of the form of Eq. (6.9).

Exercise 6.1 *Verify this for the master equation (6.6), by showing that the state*

$$|\psi_{ss}\rangle = |0, 0\rangle + \lambda\left(|1, 0\rangle - r|0, 1\rangle\right)$$

$$+ \lambda^2\left(\frac{\xi_0}{\sqrt{2}}|2, 0\rangle - \theta_0|1, 1\rangle + \frac{\eta_0}{\sqrt{2}}|0, 2\rangle\right) + O(\lambda^3) \tag{6.10}$$

is an eigenstate of $-i\hat{H} - (\kappa/2)\hat{a}^\dagger\hat{a} - (\gamma/2)\sum_k \hat{\sigma}_k^\dagger\hat{\sigma}_k$, *which generates the non-unitary evolution. Here* $|n, m\rangle$ *is a state with n photons and m excited atoms, while* $r = 2\Omega/\gamma$ *and*

$$\xi_0 = \zeta\left(1 - \frac{C}{N}\frac{2\kappa}{\kappa+\gamma}\right), \tag{6.11}$$

$$\theta_0 = -r\zeta, \tag{6.12}$$

$$\eta_0 = r^2\zeta\sqrt{1 - 1/N}, \tag{6.13}$$

$$\zeta = \frac{1 + 2C}{1 + 2C[1 - (1/N)\kappa/(\kappa+\gamma)]}, \tag{6.14}$$

$$C = 2\Omega^2/(\kappa\gamma). \tag{6.15}$$

C is known as the co-operativity parameter. Note for later that, if $N \to \infty$ with Ω fixed, then $\zeta \to 1$ and so $\xi_0 \to 1$, $\theta_0 \to -r$ and $\eta_0 \to r^2$.

Having established Eq. (6.9) as the stationary solution of the no-jump evolution, we will have obtained the desired result if we can show that the effect of the jumps is to add to ρ_{ss} terms of order λ^3 and higher. That the extra terms from the jumps *are* of order λ^3 can be seen as follows.

First, the rate of jumps for the system in state (6.9) is of order λ^2. This comes from the probability of excitation of the system, which is $O(\lambda^2)$, times the damping rates, which are $O(1)$. That is to say, jumps are rare events.

Second, the effect of a jump will be once more to create a state of the form $|\psi_0\rangle + O(\lambda)$. This is because any lowering operator destroys $|\psi_0\rangle$, acts on $\lambda|\psi_1\rangle$ to turn it into $|\psi_0\rangle$ times a constant $O(\lambda)$, and acts on $\lambda^2|\psi_2\rangle$ to turn it into a state with one excitation $O(\lambda^2)$.

Renormalizing gives the desired result: the state after the jump is different from $|\psi_{ss}\rangle$ only by an amount of order λ at most.

Third, after a jump, the system will relax back to $|\psi_{ss}\rangle$ at a rate of order unity. This is because the real part of the eigenvalues of the no-jump evolution operator will be of order the damping rates, which are of order unity. That is to say, the non-equilibrium state will persist only for a time of order unity.

On putting these together, we see that excursions from $|\psi_{ss}\rangle$ are only of order λ, and that the proportion of time the system spends making excursions is only of order λ^2. Thus Eq. (6.8) will hold, and, for the master equation Eq. (6.6), the stationary state is given by Eq. (6.10).

Exercise 6.2 *Convince yourself, if necessary, of the three points above by studying the particular example.*

6.2.2 Conditional evolution

In the actual experiment (see Fig. 6.1) only the photons emitted through one mirror are detected, and this with less than unit efficiency. The measurement operator for a photon detection is thus $\hat{M}_1 = \sqrt{\eta\kappa\,dt}\,\hat{a}$ for $\eta < 1$. From the above, we know that prior to a detection we can take the system to be in state $|\psi_{ss}\rangle$. After the detection (which we take to be at time $\tau = 0$) the state is, to $O(\lambda)$,

$$|\psi_c(\tau)\rangle = |0,0\rangle + \lambda[\xi(\tau)|1,0\rangle + \theta(\tau)|0,1\rangle]. \tag{6.16}$$

Here the conditioned cavity field evolution, $\xi(\tau)$, and the conditioned atomic polarization evolution, $\theta(\tau)$, have the initial values ξ_0 and θ_0 as defined above.

Exercise 6.3 *Verify Eq. (6.16) for $\tau = 0$.*

The subsequent no-jump evolution of $\xi(\tau)$ and $\theta(\tau)$ is governed by the coupled differential equations

$$\dot{\xi}(\tau) = -(\kappa/2)\xi(\tau) + \Omega\theta(\tau) + E/(2\lambda), \tag{6.17}$$

$$\dot{\theta}(\tau) = -(\gamma/2)\theta(\tau) - \Omega\xi(\tau), \tag{6.18}$$

where $\xi(0) = \xi_0$ and $\theta(0) = \theta_0$. As the system relaxes to equilibrium, we have from Eq. (6.10) $\xi(\infty) = 1$ and $\theta(\infty) = -r$. These equations can be found using the no-jump evolution via the pseudo-Hamiltonian $\hat{H} - \mathrm{i}(\kappa/2)\hat{a}^\dagger\hat{a} - \mathrm{i}(\gamma/2)\sum_k \hat{\sigma}_k^\dagger\hat{\sigma}_k$.

We thus see that, to lowest order in the excitation, the post-jump evolution is equivalent to two coupled harmonic oscillators with damping and driving (remember that we are in the interaction frame where the oscillation of each oscillator at frequency $\omega \gg \Omega, \kappa, \gamma$ has been removed). This evolution can be understood classically. What is quantum in this system is all in the quantum jump that results from the detection.

The quantum nature of this jump can be seen in the atomic polarization. Upon the detection of a photon from the cavity, this changes from $-r$ to θ_0. Since the system is in a pure state, the only way a measurement upon one subsystem (the cavity mode) can lead to a change in the state of the second subsystem (the atoms) is if they are entangled. We have already noted above that, if $N \to \infty$, $\theta_0 \to -r$, so there would be no change in the atomic polarization. That is because in this limit there is no difference between the atomic system coupled to a harmonic oscillator and two coupled harmonic oscillators. Two coupled harmonic oscillators, driven and damped, end up in coherent states, and so cannot be entangled.

Exercise 6.4 *Show this by substituting $\rho = |\alpha\rangle\langle\alpha| \otimes |\beta\rangle\langle\beta|$ into the master equation*

$$\dot{\rho} = \left[(E/2)(\hat{a}^\dagger - \hat{a}) + \Omega(\hat{a}^\dagger \hat{b} - \hat{a}\hat{b}^\dagger), \rho \right] + \kappa \mathcal{D}[\hat{a}]\rho + \gamma \mathcal{D}[\hat{b}]\rho, \qquad (6.19)$$

and show that $\alpha = \lambda$ and $\beta = -r\lambda$ make $\dot{\rho} = 0$. Here $\hat{a}|\alpha\rangle = \alpha|\alpha\rangle$ and $\hat{b}|\beta\rangle = \beta|\beta\rangle$.

In fact, as discussed in Section 5.4.2, there is absolutely no state change in a coherent state under photodetection, so there would be nothing at all to see in an experiment involving harmonic oscillators. Everything interesting in this experiment comes from the finite number of atoms N, which leads to an anharmonicity in the atomic oscillator (at second order in the excitation), which leads to atom–field entanglement.

The solutions to the differential equations (6.17) and (6.18), for the field and atomic excitation amplitude, respectively, are much simplified if we take $\kappa = \gamma$. This is a good approximation if $|\kappa - \gamma| \ll \Omega$, which is the case experimentally since typical values are $(\Omega, \kappa, \gamma)/(2\pi) = (48.5, 9.8, 9.1)$ MHz [RSO+04]. Under this assumption, both solutions $\xi(t)$ and $\theta(t)$ are of the form

$$f(\tau) = f_{ss} + e^{-(\kappa+\gamma)\tau/4}[A_f \cos(\Omega\tau) + B_f \sin(\Omega\tau)], \qquad (6.20)$$

The steady-state values are, as stated above, $\xi_{ss} = 1$ and $\theta_{ss} = -r$. The four constants A_ξ, A_θ, B_ξ and B_θ are given by

$$A_\xi = \xi_0 - \xi_{ss} = -B_\theta, \qquad (6.21)$$

$$A_\theta = \theta_0 - \theta_{ss} = B_\xi, \qquad (6.22)$$

so that the two functions oscillate exactly out of phase.

Exercise 6.5 *Verify the above solutions.*

Using Eqs. (6.3) and (6.16) it is easy to show that $g^{(2)}(\tau)$ is given by

$$g^{(2)}(\tau) = \xi(\tau)^2. \qquad (6.23)$$

That is, the correlation function measures the square of the conditioned field amplitude. It has an initial value of ξ_0^2, which, from Eq. (6.12), is always less than unity. This in itself is a nonclassical effect – it is the antibunching discussed in Section 4.6.1. For coherent light $g^{(2)}(0) = 1$, while for any classical light source (which can described as a statistical mixture of coherent states), $g^{(2)}(0)$ can only increase from unity, giving bunching.

Exercise 6.6 *Verify Eq. (6.23), and convince yourself that antibunching is a nonclassical effect.*

6.2.3 Quantum feedback control

In this section we outline the feedback protocol used to capture and stabilize the conditional state. From Eq. (6.16) we note that the evolution of the conditional state (to first order in λ) depends on only the two functions $\xi(\tau)$ and $\theta(\tau)$. This state can be stabilized if we can, via feedback, make $\xi(\tau) = 0$ and $\theta(\tau) = 0$, putting it into a new steady state. From Eqs. (6.17) and (6.18) there are two parameters that are easily controlled in an experiment: the feedback time $T = \tau$ and the driving strength E. The feedback protocol simply consists of applying a different driving strength, E', from certain feedback times T_n.

To calculate the feedback times we set $\dot\theta(\tau)$ in Eq. (6.18) to zero. That is, we choose a time such that the magnitude of the atomic polarization is at a maximum or minimum. This is necessary because changing E directly affects only $\xi(\tau)$ (Eq. (6.17)), not $\theta(\tau)$ (Eq. (6.18)). Doing this gives the feedback time constraint $\Omega T_n = \left(n + \frac{1}{2}\right)\pi$ where $n = 0, 1, 2, \ldots$.

Exercise 6.7 *Show this.*

The change in the driving strength is then determined by substituting $\theta(T_n)$ and $\dot\xi(\tau) = 0$ into Eq. (6.17) and solving for E. Doing this gives

$$E'/E = \xi(T_n) = 1 + e^{-(\kappa+\gamma)(n+1/2)\pi/(4\Omega)}(\theta_0 - \theta_{ss}). \tag{6.24}$$

Exercise 6.8 *Show this.*

Thus, the size of the feedback $(E' - E)$ is directly proportional to the jump in the atomic polarization upon a photodetection, which is due to the entanglement. Without entanglement in the initial state, there could be no feedback stabilization.

It might seem that the last claim has not been properly justified. We have shown that for a coherent system (driven, damped harmonic oscillators) there is no change at all in the system upon a detection, and so nothing to stabilize by feedback. However, if one introduced classical noise into such a system, then the stationary state could be mixed, with classical correlations between the two harmonic oscillators. Then the detection of a photon from the first oscillator could cause a jump in the second, and there would be oscillations as the system relaxes, which one would think could be stabilized by feedback. It turns out that this is not the case, because there are not sufficient control parameters to stabilize a mixed state. This is discussed in detail in Ref. [RSO+04].

6.2.4 Experimental results

The typical experimental values, $(\Omega, \kappa, \gamma)/(2\pi) = (48.5, 9.8, 9.1)$ MHz, give a co-operativity $C \approx 53$. It is not possible to measure N, the number of atoms in the cavity

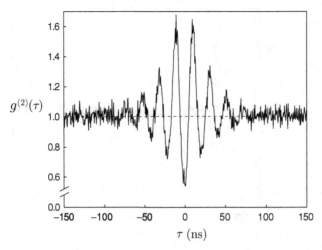

Fig. 6.2 Measured $g^{(2)}(\tau)$. $\tau = 0$ is defined by a photodetection in APD1. Data are binned into 1.0-ns bins. Figure 8 adapted with permission from J. E. Reiner *et al.*, *Phys. Rev.* A **70**, 023819, (2004). Copyrighted by the American Physical Society.

at any given time, directly. Indeed, this concept is not even well defined. First, it will fluctuate because of the random arrival times and velocities of the atoms in the beam. Second, the cavity mode is Gaussian, and so has no sharp cut-off in the transverse direction. The coupling constant g also varies longitudinally in the cavity, because it is a standing-wave mode. However, an average g can be calculated from the cavity geometry, and was found to be $g/(2\pi) = 3.7$ MHz. This implies an effective N of about 170, which is quite large.

Recall that in the limit $N \to \infty$ (with Ω fixed) there are no jumps in the system. However, from Eq. (6.11), the jump in the field amplitude scales as C/N, and C is large enough for this to be significant, with $C/N \approx 0.3$. (This parameter is known as the single-atom co-operativity.) Thus a photon detection sets up a significant oscillation in the quantum state, which is detectable by $g^{(2)}(\tau)$. A typical experimental trace of this is shown in Fig. 6.2. Referring back to Fig. 6.1, two APDs are necessary to measure $g^{(2)}(\tau)$ because the dead-time of the first detector after the detection at $\tau = 0$ (i.e. the time during which it cannot respond) is long compared with the system time-scales. Since it is very unlikely that more than two photons will be detected in the window of interest, the dead-time of the second APD does not matter.

The large value of N has a greater impact on the feedback. From Eq. (6.14), and using the approximation $\kappa \approx \gamma$, the size of the jump in the atomic polarization scales as $\sqrt{2C}/(2N) \approx 0.03$. Thus, the size of the change in the driving field in order to stabilize a conditional state, given by Eq. (6.24), is only a few per cent. Nevertheless, this small change in the driving amplitude is able to freeze the state, as shown in Fig. 6.3. When the driving is returned to its original amplitude, the relaxation of the state to $|\psi_{ss}\rangle$ resumes, so

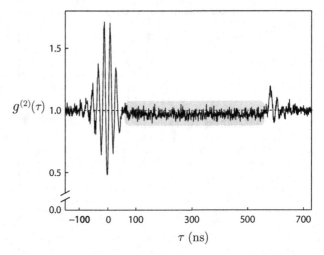

Fig. 6.3 The measured intensity correlation function with the feedback in operation. The grey box indicates the application time of the square feedback pulse, which reduced the driving amplitude by about 0.013. (This is somewhat larger than the value predicted by theory.) The pulse was turned on at $\tau = 45$ ns, in agreement with theory, and turned off 500 ns later. Data are binned into 1.0-ns bins. Figure 9 adapted with permission from J. E. Reiner *et al.*, *Phys. Rev.* A **70**, 023819, (2004). Copyrighted by the American Physical Society.

the effect of the feedback is to insert a flat line of arbitrary length, at the value $|\xi(T_n)|^2$, into the photocurrent autocorrelation function. From Eq. (6.24), this straight line can be at most of order $\sqrt{2C}/N \approx 0.06$ below the steady-state value of unity.

The smallness of the feedback signal (and the smallness of the deviation of the frozen state from the steady state) is exacerbated by the fact that the feedback cannot catch the system at the earliest time $\tau = T_1$. This is because of delays in all parts of the feedback loop, including the APDs themselves, totalling to 43 ns of delay. Since $\pi/\Omega \approx 10$ ns, this means that the earliest time at which the feedback can be applied is $T_4 \approx 45$ ns, by which time the envelope of the oscillations has decayed to about 0.25 of its original size (according to Eq. (6.24)). Even achieving feedback this fast required using electronics designed for data collection in high-energy physics experiments. This highlights the difficulties in making the delay time negligible in quantum feedback experiments.

6.3 General classical systems

Before tackling state-based feedback in general quantum systems, it is useful to review how similar concepts apply in the analogous classical systems.

6.3.1 Notation and terminology

In the remainder of this chapter we are concerned chiefly with linear systems. Hence we will be making frequent use of matrices, which we will denote by capital letters, and vectors, which we will denote by bold-font small letters. This greatly simplifies the equations we present, but necessitates a change in some of the conventions introduced in Chapter 1. In particular, there we used a capital letter to denote a random variable, and the corresponding small letter to act as a dummy argument in a probability distribution (or a ket). We maintain different notation for these distinct concepts, but use a new convention, explained below.

A precisely known classical system can be described by a list of real numbers that can be expressed as a vector $\mathbf{x}^\top = (x_1, x_2, \ldots, x_n)$. Here $\mathbf{v}^\top$ indicates the transpose of vector $\mathbf{v}$. We require that these variables form a *complete* set, by which we mean that any property o of the system is a function of (i.e. can be determined from) $\mathbf{x}$, and that none of the elements x_k can be determined from the remainder of them. (If some of the elements x_k could be determined from the remainder of them, then $\mathbf{x}$ would represent an *overcomplete* set of variables.) For example, for a Newtonian system with several physical degrees of freedom one could have $\mathbf{x}^\top = (\mathbf{q}^\top, \mathbf{p}^\top)$, where $\mathbf{q}$ is the vector of coordinates and $\mathbf{p}$ the vector of conjugate momenta. Taking $\mathbf{x}$ to be complete, we will refer to it as the *configuration* of the system, so that $\mathbb{R}^n$ is the system's configuration space. This coincides with the terminology introduced in Chapter 1.

We are interested in situations in which $\mathbf{x}$ is not known precisely, and is therefore a vector of random variables. An observer's *state of knowledge* of the system is then described by a probability density function $\wp(\mathbf{\check{x}})$. Here we use $\mathbf{\check{x}}$ to denote the argument of this function (a dummy variable) as opposed to $\mathbf{x}$, the random variable itself. The probability density is a non-negative function normalized according to

$$\int d^n\mathbf{\check{x}}\, \wp(\mathbf{\check{x}}) = 1. \tag{6.25}$$

Here $d^n\mathbf{\check{x}} \equiv \prod_{m=1}^n d\check{x}_m$, and an indefinite integral indicates integration over all of configuration space. The state defines an expectation value for any property $o(\mathbf{x})$, by

$$\mathrm{E}[o] = \int d^n\mathbf{\check{x}}\, \wp(\mathbf{\check{x}})o(\mathbf{\check{x}}). \tag{6.26}$$

If the notion of expectation value is taken as basic, we can instead use it to define the probability distribution:

$$\wp(\mathbf{\check{x}}) = \mathrm{E}[\delta^{(n)}(\mathbf{x} - \mathbf{\check{x}})]. \tag{6.27}$$

As in Chapter 1, we refer to $\wp(\mathbf{\check{x}})$ as the *state* of the system. Note that this differs from usual engineering practice, where $\mathbf{x}$ is sometimes called the state or (even more confusingly for quantum physicists) the state vector. Since we will soon be concerned with feedback control, there is another potential confusion worth mentioning: engineers use the term 'plant' to refer to the configuration $\mathbf{x}$ and its dynamics, reserving 'system' for the operation of the combined plant and controller.

6.3.2 The Kushner–Stratonovich equation

We now consider conditioning the classical state upon continuous monitoring, just as we have previously done for quantum systems in Chapter 4. In fact, we have already described such monitoring for a particular case in Section 4.8.4, in the context of a classical circuit coupled to a detector for a quantum system. There we considered conditioning upon measurement of a current that contained white noise (i.e. a Wiener process). In this subsection we generalize that theory to an arbitrary measurement of this type, performed on a classical system, including back-action. (Recall that the idea of classical back-action was introduced in Section 1.1.5). Just as in the quantum case, it is also possible to consider conditioning a classical system upon a point process. However, for linear systems, as we will specialize to in Section 6.4, conditioning upon a measurement with Wiener noise can be treated semi-analytically, whereas this is not true for conditioning upon a point process. Thus we restrict our analysis to the former.

For a system with configuration $\mathbf{x}_t$ we wish to consider a measurement result that in an infinitesimal interval $[t, t + dt)$ is a real number, defined by

$$y(t)dt = \bar{y}(\mathbf{x}_t)dt + dv(t), \tag{6.28}$$

where $\bar{y}(\mathbf{x}_t)$ is an arbitrary real function of $\mathbf{x}_t$ and $dv(t)$ is a Wiener process, defined as usual by

$$[dv(t)]^2 = dt, \tag{6.29}$$

$$E[dv(t)] = 0, \tag{6.30}$$

$$dv(t)dv(t') = 0 \text{ for } t \neq t'. \tag{6.31}$$

Note that Eqs. (6.29)–(6.30) mean that the result $y(t)$ has an infinite amount of noise in it, and so does not strictly exist. However, Eq. (6.31) means that the noise is independent from one moment to the next so that if $y(t)$ is averaged over any finite time the noise in it will be finite. Nevertheless, in this chapter we are being a little more rigorous than previously, and will always write the product $y\,dt$ (which does exist) rather than y.

Using the methods of Section 4.8.4, it is not difficult to show that the equation for the conditioned classical state (commonly known as the Kushner–Stratonovich equation) is

$$d\wp(\check{\mathbf{x}}|y) = dw(t)\{\bar{y}(\check{\mathbf{x}}) - E[\bar{y}(\mathbf{x})]\}\wp(\check{\mathbf{x}}). \tag{6.32}$$

This is a simple example of *filtering* the current to obtain information about the system, a term that will be explained in Section 6.4.3. Remember that $E[\bar{y}(\mathbf{x}_t)]$ means $\int d^n\check{\mathbf{x}}\,\wp(\check{\mathbf{x}};t)\bar{y}(\check{\mathbf{x}})$. Here $dw(t)$ is another Wiener process defined by

$$dw(t) \equiv y(t)dt - E[y(t)dt] \tag{6.33}$$

$$= y(t)dt - E[\bar{y}(\mathbf{x}_t)]dt \tag{6.34}$$

$$= dv(t) + \bar{y}(\mathbf{x}_t)dt - E[\bar{y}(\mathbf{x}_t)]dt. \tag{6.35}$$

This $dw(t)$ is known as the *innovation* or *residual*. It is the *unexpected* part of the result $y(t)dt$, which by definition is the only part that can yield information about the system.

It may appear odd to claim that dw is a Wiener process (and so has zero mean) when it is equal to another Wiener process dv plus something non-zero, namely $\bar{y}(\mathbf{x}_t)dt - E[\bar{y}(\mathbf{x}_t)]dt$. The point is that the observer (say Alice) whose state of knowledge is $\wp(\check{\mathbf{x}})$ does not know $\mathbf{x}$. There is no way therefore for her to discover the 'true' noise dv. Insofar as she is concerned $\bar{y}(\mathbf{x}_t) - E[\bar{y}(\mathbf{x}_t)]$ is a finite random variable of mean zero, so it makes no difference if this is added to dv/dt, which has an unbounded variance as stated above. Technically, dw is related to dv by a Girsanov transformation [IW89].

In general the system configuration will change in conjunction with yielding the measurement result $y(t)dt$. Allowing for deterministic change as well as a purely stochastic change, the system configuration will obey the Langevin equation

$$d\mathbf{x} = \mathbf{a}(\mathbf{x})dt + \mathbf{b}(\mathbf{x})dv \tag{6.36}$$

$$= [\mathbf{a}(\mathbf{x}) - \mathbf{b}(\mathbf{x})\bar{y}(\mathbf{x})]dt + \mathbf{b}(\mathbf{x})y(t)dt. \tag{6.37}$$

Note that the noise that appears in this SDE is *not* the innovation dw, since that is an observer-dependent quantity that has no role in the dynamics of the system configuration (unless introduced by a particular observer through feedback as will be considered later). It can be shown that these dynamics alter the SDE for the system state from the 'purely Bayesian' Kushner–Stratonovich equation (6.32) to the following:

$$d\wp(\check{\mathbf{x}}|y) = dw(t)\left\{ \bar{y}(\check{\mathbf{x}}) - \sum_k \nabla_k b_k(\check{\mathbf{x}}) - E[\bar{y}(\mathbf{x})] \right\}\wp(\check{\mathbf{x}})$$

$$- dt \sum_k \nabla_k a_k(\check{\mathbf{x}})\wp(\check{\mathbf{x}})$$

$$+ \frac{dt}{2} \sum_{k,k'} \nabla_k \nabla_{k'} b_k(\check{\mathbf{x}}) b_{k'}(\check{\mathbf{x}})\wp(\check{\mathbf{x}}). \tag{6.38}$$

Here $\nabla_m \equiv \partial/\partial \check{x}_m$ and the derivatives act on all functions of $\check{\mathbf{x}}$ to their right.

Exercise 6.9 *Derive Eq. (6.38) using the methods of Section 4.8.4.*

Note that this equation has a solution corresponding to complete knowledge: $\wp_c(\check{\mathbf{x}};t) = \delta^{(n)}(\check{\mathbf{x}} - \mathbf{x}_t)$, where $\mathbf{x}_t$ obeys Eq. (6.36). This can be seen from the analysis in Section 4.8.4. For an observer who starts with complete knowledge, dv and dw *are* identical in this case.

If one were to ignore the measurement results, the resulting evolution is found from Eq. (6.38) simply by setting $dw(t)$ equal to its expectation value of zero. Allowing for more than one source of noise so that $d\mathbf{x} = \sum_l \mathbf{b}^{(l)}dv^{(l)}$, we obtain

$$d\wp(\check{\mathbf{x}}) = -dt \sum_k \nabla_k a_k(\check{\mathbf{x}})\wp(\check{\mathbf{x}}) + \frac{dt}{2} \sum_{k,k'} \nabla_k \nabla_{k'} \bar{D}_{k,k'}(\check{\mathbf{x}})\wp(\check{\mathbf{x}}) \tag{6.39}$$

$$\equiv \mathcal{L}\wp(\check{\mathbf{x}}). \tag{6.40}$$

Here $\forall \check{\mathbf{x}}$, $\bar{D}(\check{\mathbf{x}})$ is an arbitrary positive semi-definite (PSD) matrix

$$\bar{D}(\check{\mathbf{x}}) = \sum_l \mathbf{b}^{(l)}(\check{\mathbf{x}})[\mathbf{b}^{(l)}(\check{\mathbf{x}})]^\top. \tag{6.41}$$

Equation (6.39) is known as a Fokker–Planck equation. (See Section B.5.)

If one is told that the unconditional evolution of the classical state is given by Eq. (6.39), how does that constrain the possible conditional evolution? The answer is: not much. Any 'purely Bayesian' measurement (i.e. with no back-action) can be added to the unconditional evolution without invalidating it. The unconditional evolution constrains only terms with back-action. By contrast, in the quantum case (as we will see), the unconditional evolution puts strong constraints on the conditional evolution.

6.3.3 Optimal feedback control

Before specializing to linear systems it is appropriate to make some brief comments about optimal feedback control. Control of a system means that the observer (Alice) can influence the dynamics of the system in a time-dependent fashion. Optimal control means that she implements the control so as to minimize some cost function. Feedback control means that Alice is monitoring the system and taking into account the results of that monitoring in her control.

Let the dynamical parameters that Alice can control be represented by the vector $\mathbf{u}(t)$. The dimension of $\mathbf{u}$ is independent of the dimension of the configuration $\mathbf{x}$. A completely general cost function would be the expectation value of an arbitrary functional of the time-functions $\mathbf{x}(t)$ and $\mathbf{u}(t)$. Setting a start time t_0 and a stop time t_1 for the control problem, we notate such a control cost as

$$j = \mathrm{E}\{\mathcal{I}[\{\mathbf{u}(t)\}_{t=t_0}^{t=t_1}, \{\mathbf{x}(t)\}_{t=t_0}^{t=t_1}]\}. \tag{6.42}$$

Of course, in the minimization of j it is necessary to restrict $\mathbf{u}(t)$ to being a functional of the system output for times less than t, as well as the initial state

$$\mathbf{u}(t) = \mathcal{U}\Big[\{\mathbf{y}(t')\}_{t'=t_0}^{t'=t^-}, \wp(\check{\mathbf{x}}; t_0)\Big]. \tag{6.43}$$

Consider the case in which the cost j that Alice is to minimize is of the form

$$j = \mathrm{E}\Big[\int_{t_0}^{t_1} h(\mathbf{x}(t), \mathbf{u}(t), t)\mathrm{d}t\Big], \tag{6.44}$$

which can also be written as

$$j = \int_{t_0}^{t_1} \mathrm{E}[h(\mathbf{x}(t), \mathbf{u}(t), t)]\mathrm{d}t. \tag{6.45}$$

Physically this is very reasonable since it simply says that the total cost is additive over time. In this case it is possible to show that the *separation principle* holds[1]. That is,

---

[1] In control theory texts (e.g. [Jac93]), the separation principle is often discussed only in the context of LQG systems, in which case it is almost identical to the concept of *certainty equivalence* – see Sec. 6.4.4. The concept introduced has therefore been called a *generalized* separation principle [Seg77].

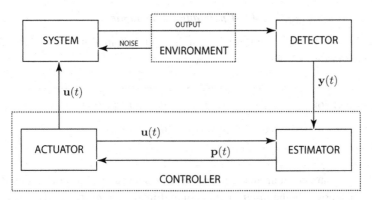

Fig. 6.4 A schematic diagram of the state-based feedback control scheme. The environment is the source of noise in the system and also mediates the system output into the detector. The controller is split into two parts: an estimator, which determines the state conditioned on the record $\mathbf{y}(t)$ from the detector, and an actuator that uses this state to control the system input $\mathbf{u}(t)$. The state, here written as $\mathbf{p}(t)$, would be the probability distribution $\wp(\mathbf{\breve{x}}; t)$ in the classical case and the state matrix $\rho(t)$ in the quantum case.

for a given function h, the optimal control strategy is

$$\mathbf{u}_{\mathrm{opt}}(t) = \mathcal{U}_h\big(\wp_{\mathrm{c}}(\mathbf{\breve{x}}; t), t\big). \tag{6.46}$$

In words, Alice should control the system on the basis of the control objective and her current knowledge of the system *and nothing else*. The control at time t is simply a function of the state at time t, though it is of course still a functional of the measurement record, as in Eq. (6.43). But all of the information in $\big\{\mathbf{y}(t')\big\}_{t'=t_0}^{t'=t}$ is irrelevant except insofar as it determines the present state $\wp_{\mathrm{c}}(\mathbf{\breve{x}}; t)$. This is illustrated in Fig. 6.4

This is a very powerful result that gives an independent definition of $\wp_{\mathrm{c}}(\mathbf{\breve{x}}; t)$. For obvious reasons, this type of feedback control is sometimes called state-based feedback, or Bayesian feedback. Determining the function $\mathcal{U}_h$ of $\wp_{\mathrm{c}}(\mathbf{\breve{x}}; t)$ is nontrivial, but can be done using the technique of dynamic programming. This involves a backwards-in-time equation called the Hamilton–Jacobi–Bellman equation (or just Bellman equation for the discrete-time case) [Jac93].

6.4 Linear classical systems

In this section we specialize to the case of linear classical systems. This entails two restrictions, defined in Sections 6.4.1 and 6.4.2. In Section 6.4.3 we introduce the concept of a stabilizing solution that is important for optimal feedback control, and Markovian feedback control, which we discuss in Sections 6.4.4 and 6.4.5, respectively. Many of the results in this section we give without proof. The interested reader is referred to Ref. [ZDG96] for more details.

6.4.1 Unconditional dynamics

The first restriction to make the system linear is explicable in terms of the dynamics of the system configuration, as follows.

(i) The system configuration obeys a linear dynamical equation

$$d\mathbf{x} = A\mathbf{x}\,dt + B\mathbf{u}(t)dt + E\,d\mathbf{v}_p(t). \tag{6.47}$$

Here A, B and E are constant matrices, while $\mathbf{u}(t)$ is a vector of arbitrary time-dependent functions. It is known as the *input* to the system. Finally, the *process noise* $d\mathbf{v}_p$ is a vector of independent Wiener processes. That is,

$$E[d\mathbf{v}_p] = \mathbf{0}, \qquad d\mathbf{v}_p(d\mathbf{v}_p)^\top = I\,dt, \tag{6.48}$$

where I is the $n \times n$ identity matrix. Thus A is square ($n \times n$), and is known as the drift matrix. The matrices B and E are not necessarily square, but can be taken to be of full column rank, so $\omega[B]$ and $\omega[E]$ can be taken to be no greater than n. (See Box 6.1.)

Strictly, the Wiener process is an example of a time-dependent function, so $\mathbf{u}(t)dt$ could be extended to include $d\mathbf{v}_p(t)$ and the matrix E eliminated. This is a common convention, but we will keep the distinction because $\mathbf{u}(t)$ will later be taken to be the feedback term, which is known by the observer, whereas $d\mathbf{v}_p(t)$ is unknown.

As explained generally in Section B.5, we can turn the Langevin equation (6.47) into an equation for the state $\wp(\check{\mathbf{x}})$. With $\mathbf{u}(t)$ known but $d\mathbf{v}_p(t)$ unknown, $\wp(\check{\mathbf{x}})$ obeys a multi-dimensional OUE (Ornstein–Uhlenbeck equation – see Section 5.6) with time-dependent driving:

$$\dot{\wp}(\check{\mathbf{x}}) = \left\{ -\nabla^\top [A\check{\mathbf{x}} + B\mathbf{u}(t)] + \tfrac{1}{2}\nabla^\top D\,\nabla \right\}\wp(\check{\mathbf{x}}), \tag{6.49}$$

where $D = EE^\top$ is called the diffusion matrix. Denote the mean system configuration by $\langle \mathbf{x} \rangle$, as usual, and the system covariance matrix $\langle \mathbf{x}\mathbf{x}^\top \rangle - \langle \mathbf{x} \rangle \langle \mathbf{x} \rangle^\top$ by V. These moments evolve as

$$\langle \dot{\mathbf{x}} \rangle = A\langle \mathbf{x} \rangle + B\mathbf{u}(t), \tag{6.50}$$

$$\dot{V} = AV + VA^\top + D. \tag{6.51}$$

Exercise 6.10 *Show this from Eq. (6.49).*
Hint: *Use integration by parts.*

Say the system state is initially a Gaussian state, which can be written in terms of the moments $\langle \mathbf{x} \rangle$ and V as

$$\wp(\check{\mathbf{x}}) = g(\check{\mathbf{x}}; \langle \mathbf{x} \rangle, V) \tag{6.52}$$

$$\equiv (2\pi \det V)^{-1/2} \exp[-(\check{\mathbf{x}} - \langle \mathbf{x} \rangle)^\top (2V)^{-1}(\check{\mathbf{x}} - \langle \mathbf{x} \rangle)]. \tag{6.53}$$

Box 6.1 Some properties of matrices

Consider a general *real* matrix A. We denote the number of columns of A by $\omega[A]$. (The vertical 'prongs' of the ω are meant to suggest columns.) Similarly, we denote the number of rows of A to be $\varepsilon[A]$. The *column space* of A is the space spanned by its column vectors, and similarly for the *row space*. The dimensionalities of these spaces are equal, and this dimensionality is called the the *rank* of A. That is, the number of *linearly independent* column vectors in A equals the number of linearly independent row vectors. Clearly

$$\text{rank}[A] \le \min\{\omega[A], \varepsilon[A]\}.$$

The matrix A is said to be full column rank iff $\text{rank}[A] = \omega[A]$, and similarly full row rank iff $\text{rank}[A] = \varepsilon[A]$.

For a *square* real matrix, we denote the set of eigenvalues by $\{\lambda(A)\}$. Note that $\{\lambda(A)\} = \{\lambda(A^\top)\}$. We define

$$\lambda_{\max}[A] \equiv \max\{\text{Re}[\lambda(A)]\}.$$

If $\lambda_{\max}[A] < 0$ then A is said to be *Hurwitz*, or, as is appropriate in the context of dynamical systems, *strictly stable*. We say A is *stable* if $\lambda_{\max}[A] \le 0$, *marginally stable* if $\lambda_{\max}[A] = 0$ and *unstable* if $\lambda_{\max}[A] > 0$. In the context of dynamical systems, we call the eigenvectors $\{\mathbf{x}_\lambda : A\mathbf{x}_\lambda = \lambda\mathbf{x}_\lambda\}$ the *dynamical modes* of the system.

For a *symmetric* real matrix, all $\lambda(A)$ are real. If they are all positive A is called positive definite, denoted $A > 0$. If they are non-negative A is called positive semidefinite (PSD), denoted $A \ge 0$. These remarks also apply to an *Hermitian* complex matrix. Any matrix P satisfying

$$P = BQB^\dagger,$$

where Q is PSD and B is an arbitrary matrix, is PSD. Also, any matrix P satisfying

$$AP + PA^\dagger + Q = 0,$$

where Q is PSD and A is *stable*, is PSD. Conversely, if P and Q are PSD then A is necessarily stable.

For an arbitrary matrix A, there exists a unique *pseudoinverse*, or *Moore–Penrose inverse*, A^+. It is defined by the four properties

1. $AA^+A = A$
2. $A^+AA^+ = A^+$
3. $(AA^+)^\dagger = AA^+$
4. $(A^+A)^\dagger = A^+A$.

If A is square and invertible then $A^+ = A^{-1}$. If A is non-square, then A^+ is also non-square, with $\varepsilon[A^+] = \omega[A]$ and $\omega[A^+] = \varepsilon[A]$. The pseudoinverse finds the 'best' solution $\mathbf{x}$ to the linear equation set $A\mathbf{x} = \mathbf{b}$, in the sense that $\mathbf{x} = A^+\mathbf{b}$ minimizes the Euclidean norm $\|A\mathbf{x} - \mathbf{b}\|^2$.

Then it can be shown that the system state will forever remain a Gaussian state, with the moments evolving as given above. This can be shown by substitution, as discussed in Exercise 5.26

The moment evolution equations can also be obtained directly from the Langevin equation for the configuration (6.47). For example, Eq. (6.50) can be derived directly from Eq. (6.47) by taking the expectation value, while

$$d(\mathbf{x}\mathbf{x}^\top) = (d\mathbf{x})\mathbf{x}^\top + \mathbf{x}(d\mathbf{x}^\top) + (d\mathbf{x})(d\mathbf{x}^\top) \tag{6.54}$$

$$= \{[A\mathbf{x}\,dt + B\mathbf{u}(t)dt + E\,d\mathbf{v}(t)]\mathbf{x}^\top + \text{m.t.}\} + dt\,EE^\top, \tag{6.55}$$

where m.t. stands for matrix transpose and we have used Eq. (6.48). Taking the expectation value and subtracting

$$d(\langle\mathbf{x}\rangle\langle\mathbf{x}\rangle^\top) = \langle d\mathbf{x}\rangle\langle\mathbf{x}\rangle^\top + \langle\mathbf{x}\rangle\langle d\mathbf{x}^\top\rangle \tag{6.56}$$

$$= \{[A\langle\mathbf{x}\rangle + B\mathbf{u}(t)]dt\langle\mathbf{x}^\top\rangle + \text{m.t.}\} \tag{6.57}$$

yields Eq. (6.51).

Stability. Consider the case $\mathbf{u} = \mathbf{0}$ – that is, no driving of the system. Then the system state will relax to a time-independent (stationary) state iff A is *strictly stable*. By this we mean that $\lambda_{max}[A] < 0$ – see Box. 6.1. For linear systems we use the terminology for the dynamics corresponding to that of the matrix A: *stable* if $\lambda_{max}(A) \leq 0$, *marginally stable* if $\lambda_{max}(A) = 0$ and *unstable* if $\lambda_{max}(A) > 0$. Note, however, that the commonly used terminology *asymptotically stable* describes the dynamics iff A is strictly stable. Returning to that case, the stationary state (ss) is then given by

$$\langle\mathbf{x}\rangle_{ss} = \mathbf{0}, \tag{6.58}$$

$$AV_{ss} + V_{ss}A^\top + D = 0. \tag{6.59}$$

The *linear matrix equation* (LME) for V_{ss} can be solved analytically for $n = 1$ (trivially) or $n = 2$, for which the solution is [Gar85]

$$V_{ss} = \frac{(\det A)D + (A - I\,\text{tr}\,A)D(A - I\,\text{tr}\,A)^\top}{2(\text{tr}\,A)(\det A)}, \tag{6.60}$$

where I is the 2×2 identity matrix. Note that here we are using tr for the trace of ordinary matrices, as opposed to Tr for the trace of operators that act on the Hilbert space for a quantum system. For $n > 2$, Eq. (6.59) can be readily solved numerically. If the dynamics is asymptotically stable then all observers will end up agreeing on the state of the system: $\wp_{ss}(\check{\mathbf{x}}) = g(\check{\mathbf{x}}; \mathbf{0}, V_{ss})$.

Stabilizability and controllability. As explained above, the above concept of stability has ready applicability to a system with noise ($E \neq 0$) but with no driving ($\mathbf{u}(t) = \mathbf{0}$). However, there is another concept of stability that has ready applicability in the opposite case – that is, when there is no noise ($E = 0$) but the driving $\mathbf{u}(t)$ may be chosen arbitrarily. In that case, if we ignore uncertain initial conditions, the system state is essentially identical to its configuration which obeys

$$\dot{\mathbf{x}} = A\mathbf{x} + B\mathbf{u}(t). \tag{6.61}$$

Since $\mathbf{u}(t)$ is arbitrary and $\mathbf{x}$ is knowable to the observer, we can consider the case $\mathbf{u} = F\mathbf{x}$ so that

$$\dot{\mathbf{x}} = (A + BF)\mathbf{x}. \tag{6.62}$$

This motivates the following definition.

The pair (A, B) is said to be *stabilizable* iff there exists an F such that $A + BF$ is strictly stable.

This ensures that the observer can control the system to ensure that $\mathbf{x} \to \mathbf{0}$ in the long-time limit. As we will see later, the concept of stabilizability is useful even in the presence of noise.

Consider, for example, the free particle of mass m, with $\mathbf{x}^\top = (q, p)$. Say the observer Alice can directly affect only the momentum of the particle, using a time-dependent linear potential.

Exercise 6.11 *Show that this corresponds to the choices*

$$A = \begin{pmatrix} 0 & 1/m \\ 0 & 0 \end{pmatrix}, \quad B = \begin{pmatrix} 0 \\ 1 \end{pmatrix}. \tag{6.63}$$

Then, with arbitrary $F = (f_q, f_p)$, we have

$$A + BF = \begin{pmatrix} 0 & 1/m \\ f_q & f_p \end{pmatrix}. \tag{6.64}$$

This matrix has eigenvalues given by $\lambda^2 - \lambda f_p - f_q/m = 0$. Clearly, for suitable f_q and f_p, $A + BF < 0$, so the system is stabilizable. This is because by affecting the momentum the observer can also indirectly affect the position, via the free evolution. By contrast, if Alice can directly affect only the position then $B = (1, 0)^\top$ and

$$A + BF = \begin{pmatrix} f_q & f_p + 1/m \\ 0 & 0 \end{pmatrix}. \tag{6.65}$$

This always has a zero eigenvalue, so the system is not stabilizable. The zero eigenvalue arises because the momentum never changes under this assumption.

A stronger property than being stabilizable is for (A, B) to be *controllable*. This allows the observer to do more: by suitable choice of $\mathbf{u}(t)$, Alice can move the configuration from

$\mathbf{x}_0$ at t_0 to any $\mathbf{x}_1$ at any $t_1 > t_0$. It can be shown that the following holds.

> The pair (A, B) is controllable iff for any $n \times n$ real matrix O there exists a matrix F such that $\{\lambda(A + BF)\} = \{\lambda(O)\}$.

An equivalent characterization of controllability is that the *controllability matrix*

$$[B \ AB \ A^2 B \ \cdots \ A^{n-1} B] \tag{6.66}$$

has full row rank (see Box. 6.1). It can be shown that this is also equivalent to the condition that $[(sI - A)B]$ has full row rank for all $s \in \mathbb{C}$. For proofs see Ref. [ZDG96].

In the above example of a free particle, the stabilizable system is also controllable because in fact the eigenvalues of Eq. (6.64) are those of an arbitrary real matrix. Note that this does *not* mean that $A + BF$ is an arbitrary real matrix – two of its elements are fixed!

6.4.2 Conditional dynamics

The second restriction necessary to obtain a linear system, in addition to Eq. (6.47), relates to the conditional dynamics.

> (ii) The system state is conditioned on the measurement result
>
> $$\mathbf{y} \, dt = C\mathbf{x} \, dt + d\mathbf{v}_{\mathrm{m}}(t). \tag{6.67}$$

This is usually known as the *output* of the system, but we will also refer to it as the measured current. Here C is not necessarily square, but can be taken to be of full row rank. The *measurement noise* $d\mathbf{v}_{\mathrm{m}}$ is a vector of independent Wiener processes. That is,

$$\mathrm{E}[d\mathbf{v}_{\mathrm{m}}] = \mathbf{0}, \qquad d\mathbf{v}_{\mathrm{m}}(d\mathbf{v}_{\mathrm{m}})^{\top} = I \, dt. \tag{6.68}$$

As explained in Section 6.3.2, the measurement noise need not be independent of the process noise (although in many control-theory texts this assumption is made). We can describe the correlations between the measurement and process noises by introducing another matrix Γ:

$$E \, d\mathbf{v}_{\mathrm{p}} \, d\mathbf{v}_{\mathrm{m}}^{\top} = \Gamma^{\top} \, dt. \tag{6.69}$$

A cross-correlation matrix Γ is compatible with a given process noise matrix E iff we can define a matrix $\tilde{E}$ such that

$$\tilde{E}\tilde{E}^{\top} = EE^{\top} - \Gamma^{\top}\Gamma. \tag{6.70}$$

That is, iff $D - \Gamma^{\top}\Gamma$ is PSD.

Using the theory presented in Section 6.3.2, the Kushner–Stratonovich equation appropriate to this conditioning is

$$d\wp_c(\check{\mathbf{x}}) = \left\{ -\nabla^\top [A\check{\mathbf{x}} + B\mathbf{u}(t)] + \tfrac{1}{2}\nabla^\top D\,\nabla \right\} \wp_c(\check{\mathbf{x}})dt$$

$$+ d\mathbf{w}^\top \{C(\check{\mathbf{x}} - \langle \mathbf{x}\rangle) - \Gamma\,\nabla\}\wp_c(\check{\mathbf{x}}). \tag{6.71}$$

Here the vector of innovations is given by

$$d\mathbf{w} = \mathbf{y}\,dt - C\langle \mathbf{x}\rangle_c\,dt = d\mathbf{v}_m + C(\mathbf{x} - \langle \mathbf{x}\rangle_c)dt. \tag{6.72}$$

It can be shown that, like the unconditional equation (6.49), the conditional equation (6.71) admits a Gaussian state as its solution. This can be shown using the Itô calculus as in Exercise 5.34. However, it is easier to derive this solution directly from the Langevin equation (6.47) and the current equation (6.67), as we now show.

The crucial fact underlying the derivation is that, if one has two estimates $\bar{\mathbf{x}}_1$ and $\bar{\mathbf{x}}_2$ for a random variable $\mathbf{x}$, and these estimates have Gaussian uncertainties described by the covariance matrices V_1 and V_2, respectively, then the *optimal* way to combine these estimates yields a new estimate $\bar{\mathbf{x}}_3$ also with a Gaussian uncertainty V_3, given by

$$V_3 = (V_1^{-1} + V_2^{-1})^{-1}, \tag{6.73}$$

$$\bar{\mathbf{x}}_3 = V_3(V_1^{-1}\bar{\mathbf{x}}_1 + V_2^{-1}\bar{\mathbf{x}}_2). \tag{6.74}$$

Here *optimality* is defined in terms of minimizing $\mathrm{tr}[M\Delta]$, where Δ is the covariance matrix $E[(\bar{\mathbf{x}}_3 - \mathbf{x})(\bar{\mathbf{x}}_3 - \mathbf{x})^\top]$ of the error in the final estimate and M is any PD matrix. This result from standard error analysis can also be derived from Bayes' rule, with $g(\mathbf{x}; \bar{\mathbf{x}}_1, V_1)$ being the prior state and $g(\mathbf{x}; \bar{\mathbf{x}}_2, V_2)$ the forward probability (or vice versa), and $g(\mathbf{x}; \bar{\mathbf{x}}_3, V_3)$ the posterior probability. The derivation of this result in the one-dimensional case was the subject of Exercise 1.5.

Before starting the derivation it is also useful to write the problem in terms of independent noise processes. It is straightforward to check from Eq. (6.69) that this is achieved by defining

$$E\,d\mathbf{v}_p = \Gamma^\top\,d\mathbf{v}_m + \tilde{E}\,d\mathbf{v}_{p:m}, \tag{6.75}$$

where $d\mathbf{v}_{p:m}$ is pure process noise, *uncorrelated* with $d\mathbf{v}_m$. This allows the system Langevin equation to be rewritten as

$$d\mathbf{x} = A\mathbf{x}\,dt + B\mathbf{u}(t)dt + \Gamma^\top(\mathbf{y} - C\mathbf{x})dt + \tilde{E}\,d\mathbf{v}_{p:m}(t). \tag{6.76}$$

Now, consider a Gaussian state $\wp(\mathbf{x}, t) = g(\mathbf{x}; \langle \mathbf{x}\rangle, V)$, and consider the effect of the observation of $\mathbf{y}(t)dt$. Let $\bar{\mathbf{x}}_1$ be an estimate for $\mathbf{x} + d\mathbf{x}$, taking into account the dynamical (back-action) effect of $\mathbf{y}$ on $\mathbf{x}$. From Eq. (6.76), this is

$$\bar{\mathbf{x}}_1 = \langle \mathbf{x}\rangle + (A - \Gamma^\top C)\langle \mathbf{x}\rangle dt + B\mathbf{u}(t)dt + \Gamma^\top \mathbf{y}\,dt. \tag{6.77}$$

The uncertainty in this estimate is quantified by the covariance matrix

$$V_1 = \langle (\mathbf{x} + d\mathbf{x} - \bar{\mathbf{x}}_1)(\mathbf{x} + d\mathbf{x} - \bar{\mathbf{x}}_1)^\top \rangle \tag{6.78}$$

$$= V + dt \{ [(A - \Gamma^\top C)V + \text{m.t.}] + \tilde{E}\tilde{E}^\top \}, \tag{6.79}$$

where the final term comes from the independent (and unknown) final noise term in Eq. (6.76). The estimate $\bar{\mathbf{x}}_1$ for $\mathbf{x} + d\mathbf{x}$ does not take into account the fact that $\mathbf{y}$ depends upon $\mathbf{x}$ and so yields information about it. Thus from Eq. (6.67) we can form another estimate. Taking C to be invertible for simplicity (the final result holds regardless),

$$\bar{\mathbf{x}}_2 = C^{-1}\mathbf{y} \tag{6.80}$$

is an extimate with a convariance matrix

$$V_2 = \langle (\mathbf{x} + d\mathbf{x} - \bar{\mathbf{x}}_2)(\mathbf{x} + d\mathbf{x} - \bar{\mathbf{x}}_2)^\top \rangle \tag{6.81}$$

$$= (C^\top C \, dt)^{-1} \tag{6.82}$$

to leading order. Strictly, $\bar{\mathbf{x}}_2$ as defined is an estimate for $\mathbf{x}$, not $\mathbf{x} + d\mathbf{x}$. However, the infinite noise in this estimate (6.82) means that the distinction is irrelevant.

Because $d\mathbf{v}_m$ is independent of $d\mathbf{v}_{p:m}$, the estimates $\bar{\mathbf{x}}_1$ and $\bar{\mathbf{x}}_2$ are independent. Thus we can optimally combine these two estimates to obtain a new estimate $\bar{\mathbf{x}}_3$ and its variance V_3:

$$V_3 = V + dt[(A - \Gamma^\top C)V + V(A - \Gamma^\top C)^\top + \tilde{E}\tilde{E}^\top - VC^\top CV], \tag{6.83}$$

$$\bar{\mathbf{x}}_3 = \langle \mathbf{x} \rangle + dt[(A - \Gamma^\top C)\langle \mathbf{x} \rangle + B\mathbf{u}(t) - VC^\top C\langle \mathbf{x} \rangle] + (VC^\top + \Gamma^\top)\mathbf{y}(t)dt. \tag{6.84}$$

Exercise 6.12 *Verify these by expanding Eqs. (6.73) and (6.74) to $O(dt)$.*

Since $\bar{\mathbf{x}}_3$ is the optimal estimate for the system configuration, it can be identified with $\langle \mathbf{x} \rangle_c (t + dt)$ and V_3 with $V_c(t + dt)$. Thus we arrive at the SDEs for the moments which define the Gaussian state $\wp_c(\mathbf{x})$,

$$d\langle \mathbf{x} \rangle_c = [A\langle \mathbf{x} \rangle_c + B\mathbf{u}(t)]dt + (V_c C^\top + \Gamma^\top)d\mathbf{w}, \tag{6.85}$$

$$\dot{V}_c = AV_c + V_c A^\top + D - (V_c C^\top + \Gamma^\top)(CV_c + \Gamma). \tag{6.86}$$

Note that the equation for V_c is actually not stochastic, and is of the form known as a *Riccati differential equation*. Equations (6.85) and (6.86) together are known as the (generalized) Kalman filter.

Detectability and observability. The concepts of stabilizability and controllability from control engineering introduced in Section 6.4.1 are defined in terms of one's ability to control a system. There is a complementary pair of concepts, detectability and observability, that quantify one's ability to acquire information about a system.

A system is said to be *detectable* if every dynamical mode that is not strictly stable is monitored. (See Box. 6.1 for the definition of a dynamical mode.) That is, given a system described by Eqs. (6.47) and (6.67), detectability means that, if the drift matrix A leads to unstable or marginally stable motion, then $\mathbf{y} \propto C\mathbf{x}$ should contain information about that motion. Mathematically, it means the following.

The pair (C, A) is detectable iff

$$C\mathbf{x}_\lambda \neq \mathbf{0} \ \forall \mathbf{x}_\lambda: \ A\mathbf{x}_\lambda = \lambda\mathbf{x}_\lambda \text{ with } \mathrm{Re}(\lambda) \geq 0. \tag{6.87}$$

Clearly, if a system is not detectable then any noise in the unstable or marginally stable modes will lead to an increasing uncertainty in those modes. That is, there cannot be a stationary conditional state for the system.

A simple example is a free particle for which only the momentum is observed. That is,

$$A = \begin{pmatrix} 0 & 1/m \\ 0 & 0 \end{pmatrix}, \qquad C = (0, c), \tag{6.88}$$

for which (C, A) is not detectable since $C(1, 0)^\top = 0$ while $A(1, 0)^\top = 0(1, 0)^\top$. No information about the position will ever be obtained, so its uncertainty can only increase with time. By contrast, a free particle for which only the position is observed, that is,

$$A = \begin{pmatrix} 0 & 1/m \\ 0 & 0 \end{pmatrix}, \qquad C = (c, 0), \tag{6.89}$$

is detectable, since $(1, 0)^\top$ is the only eigenvector of A, and $C(1, 0)^\top = c$.

A very important result is the duality between detectability and stabilizability:

$$(C, A) \text{ detectable} \iff (A^\top, C^\top) \text{ stabilizable}. \tag{6.90}$$

This means that the above definition of detectability gives another definition for stabilizability, while the definition of stabilizability in Section 6.4.1 gives another definition for detectability.

A stronger concept related to information gathering is *observability*. Like controllability, it has a simple definition for the case in which there is no process noise and, in this case, no measurement noise either (although there must be uncertainty in the initial conditions otherwise there is no information to gather). Thus the system is defined by $\dot{\mathbf{x}} = A\mathbf{x} + B\mathbf{u}(t)$ and $\mathbf{y} = C\mathbf{x}$, and this is observable iff the initial condition $\mathbf{x}_0$ can be determined with certainty from the measurement record $\{\mathbf{y}(t)\}_{t=t_0}^{t=t_1}$ in any finite interval. This can be shown to imply the following.

The pair (C, A) is observable iff

$$C\mathbf{x}_\lambda \neq 0 \ \forall \mathbf{x}_\lambda: \ A\mathbf{x}_\lambda = \lambda\mathbf{x}_\lambda. \tag{6.91}$$

That is, even strictly stable modes are monitored. For the example of the free particle above, observability and detectability coincide because there are no stable modes.

Like the detectable–stabilizable duality, there exists an observable–controllable duality:

$$(C, A) \text{ observable} \iff (A^\top, C^\top) \text{ controllable.} \tag{6.92}$$

Thus the above definition of observability gives another definition for controllability, while the two definitions of controllability in Section 6.4.1 give another two definitions for observability.

6.4.3 Stabilizing solutions

Even if the unconditioned system evolution is unstable (see Section 6.4.1), there may be a unique stable solution to the Riccati equation (6.86) for the conditioned variance. If such a solution V_c^{ss} exists, it satisfies the *algebraic* Riccati equation

$$\tilde{A} V_c^{ss} + V_c^{ss} \tilde{A}^\top + \tilde{E}\tilde{E}^\top - V_c^{ss} C^\top C V_c^{ss} = 0, \tag{6.93}$$

where

$$\tilde{A} \equiv A - \Gamma^\top C. \tag{6.94}$$

If V_c^{ss} does exist, this means that, if two observers were to start with different initial Gaussian states to describe their information about the system, they would end up with the same uncertainty, described by V_c^{ss}.

It might be thought that this is all that could be asked for in a solution to Eq. (6.93). However, it should not be forgotten that there is more to the dynamics than the conditioned covariance matrix; there is also the conditioned mean $\langle \mathbf{x} \rangle_c$. Consider two observers (Alice and Bob) with different initial knowledge so that they describe the system by different initial states, $g(\check{\mathbf{x}}; \langle \mathbf{x} \rangle^A, V^A)$ and $g(\check{\mathbf{x}}; \langle \mathbf{x} \rangle^B, V^B)$, respectively. Consider the equation of motion for the discrepancy between their means $\mathbf{d}_c = \langle \mathbf{x} \rangle_c^A - \langle \mathbf{x} \rangle_c^B$. Assuming both observers know $\mathbf{y}(t)$ and $\mathbf{u}(t)$, we find from Eq. (6.85)

$$d\mathbf{d}_c = dt\left[\tilde{A}\mathbf{d}_c - V_c^A C^\top C \langle \mathbf{x} \rangle_c^A + V_c^B C^\top C \langle \mathbf{x} \rangle_c^B \right] + \left(V_c^A - V_c^B \right) C^\top \mathbf{y}(t)dt. \tag{6.95}$$

Now say in the long-time limit V_c^A and V_c^B asymptotically approach V_c^{ss}. In this limit the equation for $\mathbf{d}_c$ becomes deterministic:

$$\dot{\mathbf{d}}_c = M\mathbf{d}_c, \tag{6.96}$$

where

$$M \equiv \tilde{A} - V_c^{ss} C^\top C = A - (V_c^{ss} C^\top + \Gamma^\top)C. \tag{6.97}$$

Thus for Alice and Bob to agree on the long-time system state it is necessary to have M strictly stable.

A solution V_c^{ss} to Eq. (6.93) that makes M strictly stable is known as a *stabilizing solution*. Because of their nice properties, we are interested in the conditions under which stabilizing solutions (rather than merely stationary solutions) to Eq. (6.93) arise. We will also introduce a new notation W to denote a stabilizing V_c^{ss}. Note that, from Eq. (6.93), if W exists then

$$-MW - WM^\top = \tilde{E}\tilde{E}^\top + WC^\top CW. \tag{6.98}$$

Now the matrix on the right-hand side is PSD, and so is W. From this it can be shown that M is necessarily stable. But to obtain a stabilizing solution we require M to be *strictly stable*.

It can be shown that a stabilizing solution exists iff $(C, \tilde{A})$ is detectable, and

$$\tilde{E}^\top \mathbf{x}_\lambda \neq \mathbf{0} \ \forall \mathbf{x}_\lambda: \ \tilde{A}^\top \mathbf{x}_\lambda = \lambda \mathbf{x}_\lambda \text{ with } \mathrm{Re}(\lambda) = 0. \tag{6.99}$$

Note that this second condition is satisfied if $(\tilde{A}, \tilde{E})$ is stabilizable (or, indeed, if $(-\tilde{A}, \tilde{E})$ is stabilizable), which also guarantees uniqueness.

Exercise 6.13 *Show that $(C, \tilde{A})$ is detectable iff (C, A) is detectable.*
Hint: *First show that $(C, \tilde{A})$ is detectable iff $(C, A - \Gamma^\top C)$ is detectable, and that the latter holds iff $\exists L: A - \Gamma^\top C + LC$ is strictly stable. Define $L' = L - \Gamma^\top$, to show that this holds iff $\exists L': A + L'C$ is strictly stable.*

The second condition (6.99) above deserves some discussion. Recall that $\tilde{E}$ is related to the process noise in the system – if there is no diffusion ($D = 0$) then $\tilde{E} = 0$. The condition means that there is process noise in all modes of $\tilde{A}$ that are marginally stable. It might seem odd that the existence of noise helps make the system more stable, in the sense of having all observers agree on the best estimate for the system configuration $\mathbf{x}$ in the long-time limit. The reason why noise can help can be understood as follows. Consider a system with a marginally stable mode x with the dynamics $\dot{x} = 0$ (i.e. no process noise). Now say our two observers begin with *inconsistent* states of knowledge, say $\wp^\alpha(\breve{x}) = \delta(\breve{x} - x^\alpha)$ with $\alpha = A$ or B and $x^A \neq x^B$. Then, with no process noise, they will never come to agreement, because the noise in $\mathbf{y}(t)$ enables each of them to maintain that their initial conditions are consistent with the measurement record. By contrast, if there is process noise in x then Alice's and Bob's states will broaden, and then conditioning on the measurement record will enable them to come into agreement.

For a system with a stabilizing solution W, the terminology 'filter' for the equations describing the conditional state is easy to explain by considering the stochastic equation for the mean. For simplicity let $\mathbf{u} = \mathbf{0}$. Then, in the long-time limit, Eq. (6.85) turns into

$$d\langle \mathbf{x} \rangle_c = M \langle \mathbf{x} \rangle_c \, dt + F^\top \mathbf{y}(t) dt. \tag{6.100}$$

Here F (a capital F) is defined as

$$F = CW + \Gamma. \tag{6.101}$$

Equation (6.100) can be formally integrated to give, in the long-time limit,

$$\langle \mathbf{x} \rangle_c (t) \rightarrow \int_{-\infty}^{t} e^{M(t-s)} \mathbf{F}^\top \mathbf{y}(s) ds. \tag{6.102}$$

Since $M < 0$, the Kalman filter for the mean is exactly a low-pass filter of the current $\mathbf{y}$.

Possible conditional steady states. For a linear system with a stabilizing solution of the algebraic Riccati equation, we have from the above analysis a simple description for the steady-state conditioned dynamics. The conditioned state is a Gaussian that jitters around in configuration space without changing 'shape'. That is, V_c is constant, while $\langle \mathbf{x} \rangle_c$ evolves stochastically. For $\mathbf{u}(t) \equiv \mathbf{0}$, the evolution of $\langle \mathbf{x} \rangle_c$ is

$$d\langle \mathbf{x} \rangle_c = A \langle \mathbf{x} \rangle_c \, dt + \mathbf{F}^\top \, d\mathbf{w}. \tag{6.103}$$

Now the stationary variance in $\mathbf{x}$ is given by

$$V_{ss} = \mathrm{E}_{ss}[\mathbf{x}\mathbf{x}^\top] \tag{6.104}$$

$$= \mathrm{E}_{ss}[(\mathbf{x} - \langle \mathbf{x} \rangle_c)(\mathbf{x} - \langle \mathbf{x} \rangle_c)^\top + \langle \mathbf{x} \rangle_c \langle \mathbf{x} \rangle_c^\top] \tag{6.105}$$

$$= W + \mathrm{E}_{ss}[\langle \mathbf{x} \rangle_c \langle \mathbf{x} \rangle_c^\top]. \tag{6.106}$$

For a system with strictly stable A, one can find $\mathrm{E}_{ss}[\langle \mathbf{x} \rangle_c \langle \mathbf{x} \rangle_c^\top]$ from the Ornstein–Uhlenbeck equation (6.103). By doing so, and remembering that W satsfies

$$AW + WA^\top + D = \mathbf{F}^\top \mathbf{F}, \tag{6.107}$$

it is easy to verify that V_{ss} as given in Eq. (6.106) does indeed satisfy the LME (6.59), which we repeat here:

$$AV_{ss} + V_{ss}A^\top + D = 0. \tag{6.108}$$

Since $\mathrm{E}_{ss}[\langle \mathbf{x} \rangle_c \langle \mathbf{x} \rangle_c^\top] \geq 0$ it is clear that

$$V_{ss} - W \geq 0. \tag{6.109}$$

That is, the conditioned state is more certain than the unconditioned state. It might be thought that for any given (strictly stable) unconditioned dynamics there would always be a way to monitor the system so that the conditional state is any Gaussian described by a covariance matrix W as long as W satisfies Eq. (6.109). That is, any conditional state that 'fits inside' the unconditional state would be a possible stationary conditional state. However, this is not the case. Since $\mathbf{F}^\top \mathbf{F} \geq 0$, it follows from Eq. (6.107) that W must satisfy the linear matrix inequality (LMI)

$$AW + WA^\top + D \geq 0, \tag{6.110}$$

which is strictly stronger than Eq. (6.109). That is, it is the unconditioned dynamics (A and D), not just the unconditioned steady state V_{ss}, that determines the possible asymptotic conditioned states.

The LMI (6.110) is easy to interpret. Say the system has reached an asymptotic conditioned state, but from time t to $t + dt$ we ignore the result of the monitoring. Then the covariance matrix for the state an infinitesimal time later is, from Eq. (6.51),

$$V(t + dt) = W + dt(AW + WA^\top + D). \qquad (6.111)$$

Now, if we had not ignored the results of the monitoring then by definition the conditioned covariance matrix at time $t + dt$ would have been W. For this to be consistent with the state unconditioned upon the result $\mathbf{y}(t)$, the unconditioned state must be a convex (Gaussian) combination of the conditioned states. In simpler language, the conditioned states must 'fit inside' the unconditioned state. This will be the case iff

$$V(t + dt) - W \geq 0, \qquad (6.112)$$

which is identical to Eq. (6.110).

As well as being a necessary condition on W, Eq. (6.110) is a sufficient condition. That is, given a W such that $AW + WA^\top + D \geq 0$, it is always possible to find a C and Γ such that $(CW + \Gamma)^\top(CW + \Gamma) = AW + WA^\top + D$. This is easy to see for the case $W > 0$, since then Γ can be dispensed with.

Exercise 6.14 *Prove the result for the case in which some of the eigenvalues of W are zero, remembering that Γ must satisfy $D - \Gamma^\top\Gamma \geq 0$. For simplicity, assume $D > 0$.*

6.4.4 LQG optimal feedback control

LQG problems. Recall from Section 6.3.3 that for a control problem with an additive cost function (6.44) the separation principle can be used to find the optimal feedback control. To make best use of the linear systems theory we have presented above, it is desirable to put some additional restrictions on the control cost, namely that the function h in Eq. (6.44) be given by

$$h(\mathbf{x}_t, \mathbf{u}_t, t) = 2\delta(t - t_1)\mathbf{x}_t^\top P_1 \mathbf{x}_t + \mathbf{x}_t^\top P \mathbf{x}_t + \mathbf{u}_t^\top Q \mathbf{u}_t. \qquad (6.113)$$

Here P_1 and P are PSD symmetric matrices, while Q is a PD symmetric matrix. In general P and Q could be time-dependent, but we will not consider that option. They represent on-going costs associated with deviation of the system configuration $\mathbf{x}(t)$ and control parameters $\mathbf{u}(t)$ from zero. The cost associated with P_1 we call the *terminal cost* (recall that t_1 is the final time for the control problem). That is, it is the cost associated with not achieving $\mathbf{x}(t_1) = \mathbf{0}$. (The factor of two before the δ-function is so that this term integrates to $\mathbf{x}_{t_1}^\top P_1 \mathbf{x}_{t_1}$.)

It is also convenient to place one final restriction on our control problem: that all noise be Gaussian. We have assumed from the start of Section 6.4 that the measurement and process

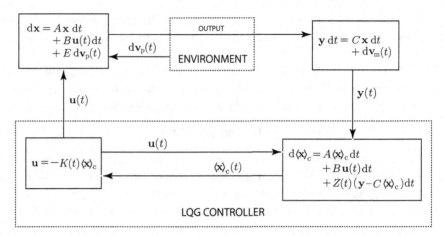

Fig. 6.5 A schematic diagram of the LQG feedback control scheme. Compare this with Fig. 6.4. Here we have introduced $Z(t) = V_c(t)C^\top - \Gamma^\top$, where $V_c(t)$ is the conditioned covariance matrix of $\mathbf{x}$ and $E\, d\mathbf{v}_p\, d\mathbf{v}_m^\top = \Gamma^\top\, dt$. The gain K depends upon the control costs for the system and actuator. Note how the Kalman filter (the equation for $d\langle\mathbf{x}\rangle$) depends upon $\mathbf{u}(t)$, the output of the actuator.

noises are Gaussian, and we now assume that the initial conditions also have Gaussian noise so that the Riccati equation (6.86) applies. With these restrictions we have defined a LQG control problem: linear dynamics for $\mathbf{x}$ and linear mapping from $\mathbf{x}$ and $\mathbf{u}$ to output $\mathbf{y}$; quadratic cost in $\mathbf{x}$ and $\mathbf{u}$, and Gaussian noise, including initial conditions.

For LQG problems the principle of *certainty equivalence* holds. This is stronger than the separation principle, and means that the optimal input $\mathbf{u}(t)$ depends upon $\wp_c(\mathbf{\check{x}}; t)$ only through the best estimate of the system configuration $\langle\mathbf{x}(t)\rangle_c$, as if there were no noise and we were certain that the system configuration was $\langle\mathbf{x}(t)\rangle_c$. Moreover, the optimal $\mathbf{u}(t)$ depends linearly upon the mean:

$$\mathbf{u}(t) = -K(t)\langle\mathbf{x}(t)\rangle_c. \tag{6.114}$$

The matrix $K(t)$ is given by

$$K(t) = Q^{-1}B^\top X(t). \tag{6.115}$$

(Recall that $Q > 0$ so that Q^{-1} always exists.) Here $X(t)$ is a symmetric PSD matrix with the final condition $X(t_1) = P_1$, which is determined for $t_0 \leq t < t_1$ by solving the time-reversed equation

$$\frac{dX}{d(-t)} = P + A^\top X + XA - XBQ^{-1}B^\top X. \tag{6.116}$$

Note that K is independent of D and C, and so is independent of the size of the process and measurement noise – this part of the feedback control problem is completely equivalent to the no-noise control problem (hence 'certainty equivalence'). The overall feedback control scheme is shown in Fig. 6.5.

Table 6.1. *Relations between stabilizing solutions of the algebraic Riccati equation (ARE) for the cases of observing and controlling a linear system. Here 's.s.' means 'stabilizing solution', 'det.' means 'detectable' and 'stab.' means 'stabilizable'. The stabilizee is the quantity which is stabilized, and 'ss' means 'steady state'.*

	Observing	Controlling
The ARE	$\tilde{A}W + W\tilde{A}^\top + \tilde{E}\tilde{E}^\top = WC^\top CW$	$A^\top Y + YA + P = YBQ^{-1}B^\top Y$
has a unique s.s.	$W: \lambda_{max}[M] < 0,$	$Y: \lambda_{max}[N] < 0,$
where	$M = \tilde{A} - WC^\top C,$	$N^\top = A^\top - YBQ^{-1}B^\top,$
iff	$(C, \tilde{A})$ det. and $(\tilde{A}, \tilde{E})$ stab.	(A, B) stab. and (P, A) det.
The stabilizee is	$\mathbf{d}_c = \langle \mathbf{x} \rangle_c^A - \langle \mathbf{x} \rangle_c^B$	$E[\langle \mathbf{x} \rangle_c]$
since in ss	$\dot{\mathbf{d}}_c = M\mathbf{d}_c.$	$d\langle \mathbf{x} \rangle_c = N\langle \mathbf{x} \rangle_c \, dt + F^\top \, d\mathbf{w}.$

Asymptotic LQG problems. Note that Eq. (6.116) has the form of a Riccati equation, like that for the conditioned covariance matrix V_c. As in that case, we are often interested in asymptotic problems in which $t_1 - t_0$ is much greater than any relevant relaxation time. Then, if the Riccati equation (6.116) has a unique stationary solution X_{ss} that is PSD, X will equal X_{ss} for much the greater part of the control interval, having relaxed there from P_1 (which is thus irrelevant). In such cases, the optimal control input (6.114) will be time-dependent through $\langle \mathbf{x} \rangle_c$ but K will be basically time-independent. It is often convenient to assume that P is positive definite, in which case X_{ss} will also be positive definite.

For such asymptotic problems it is natural to consider the stability and uniqueness of solutions. There is a close relation between this analysis of stability and uniqueness and that for the conditioned state in Section 6.4.3. In particular, the concept of a stabilizing solution applies here as well. We show these relations in Table 6.1, but first we motivate a few definitions. Just as we denote a stabilizing solution V_c^{ss} of the algebraic Riccati equation Eq. (6.93) as W, so we will denote a stabilizing solution X_{ss} of the Riccati equation (6.116) in steady state by Y. That is, Y is a symmetric PSD matrix satisfying

$$A^\top Y + YA + P - YBQ^{-1}B^\top Y = 0 \tag{6.117}$$

such that

$$N^\top = A^\top - YBQ^{-1}B^\top \tag{6.118}$$

is strictly stable. The relevance of this is that, for this optimal control, the conditioned system mean obeys, in the long-time limit, the linear equation

$$d\langle \mathbf{x} \rangle_c = N\langle \mathbf{x} \rangle_c \, dt + F^\top \, d\mathbf{w}, \tag{6.119}$$

where $F = CW + \Gamma$ as before.

Exercise 6.15 *Show this from the control law in Eq. (6.114).*

Since $\{\lambda(N)\} = \{\lambda(N^\top)\}$, a stabilizing solution Y ensures that the dynamics of the feedback-controlled system mean will be asymptotically stable.

The conditions under which Y is a stabilizing solution, given in Table 6.1, follow from those for W, using the duality relations of Section 6.4.2. Just as in the case of Section 6.4.3 with the noise E, it might be questioned why putting a lower bound on the cost, by requiring that (P, A) be detectable, should help make the feedback loop stable. The explanation is as follows. If (P, A) were not detectable, that would mean that there were some unstable or marginally stable modes of A to which no cost was assigned. Hence the optimal control loop would expend no resources to control such modes, and they would drift or diffuse to infinity. In theory this would not matter, since the associated cost is zero, but in practice any instability in the system is bad, not least because the linearization of the system will probably break down. Note that if $P > 0$ (as is often assumed) then (P, A) is always detectable.

In summary, for the optimal LQG controller to be strictly stable it is sufficient that (A, B) and $(\tilde{A}, \tilde{E})$ be stabilizable and that $(C, \tilde{A})$ and (P, A) be detectable. If we do not require the controller to be optimal, then it can be shown (see Lemma 12.1 of Ref. [ZDG96]) that stability can be achieved iff (A, B) is stabilizable and (C, A) is detectable.

The 'if' part (sufficiency) can be easily shown since without the requirement of optimality there is a very large family of stable controllers. By assumption we can choose F such that $A + BF$ is strictly stable and L such that $A + LC$ is strictly stable. Then, if the observer uses a (non-optimal) estimate $\bar{\mathbf{x}}$ for the system mean defined by

$$d\bar{\mathbf{x}} = A\bar{\mathbf{x}}\,dt + B\mathbf{u}\,dt - L(\mathbf{y} - C\bar{\mathbf{x}})dt \tag{6.120}$$

(compare this with Eq. (6.85)) and uses the control law

$$\mathbf{u} = F\bar{\mathbf{x}}, \tag{6.121}$$

the resulting controller is stable. This can be seen by considering the equations for the configuration $\mathbf{x}$ and the estimation error $\mathbf{e} = \mathbf{x} - \bar{\mathbf{x}}$ which obey the coupled equations

$$d\mathbf{x} = (A + BF)\mathbf{x}\,dt - BF\mathbf{e}\,dt + E\,d\mathbf{v}_\mathrm{p}, \tag{6.122}$$

$$d\mathbf{e} = (A + LC)\mathbf{e}\,dt + L\,d\mathbf{v}_\mathrm{m} + E\,d\mathbf{v}_\mathrm{p}. \tag{6.123}$$

Exercise 6.16 *Derive these.*

Strict stability of $A + BF$ and $A + LC$ guarantees that $\langle\mathbf{x}\rangle \to \mathbf{0}$ and that $\mathbf{e}$ has a bounded variance, so that $V = \langle\mathbf{xx}^\top\rangle$ is bounded also.

Control costs and pacifiability. For a stable asymptotic problem the stochastic dynamics in the long-time limit are governed by the Ornstein–Uhlenbeck process (6.119). This has a closed-form solution, and thus so does the controller $\mathbf{u}$ from Eq. (6.114). Hence any statistical properties of the system and controller can be determined. For example, the stationary variance of $\mathbf{x}$ is given by Eq. (6.106). From Eq. (6.119), it thus follows that

$$N(V_\mathrm{ss} - W) + (V_\mathrm{ss} - W)N^\top + \mathrm{F}^\top\mathrm{F} = 0, \tag{6.124}$$

and hence V_ss can be determined since N is strictly stable.

One quantity we are particularly interested in is the integrand in the cost function, which has the stationary expectation value

$$E_{ss}[h] = \text{tr}[P V_{ss}] + \text{tr}[Q K (V_{ss} - W) K^\top]. \tag{6.125}$$

Here the stationary expectation value means at a time long after t_0, but long before t_1, so that both the initial conditions on $\mathbf{x}$ and the final condition P_1 on the control are irrelevant. For ease of notation, we simply use K to denote the stationary value for $K(t)$. From the above results it is not too difficult to show that Eq. (6.125) evaluates to

$$E_{ss}[h] = \text{tr}[Y B Q^{-1} B^\top Y W] + \text{tr}[Y D]. \tag{6.126}$$

Note that this result depends implicitly upon A, C, Γ and P through W and Y.

Exercise 6.17 *Derive Eq. (6.125) and verify that it is equivalent to Eq. (6.126).*

It might be thought that if control is cheap ($Q \to 0$) then the gain K will be arbitrarily large, and hence $N = A - BK$ will be such that the fluctuations in $\langle \mathbf{x} \rangle_c$ will be completely suppressed. That is, from Eq. (6.124), it might be thought that the distinction between the conditioned W and unconditioned V_{ss} covariance matrix will vanish. However, this will be the case only if B allows a sufficient degree of control over the system. Specifically, it can be seen from Eq. (6.124) (or perhaps more clearly from Eq. (6.119)) that what is required is for the columns of $\mathsf{F}^\top$ to be in the column space of B (see Box 6.1). This is equivalent to the condition that

$$\text{rank}[B] = \text{rank}[B \; \mathsf{F}^\top]. \tag{6.127}$$

We will call a system that satisfies this condition, for $\mathsf{F} = CW + \Gamma$ with W a stabilizing solution, *pacifiable*.

Note that, unlike the concepts of stabilizability and controllability, the notion of pacifiability relies not only upon the unconditioned evolution (matrices B and A), but also upon the measurement via C and Γ (both explicitly in F and implicitly through W). Thus it cannot be said that pacifiability is stronger or weaker than stabilizability or controllability. However, if B is full row rank then all three notions will be satisfied. In this case, for cheap control the solution Y of Eq. (6.117) will scale as $Q^{1/2}$, and we can approximate Y by the solution to the equation

$$Y B Q^{-1} B^\top Y = P, \tag{6.128}$$

which is independent of the system dynamics.

Exercise 6.18 *Convince yourself of this.*

In this case, the second term in Eq. (6.126) scales as $Q^{1/2}$ so that

$$E_{ss}[h] \to \text{tr}[PW]. \tag{6.129}$$

Realistic control constraints. While it seems perfectly reasonable to consider minimizing a quadratic function of system variables (such as the energy of a harmonic oscillator), it might be questioned whether the quadratic cost associated with the inputs $\mathbf{u}$ is an accurate reflection of the control constraints in a given instance. For instance, in an experiment on a microscopic physical system the power consumption of the controller is typically not a concern. Rather, one tries to optimize one's control of the system within the constraints of the apparatus one has built. For example one might wish to put bounds on $\mathrm{E}[\mathbf{uu}^\top]$ in order that the apparatus does produce the desired change in the system configuration, $B\mathbf{u}\,\mathrm{d}t$, for a given input $\mathbf{u}$.[2] That is, we could require

$$J - K(V - V_c)K^\top \geq 0 \tag{6.130}$$

for some PSD matrix J. The genuinely optimal control for this physical problem would saturate the LMI (6.130). To discover the control law that achieves this optimum it would be necessary to follow an iterative procedure to find the Q that minimizes j while respecting Eq. (6.130).

Another sort of constraint that arises naturally in manipulating microscopic systems is time delays and bandwidth problems in general. This can be dealt with in a systematic manner by introducing extra variables that are included within the system configuration $\mathbf{x}$, as discussed in Ref. [BM04]. To take a simple illustration, for feedback with a delay time τ, the Langevin equation would be

$$\mathrm{d}\mathbf{x} = A\mathbf{x}\,\mathrm{d}t + B\mathbf{u}(t - \tau)\mathrm{d}t + E\,\mathrm{d}\mathbf{v}_\mathrm{p}(t). \tag{6.131}$$

To describe this exactly would require infinite order derivatives, and hence an infinite number of extra variables. However, as a crude approximation (which is expected to be reasonable for sufficiently short delays) we can make a first-order Taylor expansion, to write

$$\mathrm{d}\mathbf{x} = A\mathbf{x}\,\mathrm{d}t + B[\mathbf{u}(t) - \tau\dot{\mathbf{u}}(t)]\mathrm{d}t + E\,\mathrm{d}\mathbf{v}_\mathrm{p}(t). \tag{6.132}$$

Now we define new variables as follows:

$$\mathbf{u}'(t) = -\tau\dot{\mathbf{u}}(t), \tag{6.133}$$

$$\mathbf{x}' = \mathbf{u}(t), \tag{6.134}$$

such that $\mathbf{u}'$ is to be considered the new control variable and $\mathbf{x}'$ an extra system variable. Thus the system Langevin equation would be replaced by the pair of equations

$$\mathrm{d}\mathbf{x} = [A\mathbf{x} + B\mathbf{x}']\mathrm{d}t + B\mathbf{u}'(t)\mathrm{d}t + E\,\mathrm{d}\mathbf{v}_\mathrm{p}(t), \tag{6.135}$$

$$\mathrm{d}\mathbf{x}' = B'\mathbf{u}'(t)\mathrm{d}t, \tag{6.136}$$

where $B' = -(1/\tau)I$. Note that there is no noise in the equation of $\mathbf{x}'$, so the observer will have no uncertainty about these variables.

---

[2] Actually, it would be even more natural to put absolute bounds on $\mathbf{u}$, rather than mean-square bounds. However, such non-quadratic bounds cannot be treated within the LQG framework.

If there were no costs assigned with either $\mathbf{x}'$ or $\mathbf{u}'$ then the above procedure would be nullified, since one could always choose $\mathbf{u}'$ such that

$$\mathbf{u}'(t) = K \langle \mathbf{x} \rangle_{\mathrm{c}}(t) - \mathbf{x}', \qquad (6.137)$$

which would lead to the usual equation for LQG feedback with no delay. But note that this equation can be rewritten as

$$\tau \dot{\mathbf{x}}' = \mathbf{x}' - K \langle \mathbf{x} \rangle_{\mathrm{c}}(t). \qquad (6.138)$$

This is an unstable equation for $\mathbf{x}'$, so, as long as some suitable finite cost is assigned to $\mathbf{x}'$ and/or $\mathbf{u}'$, the choice (6.137) would be ruled out. Costs on the control $\mathbf{u}$ in the original problem translate into a corresponding cost on $\mathbf{x}'$ in the new formulation, while a cost placed on $\mathbf{u}'$ would reflect limitations on how fast the control signal $\mathbf{u}$ can be modified. In practice there is considerable arbitrariness in how the cost functions are assigned.

6.4.5 Markovian feedback

General principles. As discussed above, for an additive cost function, state-based feedback (where $\mathbf{u}(t)$ is a function of the conditional state $\wp_{\mathrm{c}}(\check{\mathbf{x}}; t)$) is optimal. In this section we consider a different (and hence non-optimal) sort of feedback: Markovian feedback. By this we mean that the system input $\mathbf{u}(t)$ is a function of the *just-recorded* current $\mathbf{y}(t)$. We have already considered such feedback in the quantum setting in the preceding chapter, in Sections 5.4 and 5.5. This sort of feedback is not so commonly considered in the classical setting, but we will see that for linear systems much of the analysis of optimal feedback also applies to Markovian feedback.

The name 'Markovian feedback' is appropriate because it leads to Markovian evolution of the system, described by a Markovian Langevin equation. For example, for the linear system we have been considering, the control law is

$$\mathbf{u}(t) = L\mathbf{y}(t), \qquad (6.139)$$

with L a matrix that could be time-dependent, but for strict Markovicity would not be. The Langevin equation for the system configuration is then

$$d\mathbf{x} = A\mathbf{x}\,dt + BL\mathbf{y}\,dt + E\,d\mathbf{v}_{\mathrm{p}}(t) \qquad (6.140)$$

$$= (A + BLC)\mathbf{x}\,dt + BL\,d\mathbf{v}_{\mathrm{m}} + E\,d\mathbf{v}_{\mathrm{p}}. \qquad (6.141)$$

Note that for Markovian feedback it is not necessary to assume or derive Eq. (6.139); any function of the instantaneous current $\mathbf{y}(t)$ that is not linear is not well defined. That is, if one wishes to have Markovian system dynamics then one can only consider what engineers call *proportional feedback*. It should be noted that $\mathbf{y}(t)$ has unbounded variation, so Markovian control is no less onerous than optimal control with unbounded $K(t)$, as occurs for zero control cost, $Q \to 0$. In both cases this is an idealization, since in any physical realisation both the measured current and the controller response would roll off in some way at high frequency.

The motivation for considering Markovian feedback is that it is much simpler than optimal feedback. Optimal feedback requires processing or filtering the current $\mathbf{y}(t)$ in an optimal way to determine the state $\wp_c(\check{\mathbf{x}})$ (or, in the LQG case, just its mean $\langle \mathbf{x} \rangle_c(t)$). Markovian feedback is much simpler to implement experimentally. One notable example of the use of Markovian feedback is the feedback-cooling of a single electron in a harmonic trap (in the classical regime) [DOG03], which we discuss below. Markovian feedback is also much simpler to describe theoretically, since it requires only a model of the system instead of a model of the system plus the estimator and the actuator.

The simplicity of Markovian feedback can be seen in that Eq. (6.141) can be turned directly into an OUE. The moment equations are as in Eqs. (6.50) and (6.51) but with drift and diffusion matrices

$$A' = A + BLC, \tag{6.142}$$

$$D' = D + BLL^\top B^\top + BL\Gamma + \Gamma^\top L^\top B^\top. \tag{6.143}$$

Exercise 6.19 *Derive these.*

Recall that the stationary covariance matrix satisfies

$$A'V_{ss} + V_{ss}A'^\top + D' = 0. \tag{6.144}$$

As for an asymptotic LQG problem with no control costs, the aim of the feedback would be to minimize $\mathrm{tr}[PV_{ss}]$ for some PSD matrix P. If B is full row rank and (C, A) is detectable, then by the definition of detectability it is possible to choose an L such that A' is strictly stable. It might be thought that the optimal Markovian feedback would have $\|L\|$ large in order to make the eigenvalues of A' as negative as possible. However, this is not the case in general, because L also affects the diffusion term, and if $\|L\| \to \infty$ then so does D' (quadratically) so that $\|V_{ss}\| \to \infty$ also. Thus there is in general an optimal value for L, to which we return after the following experimental example.

Experimental example: cooling a one-electron oscillator. The existence of an optimal feedback strength for Markovian feedback (in contrast to the case for state-based feedback) is well illustrated in the recent experiment performed at Harvard [DOG03]. Their system was a single electron in a harmonic trap of frequency $\omega = 2\pi \times 65$ MHz, coupled electromagnetically to an external circuit. This induces a current through a resistor, which dissipates energy, causing damping of the electron's motion at rate $\gamma \approx 2\pi \times 8.4$ Hz. Because the resistor is at finite temperature $T \approx 5.2$ K, the coupling also introduces thermal noise into the electron's motion, so that it comes to thermal equilibrium at temperature T. The damping rate γ is seven orders of magnitude smaller than the oscillation frequency, so a secular approximation (explained below) is extremely well justified.

We can describe the motion by the complex amplitude

$$\alpha(t) = e^{i\omega t}[x(t) - ip(t)]. \tag{6.145}$$

Here (for convenience) x is the electron *momentum* divided by $\sqrt{2m}$ and p is the electron *position* divided by $\sqrt{2/(m\omega^2)}$, where m is the electron mass. The complex exponential in the definition of α removes the fast oscillation from its evolution, so that it should obey the Langevin equation

$$d\alpha = -\frac{\gamma}{2}\alpha\,dt + \sqrt{\gamma T}[dv_1(t) + i\,dv_2(t)]/\sqrt{2}, \tag{6.146}$$

where dv_1 and dv_2 are independent Wiener increments. This equation ensures that $\wp_{ss}(\alpha)$ is a Gaussian that is independent of the phase of α, has a mean of zero, and has

$$E_{ss}[|\alpha|^2] = E_{ss}[x^2 + p^2] = 2E_{ss}[x^2] = T. \tag{6.147}$$

This is as required by the equipartition theorem since $|\alpha|^2$ equals the total energy (we are using units for which $k_B \equiv 1$). For cooling the electron we wish to reduce this mean $|\alpha|^2$.

From Eq. (6.146), the steady-state rate of energy loss from the electron due to the damping (which is balanced by energy gain from the noisy environment) is

$$P = \gamma E_{ss}[|\alpha|^2] = 2\gamma E_{ss}[x^2]. \tag{6.148}$$

This can be identified with I^2R, the power dissipated in the resistor, so if $I \propto x$ we must have

$$I = \sqrt{2\gamma/R}\,x. \tag{6.149}$$

The voltage drop across the resistor is $V = V_J + IR$, where V_J is Johnson (also known as Nyquist) noise. Taking this noise to be white, as is the usual approximation, we have in differential form [Gar85],

$$V\,dt = \sqrt{2\gamma R}\,x\,dt + \sqrt{2TR}\,dv_J. \tag{6.150}$$

But it is this voltage that drives the motion of the electron, giving a term so that the equation for x is

$$dx = -\omega p\,dt - \beta V\,dt \tag{6.151}$$

for some coupling β.

The damping of α at rate $\gamma/2$ arises from this coupling of the electron to the resistor. To derive this we must obtain a damping term $-\gamma x\,dt$ in Eq. (6.151), which requires that $\beta = \sqrt{\gamma/(2R)}$. This gives

$$dx = -\omega p\,dt - \gamma x\,dt - \sqrt{\gamma T}\,dv_J, \tag{6.152}$$

together with the position equation $dp = \omega x\,dt$. On turning these into an equation for α and making a secular approximation by dropping terms rotating at frequency ω or 2ω, we do indeed end up with Eq. (6.146). Note that the secular approximation does not allow us to drop terms like $e^{i\omega t}\,dv_J(t)$, because white noise by assumption fluctuates faster than any frequency. Instead, this term gives the complex noise $[dv_1(t) + i\,dv_2(t)]/\sqrt{2}$, as can be verified by making the secular approximation on the *correlation function* for the Gaussian noise process $e^{i\omega t}\,dv_J(t)$.

Exercise 6.20 *Verify this derivation of Eq. (6.146), and show that it has the steady-state solution with the moments (6.147).*

Note also that the equilibrium temperature T can be obtained by *ignoring the free evolution* (that is, deleting the $-\omega p\, dt$ term in Eq. (6.152)), calculating $E_{ss}[x^2]$ and defining

$$T = 2E_{ss}[x^2]. \tag{6.153}$$

That is, the same expression holds as when the $-\omega p\, dt$ is retained and the secular approximation made as in Eq. (6.147). This happy coincidence will be used later to simplify our description.

From Eq. (6.152), if the voltage were directly measured, the measurement noise would be perfectly correlated with the process noise. For the purpose of feedback, it is necessary to amplify this voltage. In practice this introduces noise into the fed-back signal. Thus, it is better to model the measured voltage as $\varepsilon/\sqrt{\gamma/(2R)}$, where

$$\varepsilon\, dt = \gamma x\, dt + \sqrt{\gamma T}\, dv_J + \sqrt{\gamma T_g}\, dv_g \tag{6.154}$$

Here the temperature T_g (as used in Ref. [DOG03]) scales the amplifier noise dv_g, and is, in their experiment, much less than T.

In this experiment it was easy to apply feedback using a second plate to which the electron motion also couples. Because the relevant system time-scale $\gamma^{-1} \sim 0.02\,s$ is so long, the Markovian approximation (ignoring feedback delay time and bandwidth constraints) is excellent. It is not necessary to feed back fast compared with the oscillation frequency – as long as the phase of the fed-back signal is matched to that of the system the effect will be the same as if there were no time delay. As we will show, the effect of the feedback is simply to modify the damping rate and noise power in the system. But, rather than deal with an equation for two real variables (or an equation for a complex variable), we can instead ignore the $-\omega p\, dt$ term in our equation for dx. This is because the equation with feedback is of the same form as Eq. (6.152), so the same argument as was presented there applies here too. That is, this procedure leads to the same results as are obtained more carefully by applying the secular approximation to the full equations of motion. Thus, after introducing a feedback gain g, the system can be modelled by the following equation for one real variable:

$$dx = -\gamma x\, dt - \sqrt{\gamma T}\, dv_J + g\varepsilon\, dt \tag{6.155}$$

$$= -(1-g)[\gamma x\, dt + \sqrt{\gamma T}\, dv_J] + g\sqrt{\gamma T_g}\, dv_g. \tag{6.156}$$

This gives a new system damping rate $\gamma_e = \gamma(1-g)$ and a new equilibrium system temperature

$$T_e \equiv 2E_{ss}[x^2] = T(1-g) + \frac{g^2 T_g}{1-g}. \tag{6.157}$$

For $T_g \ll T$, the new temperature T_e decreases linearly as g increases towards unity until a turning point at $g \simeq 1 - \sqrt{T_g/T}$, after which it increases rapidly with g. The minimal T_e,

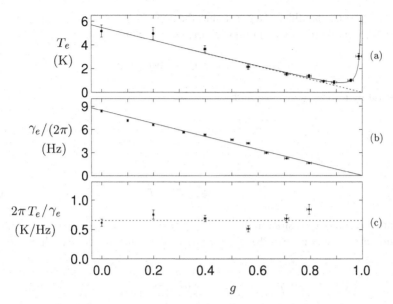

Fig. 6.6 Experimental results for cooling of a single electron by Markovian feedback [DOG03]. T_e is the equilibrium temperature, while γ_e is the measured damping rate of the electron's energy. The lines or curves are the theoretical predictions. Note the existence of an optimal gain g. Figure 5 adapted with permission from B. D'Urso et al., Phys. Rev. Lett. 90, 043001, (2003). Copyrighted by the American Physical Society.

at the optimal gain value, is $T_e \simeq 2\sqrt{T_g T}$. All of this was seen clearly in the experiment, with $T_g \approx 0.04$ K giving a minimum $T_e \approx 0.85$ K, a six-fold reduction in temperature. This is shown in Fig. 6.6. The full expression (not given in Ref. [DOG03]) for the minimum temperature with feedback is

$$T_e = 2\left(\sqrt{TT_g + T_g^2} - T_g\right).\tag{6.158}$$

It is interesting to re-examine this system from the viewpoint of conditional dynamics. Ignoring the $-\omega p\, dt$ term in Eq. (6.152), we have a one-dimensional system, so all of the matrices become scalars. From this equation and Eq. (6.154) it is easy to identify the following:

$$y(t) = [\gamma(T + T_g)]^{-1/2}\varepsilon(t),\tag{6.159}$$

$$A = -\gamma,\tag{6.160}$$

$$C = [\gamma/(T + T_g)]^{1/2},\tag{6.161}$$

$$\Gamma = -T[\gamma/(T + T_g)]^{1/2},\tag{6.162}$$

$$\tilde{A} = A - C\Gamma = -\gamma T_g/(T + T_g),\tag{6.163}$$

$$D = \gamma T,\tag{6.164}$$

$$\tilde{E}^2 = D - \Gamma^2 = \gamma TT_g/(T + T_g).\tag{6.165}$$

It is trivial to verify that this system satisfies the conditions for there to exist a stabilizing solution W for the stationary conditioned variance equation

$$\tilde{A}W + W\tilde{A} + \tilde{E}\tilde{E}^\top = WC^\top CW. \tag{6.166}$$

By substituting into the above we find

$$W^2 + 2T_g W - TT_g = 0, \tag{6.167}$$

giving

$$W = -T_g + \sqrt{T_g^2 + TT_g}. \tag{6.168}$$

Remarkably, this expression, multiplied by two, is identical to the above expression for the minimum temperature (6.158). This identity is no coincidence. Recall that in steady state

$$T = 2\mathrm{E}_{\mathrm{ss}}[x^2] = 2W + 2\mathrm{E}_{\mathrm{ss}}[\langle x \rangle_{\mathrm{c}}^2]. \tag{6.169}$$

Thus $2W$ is a lower bound for the temperature T_e of Eq. (6.157). At the optimal value of feedback gain, the feedback exactly cancels out the noise in the equation for the conditional mean. We saw the same phenomenon for the one-dimensional quantum system considered in Section 5.6. Thus with optimal Markovian feedback $\langle x \rangle_{\mathrm{c}}(t) = 0$ in steady state, and $\mathrm{E}_{\mathrm{ss}}[x^2] = W$. Rather than show this explicitly, we show now that this can be done for any linear system that has a stabilizing solution W.

Understanding Markovian feedback by conditioning. Recall that we can write the conditional mean equation for an arbitrary linear system as

$$\mathrm{d}\langle \mathbf{x} \rangle_{\mathrm{c}} = \left[A - V_{\mathrm{c}}(t)C^\top C - \Gamma^\top C\right]\langle \mathbf{x} \rangle_{\mathrm{c}}\, \mathrm{d}t + B\mathbf{u}(t)\mathrm{d}t + \left[V_{\mathrm{c}}(t)C^\top + \Gamma^\top\right]\mathbf{y}(t)\mathrm{d}t. \tag{6.170}$$

Now, if we add Markovian feedback as defined above then the equation for the covariance matrix is of course unaffected, and that of the mean becomes

$$\mathrm{d}\langle \mathbf{x} \rangle_{\mathrm{c}} = \left[A - V_{\mathrm{c}}(t)C^\top C - \Gamma^\top C\right]\langle \mathbf{x} \rangle_{\mathrm{c}}\mathrm{d}t + \left[V_{\mathrm{c}}(t)C^\top + \Gamma^\top + BL\right]\mathbf{y}(t)\mathrm{d}t. \tag{6.171}$$

Now let us assume that B is such that for all times there exists an L satisfying

$$BL(t) = -V_{\mathrm{c}}(t)C^\top - \Gamma^\top. \tag{6.172}$$

(Obviously that will be the case if B is full row rank.) Then, for this choice of L, the equation for the conditioned mean is simply

$$\mathrm{d}\langle \mathbf{x} \rangle_{\mathrm{c}} = \left[A - V_{\mathrm{c}}(t)C^\top C - \Gamma^\top C\right]\langle \mathbf{x} \rangle_{\mathrm{c}}\, \mathrm{d}t. \tag{6.173}$$

This is a *deterministic* equation. All noise in the conditional mean has been cancelled out by the feedback, so the unconditioned variance and conditioned variance are equal:

$$V(t) = V_c(t), \tag{6.174}$$

$$\dot{V}_c = AV_c + V_c A^\top + D - (V_c C^\top + \Gamma^\top)(CV_c + \Gamma). \tag{6.175}$$

Strictly, the feedback scheme is Markovian only if L is time-independent. This makes sense if we are concerned with asymptotic problems, for which we should choose

$$BL = -WC^\top - \Gamma^\top = -F^\top, \tag{6.176}$$

where W is the stationary conditioned covariance matrix, which we have assumed to be stabilizing. Such an L will exist provided that $\mathrm{rank}[B] = \mathrm{rank}[B\ F^\top]$. That is, provided that the system is pacifiable (see Section 6.4.4). Then there is a unique (observer-independent) long-time behaviour for $\langle \mathbf{x} \rangle_c$, governed by

$$d\langle \mathbf{x} \rangle_c = M \langle \mathbf{x} \rangle_c\, dt, \tag{6.177}$$

where $M = A - WC^\top C - \Gamma^\top C$. But this M is precisely the matrix defined earlier in Section 6.4.3: it is the matrix that is strictly stable iff W is a stabilizing solution. Therefore in the long-time limit $\langle \mathbf{x} \rangle_c = \mathbf{0}$ and

$$V = W. \tag{6.178}$$

In this limit the current is pure noise:

$$\mathbf{y}\, dt = d\mathbf{w}(t). \tag{6.179}$$

This may be a useful fact experimentally for fine-tuning the feedback to achieve the desired result when the system parameters are not known exactly, as shown in the experiment [BRW+06] discussed in Section 5.8.2.

To reiterate, under the conditions

1. there exists a stabilizing solution W to Eq. (6.175)
2. there are no control costs ($Q \to 0$)
3. the system is pacifiable: $\mathrm{rank}[B] = \mathrm{rank}[B\ F^\top]$, where $F = CW + \Gamma$
4. the system costs apply only to the steady state ($t_1 \to \infty$)

the optimal Markovian feedback scheme is strictly stable and performs precisely as well as does the optimal state-based feedback scheme.

In fact, we can prove that the above Markovian feedback algorithm can be derived as a limit of the optimal feedback algorithm that arises when P is positive definite and $Q \to 0$. As discussed in Section 6.4.4, in this case the optimal control is such that BK acts as an infinite positive matrix on the column space of $F^\top$. Recall that for optimal feedback the conditioned mean obeys

$$d\langle \mathbf{x} \rangle_c(t) = (M - BK)\langle \mathbf{x} \rangle_c(t) + F^\top \mathbf{y}(t), \tag{6.180}$$

because $\mathbf{u}(t) = -K\langle\mathbf{x}\rangle_c(t)$. Taking the eigenvalues of BK to positive infinity, the solution to Eq. (6.180) is simply

$$B\mathbf{u}(t) = -\mathrm{F}^\top\mathbf{y}(t), \qquad\qquad (6.181)$$

which is exactly as derived for Markovian feedback.

Exercise 6.21 *Derive Eq. (6.181).*
Hint: *Solve Eq. (6.180) by use of the Laplace transform, and expand in powers of s (the Laplace variable). Then take the $Q \to 0$ limit to eliminate all terms involving non-zero powers of s.*

The most important message is that the optimal Markovian scheme is intimately connected with the conditioned state $\wp_c(\check{\mathbf{x}};t)$. It might be thought therefore that distinguishing Markovian feedback from state-based feedback is a false dichotomy. However, there is still an important distinction between the two: Markovian feedback can be described by Langevin equations for the system configuration $\mathbf{x}$; state-based feedback requires the addition of (at minimum) the configuration of the estimator, $\langle\mathbf{x}\rangle_c$. Moreover, we will see in Section 6.6.7 that if the system is not pacifiable then Markovian feedback is generally inferior.

6.5 General quantum systems

This section mirrors Section 6.3 but for quantum systems. That is, we introduce some new notation and terminology, then consider the diffusive unravelling of general Markovian master equations, and finally make some general remarks on optimal feedback control.

6.5.1 Notation and terminology

In quantum mechanics the system configuration $\mathbf{x} = (x_1, x_2, \ldots, x_n)$ is represented not by real numbers, but rather by self-adjoint linear operators on a Hilbert space. We write $\mathfrak{L}(\mathbb{H})$ for the set of all linear operators on a Hilbert space $\mathbb{H}$, and $\mathfrak{D}(\mathbb{H}) \subset \mathfrak{L}(\mathbb{H})$ for all self-adjoint operators. Thus $\forall k$, $\hat{x}_k \in \mathfrak{D}(\mathbb{H})$. To obtain $\{\lambda(\hat{x})\} = \mathbb{R}$ it is necessary to consider an infinite-dimensional Hilbert space. As in the classical case, we assume that $\hat{\mathbf{x}}$ is a complete set of observables in the sense that any operator in $\mathfrak{D}(\mathbb{H})$ can be expressed as a function of $\hat{\mathbf{x}}$, but that this is not so if any element $\hat{x}_k$ of $\hat{\mathbf{x}}$ is omitted.

We denote the set of state matrices as $\mathfrak{S}(\mathbb{H}) \subset \mathfrak{D}(\mathbb{H})$. For an observable x with $\{\lambda(\hat{x})\} = \mathbb{R}$, the quantum state ρ defines a function

$$\wp(\check{x}) = \mathrm{Tr}[\rho\delta(\check{x} - \hat{x})], \qquad\qquad (6.182)$$

which is the probability distribution for x. Note that a multi-dimensional $\delta^{(n)}(\check{\mathbf{x}} - \hat{\mathbf{x}})$ cannot be defined as an operator in $\mathfrak{D}(\mathbb{H})$ because of the non-commutativity of the elements of $\hat{\mathbf{x}}$.

We will also introduce a new notation for the general Lindblad master equation:

$$\hbar\dot{\rho} = -i[\hat{H}, \rho] + \mathcal{D}[\hat{\mathbf{c}}]\rho. \tag{6.183}$$

Here $\hat{H} = \hat{H}^\dagger$ as before, we have defined a vector of operators $\hat{\mathbf{c}} = (\hat{c}_1, \ldots, \hat{c}_L)^\top$, and

$$\mathcal{D}[\hat{\mathbf{c}}] \equiv \sum_{l=1}^{L} \mathcal{D}[\hat{c}_l]. \tag{6.184}$$

Note that we have introduced Planck's constant $\hbar \approx 10^{-34}\,\mathrm{J\,s}$ on the left-hand side of Eq. (6.183). This is simply a matter of redefinition of units and is necessary in order to connect quantum operators with their classical counterparts. For example, in this case it is necessary if $\hat{H}$ is to correspond to the classical Hamiltonian function. We will see later in Section 6.6 that keeping $\hbar$ in the formulae, rather than setting $\hbar = 1$ as we have been doing, is useful for keeping track of what is distinctively quantum about quantum control of linear systems.

Equation (6.183) can be derived by generalizing the system–bath coupling introduced in Section 3.11 by introducing L independent baths. Then the Itô stochastic differential equation for the unitary operator that generates the evolution of the system and bath observables obeys

$$\hbar\,d\hat{U}(t, t_0) = -\left[dt(\hat{\mathbf{c}}^\dagger\hat{\mathbf{c}}/2 + i\hat{H}) + (\hat{\mathbf{c}}^\dagger\,d\hat{\mathbf{B}}_{z:=-t} - d\hat{\mathbf{B}}^\dagger_{z:=-t}\,\hat{\mathbf{c}})\right]\hat{U}(t, t_0). \tag{6.185}$$

Note here that $\hat{\mathbf{c}}^\dagger$ means the transpose of the vector as well as the Hermitian adjoint of the operators. The vector of operators $d\hat{\mathbf{B}}_{z:=-t}$ has all second-order moments equal to zero except for

$$d\hat{\mathbf{B}}_{z:=-t}\,d\hat{\mathbf{B}}^\dagger_{z:=-t} = I\hbar\,dt, \tag{6.186}$$

where I is the $L \times L$ identity. Note the appearance of $\hbar$ here also.

6.5.2 The Belavkin equation

The master equation (6.183) may be thought of as a quantum analogue of a general Fokker–Planck equation. If we restrict ourselves to diffusive unravellings (see Section 4.5.2), the resultant stochastic master equation can be considered an analogue of the Kushner–Stratonovich equation. This form of the stochastic master equation is sometimes called the Belavkin equation, since diffusive unravellings were first formulated by Belavkin [Bel88] in the mathematical physics literature. (He called it a quantum filtering equation.)

We have already presented a general diffusive unravelling in Section 4.5.2. We present it here again for two reasons. First, we allow for inefficient detection, which we ignored in Section 4.5.2. Second, we use different notation in order to be consistent with the convention in this chapter of capital letters for matrices and small bold letters (usually lower case) for vectors. The most general Belavkin equation compatible with Eq. (6.183) is

$$\hbar \, d\rho_c = dt \, \mathcal{D}[\hat{\mathbf{c}}]\rho_c + \mathcal{H}[-i\hat{H} \, dt + d\mathbf{z}^\dagger(t)\hat{\mathbf{c}}]\rho_c. \tag{6.187}$$

Here we are defining $d\mathbf{z} = (dz_1, \ldots, dz_L)^\top$, a vector of infinitesimal complex Wiener increments. Like dw, these are c-number innovations and $d\mathbf{z}^\dagger$ simply means $(d\mathbf{z}^*)^\top$. Recall that $\mathcal{H}$ is the nonlinear superoperator defined in Eq. (4.24).

The innovations vector $d\mathbf{z}$ satisfies $\mathrm{E}[d\mathbf{z}] = 0$, and has the correlations

$$d\mathbf{z} \, d\mathbf{z}^\dagger = \hbar \mathrm{H} \, dt, \qquad d\mathbf{z} \, d\mathbf{z}^\top = \hbar \Upsilon \, dt, \tag{6.188}$$

where Υ is a complex *symmetric* matrix. Here H (capital η) allows for inefficient detection. The set of allowed Hs is

$$\mathfrak{H} = \left\{ \mathrm{H} = \mathrm{diag}(\eta_1, \cdots, \eta_L) : \forall l, \, \eta_l \in [0, 1] \right\}. \tag{6.189}$$

Here η_l can be interpreted as the efficiency of monitoring the lth output channel. This allows for conditional evolution that does not preserve the purity of states when $\mathrm{H} \neq I$.

It is convenient to combine Υ and H in an *unravelling matrix*

$$U = U(\mathrm{H}, \Upsilon) \equiv \frac{1}{2} \begin{pmatrix} \mathrm{H} + \mathrm{Re}[\Upsilon] & \mathrm{Im}[\Upsilon] \\ \mathrm{Im}[\Upsilon] & \mathrm{H} - \mathrm{Re}[\Upsilon] \end{pmatrix}. \tag{6.190}$$

The set $\mathfrak{U}$ of valid Us can then be defined by

$$\mathfrak{U} = \left\{ U(\mathrm{H}, \Upsilon) : \Upsilon = \Upsilon^\top, \mathrm{H} \in \mathfrak{H}, U(\mathrm{H}, \Upsilon) \geq 0 \right\}. \tag{6.191}$$

Note the requirement that U be PSD.

The output upon which the conditioned state of Eq. (6.187) is conditioned can be written as a vector of complex currents

$$\mathbf{J}^\top \, dt = \left\langle \hat{\mathbf{c}}^\top \mathrm{H} + \hat{\mathbf{c}}^\dagger \Upsilon \right\rangle_c dt + d\mathbf{z}^\top. \tag{6.192}$$

In the Heisenberg picture, the output is represented by the following operator:

$$\hat{\mathbf{J}}^\top \, dt = d\hat{\mathbf{B}}_{\mathrm{out}}^\top \mathrm{H} + d\hat{\mathbf{B}}_{\mathrm{out}}^\dagger \Upsilon + d\hat{\mathbf{A}}^\dagger \sqrt{\mathrm{H} - \Upsilon^* \mathrm{H}^{-1} \Upsilon}$$
$$+ d\hat{\mathbf{V}}^\top \sqrt{\mathrm{H}(I - \mathrm{H})} + d\hat{\mathbf{V}}^\dagger \sqrt{\mathrm{H}^{-1} - I} \, \Upsilon. \tag{6.193}$$

This is the generalization of Eq. (4.210) to allow for inefficent detection, with *two* vectors of ancillary annihilation operators $d\hat{\mathbf{A}}$ and $d\hat{\mathbf{V}}$ understood to act on other (ancillary) baths in the vacuum state. Thus, for example,

$$[d\hat{A}_l, d\hat{A}_j^\dagger] = \hbar \delta_{lj} \, dt. \tag{6.194}$$

Exercise 6.22 *Show that all of the components of $\hat{\mathbf{J}}$ and $\hat{\mathbf{J}}^\dagger$ commute with one another, as required.*

Note that the restriction on Υ enforced by the requirement $U \geq 0$ ensures that the appearances of the matrix inverse H^{-1} in Eq. (6.193) do not cause problems even if H is not positive definite. This restriction also implies that all of the matrices under the radical signs in Eq. (6.193) are PSD, so that the square roots here can be unambiguously defined. Finally, note also that, for efficient monitoring ($H = I$), the ancillary operators $d\hat{V}$ and $d\hat{V}^{\dagger}$ are not needed, but $d\hat{A}$ still is, as in Eq. (4.210).

6.5.3 Optimal feedback control

The discussion of optimal feedback control for classical systems in Section 6.3.3 applies to quantum systems with essentially no changes. To re-establish the notation, we will use $\mathbf{y}(t)$ to represent the result of the monitoring rather than $\mathbf{J}(t)$ as above. In the Heisenberg picture, this will be an operator $\hat{\mathbf{y}}(t)$ just as $\hat{\mathbf{J}}(t)$ is in Eq. (6.193). Similarly, the feedback signal $\mathbf{u}(t)$, a functional of $\mathbf{y}(t)$, will be an operator $\hat{\mathbf{u}}(t)$ in the Heisenberg picture. Given the restrictions on quantum dynamics, we cannot postulate arbitrary terms in the evolution equation for the system configuration dependent upon $\mathbf{u}(t)$. Rather we must work within the structure of quantum dynamics, for example by postulating a feedback Hamiltonian that is a function of $\mathbf{u}(t)$:

$$\hat{H}_{\mathrm{fb}}(t) = \hat{F}\big(\mathbf{u}(t), t\big). \tag{6.195}$$

Say the aim of the control is to minimize a cost function that is additive in time. In the Heisenberg picture we can write the minimand as

$$j = \int_{t_0}^{t_1} \langle h(\hat{\mathbf{x}}, \hat{\mathbf{u}}, t)\rangle \mathrm{d}t, \tag{6.196}$$

where $\mathbf{x}$ and $\mathbf{u}$ are implicitly time-dependent as usual, and the expectation value is taken using the initial state of the system and bath, including any ancillary baths needed to define the current such as in Eq. (6.193).

In the Schrödinger picture, the current can be treated as a c-number, and we need only the system state. However, this state $\rho_{\mathrm{c}}(t)$ is conditioned and hence stochastic, so we must also take an ensemble average over this stochasticity. Also, the system variables must still be represented by operators, $\hat{\mathbf{x}}$, so that the final expression is

$$j = \int_{t_0}^{t_1} \mathrm{E}\{\mathrm{Tr}[\rho_{\mathrm{c}}(t)h(\hat{\mathbf{x}}, \mathbf{u}, t)]\}\mathrm{d}t. \tag{6.197}$$

As in Section 6.3.3, for an additive cost function the separation principle holds. This was first pointed out by Belavkin [Bel83], who also developed the quantum Bellman equation [Bel83, Bel88, Bel99] (see also Refs. [DHJ+00, Jam04] and Ref. [BvH08] for a rigorous treatment). The quantum separation principle means that the optimal control strategy is quantum-state-based:

$$\mathbf{u}_{\mathrm{opt}} = \mathcal{U}_h\big(\rho_{\mathrm{c}}(t), t\big). \tag{6.198}$$

Even in the Heisenberg picture this equation will hold, but with hats on. The conditional state $\rho_{\mathrm{c}}(t)$ is a functional (or a filter in the broad sense of the word) of the output $\mathbf{y}(t)$.

So, in the Heisenberg picture, $\hat{y}(t)$ begets $\hat{\rho}_c(t)$, which begets $\hat{u}_{opt}(t)$. Note the distinction between $\hat{\rho}_c(t)$ and $\rho_c(t)$, the latter being a state conditioned upon a c-number measurement record.

6.6 Linear quantum systems

We turn now to linear quantum systems. By linear quantum systems we mean the analogue of linear classical systems as discussed in Section 6.4. This section mirrors that classical presentation, but there are several differences that it is necessary, or at least interesting, to discuss. Thus we include additional subsections on the structure of quantum phase-space and on optimal unravellings.

Before beginning our detailed treatment, we take this opportunity to make the following point. It is commonly stated, see for example [AD01], that quantum control is a *bilinear* control problem, because even if the time-dependent Hamiltonian is linear in the control signal, such as $\hat{F}u(t)$, the equation of motion for the quantum state is not linear in $u(t)$. Rather, taking $\rho(t)$ and $u(t)$ together as describing the control loop, we have the bilinear equation

$$\hbar\dot{\rho} = -iu(t)[\hat{F}, \rho(t)]. \tag{6.199}$$

However, from our point of view, there is nothing peculiarly quantum about this situation. The classical control problem, expressed in terms of the classical state, has a completely analogous form. For example, consider a classical system, with configuration (q, p), and a classical control Hamiltonian $u(t)F(q, p)$. Then the equation for the classical state is

$$\dot{\wp}(q, p; t) = u(t)\{F(q, p), \wp(q, p; t)\}_{PB}, \tag{6.200}$$

where $\{\cdot, \cdot\}_{PB}$ denotes a Poisson bracket [GPS01].

The existence of this bilinear description of classical control of course does not preclude a linear description in terms of the system configuration. Exactly as in the classical case, a quantum system may have linear dynamics (i.e. a suitable set of observables may have linear Heisenberg equations of motion), and the measured current may be a linear function of these observables, with added Gaussian noise (which will be operator-valued). As we will see, a great deal of classical control theory for such linear systems goes over to the quantum case.

6.6.1 Quantum phase-space

In order to obtain linear dynamics we require observables with unbounded spectrum. For example, the position $\hat{q}$ of a particle has $\{\lambda(\hat{q})\} = \mathbb{R}$. Say our system has N such position observables, which all commute. To obtain a complete set of such observables, for each $\hat{q}_m$ we must include a *canonically conjugate* momentum $\hat{p}_m$. It satisfies the canonical commutation relation

$$[\hat{p}_m, \hat{q}_m] = -i\hbar \tag{6.201}$$

with its partner, but commutes with all other positions and momenta. To connect with the classical theory, we write our complete set of observables as

$$\hat{\mathbf{x}}^\top = (\hat{q}_1, \hat{p}_1, \hat{q}_2, \hat{p}_2, \ldots, \hat{q}_N, \hat{p}_N). \tag{6.202}$$

Then the commutation relations they satisfy can be written

$$[\hat{x}_{m'}, \hat{x}_m] = i\hbar \Sigma_{m',m}, \tag{6.203}$$

where Σ is a $(2N) \times (2N)$ skew-symmetric matrix with the following block-diagonal form:

$$\Sigma = \bigoplus_1^N \begin{pmatrix} 0 & 1 \\ -1 & 0 \end{pmatrix}. \tag{6.204}$$

This matrix, called the symplectic matrix, is an orthogonal matrix. That is, it satisfies $\Sigma^{-1} = \Sigma^\top = -\Sigma$. This means that $i\Sigma$ is Hermitian. In this situation, the configuration space is usually called 'phase-space' and the term 'configuration space' is reserved for the space in which $\mathbf{q}$ resides. However, we will not use 'configuration space' with this meaning.

A consequence of the canonical commutation relation is the Schrödinger–Heisenberg uncertainty relation [Sch30], which for any given conjugate pair $(\hat{q}, \hat{p})$ is

$$V_q V_p - C_{qp}^2 \geq (\hbar/2)^2. \tag{6.205}$$

Here the variances are $V_q = \langle(\Delta \hat{q})^2\rangle$ and V_p similarly, while the covariance $C_{qp} = \langle(\Delta\hat{q})(\Delta\hat{p}) + (\Delta\hat{q})(\Delta\hat{p})\rangle/2$. Note the symmetrization necessary in the covariance because of the non-commutation of the deviation terms, defined for an arbitrary observable $\hat{o}$ as $\Delta\hat{o} = \hat{o} - \langle\hat{o}\rangle$. The original Heisenberg uncertainty relation (A.10) [Hei27] is weaker, lacking the term involving the covariance. Using the matrix Σ and the covariance matrix

$$V_{m',m} = V_{m,m'} = \langle\Delta\hat{x}_{m'} \, \Delta\hat{x}_m + \Delta\hat{x}_m \, \Delta\hat{x}_{m'}\rangle/2, \tag{6.206}$$

we can write the Schrödinger–Heisenberg uncertainty relation as the linear matrix inequality [Hol82]

$$V + i\hbar\Sigma/2 \geq 0. \tag{6.207}$$

This LMI can be derived immediately from Eqs. 6.203 and 6.206, since

$$V + i\hbar\Sigma/2 = \langle(\Delta\hat{\mathbf{x}})(\Delta\hat{\mathbf{x}})^\top\rangle, \tag{6.208}$$

and the matrix on the right-hand side is PSD by construction. Since Σ is a real matrix, we can thus also define V by

$$V = \mathrm{Re}\left[\langle(\Delta\hat{\mathbf{x}})(\Delta\hat{\mathbf{x}})^\top\rangle\right]. \tag{6.209}$$

Recall that this means the real part of each element of the matrix.

Exercise 6.23 *Show from Eq. (6.207) that V (if finite) must be positive definite.*
Hint: *First show that, if* **r** *and* **h** *are the real and imaginary parts of an eigenvector of* $V + i\hbar\Sigma/2$ *with eigenvalue* λ*, then*

$$\begin{pmatrix} V & -\hbar\Sigma/2 \\ \hbar\Sigma/2 & V \end{pmatrix}\begin{pmatrix} \mathbf{r} \\ \mathbf{h} \end{pmatrix} = \lambda\begin{pmatrix} \mathbf{r} \\ \mathbf{h} \end{pmatrix}. \tag{6.210}$$

Then show that this real matrix cannot be positive if V has a zero eigenvalue, using the fact that Σ *has full rank.*

It is convenient to represent a quantum state for this type of system not as a state matrix ρ but as a Wigner function $W(\check{\mathbf{x}})$ – see Section A.5. This is a pseudo-probability distribution over a classical configuration corresponding to the quantum configuration $\hat{\mathbf{x}}$. It is related to ρ by

$$W(\check{\mathbf{x}}) = \langle \delta_W(\check{\mathbf{x}} - \hat{\mathbf{x}}) \rangle = \text{Tr}[\rho\delta_W(\check{\mathbf{x}} - \hat{\mathbf{x}})] \tag{6.211}$$

(*cf.* Eq. (6.27)), where

$$\delta_W(\check{\mathbf{x}} - \hat{\mathbf{x}}) = \int d^{2N}\mathbf{k}\, \exp\big[2\pi i\mathbf{k}^{\top}(\check{\mathbf{x}} - \hat{\mathbf{x}})\big]. \tag{6.212}$$

If the observables commuted then $\delta_W(\check{\mathbf{x}} - \hat{\mathbf{x}})$ would become a Dirac δ-function $\delta^{(2N)}(\check{\mathbf{x}} - \hat{\mathbf{x}})$. Recall from Section 6.5.1 that the latter object does not exist, which is why the state of a quantum system cannot in general be represented by a true probability distribution over the values of its observables. The Wigner function evades this because $\delta_W(\check{\mathbf{x}} - \hat{\mathbf{x}})$ is not PSD. Thus, for some states, the Wigner function will take negative values for some $\check{\mathbf{x}}$. Nevertheless, it is easy to verify that for any subset of N *commuting* observables the marginal Wigner function is a true probability distribution for those observables. For example,

$$\wp(\check{q}_1, \check{q}_2, \ldots, \check{q}_N) = \int d^N\check{\mathbf{p}}\, W(\check{q}_1, \check{p}_1, \ldots, \check{q}_N, \check{p}_N) \tag{6.213}$$

is the true probability density for finding the system positions $\mathbf{q}$ to be equal to $\check{\mathbf{q}}$. Moreover, any moments calculated using the Wigner function equal the corresponding *symmetrized* moments of the quantum state. For example, with a two-dimensional phase-space,

$$\int d\check{q}\, d\check{p}\, W(\check{q}, \check{p})\check{q}^2\check{p} = \text{Tr}\big[\rho(\hat{q}^2\hat{p} + 2\hat{q}\hat{p}\hat{q} + \hat{p}\hat{q}^2)/4\big]. \tag{6.214}$$

Having defined the Wigner function, we can now define a *Gaussian state* to be a state with a Gaussian Wigner function. That is, $W(\check{\mathbf{x}})$ is of the form of Eq. (6.53), with mean vector $\langle \hat{\mathbf{x}} \rangle$ and covariance matrix V as defined in Eq. (6.206). Such a Wigner function is of course positive everywhere, and so has the form of a classical distribution function. Thus, if one restricts one's attention to the observables symmetrized in $\check{\mathbf{x}}$, Gaussian states have a classical analogue. Note that the vacuum state of a bosonic field, which is the bath state we assumed in Section 6.5.2, is a Gaussian state (see Section A.3.3). For Gaussian states, the Schrödinger–Heisenberg uncertainty relation (6.207) is a *sufficient* as well as necessary condition on V for it to describe a valid quantum state. This, and the fact that (6.207) is a LMI in V, will be important later.

Exercise 6.24 *Show that the purity $p = \mathrm{Tr}[\rho^2]$ of a state with Wigner function $W(\breve{\mathbf{x}})$ is given by*

$$p = (2\pi\hbar)^N \int d^{2N}\breve{\mathbf{x}}[W(\breve{\mathbf{x}})]^2. \tag{6.215}$$

Hint: *First generalize Eq. (A.117) to N dimensions.*
Then show that for a Gaussian state this evaluates to

$$p = \det[V]/(\hbar/2)^{2N}. \tag{6.216}$$

6.6.2 Linear unconditional dynamics

Having delineated the structure of quantum phase-space, we can now state the first restriction we require in order to obtain a linear quantum system (in the control theory sense).

(i) The system configuration obeys a linear dynamical equation

$$d\hat{\mathbf{x}} = A\hat{\mathbf{x}}\,dt + B\mathbf{u}(t)dt + E\,d\hat{\mathbf{v}}_{\mathrm{p}}(t). \tag{6.217}$$

This is precisely the same as the classical condition. Unlike in that case, the restrictions on quantum dynamics mean that the matrices A and E cannot be specified independently. To see this, we must derive Eq. (6.217) from the quantum Langevin equation generated by Eq. (6.185). The QLE for $\hat{\mathbf{x}}$, generalizing that of Eq. (3.172) to multiple baths, is

$$\hbar\,d\hat{\mathbf{x}} = dt\left(i[\hat{H}, \hat{\mathbf{x}}] + \hat{\mathbf{c}}^\dagger\hat{\mathbf{x}}\hat{\mathbf{c}} - \{\hat{\mathbf{c}}^\dagger\hat{\mathbf{c}}/2, \hat{\mathbf{x}}\}\right) + [\hat{\mathbf{c}}^\dagger\,d\hat{\mathbf{B}}_{\mathrm{in}}(t), \hat{\mathbf{x}}] - [d\hat{\mathbf{B}}_{\mathrm{in}}^\dagger(t)\hat{\mathbf{c}}, \hat{\mathbf{x}}]. \tag{6.218}$$

Here $\hat{\mathbf{c}}^\dagger\hat{\mathbf{x}}\hat{\mathbf{c}}$ is to be interpreted as $\sum_l \hat{c}_l^\dagger\hat{\mathbf{x}}\hat{c}_l$, and $d\hat{\mathbf{B}}_{\mathrm{in}}(t) = d\hat{\mathbf{B}}_{z:=-t}(t)$ is the vectorial generalization of Eq. (3.167).

We can derive Eq. (6.217) from Eq. (6.218) if we choose the system Hamiltonian to be quadratic in $\hat{\mathbf{x}}$, with the form

$$\hat{H} = \tfrac{1}{2}\hat{\mathbf{x}}^\top G\hat{\mathbf{x}} - \hat{\mathbf{x}}^\top \Sigma B\mathbf{u}(t), \tag{6.219}$$

with G real and symmetric, and the vector of Lindblad operators to be linear in $\hat{\mathbf{x}}$:

$$\hat{\mathbf{c}} = \tilde{C}\hat{\mathbf{x}}. \tag{6.220}$$

In terms of these we then have

$$A = \Sigma(G + \mathrm{Im}[\tilde{C}^\dagger\tilde{C}]), \tag{6.221}$$

$$E\,d\hat{\mathbf{v}}_{\mathrm{p}}(t) = [-i\,d\hat{\mathbf{B}}_{\mathrm{in}}^\dagger(t)\tilde{C}\Sigma + i\,d\hat{\mathbf{B}}_{\mathrm{in}}^\top(t)\tilde{C}^*\Sigma]^\top. \tag{6.222}$$

The second expression can be interpreted as Gaussian quantum process noise, because of the stochastic properties of $d\hat{\mathbf{B}}_{in}(t)$ as defined by Eq. (6.186).

Exercise 6.25 *Derive Eqs. (6.221) and (6.222) from Eq. (6.218).*

We have not specified separately E and $d\hat{\mathbf{v}}_p(t)$ because the choice would not be unique. All that is required (for the moment) is that the above expression for $E \, d\hat{\mathbf{v}}_p(t)$ gives the correct diffusion matrix:

$$D \, dt = \text{Re}[E \, d\hat{\mathbf{v}}_p(t) d\hat{\mathbf{v}}_p(t)^\top E^\top] \tag{6.223}$$

$$= \hbar \Sigma \, \text{Re}[\tilde{C}^\dagger \tilde{C}] \Sigma^\top dt \tag{6.224}$$

$$= \hbar \Sigma [\bar{C}^\top \bar{C}] \Sigma^\top dt. \tag{6.225}$$

In Eq. (6.223) we take the real part for the same reason as in Eq. (6.209): to determine the symmetrically ordered moments. In Eq. (6.225) we have introduced a new matrix,

$$\bar{C}^\top \equiv (\text{Re}[\tilde{C}^\top], \text{Im}[\tilde{C}^\top]). \tag{6.226}$$

Using this matrix, we can also write the drift matrix as

$$A = \Sigma(G + \bar{C}^\top S \bar{C}), \tag{6.227}$$

where, in terms of the blocks defined by Eq. (6.226),

$$S = \begin{pmatrix} 0 & I \\ -I & 0 \end{pmatrix}. \tag{6.228}$$

Exercise 6.26 *Verify that Eq. (6.225) is the correct expression for the diffusion matrix D.* **Hint:** *Calculate the moment equations (6.50) and (6.51) for $\langle \hat{\mathbf{x}} \rangle$ and V using the Itô calculus from the quantum Langevin equation (6.218).*

The calculations in the above exercise follow exactly the same form as for the classical Langevin equation. The non-commutativity of the noise operators actually plays no important role here, because of the linearity of the dynamics. Alternatively, the moment equations can be calculated (as in the classical case) directly from the equation for the state:

$$\hbar \dot{\rho} = -i[\hat{H}, \rho] + \mathcal{D}[\hat{\mathbf{c}}]\rho. \tag{6.229}$$

To make an even closer connection to the classical case, this master equation for the state matrix can be converted into an evolution equation for the Wigner function using the operator correspondences in Section A.5. This evolution equation has precisely the form of the OUE (6.49). Thus, the Wigner function has a Gaussian solution if it has a Gaussian initial state. As explained above, this means that there is a classical analogue to the quantum state, which is precisely the probability distribution that arises from the classical Langevin equation (6.47).

Fluctuation–dissipation relations. If we restrict ourselves to considering symmetrized moments of the quantum system, then we can go further and say that there is a classical system that is *equivalent* to our quantum system. This is because the linearity of the dynamics means that the quantum configuration ends up being a linear combination of the initial quantum configuration (assumed to have a Gaussian state) and the bath configuration (also assumed to have a Gaussian state, the vacuum). Thus any symmetrized function of the system configuration will be symmetrized in the initial system and bath configuration, and corresponds to a function of a classical random variable with a Gaussian probability distribution.

It should not be thought, however, that there are no quantum–classical differences in the unconditional dynamics of linear systems. As mooted above, the conditions of unitarity place restrictions on quantum evolutions that are not present classically. Specifically, the drift and diffusion matrices, A and D, respectively, cannot be specified independently of one another because both are related to $\tilde{C}$ (or $\bar{C}$). This is despite the fact that, considered on their own, neither A nor D is restricted by quantum mechanics. That is, the drift matrix A is an arbitrary real matrix. To see this, recall that G is arbitrary real symmetric, whereas $\bar{C}^{\mathsf{T}} S \bar{C}$ is an arbitrary real skew-symmetric matrix, and Σ is invertible. Also, since Σ is orthogonal, the diffusion matrix D is an arbitrary real PSD matrix.

The relation between D and A can be seen by noting that

$$\Sigma^{-1} D \Sigma + \frac{i\hbar}{2}\left[\Sigma^{-1} A - \left(\Sigma^{-1} A\right)^{\mathsf{T}}\right] = \hbar \tilde{C}^{\dagger} \tilde{C} \geq 0. \tag{6.230}$$

Here we have used Eqs. (6.221) and (6.224). Thus

$$D - i\hbar\left(A\Sigma - \Sigma^{\mathsf{T}} A^{\mathsf{T}}\right)/2 \geq 0. \tag{6.231}$$

As well as being a necessary condition, this LMI is also a sufficient condition on D for a given drift matrix A. That is, it guarantees that $V(t) + i\hbar\Sigma/2 \geq 0$ for all $t > t_0$, provided that it is true at $t = t_0$. This is because the invertibility of Σ allows us to construct a Lindblad master equation explicitly from the above equations given valid A and D matrices.

The LMI (6.231) can be interpreted as a generalized fluctuation–dissipation relation for open quantum systems. A dissipative system is one that loses energy so that the evolution is strictly stable (here we are assuming that the energy is bounded from below; that is, $G \geq 0$). As discussed in Section 6.4.1, this means that the real parts of the eigenvalues of A must be negative. Any strictly stable A must have a non-vanishing value of $A\Sigma - \Sigma^{\mathsf{T}} A^{\mathsf{T}}$ and in this case the LMI (6.231) places a lower bound on the fluctuations D about equilibrium. Note that in fact exactly the same argument holds for a strictly unstable system (i.e. one for which all modes are unstable).

By contrast, it is easy to verify that the contribution to A arising from the Hamiltonian $\hat{H}$ places no restriction on D. This is because energy-conserving dynamics cannot give rise to dissipation. To see this, note that $\Sigma^{-1}(\Sigma G)\Sigma = G\Sigma = -(\Sigma G)^{\mathsf{T}}$. This implies that ΣG has the same eigenvalues as the negative of its transpose, which is to say, the same eigenvalues as the negative of itself. Thus, if λ is an eigenvalue then so is $-\lambda$, and therefore it is impossible for all the eignevalues of ΣG to have negative real parts. That is, $A = \Sigma G$ cannot be a strictly stable system.

It might be questioned whether it is appropriate to call Eq. (6.231) a fluctuation–dissipation relation, because in equilibrium thermodynamics this term is used for a relation that precisely specifies (not merely bounds) the strength of the fluctuations for a given linear dissipation [Nyq28, Gar85]. The reason why our relation is weaker is that we have not made the assumption that our system is governed by an evolution equation that will bring it to thermal equilibrium at any particular temperature (including zero temperature). Apart from the Markovicity requirement, we are considering completely general linear evolution. Our formalism can describe the situation of thermal equilibrium, but also a situation of coupling to baths at different temperatures, and even more general situations. Thus, just as the Schrödinger–Heisenberg uncertainty relation can provide only a lower bound on uncertainties in the system observables, our fluctuation–dissipation relation can provide only a lower bound on fluctuations in their evolution.

Stabilizability and controllability. The concepts of stabilizability and controllability for linear quantum systems can be brought over without change from the corresponding classical definitions in Section 6.4.1. However, the term 'controllability' is also used in the context of control of Hamiltonian quantum systems [RSD+95], with a different meaning. To appreciate the relation, let us write the system Hamiltonian, including the control term, as

$$\hat{H} = \hat{H}_0 + \sum_j \hat{H}_j u_j(t), \qquad (6.232)$$

where

$$\hat{H}_0 = \tfrac{1}{2}\hat{\mathbf{x}}^\top G \hat{\mathbf{x}}, \qquad (6.233)$$

$$\hat{H}_j = -\hat{\mathbf{x}}^\top \Sigma B \mathbf{e}_j, \qquad (6.234)$$

where the $\mathbf{e}_j$s are orthonormal vectors such that $\mathbf{u}(t) = \sum_j u_j(t)\mathbf{e}_j$. Note that j in this section is understood to range from 1 to $\omega[B]$, the number of columns of B. From these operators we can form the following quantities:

$$\hat{H}_j = -(\hat{\mathbf{x}}^\top \Sigma) B \mathbf{e}_j, \qquad (6.235)$$

$$[\hat{H}_0, \hat{H}_j]/(i\hbar) = -(\hat{\mathbf{x}}^\top \Sigma) A B \mathbf{e}_j, \qquad (6.236)$$

$$[\hat{H}_0, [\hat{H}_0, \hat{H}_j]]/(i\hbar)^2 = -(\hat{\mathbf{x}}^\top \Sigma) A^2 B \mathbf{e}_j \qquad (6.237)$$

and so on. Here we have used $A = \Sigma G$ as appropriate for Hamiltonian systems. The complete set of these operators, plus $\hat{H}_0$, plus real linear combinations thereof, is known as the Lie algebra generated by the operators $\{\hat{H}_0, \hat{H}_1, \hat{H}_2, \ldots, \hat{H}_{\omega[B]}\}$. (See Box 6.2.) It is these operators (divided by $\hbar$ and multiplied by the duration over which they act) which generate the Lie group of unitary operators which can act on the system. Note that there is

---

Box 6.2 Groups, Lie groups and Lie algebras

A group is a set $\mathfrak{G}$ with a binary operation '·' satisfying the four axioms

1. closure: $\forall A, B \in \mathfrak{G}, A \cdot B \in \mathfrak{G}$
2. associativity: $\forall A, B, C \in \mathfrak{G}, (A \cdot B) \cdot C = A \cdot (B \cdot C)$
3. existence of an identity element $I \in \mathfrak{G}$: $\forall A \in \mathfrak{G}, A \cdot I = I \cdot A = A$
4. existence of an inverse: $\forall A \in \mathfrak{G}, \exists A^{-1} \in \mathfrak{G}$: $A \cdot A^{-1} = A^{-1} \cdot A = I$.

For example, the set of real numbers, with · being addition, forms a group. Also, the set of positive real numbers, with · being multiplication, forms a group. Both of these examples are Abelian groups; that is, $\forall A, B \in \mathfrak{G}, A \cdot B = B \cdot A$. An example of a *non-Abelian* group is the set of unitary matrices in some dimensions $d > 1$, with · being the usual matrix product.

In physics, it is very common to consider groups that are continuous, called *Lie groups*. A common example is a group of matrices (that may be real or complex):

$$\mathfrak{G} = \{\exp(-iY): Y \in \mathfrak{g}\}. \qquad (6.238)$$

Here $\mathfrak{g}$ is the *Lie algebra* for the Lie group $\mathfrak{G}$. This set is called a Lie algebra because (i) it forms a vector space with a concept of multiplication (in this case, the usual matrix multiplication) that is distributive and (ii) it is closed under a particular sort of binary operation called the *Lie bracket* (in this case, equal to $-i$ times the commutator $[Y, Z] \equiv YZ - ZY$). Closure means that

$$\forall Y, Z \in \mathfrak{g}, -i[Y, Z] \in \mathfrak{g}. \qquad (6.239)$$

The *generator* of a Lie algebra $\mathfrak{g}$ is a set $\mathbb{X}^0 = \{X_k: k\}$ of elements of $\mathfrak{g}$ that generate the whole algebra using the Lie bracket. This means the following. We introduce the recursive definition $\mathbb{X}^{n+1} = \mathbb{X}^n \bigcup \{-i[Y, Z]: Y, Z \in \mathbb{X}^n\}$. Then

$$\mathfrak{g} = \text{span}(\mathbb{X}^\infty), \qquad (6.240)$$

where span($\mathbb{S}$) is the set of all real linear combinations of matrices in the set $\mathbb{S}$. In many cases (for example when the X_k are finite-dimensional matrices) the recursive definition will cease to produce distinct sets after some finite number of iterations, so that $\mathbb{X}^\infty = \mathbb{X}^N$ for some N.

---

no point in considering commutators containing more than one $\hat{H}_j$ ($j \neq 0$) since they will be proportional to a constant.

Now, the criterion for controllability for a linear system is that the controllability matrix (6.66) has full row rank. By inspection, this is equivalent to the condition that, out of the $2N \times \omega(B)$ Hilbert-space operators in the row-vector

$$(\hat{\mathbf{x}}^\top \Sigma)[B \; AB \; A^2B \; \cdots \; A^{2N-1}B], \qquad (6.241)$$

$2N$ are linearly independent combinations of the $2N$ canonical phase-space variables $\hat{x}_k$. From the above relations, we can thus see that controllability in the sense appropriate to linear systems can be restated as follows.

A linear quantum system is controllable iff the Lie algebra generated by $\{\hat{H}_0, \hat{H}_1, \hat{H}_2, \ldots, \hat{H}_{\omega[B]}\}$ includes a complete set of observables.

(See Section 6.5.1 for the definition of a complete set of observables.) The significance of this concept of controllability is that, as defined in Section 6.4.1, the centroid in phase space $\langle \hat{x} \rangle$ can be arbitrarily displaced as a function of time for a suitable choice of $\mathbf{u}(t)$. Note that this concept of controllability does *not* mean that it is possible to prepare an arbitrary quantum state of the system.

Having formulated our sense of controllability in Lie-algebraic terms, we can now compare this with the other sense used in Refs. [Bro73, HTC83, Alt02, RSD⁺95] and elsewhere. This sense applies to arbitrary quantum systems, and is much stronger than the notion used above. We consider a Hamiltonian of the form of Eq. (6.232), but with no restrictions on the forms of $\hat{H}_0$ and the $\hat{H}_j$. Then Ref. [RSD⁺95] defines *operator-controllability* as follows.

A quantum system is operator-controllable iff the Lie algebra generated by $\{\hat{H}_0, \hat{H}_1, \hat{H}_2, \ldots, \hat{H}_{\omega[B]}\}$ is equal to $\mathfrak{D}(\mathbb{H})$.

(See Section 6.5.1 for the definition of $\mathfrak{D}(\mathbb{H})$.) The significance of this concept of controllability is that any unitary evolution $\hat{U}$ can be realized by some control vector $\mathbf{u}(t)$ over some time interval $[t_0, t_1]$. Hence, operator-controllability means that from an initial pure state it is possible to prepare an arbitrary quantum state of the system.

6.6.3 Linear conditional dynamics

As for the first restriction of Section 6.6.2, the second restriction necessary to obtain a linear quantum system is again identical to the classical case (apart from hats on all of the quantities).

(ii) The system state is conditioned on the measurement result

$$\hat{\mathbf{y}} \, dt = C\hat{\mathbf{x}} \, dt + d\hat{\mathbf{v}}_m(t). \qquad (6.242)$$

This follows automatically from the assumption (6.220) that we have already made, provided that we use a Wiener-process unravelling to condition the system. Once again, however, quantum mechanics places restrictions on the matrix C and the correlations of the measurement noise $d\hat{\mathbf{v}}_m(t)$.

We saw in Section 6.5.2 that the most general output of a quantum system with Wiener noise is a vector of complex currents $\hat{\mathbf{J}}$ defined in Eq. (6.193). This can be turned into a real vector by defining

$$\hat{\mathbf{y}} = T^+ \begin{pmatrix} \mathrm{Re}\,\hat{\mathbf{J}} \\ \mathrm{Im}\,\hat{\mathbf{J}} \end{pmatrix} = C\hat{\mathbf{x}} + \frac{d\hat{\mathbf{v}}_m}{dt}, \tag{6.243}$$

with

$$C = 2T^\top \bar{C}/\hbar. \tag{6.244}$$

Here T is, in general, a *non-square* matrix, with $\varepsilon[\bar{C}]$ rows and $\varepsilon[C]$ columns, such that

$$TT^\top = \hbar U, \tag{6.245}$$

where U is the unravelling matrix as usual. In Eq. (6.243) T^+ is the pseudoinverse, or Moore–Penrose inverse [CM91, ZDG96] of T (see Box. 6.1). Note that the numbers of columns of $\bar{C}$ and of C are equal: $\omega[C] = \omega[\bar{C}] = 2N$. The number of rows of C, $\varepsilon[C]$ (also equal to the dimension of $\hat{\mathbf{y}}$), is equal to the rank of U. The number of rows of $\bar{C}$, $\varepsilon[\bar{C}]$ (also equal to twice the dimension of $\hat{\mathbf{c}}$), is equal to the number of rows (or columns) of U. This guarantees that the matrix T exists.

In Eq. (6.243) we have defined

$$d\hat{\mathbf{v}}_m = T^+ \begin{pmatrix} \mathrm{Re}\,\delta\hat{\mathbf{J}} \\ \mathrm{Im}\,\delta\hat{\mathbf{J}} \end{pmatrix} dt, \tag{6.246}$$

where (*cf.* Eq. (6.193))

$$\delta\hat{\mathbf{J}}^\top dt = d\hat{\mathbf{B}}_{in}^\top H + d\hat{\mathbf{B}}_{in}^\dagger \Upsilon + d\hat{\mathbf{A}}^\dagger \sqrt{H - \Upsilon^* H^{-1} \Upsilon}$$
$$+ d\hat{\mathbf{V}}^\top \sqrt{H(I - H)} + d\hat{\mathbf{V}}^\dagger \sqrt{H^{-1} - I}\,\Upsilon. \tag{6.247}$$

The measurement noise operator $d\hat{\mathbf{v}}_m(t)$ has the following correlations:

$$d\hat{\mathbf{v}}_m\, d\hat{\mathbf{v}}_m^\top = I\, dt, \tag{6.248}$$
$$\mathrm{Re}[E\, d\hat{\mathbf{v}}_p\, d\hat{\mathbf{v}}_m^\top] = \Gamma^\top dt, \tag{6.249}$$

just as in the classical case, where here

$$\Gamma = -T^\top S\bar{C}\Sigma^\top. \tag{6.250}$$

Exercise 6.27 *Verify these correlations from the above definitions.*

The evolution conditioned upon measuring $\mathbf{y}(t)$ is the same as in the classical case. That is, the conditioned state is Gaussian, with mean vector and covariance matrix obeying Eqs. (6.85) and (6.86), respectively. In the stochastic equation for the mean, the innovation is, as expected,

$$dw = d\hat{v}_m + C(\hat{x} - \langle \hat{x} \rangle_c). \tag{6.251}$$

These quantum Kalman-filter equations can also be derived from the quantum version of the Kushner–Stratonovich equation, Eq. (6.187). Indeed, by using the Wigner function to represent the quantum state, the evolution can be expressed precisely as the Kushner–Stratonovich equation (6.71), involving the matrices A, B, D, C and Γ.

Fluctuation–observation relations. The above analysis shows that, even including measurement, there is a linear classical system with all the same properties as our linear quantum system. As in the case of the unconditioned dynamics, however, the structure of quantum mechanics constrains the possible conditional dynamics. In the unconditioned case this was expressed as a fluctuation–dissipation relation. That is, any dissipation puts lower bounds on the fluctuations. In the present case we can express the constraints as a *fluctuation–observation relation*. That is, for a quantum system, any information gained puts lower bounds on fluctuations in the conjugate variables, which are necessary in order to preserve the uncertainty relations.

Recall that the Riccati equation for the conditioned covariance matrix can be written as

$$V_c = \tilde{A} V_c + V_c \tilde{A}^\top + \tilde{E} \tilde{E}^\top - V_c C^\top C V_c, \tag{6.252}$$

where in the quantum case

$$\tilde{A} = A - \Gamma^\top C = \Sigma[G + \bar{C}^\top S(I - 2U)\bar{C}], \tag{6.253}$$

$$\tilde{E}\tilde{E}^\top = D - \Gamma^\top \Gamma = \hbar \Sigma \bar{C}^\top [I - S^\top U S] \bar{C} \Sigma^\top. \tag{6.254}$$

From Eq. (6.252) we see that $\tilde{E}\tilde{E}^\top$ always increases the uncertainty in the system state. This represents fluctuations. By contrast, the term $-V_c C^\top C V_c$ always decreases the uncertainty. This represents information gathering, or observation. The fluctuation–observation relation is expressed by the LMI

$$\tilde{E}\tilde{E}^\top - \frac{\hbar^2}{4}\Sigma C^\top C \Sigma^\top \geq 0. \tag{6.255}$$

Exercise 6.28 *Show this, by showing that the left-hand side evaluates to the following matrix which is clearly PSD:*

$$\hbar \Sigma \bar{C}^\top \begin{pmatrix} I - \mathrm{H} & 0 \\ 0 & I - \mathrm{H} \end{pmatrix} \bar{C} \Sigma^\top. \tag{6.256}$$

The first thing to note about Eq. (6.255) is that it is quantum in origin. If $\hbar$ were zero, there would be no lower bound on the fluctuations. The second thing to notice is that observation of one variable induces fluctuations in the conjugate variable. This follows from the presence of the matrix $\Sigma^\top$ that postmultiplies C in Eq. (6.255). It is most easily seen in the case of motion in one dimension ($N = 1$). Say we observe the position q, so that

$$y \, dt = \sqrt{\kappa} \langle \hat{q} \rangle_c \, dt + dw, \tag{6.257}$$

where here κ is a scalar expressing the measurement strength. Then Eq. (6.255) says that

$$\tilde{E}\tilde{E}^\mathsf{T} - \frac{\hbar^2}{4} \begin{pmatrix} 0 & 0 \\ 0 & \kappa \end{pmatrix} \geq 0. \tag{6.258}$$

That is, there is a lower bound of $(\hbar\sqrt{\kappa}/2)^2$ on the spectral power of momentum fluctuations. The third thing to note about Eq. (6.255) is that, since $D = \tilde{E}\tilde{E}^\mathsf{T} + \Gamma^\mathsf{T}\Gamma$, our relation implies the *weaker* relation

$$D - \frac{\hbar^2}{4} \Sigma C^\mathsf{T} C \Sigma^\mathsf{T} \geq 0. \tag{6.259}$$

As well as being a necessary condition on $\tilde{E}$ given C, Eq. (6.255) is also a sufficient condition.

Detectability and observability. The definitions of detectability and observability for linear quantum systems replicate those for their classical counterparts – see Section 6.4.2. However, there are some interesting points to make about the quantum case.

First, we can define the notion of *potential detectability*. By this we mean that, given the unconditioned evolution described by A and D, there exists a matrix C such that (C, A) is detectable. Classically this is always the case because C can be specified independently of A and D, so this notion would be trivial, but quantum mechanically there are some evolutions that are not potentially detectable; Hamiltonian evolution is the obvious example.

We can determine which unconditional evolutions are potentially detectable from A and D as follows. First note that from Eq. (6.244) the existence of an unravelling U such that (C, A) is detectable is equivalent to $(\bar{C}, A)$ being detectable. Indeed, $C \propto \bar{C}$ results from the unravelling $U = I/2$, so a system is potentially detectable iff the $U = I/2$ unravelling is detectable. Now, $(\bar{C}, A)$ being detectable is equivalent to $(\bar{C}^\mathsf{T}\bar{C}, A)$ being detectable. But, from Eq. (6.225), $\bar{C}^\mathsf{T}\bar{C} = \Sigma^\mathsf{T} D \Sigma/\hbar$. Since the above arguments, *mutatis mutandis*, also apply for potential observability, we can state the following.

A quantum system is potentially detectable (observable) iff $(\Sigma^\mathsf{T} D \Sigma, A)$ is detectable (observable).

The second interesting point to make in this section is that, for quantum systems, if $(C, \tilde{A})$ is detectable then $(-\tilde{A}, \tilde{E})$ is stabilizable.[3] Consider for simplicity the case of efficient detection, where $H = I$. Then the left-hand side of Eq. (6.255) is zero, and we can

---

[3] Note that in Ref. [WD05] it was incorrectly stated that $(\tilde{A}, \tilde{E})$ was stabilizable, but the conclusion drawn there, discussed in Section 6.6.4, is still correct.

choose

$$\tilde{E} = \frac{\hbar}{2} \Sigma C^\top. \tag{6.260}$$

Moreover, $\tilde{A} = \Sigma \tilde{G}$, where

$$\tilde{G} = G + \bar{C}^\top \begin{pmatrix} -\mathrm{Im}\,\Upsilon & \mathrm{Re}\,\Upsilon \\ \mathrm{Re}\,\Upsilon & \mathrm{Im}\,\Upsilon \end{pmatrix} \bar{C} \tag{6.261}$$

is a symmetric matrix, so that $\frac{1}{2}\hat{\mathbf{x}}^\top \tilde{G}\hat{\mathbf{x}}$ is a pseudo-Hamiltonian that generates the part of the drift of the system which is independent of the record $\mathbf{y}$. Now, since Σ is invertible, we can replace $\mathbf{x}_\lambda$ by $\Sigma \mathbf{x}_\lambda$ everywhere in the definition (6.87) of detectability. It then follows that $(C, \tilde{A})$ detectable is equivalent to $(C\Sigma^\top, \Sigma\tilde{A}\Sigma^\top)$ detectable (remember that $\Sigma^{-1} = \Sigma^\top$). But $\Sigma\tilde{A}\Sigma^\top = \Sigma^2\tilde{G}\Sigma^\top = -\tilde{G}\Sigma^\top = -\tilde{A}^\top$, while $C\Sigma^\top \propto \tilde{E}^\top$. Thus, by virtue of the detectable–stabilizable duality, we have $(-\tilde{A}, \tilde{E})$ stabilizable. Now for inefficient detection the fluctuations in the system are greater, so $(-\tilde{A}, \tilde{E})$ will also be stabilizable in this case.

Third, we note that, as was the case for controllability, we can give a Lie-algebraic formulation for observability, at least for the non-dissipative case in which $\tilde{C}$ can be taken to be real, so that $\mathrm{Im}[\tilde{C}^\dagger \tilde{C}] = 0$ and $A = \Sigma G$. From Section 6.4.2, observability for the linear system is equivalent to the matrix

$$[C^\top \ A^\top C^\top \ (A^\top)^2 C^\top \ \cdots \ (A^\top)^{2N-1} C^\top] \tag{6.262}$$

having full row rank. Now, for this non-dissipative case (so called because the drift evolution is that of a Hamiltonian system), the method of Section 6.6.2 can be applied to give the following new formulation.

A non-dissipative linear quantum system is observable iff the Lie algebra generated by $\{\hat{H}_0, \hat{o}_1, \hat{o}_2, \ldots, \hat{o}_{\varepsilon[C]}\}$ includes a complete set of observables.

Here $\hat{H}_0$ is as above, while $\hat{o}_l = \mathbf{e}_l^\top C\hat{\mathbf{x}}$.

This definition does not generalize naturally to other sorts of quantum systems in the way that the definition of controllability does. If $\hat{H}_0$ and $\{\hat{o}_l\}$ are arbitrary operators, then the above Lie algebra does not correspond to the operators the observer obtains information about as the system evolves conditionally. Moreover, unlike in the linear case, the observability of a general system can be enhanced by suitable application of the control Hamiltonians $\{\hat{H}_j\}$. Indeed, Lloyd [Llo00] has defined observability for a general quantum system such that it is achievable iff it is operator-controllable (see Section 6.6.2) and the observer can make at least one nontrivial projective measurement. (His definition of observability is essentially that the observer can measure any observable, and he does not consider continuous monitoring.)

6.6.4 Stabilizing solutions

As in the classical case, for the purposes of feedback control we are interested in observed systems for which the Riccati equation (6.252) has a stationary solution that is stabilizing. Again, we will denote that solution by W, so we require that

$$- M W_U - W_U M^\top = \tilde{E} \tilde{E}^\top + W_U C^\top C W_U \qquad (6.263)$$

has a unique solution such that

$$M \equiv \tilde{A} - W_U C^\top C \qquad (6.264)$$

is strictly stable. In the above we have introduced a subscript U to emphasize that the stationary conditioned covariance matrix depends upon the unravelling U, since all of the matrices $\tilde{A}$, $\tilde{E}$ and C depend upon U. We call an unravelling stabilizing if Eq. (6.263) has a stabilizing solution.

As discussed in Section 6.4.3, a solution is stabilizing iff $(C, \tilde{A})$ (or (C, A)) is detectable and condition (6.99) is satisfied. As stated there, the second condition is satisfied if $(-\tilde{A}, \tilde{E})$ is stabilizable. But, as we saw in the preceding section, in the quantum case, this follows automatically from the first condition. That is, *quantum mechanically, the conditions for the existence of a stabilizing solution are weaker than classically.* Detectability of (C, A) is all that we require to guarantee a stabilizing solution.

In the quantum case we can also apply the notion of potential detectability from the preceding section. It can be shown [WD05] that *if the system is potentially detectable then the stabilizing unravellings form a dense subset*[4] *of the set of all unravellings.* Now, for detectable unravellings, the solutions to the algebraic MRE (6.263) are continuous in $\tilde{A}$, $\tilde{E}$ and C [LR91]. But these matrices are continuous in U, and hence W_U is continuous in U. Thus, as long as (i) one restricts oneself to a compact set of W_Us (e.g. a set of bounded W_Us); (ii) one is interested only in continuous functions of W_U; and (iii) the system is potentially detectable, then one can safely assume that any such W_U is a stabilizing solution.

Possible conditional steady states. We showed in the classical case that the only restriction on the possible stationary conditioned covariance matrices that a system described by A and D can have is the LMI

$$A W_U + W_U A^\top + D \geq 0. \qquad (6.265)$$

This is also a necessary condition in the quantum case, by exactly the same reasoning, although in this case we also have another necessary condition on the covariance matrix given by the uncertainty relation (6.207), which we repeat here:

$$W_U + i\hbar \Sigma / 2 \geq 0. \qquad (6.266)$$

If W_U is the covariance matrix of a pure state, then Eq. (6.265) is also a sufficient condition for W_U to be a realizable stationary conditioned covariance matrix. This can

---

[4] If a set A is a dense subset of a set B, then for every element of B there is an element of A that is arbitrarily close, by some natural metric.

be seen from the phenomenon of pure-state steering as discovered by Schrödinger (see Section 3.8.1). This was first pointed out in Ref. [WV01]. Say the system at time t has a covariance matrix W_U corresponding to a pure state. Recall, from Eq. (6.216), that this is true iff

$$\det[W_U] = (\hbar/2)^{2N}. \tag{6.267}$$

Then, if Eq. (6.265) is satisfied, the system at an infinitesimally later time $t + dt$ will be a mixture of states, all with covariance matrix W_U, and with Gaussian-distributed means, as explained in Section 6.4.3. We call such an ensemble a uniform Gaussian pure-state ensemble. Then, by virtue of the Schrödinger–HJW theorem, there will be some way of monitoring the environment – that part of the bath that has become entangled with the system in the interval $[t, t + dt)$ – such that the system is randomly collapsed to one of the pure state elements of this ensemble, with the appropriate Gaussian weighting. That is, a pure state with covariance matrix W_U can be reprepared by continuing the monitoring, and therefore W_U must be the stationary conditioned covariance matrix under some monitoring scheme.

Thus we have a necessary condition (Eqs. (6.265) and (6.266)) and a sufficient condition (Eqs. (6.265) and (6.267)) for W_U to be the steady-state covariance matrix of some monitoring scheme.[5] If W_U is such a covariance matrix, then there is an unravelling matrix U that will generate the appropriate matrices $\tilde{A}$, $\tilde{E}$ and C so that W_U is the solution of Eq. (6.263). Moreover, as argued above, as long as W_U is bounded and $(\bar{C}, A)$ is detectable, this W_U can be taken to be stabilizing.

To find an unravelling (which may be non-unique) generating W_U as a stationary conditioned covariance matrix, it is simply necessary to put the U-dependence explicitly in Eq. (6.263). This yields the LME for U:

$$\hbar R^\top U R = D + A W_U + W_U A^\top, \tag{6.268}$$

where $R = 2\bar{C}W_U/\hbar + S\bar{C}\Sigma$. This can be solved efficiently (that is, in a time polynomial in the size of the matrices). It does not matter whether this equation has a non-unique solution U, because in steady state the conditional state and its dynamics will be the same for all U satisfying Eq. (6.268) for a given W_U. This can be seen explicitly as follows. The shape of the conditioned state in the long-time limit is simply W_U. The stochastic dynamics of the mean $\langle \hat{\mathbf{x}} \rangle_c$ is given by

$$d\langle \hat{\mathbf{x}} \rangle_c = [A\langle \hat{\mathbf{x}} \rangle_c + B\mathbf{u}(t)]dt + F^\top d\mathbf{w}, \tag{6.269}$$

which depends upon U only through the stochastic term. Recall that $F = CW_U + \Gamma$, which depends on U through C and Γ as well as W_U. However, statistically, all that matters is the

---

[5] In Ref. [WD05] it was incorrectly stated that Eqs. (6.265) and (6.266) form the necessary and sufficient conditions. However, this does not substantially affect the conclusions of that work, since other constraints ensure that the states under consideration will be pure, as will be explained later.

covariance of the noise in Eq. (6.269), which, from Eq. (6.263), is given by

$$F^\top \, d\mathbf{w} \, d\mathbf{w}^\top \, F = dt(AW_U + W_U A^\top + D), \tag{6.270}$$

which depends on U only through W_U.

Consider the case in which A is strictly stable so that for $\mathbf{u} = \mathbf{0}$ an unconditional steady state exists, with covariance matrix satisfying

$$AV_{ss} + V_{ss}A^\top + D = 0. \tag{6.271}$$

As noted in the classical case, there exist states with covariance matrix W satisfying $V_{ss} - W \geq 0$ and yet not satisfying Eq. (6.265). This is also true in the quantum case, even with the added restriction that W correspond to a pure state by satisfying (6.267). That is, there exist uniform Gaussian pure-state ensembles that represent the stationary solution ρ_{ss} of the quantum master equation but cannot be realized by any unravelling. In saying that the uniform Gaussian ensemble represents ρ_{ss} we mean that

$$\rho_{ss} = \int d^{2N}\langle \hat{\mathbf{x}} \rangle_c \, \wp(\langle \hat{\mathbf{x}} \rangle_c) \rho^W_{\langle \hat{\mathbf{x}} \rangle_c}, \tag{6.272}$$

where $\rho^W_{\langle \hat{\mathbf{x}} \rangle_c}$ has the Gaussian Wigner function $W_c(\check{\mathbf{x}}) = g(\check{\mathbf{x}}; \langle \hat{\mathbf{x}} \rangle_c, W)$, and the Gaussian distribution of means is

$$\wp(\langle \hat{\mathbf{x}} \rangle_c) = g(\langle \hat{\mathbf{x}} \rangle_c; \mathbf{0}, V_{ss} - W). \tag{6.273}$$

In saying that the ensemble cannot be realized we mean that there is no way an observer can monitor the output of the system so as to know that the system is in the state $\rho^W_{\langle \hat{\mathbf{x}} \rangle_c}$, such that W remains fixed in time but $\langle \hat{\mathbf{x}} \rangle_c$ varies so as to sample the Gaussian distribution (6.273) over time. On the other hand, there are certainly some ensembles that satisfy both Eq. (6.265) and Eq. (6.267), which thus are physically realizable (PR) in this sense. This existence of some ensembles representing ρ_{ss} that are PR and some that are not is an instance of the *preferred-ensemble fact* discussed in Section 3.8.2.

Example: on-threshold OPO. To illustrate this idea, consider motion in one dimension with a single output channel ($N = L = 1$), described by the master equation

$$\hbar\dot{\rho} = -i[(\hat{q}\hat{p} + \hat{p}\hat{q})/2, \rho] + \mathcal{D}[\hat{q} + i\hat{p}]\rho, \tag{6.274}$$

where the output arising from the second term may be monitored. This could be realized in quantum optics as a damped cavity (a harmonic oscillator in the rotating frame) containing an on-threshold parametric down-converter, also known as an optical parametric oscillator (OPO). Here p would be the squeezed quadrature and q the anti-squeezed quadrature. The monitoring of the output could be realized by techniques such as homodyne or heterodyne detection.

Exercise 6.29 *Show that in this case we have*

$$G = \begin{pmatrix} 0 & 1 \\ 1 & 0 \end{pmatrix}, \qquad \tilde{C} = (1, i) \tag{6.275}$$

and that the drift and diffusion matrices evaluate to

$$A = \begin{pmatrix} 0 & 0 \\ 0 & -2 \end{pmatrix}, \qquad D = \hbar \begin{pmatrix} 1 & 0 \\ 0 & 1 \end{pmatrix}. \tag{6.276}$$

Since A is not strictly stable, there is no stationary unconditional covariance matrix. However, in the long-time limit

$$V \to \hbar \begin{pmatrix} \infty & 0 \\ 0 & 1/2 \end{pmatrix}, \tag{6.277}$$

and we come to no harm in defining this to be V_{ss}. Writing the conditional steady-state covariance matrix as

$$W_U = \frac{\hbar}{2} \begin{pmatrix} \alpha & \beta \\ \beta & \gamma \end{pmatrix}, \tag{6.278}$$

the LMIs (6.265) and (6.266) become, respectively,

$$\begin{pmatrix} \alpha & \beta + i \\ \beta - i & \gamma \end{pmatrix} \geq 0, \tag{6.279}$$

$$\begin{pmatrix} 1 & -\beta \\ -\beta & 1 - 2\gamma \end{pmatrix} \geq 0. \tag{6.280}$$

The first of these implies $\alpha > 0$, $\gamma > 0$ and $\alpha\gamma \geq 1 + \beta^2$. The second then implies $\gamma \leq (1 - \beta^2)/2$.

In Fig. 6.7 we show four quantum states $\rho_{\langle\hat{x}\rangle_c}^W$ that are pure (they saturate Eq. (6.279)) and satisfy $V_{ss} - W \geq 0$. That is, they 'fit inside' ρ_{ss}. However, one of them does not satisfy Eq. (6.280). We see the consequence of that when we show the mixed states that these four pure states evolve into after a short time $\tau = 0.2$ in Fig. 6.8. (We obtain this by analytically solving the moment equation (6.51), starting with $\langle\hat{x}\rangle = \mathbf{0}$ for simplicity.) This clearly shows that, for the initial state that fails to satisfy Eq. (6.280), the mixed state at time τ can no longer be represented by a mixture of the original pure state with random displacements, because the original state does not 'fit inside' the evolved state. The ensemble formed from these states is not physically realizable. We will see later, in Section 6.6.6, how this has consequences in quantum feedback control.

6.6.5 LQG optimal feedback control

We can now consider a cost function (6.197) for the quantum system and controller where h is quadratic, as in Eq. (6.113), but with $\mathbf{x}$ replaced by $\hat{\mathbf{x}}$. This can be justified in the quantum case for the same sorts of reasons as in the classical case; for instance, in linear systems, the free energy is a quadratic function of $\hat{\mathbf{x}}$. The resulting optimization (cost minimization) problem has exactly the same solution as in the classical case, so all of the discussion on LQG control in Section 6.4.4 applies here.

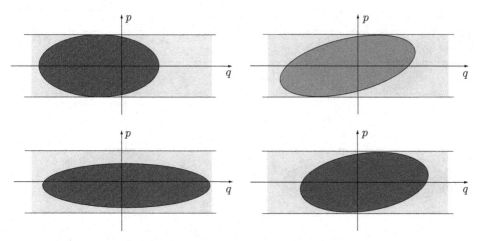

Fig. 6.7 Representation of states in phase-space for the system described in the text. The horizontal and vertical axes are q and p, respectively, and the curves representing the states are one-standard-deviation contours of the Wigner function $W(q, p)$. That is, they are curves defined by the parametric equation $W(q, p) = \mathrm{e}^{-1} W(\bar{q}, \bar{q})$, where $(\bar{q}, \bar{p})^{\top}$ is the centroid of the state. The stationary unconditioned state has a p-variance of $\hbar/2$ and an unbounded q-variance. A short segment of the Wigner function of this state is represented by the lightly shaded region between the horizontal lines at $p = \pm\sqrt{\hbar/2}$. The ellipses represent pure states, with area $\pi\hbar$. They are possible conditioned states of the system, since they 'fit inside' the stationary state. For states realizable by continuous monitoring of the system output, the centroid of the ellipses wanders stochastically in phase-space, which is indicated in the diagram by the fact that the states are not centred at the origin. The state in the top-right corner is shaded differently from the others because it cannot be physically realized in this way, as Fig. 6.8 demonstrates.

One new feature that arises in the quantum case is the following. Classically, for a minimally disturbing measurement, the stronger the measurement, the better the control. Consider the steady-state case for simplicity. If we say that $C = \sqrt{\kappa}C_1$, with C_1 fixed, then the stationary conditioned covariance matrix W is given by the ARE

$$AW + WA^{\top} + D = \kappa W C_1{}^{\top} C_1 W. \qquad (6.281)$$

(Note that we have set $\Gamma = 0$ as appropriate for a minimally disturbing measurement.)

Exercise 6.30 *Convince yourself that, for A, D and C_1 fixed, the eigenvalues of W decrease monotonically as κ increases.*

Thus the integrand in the cost function

$$E_{\mathrm{ss}}[h] = \mathrm{tr}[Y B Q^{-1} B^{\top} Y W] + \mathrm{tr}[Y D] \qquad (6.282)$$

is monotonically decreasing with κ. By contrast, in the quantum case it is not possible to say that D is fixed as κ increases. Rather, for a minimally disturbing quantum measurement, the measurement necessarily contributes to D a term $\hbar^2 \kappa \Sigma C_1{}^{\top} C_1 \Sigma^{\top}/4$, according to the

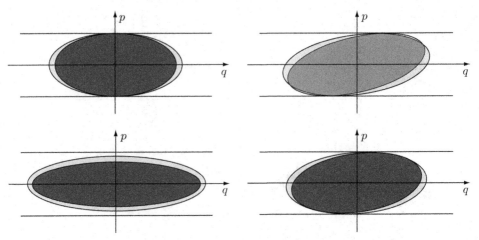

Fig. 6.8 Representation of the four pure states from Fig. 6.7, plus the mixed states they evolve into after a time $\tau = 0.2$. For ease of comparison we have centred each of these states at the origin, and we have omitted the shading for the stationary state, but the other details are the same as for the preceding figure. Note that, apart from the top-right state, the initial states (heavy shading) all 'fit inside' the evolved states (light shading). Hence they are all physically realizable (PR). The top-right initial state is unshaded, as is apparent from the parts that do not 'fit inside' the evolved state (light shading), and so is not PR. (The part that does 'fit inside' appears with medium shading, as in Fig. 6.7). The four initial states that appear here are defined as follows. Top-left: the state with minimum q-variance that fits inside the stationary state. Bottom-left: the state arising from the $U = I/2$ unravelling. Top-right: the state with minimum $(q - p)^2$ that fits inside the stationary state. Bottom-right: the state with minimum $(q - p)^2$ that is PR.

fluctuation–observation relation. Thus with D_0 fixed we have the ARE

$$AW + WA^\top + D_0 + \hbar^2 \kappa \Sigma C_1{}^\top C_1 \Sigma^\top / 4 = \kappa W C_1{}^\top C_1 W. \qquad (6.283)$$

Here, the eigenvalues of W are not monotonically decreasing with κ, and neither (in general) is $E_{ss}[h]$. The cost may actually monotonically increase with κ, or there may be some optimum κ that minimizes $E_{ss}[h]$.

Example: the harmonic oscillator. We can illustrate the above idea, as well as other concepts in LQG control, using the example of the harmonic oscillator with position measurement and controlled by a spatially invariant (but time varying) force. This was considered by Doherty and Jacobs [DJ99], who also discussed a physical realization of this system in cavity QED. We do not assume that the oscillator frequency is much larger than the measurement rate κ, so it is not appropriate to work in the interaction frame. Indeed, we take the oscillator frequency to be unity, and for convenience we will also take the particle mass to be unity. We can model this by choosing

$$G = \begin{pmatrix} 1 & 0 \\ 0 & 1 \end{pmatrix}, \qquad \tilde{C} = \sqrt{\kappa}\,(1, 0), \qquad U = \begin{pmatrix} 1 & 0 \\ 0 & 0 \end{pmatrix}. \qquad (6.284)$$

Exercise 6.31 *Show that this gives* $\Gamma = 0$ *and*

$$A = \begin{pmatrix} 0 & 1 \\ -1 & 0 \end{pmatrix}, \quad D = \hbar \begin{pmatrix} 0 & 0 \\ 0 & \kappa \end{pmatrix}, \quad C_1 = (2,0)/\sqrt{\hbar}. \tag{6.285}$$

In the above we have assumed that $D_0 = 0$ (that is, that there are no noise sources apart from the measurement back-action). This allows a simple solution to the algebraic MRE (6.283):

$$W = \frac{\hbar}{4\kappa} \begin{pmatrix} \sqrt{2v} & v \\ v & (1+v)\sqrt{2v} \end{pmatrix}, \tag{6.286}$$

where $v = \sqrt{1 + 4\kappa^2} - 1$.

Exercise 6.32 *Show that, as well as solving Eq. (6.283), this W saturates the LMI (6.266), and hence corresponds to a pure state.*

When $\kappa \ll 1$, the measurement of position is slow compared with the oscillation of the particle. In this limit, $v \to 2\kappa^2$ and $W \to (\hbar/2)I$. That is, the conditioned state is a coherent state of the oscillator, and the conditioned variance in position is $\hbar/2$. In physical units, this (the standard quantum limit for the position variance) is $\hbar/(2m\omega)$.

Now consider feedback control of the oscillator, for the purpose of minimizing the energy in steady state. That is, we choose the cost function $P = I$, and (from the control constraint mentioned above), $B = (0,1)^\top$, so that Q is just a scalar. Then Eq. (6.117) for Y becomes

$$\begin{pmatrix} 0 & -1 \\ 1 & 0 \end{pmatrix} Y + Y \begin{pmatrix} 0 & 1 \\ -1 & 0 \end{pmatrix} + \begin{pmatrix} 1 & 0 \\ 0 & 1 \end{pmatrix} = Y \begin{pmatrix} 0 & 0 \\ 0 & Q^{-1} \end{pmatrix} Y. \tag{6.287}$$

Exercise 6.33 *Show that, for $Q \ll 1$, this ARE has the approximate solution*

$$Y \simeq \begin{pmatrix} 1 & \sqrt{Q} \\ \sqrt{Q} & \sqrt{Q} \end{pmatrix}. \tag{6.288}$$

Also show that it is a stabilizing solution.
Hint: *For the second part, show first that*

$$BQ^{-1}B^\top Y \simeq -\frac{1}{\sqrt{Q}} \begin{pmatrix} 0 & 0 \\ 1 & 1 \end{pmatrix}. \tag{6.289}$$

Thus the optimal feedback, which adds to the equations of motion the term

$$\frac{d}{dt} \begin{pmatrix} \langle \hat{q} \rangle \\ \langle \hat{p} \rangle \end{pmatrix}_{\text{fb}} = -\frac{1}{\sqrt{Q}} \begin{pmatrix} 0 \\ \langle \hat{q} \rangle + \langle \hat{p} \rangle \end{pmatrix}, \tag{6.290}$$

is asymptotically stable.

Now for this problem we have

$$F = CW = \sqrt{\frac{\hbar}{4\kappa}}(\sqrt{2v}, v). \tag{6.291}$$

Therefore, under the optimal feedback, the approximate (for $Q \ll 1$) equation for the unconditioned variance (6.124) is

$$\left\{ \begin{pmatrix} 0 & 1 \\ Q^{-1/2} & Q^{-1/2} \end{pmatrix} (V_{\mathrm{ss}} - W) + \mathrm{m.t.} \right\} = -\frac{\hbar}{4\kappa} \begin{pmatrix} 2v & v\sqrt{2v} \\ v\sqrt{2v} & v^2 \end{pmatrix}. \tag{6.292}$$

In order to counter the largeness of $Q^{-1/2}$, we must have

$$V_{\mathrm{ss}} - W = \epsilon \begin{pmatrix} 1 & -1 \\ -1 & 1 \end{pmatrix} + O(Q^{1/2}), \tag{6.293}$$

for some positive constant ϵ.

Exercise 6.34 *Prove this by considering an asymptotic expansion of $V_{\mathrm{ss}} - W$ in powers of $Q^{1/2}$.*

On substituting Eq. (6.293) into Eq. (6.292), we find from equating the top-left element of both sides that $\epsilon = \hbar v/(4\kappa) + O(Q^{1/2})$. Thus we have finally (remember that $v = \sqrt{1 + 4\kappa^2} - 1$)

$$V_{\mathrm{ss}} = \frac{\hbar}{4\kappa} \begin{pmatrix} v + \sqrt{2v} & 0 \\ 0 & v + (1 + v)\sqrt{2v} \end{pmatrix} + O(Q^{1/2}). \tag{6.294}$$

Note that, even though we have set the control cost Q to zero, V_{ss} does not approach W. The 'classical' fluctuations (6.293) are of the same order ($\hbar$) as the quantum noise W. This is because the control constraint, that $B = (0, 1)^{\top}$, means that the system is not pacifiable. This follows from Eq. (6.127), since rank$[B]$ is one, but rank$[B\ \mathrm{F}^{\top}]$ is two.

Under this optimal feedback control, the integrand in the cost function (6.126) evaluates to

$$\mathrm{E}_{\mathrm{ss}}[h] = \mathrm{tr}[YBQ^{-1}B^{\top}YW] + \mathrm{tr}[YD] \tag{6.295}$$

$$= \hbar\sqrt{\frac{2v}{\kappa}}\left(1 + \sqrt{v/2} + v/2\right) + \hbar O(Q^{1/2}). \tag{6.296}$$

Considered as a function of κ, this cost has a minimum of $\hbar$ as $\kappa \to 0$ (since then $v \to \kappa/2$). That is, the optimal measurement strength is zero, and the cost rises monotonically with κ. Of course, the measurement strength must be non-zero in order for it be possible to stabilize the system at all. Moreover, the time-scale for the conditioned system covariance matrix V_{c} to reach its equilibrium value W is of order κ^{-1}, so in practice κ cannot be set too small. Finally, in a realistic system there will be other sources of noise. That is, D_0 will not be zero. The full solution in that case is considerably more complicated, but it is not difficult to see that in general there will be an optimal non-zero value of κ that depends upon D_0.

Pacifiability revisited. In the limit $\kappa \to 0$ as we have just been considering, $v \to 0$ and so $V_{\mathrm{ss}} - W \to 0$. That is, the feedback-stabilized system has no excess variance above the conditioned quantum state. This is what we expect for a pacifiable system, but we just showed above that the system is not pacifiable. There are a couple of ways to understand this conundrum. First, in the limit $\kappa \to 0$, the matrix F in Eq. (6.291) has one element much larger than the other. Thus to leading order $F \simeq (\sqrt{\hbar\kappa}, 0)$, so that the measurement-induced noise in the conditioned mean position is much larger than that for the conditioned mean momentum. Using this approximation, the system is pacifiable from the definition (6.127).

The second way to understand how the system is effectively pacifiable in the $\kappa \to 0$ limit is to make a rotating-wave (or secular) approximation. The weak measurement limit is the same as the rapid oscillation limit, so it makes sense to move to a rotating frame at the (unit) oscillator frequency and discard rotating terms. There are many ways to do this: using the Langevin equations (as discussed in Section 6.4.5), the Belavkin equation, or the Kalman filter. Here we do it using the Belavkin equation (the SME), which is

$$\hbar \, d\rho_c = -i[(\hat{q}^2 + \hat{p}^2)/2 + u(t)\hat{p}, \rho_c]dt + \kappa \, dt \, \mathcal{D}[\hat{q}]\rho_c + \sqrt{\hbar\kappa} \, dw(t)\mathcal{H}[\hat{q}]\rho_c. \quad (6.297)$$

On moving to the interaction frame with respect to $\hat{H}_0 = \hbar(\hat{q}^2 + \hat{p}^2)/2$ we have

$$\hbar \, d\rho_c = -i[u(t)(\hat{p}\cos t - \hat{q}\sin t), \rho_c]dt + \kappa \, dt \, \mathcal{D}[\hat{q}\cos t + \hat{p}\sin t]\rho_c$$
$$+ \sqrt{\hbar\kappa} \, dw(t)\mathcal{H}[\hat{q}\cos t + \hat{p}\sin t]\rho_c. \quad (6.298)$$

Under the secular approximation, $\mathcal{D}[\hat{q}\cos t + \hat{p}\sin t] \to \frac{1}{2}\mathcal{D}[\hat{q}] + \frac{1}{2}\mathcal{D}[\hat{p}]$. Recall from Section 6.4.5 that we cannot average oscillating terms that multiply $dw(t)$. Rather, we must consider the average of the correlation functions of $dw_1(t) = \sqrt{2} \, dw(t)\cos t$ and $dw_2(t) = \sqrt{2} \, dw(t)\sin t$, namely $dw_i(t) \, dw_j(t) = \delta_{ij} \, dt$. Similarly, we cannot assume that $u(t)$ is slowly varying and average over oscillating terms that multiply $u(t)$. Instead we should define $u_1(t) = u(t)\cos t$ and $u_2(t) = -u(t)\sin t$ (and we expect that these will have slowly varying parts). Thus we obtain the approximate SME

$$\hbar \, d\rho_c = -i[u_1(t)\hat{p} - u_2(t)\hat{q}, \rho_c]dt + (\kappa/2)dt(\mathcal{D}[\hat{q}] + \mathcal{D}[\hat{p}])\rho_c$$
$$+ \sqrt{\hbar\kappa/2} \, dw_1 \, \mathcal{H}[\hat{q}]\rho_c + \sqrt{\hbar\kappa/2} \, dw_2 \, \mathcal{H}[\hat{p}]\rho_c. \quad (6.299)$$

Exercise 6.35 *Show that for this system we have $C = \sqrt{2\kappa/\hbar} \, I$, $A = 0$, $D = (\hbar\kappa/2)I$ and $B = I$. Thus verify that $W = (\hbar/2)I$ and that the system is pacifiable.*

6.6.6 Optimal unravellings

The preceding section showed that, as a consequence of the fluctuation–observation relation, quantum feedback control differs from classical feedback control in that it is often the case that it is not optimal to increase the measurement strength without limit. However, even for a given (fixed) measurement strength, there are questions that arise in quantum control that are meaningless classically. In particular, given a linear system with dynamics described by the drift A and diffusion D matrices, what is the optimal way to monitor the bath to minimize some cost function? Classically, the unconditioned evolution described by A and D would not proscribe the measurements that can be made on the system in any way. But for quantum systems the fluctuation–observation relation means that the stationary conditioned covariance matrix W_U will be positive definite. Thus the control cost associated with the system will always be non-zero, and will depend upon the unravelling U.

Consider an asymptotic LQG problem. Then the cost to be minimized (by choice of unravelling) is

$$m = E_{ss}[h] = \text{tr}[YBQ^{-1}B^\top YW_U] + \text{tr}[YD], \qquad (6.300)$$

where Y, B, Q and D are constant matrices (independent of the unravelling U). If Y is a stabilizing solution then $YBQ^{-1}B^\top Y$ will be positive definite – if it were not then the optimal control could allow the uncertainty in some system modes to grow to infinity. Because of this the optimal solution W_U will always be found to correspond to a pure state, since that of a mixed state would necessarily give a larger value for $\text{tr}[YBQ^{-1}B^\top YW_U]$. Thus simply minimizing Eq. (6.300), subject to the condition that W_U correspond to a quantum state, will guarantee that W_U corresponds to pure state. Recall that W_U corresponds to a quantum state provided that it satisfies the LMI

$$W_U + i\hbar\Sigma/2 \geq 0. \qquad (6.301)$$

Recall also from Section 6.6.4 that there is a sufficient condition on a pure-state W_U for it to be physically realizable, namely that it satisfy the second LMI

$$AW_U + W_U A^\top + D \geq 0. \qquad (6.302)$$

Now the problem of minimizing a linear function (6.300) of a matrix (here W_U) subject to the restriction of one or more LMIs for that matrix is a well-known mathematical problem. Significantly, it can be solved numerically using the efficient technique of semi-definite programming [VB96]. This is a generalization of linear programming and a specialization of convex optimization. Note that here 'efficient' means that the execution time for the semi-definite program scales polynomially in the system size n. As pointed out earlier, an unravelling U that gives any particular permissible W_U can also be found efficiently by solving the linear matrix equation (6.268).

Example: on-threshold OPO. We now illustrate this with an example. Consider the system described in Section 6.6.4, a damped harmonic oscillator at threshold subject to dyne detection (such as homodyne or heterodyne). Since optimal performance will always be obtained for efficient detection, such detection is parameterized by the complex number v, such that $|v| \leq 1$, with the unravelling matrix given by

$$U = \frac{1}{2}\begin{pmatrix} 1 + \text{Re}\,v & \text{Im}\,v \\ \text{Im}\,v & 1 - \text{Re}\,v \end{pmatrix}. \qquad (6.303)$$

Homodyne detection of the cavity output corresponds to $v = e^{2i\theta}$, with θ the phase of the measured quadrature,

$$\hat{x}_\theta = \hat{q}\cos\theta - \hat{p}\sin\theta. \qquad (6.304)$$

That is, $\theta = 0$ corresponds to obtaining information only about q, while $\theta = \pi/2$ corresponds to obtaining information only about p. In heterodyne detection information about both quadratures is obtained equally, and $\upsilon = 0$ so that $U = I/2$.

Now let us say that the aim of the feedback control is to produce a stationary state where $q = p$ as nearly as possible. (There is no motivation behind this aim other than to illustrate the technique.) The quadratic cost function to be minimized is thus $\langle(\hat{q} - \hat{p})^2\rangle_{ss}$. That is,

$$P = \begin{pmatrix} 1 & -1 \\ -1 & 1. \end{pmatrix} \tag{6.305}$$

In this optical example it is simple to displace the system in its phase space by application of a coherent driving field. That is, we are justified in taking B to be full row rank, so that the system will be pacifiable.

Any quadratic cost function will be minimized for a pure state, so we may assume that Eq. (6.301) is saturated, with $\alpha\gamma = 1 + \beta^2$. Ignoring any control costs, we have $Q \to 0$. Thus, from Eq. (6.129), the minimum cost m achievable by optimal control is

$$m = E_{ss}[h] = \mathrm{tr}[PW_U], \tag{6.306}$$

constrained only by $0 < \gamma \le (1 - \beta^2)/2$. The minimum is found numerically to be $m^\star \approx 1.12\hbar$ at $\beta^\star \approx 0.248$ and $\gamma^\star = [1 - (\beta^\star)^2]/2$. Note that for this simple example we do not need semi-definite programming to find this optimum, but for larger problems it would be necessary.

Having found the optimal $W_U^\star$, we now use Eq. (6.268) to find the optimal unravelling:

$$U^\star = \begin{pmatrix} \cos^2\theta & \cos\theta\sin\theta \\ \cos\theta\sin\theta & \sin^2\theta \end{pmatrix} \text{ for } \theta \approx 0.278\pi. \tag{6.307}$$

This corresponds to homodyne detection with θ being the phase of the measured quadrature (θ above). Naively, since one wishes to minimize $(q - p)^2$, one might have expected that it would be optimal to obtain information only about $q - p$. That is, from Eq. (6.304), one might have expected the optimal θ to be $\pi/4$. The fact that the optimal θ is different from this points to the nontriviality of the problem of finding the optimal unravelling in general, and hence the usefulness of an efficient numerical technique for achieving it.

6.6.7 Markovian feedback

Recall from Section 6.4.5 that classically, under the conditions that there exists a stabilizing solution W, that there are no control costs, that the system is pacifiable, and that we are interested in steady-state performance only, the optimal control problem can be solved by Markovian feedback. Exactly the same analysis holds in the quantum case. The required feedback Hamiltonian is

$$\hat{H}_{fb}(t) = \hbar \hat{\mathbf{f}}^\top \hat{\mathbf{y}}(t), \tag{6.308}$$

where

$$\hat{\mathbf{f}}^{\mathsf{T}} = -\hat{\mathbf{x}}^{\mathsf{T}} \Sigma BL/\hbar. \tag{6.309}$$

Generalizing the analysis of Section 5.5, the ensemble-average evolution including the feedback is described by the master equation

$$\hbar\dot{\rho} = -\mathrm{i}[\hat{H}, \rho] + \mathcal{D}[\hat{\mathbf{c}}]\rho + \hbar\mathcal{D}[\hat{\mathbf{f}}]\rho$$
$$+ \left\{ \mathrm{i}[(\hat{\mathbf{c}}^{\mathsf{T}}, -\mathrm{i}\hat{\mathbf{c}}^{\mathsf{T}})T\rho\hat{\mathbf{f}} + \rho(\hat{\mathbf{c}}^{\dagger}, \mathrm{i}\hat{\mathbf{c}}^{\dagger})T\hat{\mathbf{f}}] + \text{H.c.} \right\}. \tag{6.310}$$

Remember that the matrix T is defined such that $TT^{\mathsf{T}} = \hbar U$. Equation (6.310) is not limited to linear systems. That is, it is valid for any $\hat{\mathbf{c}}$ with $\hat{c}_l \in \mathcal{L}(\mathbb{H})$, any $\hat{H} \in \mathcal{D}(\mathbb{H})$, any $\hat{\mathbf{f}}$ with $\hat{f}_l \in \mathcal{D}(\mathbb{H})$ and any $U \in \mathfrak{U}$ given by Eq. (6.190).

Exercise 6.36 *Referring back to Section 5.5, convince yourself of the correctness of Eq. (6.310) and show that it is of the Lindblad form.*

For linear systems, the master equation (6.310) can be turned into an OUE for the Wigner function, as could be done for the original master equation as explained in Section 6.6.2. However, just as for the original evolution (with no feedback), it is easier to calculate the evolution of $\hat{\mathbf{x}}$ in the Heisenberg picture, including the feedback Hamiltonian (6.308). The result is precisely Eq. (6.141), with hats placed on the variables. Thus the classical results for Markovian feedback all hold for the quantum case.

Under the conditions stated at the beginning of this section, it is thus clear that the optimal measurement sensitivity (if it exists) and the optimal unravelling are the same for Markovian feedback as for state-based feedback. The optimal unravelling is found by solving the semi-definite program of minimizing

$$m = \mathrm{E}_{ss}[h] = \mathrm{tr}[PW_U] \tag{6.311}$$

subject to the LMIs (6.302) and (6.301). Recall that the feedback-modified drift matrix is

$$M = A + BLC = A - W_U C^{\mathsf{T}} C - \Gamma^{\mathsf{T}} C. \tag{6.312}$$

For the example considered in the preceding section,

$$\hbar C/2 = -\Gamma = T^{\mathsf{T}}. \tag{6.313}$$

Thus $M = A - 4W_U U/\hbar + 2U$. For the optimal unravelling (6.307),

$$M^{\star} \approx \begin{pmatrix} -1.29 & -1.53 \\ 0.32 & -1.62 \end{pmatrix}. \tag{6.314}$$

Exercise 6.37 *Show that this is strictly stable, as it should be.*

Although it is natural to consider these ideal conditions under which state-based and Markovian feedback are equally effective, it is important to note that there are common circumstances for which these conditions do not hold. In particular, there are good reasons why the control matrix B might not have full row rank. If the ps and qs correspond to momenta and positions of particles, then it is easy to imagine implementing a time-dependent potential linear in the qs (i.e. a time-dependent but space-invariant force), but not so for a time-dependent Hamiltonian term linear in the ps. In such circumstances state-based feedback may be strictly superior to Markovian feedback.

This can be illustrated by the harmonic oscillator with position measurement, as considered in Section 6.6.5. Say $B = (0, 1)^\top$, describing the situation in which only a position-dependent potential can be controlled. Taking $m = \omega = 1$ as before, the feedback-modified drift matrix is

$$A' = A + BLC \tag{6.315}$$

$$= \begin{pmatrix} 0 & 1 \\ -1 & 0 \end{pmatrix} + \begin{pmatrix} 0 \\ 1 \end{pmatrix}(L)(C \quad 0) \tag{6.316}$$

$$= \begin{pmatrix} 0 & 1 \\ -1 + LC & 0. \end{pmatrix}. \tag{6.317}$$

Thus the only effect the feedback can have in this situation is to modify the frequency of the oscillator from unity to $\sqrt{1 - LC}$. It cannot damp the motion of the particle at all.

How do we reconcile this analysis with the experimental result, discussed in Section 5.8.2, demonstrating cooling of an ion using Markovian feedback? The answer lies in the secular approximation, as used in Section 6.6.5 for this sort of system. The rapid ($\nu = 1$ MHz) oscillation of the ion means that the signal in the measured current $y(t)$ also has rapid sinusoidal oscillations. In the experiment the current was filtered through a narrow ($B = 30$ kHz) band-pass filter centred at the ion's oscillation frequency. This gives rise to two currents – the cosine and the sine components of the original $y(t)$. The innovations in these currents correspond exactly to the two noise terms dw_1 and dw_2 in the SME (6.299) under the secular approximation. As shown in that section, the system in the secular approximation is pacifiable. Moreover, because the bandwidth B was much greater than the characteristic relaxation rate of the ion ($\Gamma = 400$ Hz) it is natural (in this rotating frame) to regard these current components as raw currents $y_1(t)$ and $y_2(t)$ that can be fed back directly, implementing a Markovian feedback algorithm. Thus we see that the limitations of Markovian feedback can sometimes be overcome if one is prepared to be lenient in one's definition of the term 'Markovian'.

6.7 Further reading

6.7.1 Approximations to LQG quantum feedback

No system is exactly linear, hence the LQG control theory discussed here is an idealization. Nevertheless, LQG control theory can be simply adapted to deal with even quite nonlinear systems. The optimal approach with nonlinear systems is to use the full nonlinear filtering equations (the Kushner–Stratonovich equation and Bellman equations classically, or their quantum equivalent). This is often not practical, because of the difficulty of solving these nonlinear equations in real time. Hence it is attractive to consider a suboptimal approach based on LQG control. The basic idea is to linearize the system around its mean configuration in phase-space, use LQG theory to control the system and to update one's estimates of the mean vector and covariance matrix for a short time, and then relinearize around the new (approximate) mean configuration. As long as the 'true' (i.e. optimal) conditioned system state remains approximately Gaussian, this procedure works reasonably well. Doherty *et al.* [DHJ$^+$00] demonstrate theoretically that it can be used to control a quantum particle in a double-well potential, forcing it to occupy one or the other well.

A more immediate application for quantum feedback control is in the cooling of oscillators subject to position monitoring. When a rotating-wave approximation can be made, Markovian feedback works well, as discussed in Section 6.6.7. However, for systems with a relatively low oscillation frequency ω, such that the feedback-induced damping rate is comparable to ω, state-based control such as LQG is required. The theory of feedback cooling of nano-mechanical resonators using a simplified version of LQG control was done by Hopkins *et al.* [HJHS03], and recent experiments suggest that it should be possible to implement this scheme [LBCS04].

Another example is the feedback cooling of atoms in a standing wave, as analysed by Steck *et al.* [SJM$^+$04]. In this case the dynamics is nonlinear, and linearization was used to derive approximate equations for the mean vector and covariance matrix as described above. However, in this case the feedback control signal was derived from considering the exact equations, and was a function of both the mean and the covariance (unlike with LQG control, where it is always a function of the mean only). It was shown that the atom could be cooled to within one oscillator quantum of its ground-state energy.

6.7.2 State-based quantum feedback control in finite-dimensional systems

In a series of papers [Kor01b, RK02, ZRK05], Korotkov and co-workers have considered the use of state-based quantum feedback control in a solid-state setting. They show that, by such control, Rabi oscillations of a solid-state qubit may be maintained indefinitely (although imperfectly) even in the presence of environmental noise. The basic idea is to compare the computed phase of the qubit state (as computed from the measurement results) with the time-dependent phase required for the desired Rabi oscillations, and to alter the qubit Hamiltonian in order to reduce the discrepancy. Korotkov has also shown that a more feasible algorithm, which does not involve computing the conditioned state from the measurement record, works almost as well [Kor05]. This approach, in which the observed current is filtered through a simple circuit before being fed back, is more like the current-based feedback considered in Chapter 5.

State-based control of a different two-level system, an atom, has also been considered. Here the measurement record is assumed to arrive from the spontaneous emission of the atom. Markovian feedback in this system was considered first [HHM98, WW01, WWM01]. It was shown that, by controlling the amplitude of a coherent driving field, the atom could be stabilized in almost any pure state (for efficient detection). The exceptions were states on the equator of the Bloch sphere, for which the Markovian feedback algorithm produced a completely mixed state in steady state. This deficiency can be overcome using state-based feedback [WMW02]. Moreover, it was proven rigorously (i.e. without reliance on numerical evidence from stochastic simulations) that state-based feedback is superior to Markovian feedback in the presence of imperfections such as inefficient detection or dephasing.

A final application of state-based control is in deterministic Dicke-state preparation. As discussed in Section 5.7, Markovian feedback can (in principle) achieve deterministic spin-squeezing close to the Heisenberg limit. This is so despite the fact that the approximations behind the feedback algorithm [TMW02b], which are based on linearizing about the mean spin vector and treating the two orthogonal spin components as continuous variables, break down in the Heisenberg limit. The breakdown is most extreme when the state collapses to an eigenstate of $\hat{J}_z$ (a Dicke state) with eigenvalue zero. This can be visualized as the equatorial ring around the spin-J Bloch sphere of Fig. 5.4, for which the spin vector has zero mean. Without feedback, the QND measurement alone will eventually collapse the state into a Dicke state, but one that can be neither predicted nor controlled. However, Stockton *et al.* show using stochastic simulations that state-based feedback does allow the deterministic production of a $J_z = 0$ Dicke state in the long-time limit [SvHM04].

Applications of state-based quantum feedback control in quantum information will be considered in Chapter 7.

6.7.3 Beyond state-based control

There are reasons to consider cost functions that are not additive in time. Considering the classical case to start, this means cost functions not of the form of Eq. (6.44) (a time-integral). One reason is found in 'risk-sensitive' control [Whi81, DGKF89], in which small excursions from the desired outcome are tolerated more, large excursions less. Such control tends to be more robust with respect to errors in the equations describing the system dynamics. In such cases it can be shown that $\wp_{\rm c}(\check{\mathbf{x}}; t)$ is not sufficient to specify the optimal control law. Interestingly, sometimes the optimal control law is a function of a *different* state, $\wp_{\rm c}'(\check{\mathbf{x}}; t)$. That is, the separation principle still applies, but for a state (a normalized probability distribution) that is *differently* conditioned upon $\{\mathbf{y}(t')\}_{t'=t_0}^{t'=t}$, and so is not an optimal predictor for the properties of the system. An example of a risk-sensitive cost function that yields such a state is an *exponential* of a time-integral.

James [Jam04, Jam05] recently derived a quantum equivalent to this type of control, involving a differently conditioned quantum state $\rho_{\rm c}'(t)$. Here care must be taken in defining the cost function, because system variables at different times will not commute. Considering the case of no terminal costs for simplicity, James defines the cost function to be $\langle \hat{R}(T) \rangle$, where $\hat{R}(t)$ is the solution of the differential equation

$$\frac{\mathrm{d}\hat{R}}{\mathrm{d}t} = \mu \hat{C}(t)\hat{R}(t) \tag{6.318}$$

satisfying $\hat{R}(0) = \hat{1}$. Here $\hat{C}(t) = \int_0^t \hat{h}(s)\mathrm{d}s$, where $\hat{h}(t)$ is a function of observables at time t, while $\mu > 0$ is a risk parameter. In the limit $\mu \to 0$, $[\hat{R}(T) - \hat{1}]/\mu \to \hat{C}(T)$, so the problem reduces to the usual ('risk-neutral') sort of control problem.

A useful and elegant example of risk-sensitive control is LEQG [Whi81]. This is akin to the LQG control discussed above (an example of risk-neutral control), in that it involves linear dynamics and Gaussian noise. But, rather than having a cost function that is the expectation of a time-integral of a quadratic function of system and control variables, it has a cost function that is the exponential of a time-integral of a quadratic function. This fits easily in James' formalism, on choosing $\hat{h}(s)$ to be a quadratic function of system observables and control variables (which are also observables in the quantum Langevin treatment [Jam05]). Just as for the LQG case, many results from classical LEQG theory follow over to quantum LEQG theory [Yam06]. This sort of risk-sensitive control is particularly useful because the linear dynamics (in either LQG or LEQG) is typically an approximation to the true dynamics. Because risk-sensitive control avoids large excursions, it can ensure that the system does not leave the regime where linearization is a good approximation. That is, the risk-sensitive nature of the control helps ensure its validity.

A different approach to dealing with uncertainties in the dynamics of systems is the robust estimator approach adopted by Yamamoto [Yam06]. Consider quantum LQG control, but with bounded uncertainties in the matrices A and C. Yamamoto finds a non-optimal linear filter such that the mean square of the estimation error is guaranteed to be within a certain bound. He then shows by example that linear feedback based on this robust observer results in stable behaviour in situations in which both standard (risk-neutral) LQG and (risk-sensitive) LEQG become unstable. Yet another approach to uncertainties in dynamical parameters is to describe them using a probability distribution. One's knowledge of these parameters is then updated simultaneously, and in conjuction, with one's knowledge of the system. The interplay between knowledge about the system and knowledge about its dynamics leads to a surprising range of behaviour under different unravellings. This is investigated for a simple quantum system (resonance fluorescence with an uncertain Rabi frequency) in Ref. [GW01].

7

Applications to quantum information processing

7.1 Introduction

Any technology that functions at the quantum level must face the issues of measurement and control. We have good reasons to believe that quantum physics enables communication and computation tasks that are either impossible or intractable in a classical world [NC00]. The security of widely used classical cryptographic systems relies upon the difficulty of certain computational tasks, such as breaking large semi-prime numbers into their two prime factors in the case of RSA encryption. By contrast, quantum cryptography can be *absolutely* secure, and is already a commercial reality. At the same time, the prospect of a quantum computer vastly faster than any classical computer at certain tasks is driving an international research programme to implement quantum information processing. Shor's factoring algorithm would enable a quantum computer to find factors exponentially faster than any known algorithm for classical computers, making classical encryption insecure. In this chapter, we investigate how issues of measurement and control arise in this most challenging quantum technology of all, quantum computation.

The subjects of information theory and computational theory at first sight appear to belong to mathematics rather than physics. For example, communication was thought to have been captured by Shannon's abstract theory of information [SW49, Sha49]. However, physics must impact on such fundamental concepts once we acknowledge the fact that information requires a physical medium to support it. This is a rather obvious point; so obvious, in fact, that it was only recently realized that the conventional mathematics of information and computation are based on an implicit classical intuition about the physical world. This intuition unnecessarily constrains our view of what tasks are tractable or even possible.

Shannon's theory of information and communication was thoroughly grounded in classical physics. He assumed that the fundamental unit of information is a classical 'bit', which is definitely either in state 'zero' or in state 'one', and that the process of sending bits through channels could be described in an entirely classical way. This focus on the classical had important practical implications. For example, in 1949 Shannon used his formulation of information theory to 'prove' [Sha49] that it is impossible for two parties to communicate with perfect privacy, unless they have pre-shared a random key as long as the message they wish to communicate.

Insofar as Shannon's theory is concerned, any physical quantity that can take one of two distinct values can support a bit. One physical instantiation of a bit is as good as any other – we might say that bits are *fungible*. Clearly, bits can exist in a quantum world. There are many quantum systems that are adequate to the task: spin of a nucleus, polarization of a photon, any two stationary states of an atom etc., but, as the reader well knows, there is a big difference between a classical bit and a two-level quantum system: the latter can be in an arbitrary *superposition* of its two levels.

One might think that such a superposition is not so different from a classical bit in a mixture, describing a lack of certainty as to whether it is in state zero or one, but actually the situations are quite different. The entropy of the classical state corresponding to an uncertain bit value is non-zero, whereas the entropy of a pure quantum superposition state is zero. To capture this difference, Schumacher coined the term *qubit* for a quantum bit [Sch95]. Like bits, qubits are fungible and we can develop quantum information theory without referring to any particular physical implementation. This theory seeks to establish abstract principles for communication and computational tasks when information is encoded in qubits. For a thorough introduction to this subject we refer the reader to the book by Nielsen and Chuang [NC00].

It will help in what follows to state a few definitions. In writing the state of a qubit, we typically use some preferred orthonormal basis, which, as in Chapter 1, we denote $\{|0\rangle, |1\rangle\}$ and call the *logical basis* or *computational basis*. The qubit Hilbert space could be the entire Hilbert space of the system or just a two-dimensional subspace of the total Hilbert space. In physical terms, the logical basis is determined by criteria such as ease of preparation, ease of measurement and isolation from sources of decoherence (as in the pointer basis of Section 3.7). For example, if the qubit is represented by a spin of a spin-half particle in a static magnetic field, it is convenient to regard the computational basis as the eigenstates of the component of spin in the direction of the field, since the spin-up state can be prepared to a good approximation by allowing the system to come to thermal equilibrium in a large enough magnetic field. If the physical system is a mesoscopic superconducting system (see Section 3.10.2), the computational basis could be two distinct charge states on a superconducting island, or two distinct phase states, or some basis in between these. A charge qubit is very difficult to isolate from the environment and thus it may be preferable to use the phase basis. On the other hand, single electronics can make the measurement of charge particularly easy. In all of these cases the qubit Hilbert space is only a two-dimensional subspace of an infinite-dimensional Hilbert space describing the superconducting system.

Once the logical basis has been fixed, we can specify three Pauli operators, X, Y and Z, by their action on the logical states $|z\rangle$, $z \in \{0, 1\}$:

$$Z|z\rangle = (-1)^z |z\rangle, \tag{7.1}$$

$$Y|z\rangle = i(-1)^z |1 - z\rangle, \tag{7.2}$$

$$X|z\rangle = |1 - z\rangle. \tag{7.3}$$

Here, we are following the convention common in the field of quantum information [NC00]. Note the different notation from what we have used previously (see Box 3.1) of $\hat{\sigma}_x$, $\hat{\sigma}_y$ and $\hat{\sigma}_z$. In particular, here we do not put hats on X, Y and Z, even though they are operators. When in this chapter we do use $\hat{X}$ and $\hat{Y}$, these indicate operators with continuous spectra, as in earlier chapters. Another convention is to omit the tensor product between Pauli operators. Thus, for a two-qubit system, ZX means $Z \otimes X$. Note that the square of any Pauli operator is unity, which we denote I.

This chapter is structured as follows. Section 7.2 introduces a widely used primitive of quantum information processing: teleportation of a qubit. This involves discrete (in time) measurement and feedforward. In Section 7.3, we consider the analogous protocol for variables with continuous spectra. In Section 7.4, we introduce the basic ideas of quantum errors, and how to protect against them by quantum encoding and error correction. In Section 7.5 we relate error correction to the quantum feedback control of Chapter 5 by considering continuously detected errors. In Section 7.6 we consider the conventional error model (i.e. undetected errors), but formulate the error correction as a control problem with continuous measurement and Hamiltonian feedback. In Section 7.7 we consider the same problem (continuous error correction) but without an explicit measurement step; that is, we treat the measurement and control apparatus as a physical system composed of a small number of qubits. In Section 7.8 we turn to quantum computing, and show that discrete measurement and control techniques can be used to engineer quantum logic gates in an optical system where the carriers of the quantum information (photons) do not interact. In Section 7.9, we show that this idea, called linear optical quantum computation, can be augmented using techniques from continuous measurement and control. In particular, adaptive phase measurements allow one to create, and perform quantum logic operations upon, qubits comprising arbitrary superpositions of zero and one photon. We conclude as usual with suggestions for further reading.

7.2 Quantum teleportation of a qubit

We begin with one of the protocols that set the ball rolling in quantum information: quantum teleportation of a qubit [BBC$^+$93]. This task explicitly involves both quantum measurement and control. It also requires an entangled state, which is shared by two parties, the sender and the receiver. The sender, Alice, using only classical communication, must send an unknown qubit state to a distant receiver, Bob. She can do this in such a way that neither of them learns anything about the state of the qubit. The protocol is called teleportation because the overall result is that the qubit is transferred from Alice to Bob even though there is no physical transportation of any quantum system from Alice to Bob. It is illustrated in Fig. 7.1 by a *quantum circuit diagram*, the first of many in this chapter.

7.2.1 The protocol

The key resource (which is consumed) in this quantum teleportation protocol is the bipartite entangled state. Alice and Bob initially each have one qubit of a two-qubit maximally

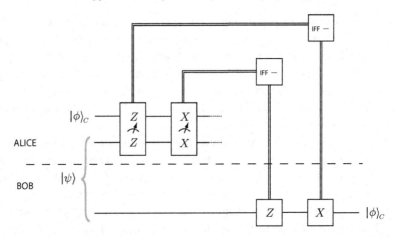

Fig. 7.1 A quantum circuit diagram for quantum teleportation of an arbitrary qubit state $|\phi\rangle_C$ from Alice to Bob, using an entangled Bell state $|\psi\rangle$ shared by Alice and Bob. The single lines represent quantum information in qubits, with time increasing from left to right. The two boxes containing dials represent a measurement of the operator contained within (ZZ and XX, respectively), with possible outcomes ± 1. The double lines represent classical bits: the outcomes of the measurements and the controls which implement (or not) the quantum gates X and Z, respectively. For details see the text.

entangled state such as

$$|\psi\rangle = \frac{1}{\sqrt{2}}(|0\rangle_A|0\rangle_B + |1\rangle_A|1\rangle_B). \tag{7.4}$$

This is often known as a Bell state, because of the important role such states play in Bell's theorem [Bel64] (see Section 1.2.1). In addition, Alice has in her possession another qubit, which we will refer to as the *client*, prepared in an arbitrary state (it could even be entangled with other systems). For ease of presentation, we will assume that the client qubit is in a pure state

$$|\phi\rangle_C = \alpha|0\rangle_C + \beta|1\rangle_C. \tag{7.5}$$

This state is unknown to Alice and Bob; it is known only to the client who has entrusted it to Alice for delivery to Bob. The total state of the three systems is then

$$|\Psi\rangle = \frac{1}{\sqrt{2}}(\alpha|0\rangle_C + \beta|1\rangle_C)(|0\rangle_A|0\rangle_B + |1\rangle_A|1\rangle_B). \tag{7.6}$$

At this stage of the protocol, Alice has at her location two qubits, the client qubit, in an unknown (to her) state, and one of an entangled pair of qubits. The other entangled qubit is held at a distant location by Bob.

The next stage requires Alice to measure two physical quantities, represented by commuting operators, on her two qubits. These quantities are joint properties of her two qubits, with operators $Z_A \otimes Z_C$ and $X_A \otimes X_C$.

Exercise 7.1 *Show that these operators commute, that they both have eigenvalues ±1 and that the simultaneous eigenstates are*

$$\sqrt{2}|+;+\rangle = |00\rangle + |11\rangle, \tag{7.7}$$

$$\sqrt{2}|+;-\rangle = |00\rangle - |11\rangle, \tag{7.8}$$

$$\sqrt{2}|-;+\rangle = |01\rangle + |10\rangle, \tag{7.9}$$

$$\sqrt{2}|-;-\rangle = |01\rangle - |10\rangle. \tag{7.10}$$

Here the first ± label refers to the eigenvalue for ZZ and the second ± label to the eigenvalue of XX, and the order of the qubits is AC as above.

This is known as a Bell measurement, because the above eigenstates are Bell states.

On rewriting the state of the three qubits, Eq. (7.6), in terms of these eigenstates for qubits A and C, we find

$$|\Psi\rangle = \frac{1}{2}[|+;+\rangle(\alpha|0\rangle_B + \beta|1\rangle_B) + |+;-\rangle(\alpha|0\rangle_B - \beta|1\rangle)$$

$$+ |-;+\rangle(\alpha|1\rangle_B + \beta|0\rangle_B) + |-;-\rangle(-\alpha|1\rangle_B + \beta|0\rangle_B)]. \tag{7.11}$$

Exercise 7.2 *Verify this.*

Remember that $|+;+\rangle$ etc. refer to entangled states of the A and C qubits held locally by Alice. It is now clear that the four possible results for the two joint measurements that Alice must make are equally probable. The results of Alice's measurement thus give two bits of information. Furthermore, we can simply read off the conditional state of Bob's qubit. For example, if Alice obtains the result $(+;+)$ then Bob's qubit must be in the state $|\phi\rangle_B$. That is, it is in the same state as the original client qubit held by Alice. Of course, until Bob knows the outcome of Alice's measurement he cannot describe the state of his qubit in this way. Meanwhile Alice's final state is unrelated to the original state of the client qubit because the Bell measurement is a *complete* measurement (see Section 1.4.2).

The final step of the protocol requires Alice to send the results of her measurements to Bob by *classical communication* (e.g. telephone or email). Once Bob has this information, he may, using a local unitary transformation conditional on Alice's results, transform his qubit into the same state as the original client qubit held by Alice. As we have seen, if Alice gets the result $(+;+)$, then Bob need do nothing. If Alice gets the result $(-;+)$, then Bob, upon receiving this information, should act upon his local system with the unitary transformation X_B to change his state into $|\phi\rangle_B$. Similarly, if Alice gets $(+;-)$ then Bob should act with Z_B, and if $(-;-)$, then with $Y_B \propto Z_B X_B$. At no time does Alice or Bob learn anything about the state of the client system; as shown above, the results of Alice's measurement are completely random. Note also that the communication from Alice to Bob is limited to the speed of light, so the teleportation protocol does not transfer the quantum state faster than light.

Exercise 7.3 *Suppose the client state is itself entangled with another system, Q. Convince yourself that, after teleportation, this will result in Bob's qubit being entangled in the same way with Q.*

Clearly the teleportation protocol just described is just a rather simple form of measurement-based control in which the results of measurement upon a part of the total system are used to effect a local unitary transformation on another part of the system. While Alice and Bob share entangled qubits they must always be regarded as acting on a single quantum system, no matter how distant they are in space. Only at the end of the protocol can Bob's qubit be regarded as an independent quantum system.

7.2.2 A criterion for demonstrating qubit teleportation

In any real experiment, every part of the teleportation will be imperfect: the preparation of the entanglement resource, Alice's measurement and Bob's control. As a result, the teleportation will not work perfectly, so Bob will end up with a state ρ different from the desired state $|\phi\rangle\langle\phi|$. The quality of the teleportation can be quantified by the *fidelity*,

$$F = \langle\phi|\rho|\phi\rangle, \tag{7.12}$$

which is the probability for the client to find Bob's system in the desired state $|\phi\rangle$, if he were to check.

Exercise 7.4 *Show that $F = 1$ iff $\rho = |\phi\rangle\langle\phi|$.*

How much less than unity can the fidelity be before we stop calling this process *quantum teleportation*? To turn the question around, what is the maximum fidelity that can be obtained without using a quantum resource (i.e. an entangled state).

It turns out that the answer to this question hangs on what it means to say that the client state is *unknown* to Alice and Bob. One answer to this question has been given by Braunstein *et al.* [BFK00] by specifying the ensemble from which client states are drawn. To make Alice's and Bob's task as difficult as possible, we take the ensemble to weight all pure states equally.

Exercise 7.5 *Convince yourself that the task of Alice and Bob is easier if any other ensemble is chosen. In particular, if the ensemble comprises two orthogonal states (known to Alice and Bob), show that they can achieve a fidelity of unity without any shared entangled state.*

We may parameterize qubit states on the Bloch sphere (see Box 3.1) by $\Omega = (\theta, \phi)$ according to

$$|\Omega\rangle = \cos\left(\frac{\theta}{2}\right)|0\rangle + e^{i\phi}\sin\left(\frac{\theta}{2}\right)|1\rangle. \tag{7.13}$$

The uniform ensemble of pure states then has the probability distribution $[1/(4\pi)]d\Omega = [1/(4\pi)]d\phi \sin\theta\, d\theta$.

For this ensemble, there are various ways of achieving the best possible classical teleportation (that is, without using entanglement). One way is for Alice to measure Z_A and tell Bob the result, and for Bob to prepare the corresponding eigenstate. From Eq. (7.13), the probabilities for Alice to obtain the results ± 1 are $\cos^2(\theta/2)$ and $\sin^2(\theta/2)$, respectively. Thus, the state that Bob will reconstruct is, on average,

$$\rho_\Omega = \cos^2\left(\frac{\theta}{2}\right)|0\rangle\langle 0| + \sin^2\left(\frac{\theta}{2}\right)|1\rangle\langle 1|. \tag{7.14}$$

Exercise 7.6 *Show that the same state results if Alice and Bob follow the quantum teleportation protocol specified in Section 7.2.1, but with their entangled state $|\psi\rangle$ replaced by the classically correlated state*

$$\rho_{AB} = \frac{1}{2}(|00\rangle\langle 00| + |11\rangle\langle 11|). \tag{7.15}$$

From Eq. (7.14), the *average fidelity* for the teleportation is

$$\bar{F} = \int \frac{d\Omega}{4\pi} \langle\Omega|\rho_\Omega|\Omega\rangle. \tag{7.16}$$

Exercise 7.7 *Show that this integral evaluates to $2/3$.*

Thus if, in a series of experimental runs, we find an average fidelity greater than 0.67, we can be confident that some degree of entanglement was present in the resource used and that the protocol used was indeed quantum. This has now been demonstrated in a number of different experimental settings, perhaps most convincingly using trapped-ion qubits, with Wineland's group achieving a fidelity of 0.78 [BCS+04].

As it stands, teleporting qubits, while certainly a fascinating aspect of quantum information theory, does not seem enormously useful. After all, it requires shared entanglement, which requires a quantum channel between the two parties to set up. If that quantum channel can be kept open then it would be far easier to send the qubit directly down that channel, rather than teleporting it. However, we will see in Section 7.8.3 that quantum teleportation has an essential role in the field of measurement-based quantum computing.

7.3 Quantum teleportation for continuous variables

Entangled qubit states are a particularly simple way to see how teleportation works. However, we can devise a teleportation protocol for quantum systems of any dimension, even infinite-dimensional ones. In fact, the infinite-dimensional case is also simple to treat [Vai94], and has also been demonstrated experimentally [FSB+98]. It is usually referred to as continuous-variable (CV) quantum teleportation, because operators with continuous spectra play a key role in the protocol. As noted in the introduction, here we denote such operators as $\hat{X}$ or $\hat{Y}$, with the hats to differentiate these from Pauli operators.

7.3.1 The ideal protocol

The basic procedure is the same as in the qubit case. Alice has an unknown (infinite-dimensional) client state $|\phi\rangle_C$ and she shares with Bob an entangled state $|\psi\rangle_{AB}$. For perfect teleportation of an arbitrary state $|\phi\rangle_C$, the state $|\psi\rangle_{AB}$ must contain an infinite amount of entanglement (see Section A.2.2). Let us define CV quadrature operators $\hat{X}_A$ and $\hat{Y}_A$ for Alice, and similarly for Bob and for the client. These obey

$$[\hat{X}_\nu, \hat{Y}_\mu] = 2i\delta_{\nu,\mu} \text{ for } \nu, \mu \in \{A, B, C\}, \tag{7.17}$$

and are assumed to form a complete set of observables (see Section 6.6). This allows us to define a particularly convenient choice of entangled state for Alice and Bob:

$$|\psi\rangle_{AB} = e^{-i\hat{Y}_A\hat{X}_B/2}|X := X_0\rangle_A|Y := Y_0\rangle_B. \tag{7.18}$$

Here we are following our usual convention so that $|X := X_0\rangle_A$ is the eigenstate of $\hat{X}_A$ with eigenvalue X_0 etc.

Exercise 7.8 *Show that* $|\psi\rangle_{AB}$ *is a joint eigenstate of* $\hat{X}_A - \hat{X}_B$ *and* $\hat{Y}_A + \hat{Y}_B$, *with eigenvalues* X_0 *and* Y_0.
Hint: *First show that* $e^{i\hat{Y}_A\hat{X}_B/2}\hat{X}_A e^{-i\hat{Y}_A\hat{X}_B/2} = \hat{X}_A + \hat{X}_B$, *using Eq. (2.109).*

For the case $X_0 = Y_0 = 0$ (as we will assume below), this state is known as an EPR state, because it was first introduced in the famous paper by Einstein, Podolsky and Rosen [EPR35]. Note that the entanglement in this state is also manifest in correlations between other pairs of observables, such as number and phase [MB99].

In the protocol for teleportation based on this state, Alice now makes joint measurements of $\hat{X}_C - \hat{X}_A$ and $\hat{Y}_C + \hat{Y}_A$ on the two systems in her possession. This yields two real numbers, X and Y, respectively. The conditional state resulting from this joint quadrature measurement is described by the projection onto the state $e^{-i\hat{Y}_C\hat{X}_A/2}|X\rangle_C|Y\rangle_A$. Thus the conditioned state of Bob's system is

$$|\phi^{XY}\rangle_B \propto {}_C\langle X|_A\langle Y|e^{i\hat{Y}_C\hat{X}_A/2}e^{-i\hat{Y}_A\hat{X}_B/2}|\phi\rangle_C|X := 0\rangle_A|Y := 0\rangle_B. \tag{7.19}$$

Calculating this in the eigenbasis of $\hat{X}_B$ gives

$$_B\langle x|\phi^{XY}\rangle_B \propto e^{-iYx/2}{}_C\langle X + x|\phi\rangle_C. \tag{7.20}$$

Exercise 7.9 *Show this, by first using the Baker–Campbell–Hausdorff theorem (A.118) to show that*

$$e^{i\hat{Y}_C\hat{X}_A/2}e^{-i\hat{Y}_A\hat{X}_B/2} = e^{-i\hat{Y}_A\hat{X}_B/2}e^{i\hat{Y}_C\hat{X}_A/2}e^{i\hat{Y}_C\hat{X}_B/2} \tag{7.21}$$

and then recalling that $e^{-i\hat{Y}x/2}|X\rangle = |X + x\rangle$.

Using the last part of this exercise a second time, we can write Eq. (7.20) in a basis-independent manner as

$$|\phi^{XY}\rangle_B \propto e^{-iY\hat{X}_B/2}e^{iX\hat{Y}_B/2}|\phi\rangle_B. \tag{7.22}$$

Thus, up to two simple unitary transformations (displacements of the canonical variables), the conditional state of Bob's system is the same as the initial unknown client state. If Alice now sends the results of her measurements (X, Y) to Bob, the two unitary transformations can be removed by local operations that correspond to a displacement in phase-space of $\hat{X}$ by X and of $\hat{Y}$ by Y. Thus the initial state of the client has successfully been teleported to Bob. Note that in this case an infinite amount of information must be communicated by Alice, because X and Y are two real numbers. It is not difficult to verify that, just as in the qubit case, no information about $|\phi\rangle$ is contained in this communication.

This whole procedure is actually far simpler to see in the Heisenberg picture. Alice measures $\hat{X}_C - \hat{X}_A = \hat{X}$ and $\hat{Y}_C + \hat{Y}_A = \hat{Y}$. Note that here we are using the operators $\hat{X}$ and $\hat{Y}$ to denote the measurement results (see Section 1.3.2), but in the EPR state, $\hat{X}_A = \hat{X}_B$ and $\hat{Y}_A = -\hat{Y}_B$. Therefore Alice knows $\hat{X} = \hat{X}_C - \hat{X}_B$ and $\hat{Y} = \hat{Y}_C - \hat{Y}_B$. When she sends this information to Bob, Bob translates $\hat{X}_B \to \hat{X}'_B = \hat{X}_B + \hat{X} = \hat{X}_C$ and $\hat{Y}_B \to \hat{Y}'_B = \hat{Y}_B + \hat{Y} = \hat{Y}_C$. Now, since $\hat{X}_B$ and $\hat{Y}_B$ are by assumption a complete set of observables, all operators for Bob's system can be written as functions of $\hat{X}_B$ and $\hat{Y}_B$. Thus, since Bob's new observables are the same as the original observables for the client, it follows that all properties of Bob's system are the same as those of the client's original system. In other words, the client's system has been teleported to Bob.

Exercise 7.10 *Perform a similar Heisenberg-picture analysis for the case of qubit teleportation.*
Hint: *You may find the operation $\oplus$ (see Section 1.3.3) useful.*

7.3.2 A more realistic protocol

The EPR state (7.18) is not a physical state because it has infinite uncertainty in the local quadratures. Thus, if it were realized as a state for two harmonic oscillators, it would contain infinite energy. In a realistic protocol we must use a state with finite mean energy. In an optical setting, an approximation to the EPR state is the two-mode squeezed vacuum state [WM94a]. This is an entangled state for two modes of an optical field. It is defined in the number eigenstate basis for each oscillator as

$$|\psi\rangle_{AB} = \sqrt{1 - \lambda^2} \sum_{n=0}^{\infty} \lambda^n |n\rangle_A \otimes |n\rangle_B, \tag{7.23}$$

where $\lambda \in [0, 1)$. This state is generated from the ground (vacuum) state $|0, 0\rangle$ by the unitary transformation

$$\hat{U}(r) = e^{r(\hat{a}^\dagger \hat{b}^\dagger - \hat{a}\hat{b})}, \tag{7.24}$$

where $\lambda = \tanh r$ and $\hat{a}$ and $\hat{b}$ are the annihilation operators for Alice's and Bob's mode, respectively. Compare this with the unitary transformation defining the one-mode squeezed state (A.103).

The two-mode squeezed state (7.23) approximates the EPR state in the limit $\lambda \to 1$ ($r \to \infty$). This can be seen from the expression for $|\psi\rangle_{AB}$ in the basis of $\hat{X}_A$ and $\hat{X}_B$:

$$\psi(x_A, x_B) = {}_A\langle x_A|_B\langle x_B|\psi\rangle \tag{7.25}$$

$$= (2\pi)^{-1/2} \exp\left[-\frac{e^{2r}}{8}(x_A - x_B)^2 - \frac{e^{-2r}}{8}(x_A + x_B)^2\right]. \tag{7.26}$$

This should be compared with the corresponding equation for the EPR state (7.18),

$$\psi(x_A, x_B) \propto {}_A\langle x_A|_B\langle x_B|e^{-i\hat{Y}_A\hat{X}_B/2}|X := 0\rangle_A|Y := 0\rangle_B \tag{7.27}$$

$$\propto \delta(x_A - x_B). \tag{7.28}$$

From Eq. (7.26) it is not difficult to show that

$$\text{Var}(\hat{X}_A - \hat{X}_B) = \text{Var}(\hat{Y}_A + \hat{Y}_B) = 2e^{-2r}, \tag{7.29}$$

so that in the limit $r \to \infty$ the perfect EPR correlations are reproduced. This result can be more easily derived in a pseudo-Heisenberg picture.

Exercise 7.11 *Consider the unitary operator $\hat{U}(r)$ as an evolution operator, with r as a pseudo-time. Show that, in the pseudo-Heisenberg picture,*

$$\frac{\partial}{\partial r}(\hat{X}_A - \hat{X}_B) = -(\hat{X}_A - \hat{X}_B), \tag{7.30}$$

$$\frac{\partial}{\partial r}(\hat{Y}_A + \hat{Y}_B) = -(\hat{Y}_A + \hat{Y}_B). \tag{7.31}$$

Hence, with $r = 0$ corresponding to the vacuum state $|0, 0\rangle$, show that, in the state (7.23), the correlations (7.29) result.

If we use the finite resource (7.23), but follow the same teleportation protocol as for the ideal EPR state, the final state for Bob is still pure, and has the wavefunction (in the X_B representation)

$$\phi_B^{(X,Y)}(x) = \int_{-\infty}^{\infty} dx' \, e^{-\frac{i}{2}x'Y} \psi(x, x')\phi_C(X + x'), \tag{7.32}$$

where $\phi_C(x)$ is the wavefunction for the client state and $\psi(x, x')$ is given by Eq. (7.26). Clearly in the limit $r \to \infty$ the teleportation works as before.

Exercise 7.12 *Show that when the client state is an oscillator coherent state $|\alpha\rangle$, with $\alpha \in \mathbb{R}$, the teleported state at B is*

$$\phi_B^{(X,Y)}(x) = (2\pi)^{-1/4} \exp\left[-\frac{1}{4}(x - \tanh r(2\alpha - X))^2 - \frac{ixY}{2}\tanh r\right]. \tag{7.33}$$

For finite squeezing the state at B is not (even after the appropriate displacements in phase space) an exact replica of the client state. We are interested in the *fidelity*,

$$F = |\langle \phi|e^{\frac{i}{2}gY\hat{X}_B}e^{-\frac{i}{2}gX\hat{Y}_B}|\phi^{(X,Y)}\rangle|^2. \tag{7.34}$$

In the ideal teleportation $g = 1$, but here we allow for the *gain* g to be non-unity. For finite squeezing, it is in fact usually optimal to choose $g \neq 1$.

Exercise 7.13 *Show that for the client state a coherent state, $|\alpha\rangle$, the optimal choice of the gain is $g = \tanh r$, in which case the fidelity is given by*

$$F = e^{-(1-g)^2|\alpha|^2}. \tag{7.35}$$

7.3.3 A criterion for demonstrating CV teleportation

In a real experiment, the fidelity is going to be less than Eq. (7.35), because of imperfections in the preparation, measurement and control. Just as in the qubit case, we are interested in the following question: what is the minimum average fidelity (over some ensemble of client states) for the protocol to be considered quantum? In this case, we will consider the ensemble of all possible coherent states, because these are easy states to generate and it allows one to obtain an analytical result [BK98].

Suppose A and B share no entanglement at all. In that case the best option for A is to make a simultaneous measurement of $\hat{X}_C$ and $\hat{Y}_C$, obtaining the results X and Y [BK98, HWPC05]. These are then sent to B over a classical noiseless channel. Upon obtaining the results, B can displace an oscillator ground state to produce the coherent state $|\beta\rangle$, with $\beta = (X + iY)/2$. Of course, from run to run β fluctuates, so the state that describes experiments over many runs is actually

$$\rho_B = \int d^2\beta \, \wp(\beta)|\beta\rangle\langle\beta|. \tag{7.36}$$

As discussed in Example 4 in Section 1.2.5,[1] the probability distribution for β when the client state is a coherent state, $|\alpha\rangle$ is

$$\wp(\beta) = \frac{1}{\pi}|\langle\alpha|\beta\rangle|^2 = \frac{1}{\pi}e^{-|\alpha-\beta|^2}. \tag{7.37}$$

From this expression it is clear that the fidelity is the same for all coherent states under this classical protocol. Thus the average fidelity would then be given by

$$\bar{F} = \langle\alpha|\left[\int d^2\beta \, \wp(\beta)|\beta\rangle\langle\beta|\right]|\alpha\rangle \tag{7.38}$$

$$= \int \frac{d^2\beta}{\pi} e^{-2|\alpha-\beta|^2}. \tag{7.39}$$

Exercise 7.14 *Show that this evaluates to give $\bar{F} = \frac{1}{2}$.*
Hint: *For simplicity set $\alpha = 0$ and write β in polar coordinates.*

Thus $\bar{F} = 0.5$ is the classical boundary for teleportation of a coherent state. To be useful, a quantum protocol would need to give an average fidelity greater than 0.5.

---

[1] Note the difference in definition of the quadratures by a factor of $\sqrt{2}$ between this chapter and the earlier chapter.

Strictly, it would be impossible to demonstrate an average fidelity greater than 0.5 for the coherent-state ensemble using the quantum teleportation protocol of Section 7.3.2. The reason for this is that for that protocol the teleportation fidelity depends upon the coherent amplitude $|\alpha|$ as given by Eq. (7.35). Because this decays exponentially with $|\alpha|^2$, if one averaged over the entire complex (α) plane, one would obtain a fidelity close to zero. In practice (as discussed in the following subsection) only a small part of the complex plane near the vacuum state ($|\alpha| = 0$) was sampled. For a discussion of how decoherence of the entangled resource due to phase fluctuations will affect Eq. (7.35), see Ref. [MB99]. For a discussion of other criteria for characterizing CV quantum teleportation, see Refs. [RL98, GG01].

7.3.4 Experimental demonstration of CV teleportation

Quantum teleportation of optical coherent states was first demonstrated by the group of Kimble [FSB$^+$98]. In order to understand the experiment we must consider how some of the formal steps in the preceding analysis are done in the laboratory. The initial entangled resource $|\psi\rangle_{AB}$, shared by the sender Alice and the receiver Bob, is a two-mode optical squeezed state as discussed above. To a very good approximation such states are produced in the output of non-degenerate parametric down-conversion using a crystal with a second-order optical polarizability [WM94a].

The joint measurement of the quadrature operators $\hat{X}_C - \hat{X}_A$, $\hat{Y}_C + \hat{Y}_A$ on the client and sender mode can be thought of as coupling modes A and C using a unitary transformation followed by a measurement of the quadratures $\hat{X}_C$ and $\hat{Y}_A$. In quantum optics the coupling can be simply effected using a 50:50 beam-splitter. This is represented by the unitary transformation

$$U_{bs} \begin{pmatrix} \hat{X}_A \\ \hat{Y}_A \\ \hat{X}_C \\ \hat{Y}_C \end{pmatrix} U_{bs}^\dagger = \frac{1}{\sqrt{2}} \begin{pmatrix} \hat{X}_A - \hat{X}_C \\ \hat{Y}_A - \hat{Y}_C \\ \hat{X}_C + \hat{X}_A \\ \hat{Y}_C + \hat{Y}_A \end{pmatrix}. \tag{7.40}$$

From this it is clear that the post-beam-splitter quadrature measurements described above are equivalent to a pre-beam-splitter measurement of $\hat{X}_C - \hat{X}_A$ and $\hat{Y}_C + \hat{Y}_A$. These quadratures can be measured using homodyne detection, as discussed in Section 4.7.6. Such measurements of course absorb all of the light, leaving Alice with only classical information (the measurement results). In a realistic device, inefficiency and dark noise introduce extra noise into these measurements, as discussed in Section 4.8.

On receipt of Alice's measurement results, Bob must apply the appropriate unitary operator, a displacement, to complete the protocol. Displacements are easy to apply in quantum optics using another mode, prepared in a coherent state with large amplitude, and a beam-splitter with very high reflectivity for mode B. This is is discussed in Section 4.4.1. In the experiment the two modes used were actually at different frequencies, and the role of the beam-splitter was played by an electro-optic modulator (EOM), which coherently mixes light at the two frequencies. The phase and amplitude of the modulation in the EOM were determined appropriately using Alice's results.

The experiment included an additional step to verify to what extent the state received by Bob faithfully reproduced the state of the client field. In this experiment the state of the client was a coherent state. In essence another party, Victor, is verifying the fidelity of the teleportation. This was done using homodyne detection to monitor the quadrature variances of the teleported state. Since the experiment dealt with broad-band fields the single-mode treatment we have discussed must be extended to deal with this situation. Without going into details, the basic technique is simple to understand. The noise power spectrum of the homodyne current obtained by Victor directly measures the quadrature operator variances as a function of frequency. (See Section 5.2.1.) Thus at any particular frequency Victor effectively selects a single mode.

The key feature that indicates success of the teleportation is a drop in the quadrature variance seen by Victor when Bob applies the appropriate displacement to his state. This is done by varying the gain g. If Bob does nothing to his state ($g = 0$), Victor gets one half of a two-mode squeezed state. Such a state has a quadrature noise level well above the vacuum level of the coherent state that the parties are trying to teleport. As Bob varies his gain, Victor will see the quadrature noise level fall until at precisely the right gain the teleportation is effected and the variance falls to the vacuum level of a coherent state (in the limit of high squeezing). In reality, the finite squeezing in the entangled state as well as extra sources of noise introduced in the detectors and control circuits will make the minimum higher than this. Furusawa *et al.* [FSB$^+$98] observed a minimum quadrature variance of 2.23 ± 0.03 times the vacuum level. This can be shown to correspond to a fidelity of $F = 0.58 \pm 0.02$. As discussed previously, this indicates that entanglement is an essential part of the protocol.

7.4 Errors and error correction

Information storage, communication and processing are physical processes and are subject to corruption by process noise. In the quantum case, this corruption can be identified with decoherence, as has been discussed in detail in Chapter 3. We say that noise or decoherence introduces *errors* into the information. A major part of the field of quantum information is methods to deal with such errors, most notably *quantum error correction*. In this section we introduce some of the basic concepts of quantum errors and error correction.

7.4.1 Types of quantum errors

To begin, consider errors in a classical bit B. To introduce errors, we couple the bit to another system also described by a binary variable Ξ, which we will refer to as the environment.

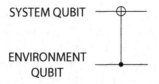

Fig. 7.2 Circuit diagram for a C-NOT interaction, which here represents the interaction of a two-state system with a two-state environment. In this case the value of the environment bit controls (·) a bit-flip error (⊕) on the target (the system bit). That is, if the environment bit has value 1, the value of the system bit changes, otherwise nothing happens. As discussed later, the same interaction or 'gate' can be applied to quantum bits, and this figure follows the conventions of Fig. 7.1.

Let the nature of the coupling be such as to transform the variables according to

$$B \to B \oplus \Xi, \tag{7.41}$$

$$\Xi \to \Xi. \tag{7.42}$$

The state of the environment is specified by the probability distribution

$$\wp(\xi := 1) = \mu = 1 - \wp(\xi := 0), \tag{7.43}$$

while the state of the system $\{\wp(b)\}$ is arbitrary. (See Section 1.1.2 for a review of notation.) This interaction or 'logic gate' is depicted in Fig. 7.2. Distinct physical systems (bits or qubits) are depicted as horizontal lines and interactions are depicted by vertical lines. In this case the interaction is referred to as a *controlled-NOT* or C-NOT gate, because the state of the lower system (environment) controls the state of the upper system according to the function defined in Eq. (7.41). The environment variable Ξ is unchanged by the interaction; see Eq. (7.42).

This model becomes the *binary symmetric channel* of classical information theory [Ash90] when we regard B as the input variable to a communication channel with output variable $B \oplus \Xi$. The received variable $B \oplus \Xi$ will reproduce the source variable B iff $\Xi = 0$. Iff $\Xi = 1$, the received variable has undergone a 'bit-flip' error. This occurs with probability μ, due to the noise or uncertainty in the environmental variable.

The same model can be used as a basis for defining errors in a qubit. The system variable B is analogous to $(I + Z)/2$, where Z is the system Pauli operator Z_S. Likewise the environment variable Ξ is analogous to $(I + Z)/2$, where here Z is the environment Pauli operator Z_E. The state of the environment is then taken as the mixed state

$$\rho_E = (1 - \mu)|0\rangle\langle0| + \mu|1\rangle\langle1|. \tag{7.44}$$

A bit-flip error on the system is analagous to swapping the eigenstates of Z, which can be achieved by applying the system Pauli operator X. Thus we take the interaction to be specified by the unitary transformation

$$\hat{U} = X \otimes \frac{I - Z}{2} + I \otimes \frac{I + Z}{2} = \frac{1}{2}(XI - XZ + II + IZ). \tag{7.45}$$

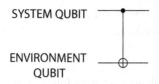

SYSTEM QUBIT

ENVIRONMENT
QUBIT

Fig. 7.3 Quantum circuit diagram for a C-NOT interaction that represents the interaction of a two-state system with a two-state environment. In this case the environment acts to produce (or not) a phase-flip error on the system qubit. Like all our quantum circuit diagrams, this figure follows the conventions of Fig. 7.1.

Here the order of operators is system then environment, and in the second expression we have dropped the tensor product, as discussed in Section 7.1.

Exercise 7.15 *Show that the state (7.44) of the environment is left unchanged by this interaction, in analogy with Eq. (7.42).*

The system qubit after the interaction is given by

$$\rho_S' = \mathrm{Tr}_E \left[\hat{U}(\rho_S \otimes \rho_E)\hat{U}^\dagger \right] \qquad (7.46)$$

$$= \mu X \rho_S X + (1 - \mu)\rho_S, \qquad (7.47)$$

where ρ_S is the state of the system qubit before the interaction.

Exercise 7.16 *Show this, and show that it also holds if the environment is prepared in the pure state $\sqrt{1-\mu}\,|0\rangle + \sqrt{\mu}\,|1\rangle$.*

In this form the interpretation of the noisy channel as an error process is quite clear: ρ_S' is the ensemble made up of a fraction μ of qubits that have suffered a bit-flip error and a fraction $1 - \mu$ that have not.

Exercise 7.17 *Show that the unitary operator in Eq. (7.45) can be generated by the system–environment interaction Hamiltonian*

$$\hat{H} = \frac{\kappa}{4}(I - X)(I + Z) \qquad (7.48)$$

for times $\kappa t = \pi$.

From the discussion so far, it might seem that there is no distinction between errors for classical bits and qubits. This is certainly not the case. A new feature arises in the quantum case on considering the example depicted in Fig. 7.3. This is the same as the previous example in Fig. 7.2, except that the direction of the C-NOT gate has been reversed. In a classical description this would do nothing at all to the system bit. The quantum case is different. The interaction is now described by the unitary operator

$$\hat{U} = \frac{1}{2}(IX - ZX + II + ZI). \qquad (7.49)$$

As in the previous example, we can take the initial state of the environment to be such that it is left unchanged by the interaction,

$$\rho_E = (1 - \mu)|+\rangle\langle+| + \mu|-\rangle\langle-|, \tag{7.50}$$

where $X|\pm\rangle = \pm1|\pm\rangle$. Equivalently (for what follows) we could take it to be the superposition $\sqrt{1-\mu}\,|+\rangle + \sqrt{\mu}\,|-\rangle$.

The reduced state of the system at the output is now seen to be

$$\rho'_S = \mu Z \rho_S Z + (1 - \mu)\rho_S. \tag{7.51}$$

We can interpret this as describing an ensemble in which a fraction $1 - \mu$ of systems remains unchanged while a fraction μ suffers the action of a Z-error. We will call this a phase-flip error since it changes the relative phase between matrix elements in the logical basis. In the logical basis the transformation is

$$\rho'_S = \begin{pmatrix} \rho_{00} & (1 - 2\mu)\rho_{01} \\ (1 - 2\mu)\rho_{10} & \rho_{11} \end{pmatrix}, \tag{7.52}$$

where $\rho_{kl} = \langle k|\rho_S|l\rangle$. The diagonal matrix elements in the logical basis are not changed by this transformation. This is a reflection of the classical result that the state of the system bit is unchanged. However, clearly the state is changed, and, for $0.5 < \mu < 1$, decoherence occurs: the magnitudes of off-diagonal matrix elements in the logical basis are decreased between input and output. Indeed, when $\mu = 0.5$, the state in the logical basis is completely decohered, since the off-diagonal matrix elements are zero at the output.

Having seen bit-flip errors, and phase-flip errors, it is not too surprising to learn that we can define a final class of qubit error by a channel that transforms the input system qubit state according to

$$\rho'_S = \mu Y \rho_S Y + (1 - \mu)\rho_S, \tag{7.53}$$

where $Y = -iXZ$ (here this product is an ordinary matrix product, not a tensor product). This error is a simultaneous bit-flip and phase-flip error. All errors can be regarded as some combination of these elementary errors. In reality, of course, a given decoherence process will not neatly fall into these categories of bit-flip or phase-flip errors. However, the theory of quantum error correction shows that if we can correct for these elementary types of errors then we can correct for an arbitrary single-qubit decoherence process [NC00].

7.4.2 Quantum error correction

The basic idea behind quantum error correction (QEC) is to encode the state of a logical qubit into more than one physical qubit. This can be understood most easily in the case of a single type of error (e.g. bit-flip errors). In that case, an encoding system that is similar to the simple classical redundancy encoding may be used. In the classical case, we simply copy the information ($X = 0$ or 1) in one bit into (say) two others. Then, after a short time, a bit-flip error may have occurred on one bit, but it is very unlikely to have occurred

Table 7.1. *The three-qubit bit-flip code*

ZZI	IZZ	Error	Correcting unitary
+1	+1	None	None
−1	+1	On qubit 1	XII
+1	−1	On qubit 3	IIX
−1	−1	On qubit 2	IXI

on two and even less likely to have occurred on all three. (A crucial assumption here is the independence of errors across the different bits. Some form of this assumption is also necessary for the quantum case.) The occurrence of an error can be detected by measuring the parity of the bit values, that is, whether they are all the same or not. If one is different, then a majority vote across the bits as to the value of X is very likely to equal the original value, even if an error has occurred on one bit. This estimate for X can then be used to change the value of the minority bit. This is the process of error correction.

These ideas can be translated into the quantum case as follows. We encode the qubit state in a two-dimensional subspace of the multi-qubit tensor-product space, known as the code space. The basis states for the code space, known as code words, are entangled states in general. For a three-qubit code to protect against bit-flip errors we can choose the code words to be simply

$$|0\rangle_L = |000\rangle, \qquad |1\rangle_L = |111\rangle. \tag{7.54}$$

An arbitrary pure state of the logical qubit then has the form $|\psi\rangle_L = \alpha|000\rangle + \beta|111\rangle$.

Suppose one of the physical qubits undergoes a bit-flip. It is easy to see that, no matter which qubit flips, the error state is always orthogonal to the code space and simultaneously orthogonal to the other two error states.

Exercise 7.18 *Show this.*

This is the crucial condition for the error to be detectable and correctable, because it makes it possible, in principle, to detect which *physical* qubit has flipped, without learning anything about the *logical* qubit, and to rotate the error state back to the code space. Unlike in the classical case, we cannot simply read out the qubits in the logical basis, because that would destroy the superposition. Rather, to detect whether and where the error occurred, we must measure the two commuting operators ZZI and IZZ. (We could also measure the third such operator, ZIZ, but that would be redundant.) The result of this measurement is the *error syndrome*. Clearly there are two possible outcomes for each operator (± 1) to give four error syndromes. These are summarized in Table 7.1.

The above encoding is an example of a *stabilizer code* [Got96]. In general this is defined as follows. First, we define the Pauli group for n qubits as

$$\mathfrak{P}_n = \{\pm 1, \pm i\} \otimes \{I, X, Y, Z\}^{\otimes n}. \tag{7.55}$$

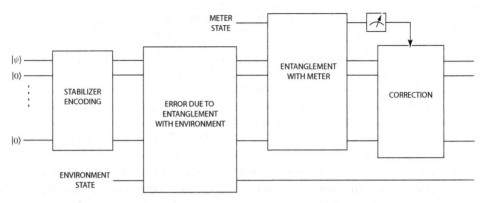

Fig. 7.4 The conventional error-correction protocol using the stabilizer formalism. After the state has been encoded, an error occurs through coupling with the environment. To correct this error, the encoded state is entangled with a meter in order to measure the stabilizer generators, and then feedback is applied on the basis of those measurements. Figure 1 adapted with permission from C. Ahn *et al., Phys. Rev.* A **67**, 052310, (2003). Copyrighted by the American Physical Society.

That is, any member may be denoted as a concatenation of letters (such as ZZI above for $n = 3$) times a phase factor of ± 1 or $\pm i$. Note that this is a discrete group (here a set of operators closed under mutiplication), not a Lie group – see Box 6.2. It can be shown that there exist subgroups of 2^{n-k} commuting Pauli operators $\mathfrak{S} \in \mathfrak{P}_n$ for all $n \geq k \geq 0$. Say that $-I$ is not an element of $\mathfrak{S}$ and that $k \geq 1$. Then it can be shown that $\mathfrak{S}$ defines the *stabilizer* of a nontrivial quantum code. The code space $C(\mathfrak{S})$ is the simultaneous $+1$ eigenspace of all the operators in $\mathfrak{S}$. Then the subspace stabilized is nontrivial, and the dimension of $C(\mathfrak{S})$ is 2^k. Hence this system can encode k logical qubits in n physical qubits. In the above example, we have $n = 3$ and $k = 1$.

The generators of the stabilizer group are defined to be a subset of this group such that any element of the stabilizer can be described as a product of generators. Note that this terminology differs from that used to define generators for Lie groups – see Box 6.2. It can be shown that $n - k$ generators suffice to describe the stabilizer group $\mathfrak{S}$. In the above example, we can take the generators of $\mathfrak{S}$ to be ZZI and IZZ, for example. As this example suggests, the error-correction process consists of measuring the stabilizer. This projection discretizes whatever error has occurred into one of 2^{n-k} error syndromes labelled by the 2^{n-k} possible outcomes of the stabilizer generator measurements. This information is then used to apply a unitary recovery operator that returns the state to the code space. A diagram of how such a protocol would be implemented in a physical system is given in Fig. 7.4.

To encode a single ($k = 1$) logical qubit against bit-flip errors, only three ($n = 3$) physical qubits are required. However, to encode against arbitrary errors, including phase-flips, a larger code must be used. The smallest *universal* encoding uses code words of length $n = 5$ [LMPZ96]. Since this has $k = 1$, the stabilizer group has four generators, which can be chosen to be

$$XZZXI, IXZZX, XIXZZ, ZXIXZ. \tag{7.56}$$

However, unlike the above example, this is not based on the usual classical codes (called linear codes), which makes it hard to generalize. The smallest universal encoding based on combining linear codes is the $n = 7$ Steane code [Ste96].

7.4.3 Detected errors

It might be thought that if one had direct knowledge of whether an error occurred, and precisely what error it was, then error correction would be trivial. Certainly this is the case classically: if one knew that a bit had flipped then one could just flip it back; no encoding is necessary. The same holds for the reversible (unitary) errors we have been considering, such as bit-flip (X), phase-flip (Z) or both (Y). For example, if one knew that a Z-error had occurred on a particular qubit, one would simply act on that qubit with the unitary operator Z. This would completely undo the effect of the error since $Z^2 = I$; again, no encoding is necessary. From the model in Section 7.4.1, one can discover whether or not a Z-error has occurred simply by measuring the state of the environment in the logical basis.

However, we know from earlier chapters to be wary of interpreting the ensemble resulting from the decoherence process (7.47) in only one way. If we measure the environment in the $|\pm\rangle$ basis then we do indeed find a Z-error with probability μ, but if we measure the environment in a different basis (which may be forced upon us by its physical context, as described in Chapter 3) then a different sort of error will be found. In particular, for the case $\mu = 1/2$ and the environment initially in a superposition state, we reproduce exactly the situation of Section 1.2.6. That is, if we measure the environment in the logical basis (which is conjugate to the $|\pm\rangle$ basis), then we discover not whether or not the qubit underwent a phase-flip, but rather which logical state the qubit is in.

Exercise 7.19 *Verify this.*

Classically such a measurement does no harm of course, but in the quantum case it changes the system state irreversibly. That is, there is no way to go back to the (unknown) pre-measurement state of the qubit. Moreover, there are some sorts of errors, which we will consider in Section 7.4.4, that are inherently irreversible. That is, there is no way to detect the error without obtaining information about the system and hence collapsing its state.

These considerations show that the effect of detected errors is nontrivial in the quantum case. Of course, we can correct any errors simply by ignoring the result of the measurement of the environment and using a conventional quantum error correcting protocol, as explained in Section 7.4.2. However, we can do better if we use quantum encoding *and* make use of the measurement results. That is, we can do the encoding using fewer physical qubits. The general idea is illustrated in Fig. 7.5. A simple example is Z-measurement as discussed above. Since this is equivalent to phase-flip errors, it can be encoded against using the three-qubit code of Eq. (7.54). However, if we record the results of the Z-measurements, then we can correct these errors using just a two-qubit code, as we now show. This is not just a hypothetical case; accidental Z-measurements are intrinsic to various schemes for linear optical quantum computing [KLM01].

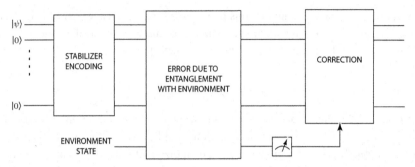

Fig. 7.5 A modified error-correction protocol using the stabilizer formalism but taking advantage of the information obtained from measuring the environment. That is, in contrast to Fig. 7.4, the error and measurement steps are the same. The correction is, of course, different from Fig. 7.4 also. Figure 1 adapted with permission from C. Ahn *et al.*, *Phys. Rev.* A **67**, 052310, (2003). Copyrighted by the American Physical Society.

We can still use the stabilizer formalism introduced above to deal with the case of detected errors. For Z-measurements, we have $n = 2$ and $k = 1$, so there is a single stabilizer generator, which can be chosen to be XX. This gives a code space spanned by the code words

$$\sqrt{2}|0\rangle_L = |00\rangle + |11\rangle, \tag{7.57}$$

$$\sqrt{2}|1\rangle_L = |01\rangle + |10\rangle. \tag{7.58}$$

Exercise 7.20 *Show that in this case* $\mathfrak{S} = \{XX, II\}$ *and verify that the code words are* $+1$ *eigenstates of both of the operators in* $\mathfrak{S}$.

Suppose now that we have an arbitrary qubit state, $|\psi\rangle = \alpha|0\rangle_L + \beta|1\rangle_L$. If the first qubit is accidentally measured in the logical basis, with result 0, the conditional state is $\alpha|00\rangle + \beta|01\rangle = |0\rangle(\alpha|0\rangle + \beta|1\rangle)$. Thus the logical qubit state is still preserved in the state of the second qubit.

Exercise 7.21 *Show that the state with this error is rotated back to the correct encoded state using the operator*

$$\hat{U} = \frac{1}{\sqrt{2}}(XX + ZI). \tag{7.59}$$

In general, if an accidental measurement of qubit 1 or 2 gives result x, the encoded state is recovered if we apply the respective corrections

$$\hat{U}_1^{(x)} = \frac{1}{\sqrt{2}}(XX + (-1)^x ZI), \tag{7.60}$$

$$\hat{U}_2^{(x)} = \frac{1}{\sqrt{2}}(XX + (-1)^x IZ). \tag{7.61}$$

7.4.4 Dynamical decoherence

The discussion so far has been largely unconcerned with dynamics. QEC arose in connection with quantum computing where the dynamics is reduced to a set of unitary operations (gates), applied at discrete times, with no dynamics in between. In the early studies of error correction, errors were modelled as occurring, probabilistically, at discrete times. The detection of a syndrome and subsequent correction were also applied, like gates, at discrete times. That is, the measurement of the error syndrome was assumed to be ideal and instantaneous. This is not a very realistic scenario. Errors are often due to continuous interactions with other systems and thus are dynamical stochastic processes occurring continuously in time. Likewise the measurement of an error syndrome must take some time and is a dynamical process that may have extra noise. Finally, one might expect that an error-correction operation may itself take some time to implement.

We have discussed in Chapter 3 how decoherence can be described as a dynamical process, the simplest model being a Markov master equation. For any particular physical implementation of quantum computation a detailed experimental study of the likely sources of decoherence needs to be made and an appropriate model defined. This is easier for some schemes than for others. Ion-trap implementations have well-characterized decoherence processes [Ste07], as do quantum-optical schemes [LWP+05], while for many solid-state schemes these studies have only just begun [SAB+06, TMB+06]. In this chapter we will consider a simple model of continuous errors that leads directly to a Markov master equation.

Consider the case of phase-flip errors. Let us suppose that, at Poisson-distributed times, the state of the system is transformed according to $\rho \to Z\rho Z$. That is, the rate of the process is a constant, say γ, that is independent of the state of the system. Then, in an infinitesimal interval of time dt, the change in the state is given by the mixture of states that have been so transformed and those that have not:

$$\rho(t + dt) = \gamma \, dt \, Z\rho(t)Z + (1 - \gamma \, dt)\rho(t), \tag{7.62}$$

from which it follows that

$$\dot{\rho} = \gamma(Z\rho Z - \rho) = \gamma \mathcal{D}[Z]\rho. \tag{7.63}$$

Exercise 7.22 *Show that this equation also describes monitoring of whether the system is in logical state 0, for example.*
Hint: *Note that $Z + I = 2\pi_0$, with $\pi_0 = |0\rangle\langle 0|$, and show that Eq. (7.63) can be unravelled using the measurement operators*

$$\hat{M}_1 = \sqrt{4\gamma \, dt} \, \pi_0, \tag{7.64}$$

$$\hat{M}_0 = 1 - 2\gamma \, dt \, \pi_0. \tag{7.65}$$

Another example of continuously occurring errors – which *cannot* be thought of as Poisson-distributed unitary errors – is spontaneous emission. Consider a register of n qubits, each coupled to an independent bath. The measurement operator for a spontaneous emission event on the jth qubit, in an infinitesimal time interval, takes the form

$$\hat{M}_1^j(\mathrm{d}t) = \sqrt{\kappa_j\,\mathrm{d}t}\,(X_j - \mathrm{i}Y_j)/2 \equiv \sqrt{\kappa_j\,\mathrm{d}t}\,L_j, \qquad (7.66)$$

where κ_j is the decay rate for the jth qubit, and we have defined the *lowering operator* $L_j = |1\rangle_j\langle 0|_j$. The corresponding no-jump measurement operator is

$$\hat{M}_0(\mathrm{d}t) = 1 - \sum_j (\kappa_j/2) L_j^\dagger L_j \,\mathrm{d}t - \mathrm{i}\hat{H}\,\mathrm{d}t, \qquad (7.67)$$

where, as in Section 4.2, we have allowed for the possibility of some additional Hamiltonian dynamics. The master equation for the n-qubit system is thus

$$\dot{\rho} = \sum_j \kappa_j \mathcal{D}[L_j]\rho - \mathrm{i}[\hat{H}, \rho]. \qquad (7.68)$$

Exercise 7.23 *Show that, for $\hat{H} = 0$, the coherence of the jth qubit, as measured by $\langle X_j(t)\rangle$ or $\langle Y_j(t)\rangle$, decays exponentially with lifetime $T_2 = 2/\kappa_j$, while the probability of its occupying logical state $|0\rangle$, as measured by $\langle Z_j(t) + 1\rangle /2$, decays exponentially with lifetime $T_1 = 1/\kappa_j$. Here the lifetime is defined as the time for the exponential to decay to e^{-1}.*

In the following section we will show that the techniques of correcting detected errors introduced in Section 7.4.3 can be adapted to deal with continuous detections, whether non-demolition, as in Eq. (7.64), or demolition, as in Eq. (7.66).

7.5 Feedback to correct continuously detected errors

In this section we consider monitored errors, that is, continuously detected errors. We assume initially that there is only one error channel for each physical qubit. We show that, for any error channel and any method of detection (as long as it is efficient), it is possible to correct for these errors using an encoding that uses only one excess qubit. That is, using n physical qubits it is possible to encode $n - 1$ logical qubits, using the stabilizer formalism. This section is based upon Ref. [AWM03].

7.5.1 Feedback to correct spontaneous emission jumps

Rather than considering encoding and error correction for an arbitrary model of continuously detected errors, we begin with a specific model and build up its generality in stages. The simple model we begin with is for encoding one logical qubit against detected spontaneous emission events. This was considered first by Mabuchi and Zoller [MZ96], for a general physical system, and subsequently by several other authors in the context of encoding in several physical qubits [PVK97, ABC+01, ABC+03]. The model we present here gives the most efficient encoding, because we allow a constant Hamiltonian in addition to the feedback Hamiltonian. Specifically, we show that by using a simple two-qubit code we can protect a one-qubit code space perfectly, provided that the spontaneously emitting qubit is known and a correcting unitary is applied instantaneously.

The code words of the code were previously introduced in Eqs. (7.57) and (7.58). If the emission is detected, such that the qubit j from which it originated is known, it is possible to correct back to the code space without knowing the state. This is because the code and error fulfil the necessary and sufficient conditions for appropriate recovery operations [KL97]:

$$\langle \mu | \hat{E}^{\dagger} \hat{E} | \nu \rangle = \Lambda_E \delta_{\mu\nu}. \tag{7.69}$$

Here $\hat{E}$ is the operator for the measurement (error) that has occurred and Λ_E is a constant. The states $|\mu\rangle$ form an orthonormal basis for the code space (they could be the logical states, such as $|0\rangle_L$ and $|1\rangle_L$ in the case of a single logical qubit). These conditions differ from the usual condition only by taking into account that we *know* a particular error $\hat{E} = L_j$ has occurred, rather than having to sum over all possible errors.

Exercise 7.24 *Convince yourself that error recovery is possible if and only if Eq. (7.69) holds for all measurement (error) operators $\hat{E}$ to which the system is subject.*

More explicitly, if a spontaneous emission on the first qubit occurs, $|0\rangle_L \to |01\rangle$ and $|1\rangle_L \to |00\rangle$. Since these are orthogonal states, this fulfills the condition given in (7.69), so a unitary exists that will correct this spontaneous emission error. One choice for the correcting unitary is

$$\hat{U}_1 = (XI + ZX)/\sqrt{2}, \tag{7.70}$$

$$\hat{U}_2 = (IX + XZ)/\sqrt{2}. \tag{7.71}$$

Exercise 7.25 *Verify that these are unitary operators and that they correct the errors as stated.*

As discussed above, in this jump process the evolution between jumps is non-unitary, and so also represents an error. For this two-qubit system the no-jump infinitesimal measurement operator Eq. (7.67) is

$$\hat{M}_0(dt) = 1 - \frac{\kappa_1}{2} L_1^{\dagger} L_1 \, dt - \frac{\kappa_2}{2} L_2^{\dagger} L_2 \, dt - i\hat{H} \, dt \tag{7.72}$$

$$= II - dt[(\kappa_1 + \kappa_2)II + \kappa_1 ZI + \kappa_2 IZ + i\hat{H}]. \tag{7.73}$$

The non-unitary part of this evolution can be corrected by assuming a driving Hamiltonian of the form

$$\hat{H} = -(\kappa_1 YX + \kappa_2 XY). \tag{7.74}$$

This result can easily be seen by plugging (7.74) into (7.73) with a suitable rearrangement of terms:

$$\hat{M}_0(dt) = II[1 - (\kappa_1 + \kappa_2)dt] - \kappa_1 \, dt \, ZI(II - XX) - \kappa_2 \, dt \, IZ(II - XX). \tag{7.75}$$

Exercise 7.26 *Verify this.*

Since $II - XX$ acts to annihilate the code space, $\hat{M}_0$ acts trivially on the code space.

Including the unitary feedback and the driving Hamiltonian, we then have the following master equation for the evolution of the system:

$$d\rho = \hat{M}_0(dt)\rho\hat{M}_0^\dagger(dt) - \rho + dt \sum_{j=1}^{2} \kappa_j \hat{U}_j L_j \rho L_j^\dagger \hat{U}_j^\dagger. \tag{7.76}$$

On writing this in the Lindblad form, we have

$$\dot{\rho} = \sum_{j=1}^{2} \kappa_j \mathcal{D}[\hat{U}_j L_j]\rho + i[\kappa_1 YX + \kappa_2 XY, \rho]. \tag{7.77}$$

From Section 5.4.2, the unitary feedback can be achieved by a feedback Hamiltonian of the form

$$\hat{H}_{\text{fb}} = I_1(t)\hat{V}_1 + I_2(t)\hat{V}_2. \tag{7.78}$$

Here $I_j(t) = dN_j(t)/dt$ is the observed photocurrent from the emissions by the jth qubit, while $\hat{V}_j$ is an Hermitian operator such that $\exp(-i\hat{V}_j) = \hat{U}_j$.

Exercise 7.27 *Show that choosing $\hat{V}_j = (\pi/2)\hat{U}_j$ works.*
Hint: *Show that $\hat{U}_j^2 = I$, like a Pauli operator.*

This code is optimal in the sense that it uses the smallest possible number of qubits required to perform the task of correcting a spontaneous emission error, since we know that the information stored in one unencoded qubit is destroyed by spontaneous emission.

7.5.2 Feedback to correct spontaneous-emission diffusion

So far we have considered only one unravelling of spontaneous emission, by direct detection giving rise to quantum jumps. However, as emphasized in Chapter 4, other unravellings are possible, giving rise to quantum diffusion for example. In this subsection we consider homodyne detection (which may be useful experimentally because it typically has a higher efficiency than direct detection) and show that the same encoding allows quantum diffusion also to be corrected by feedback.

As shown in Section 4.4, homodyne detection of radiative emission of the two qubits gives rise to currents with white noise,

$$J_j(t)dt = \kappa_j\langle e^{-i\phi_j}L_j + e^{i\phi_j}L_j^\dagger\rangle dt + \sqrt{\kappa_j}\,dW_j(t). \tag{7.79}$$

Choosing the Y-quadratures ($\phi_j = \pi/2 \,\forall j$) for definiteness, the corresponding conditional evolution of the system is

$$d\rho_{\bar{J}}(t) = -i[\hat{H}, \rho_{\bar{J}}]dt + \sum_{j=1}^{2} \kappa_j \mathcal{D}[L_j]\rho_{\bar{J}}\,dt + \sum_{j=1}^{2} \sqrt{\kappa_j}\mathcal{H}[-iL_j]\rho_{\bar{J}}\,dW_j(t). \tag{7.80}$$

We can now apply the homodyne mediated feedback scheme introduced in Section 5.5. With the feedback Hamiltonian

$$\hat{H}_{\text{fb}} = \sqrt{\kappa_1}\hat{F}_1 J_1(t) + \sqrt{\kappa_2}\hat{F}_2 J_2(t), \tag{7.81}$$

the resulting Markovian master equation is

$$\dot{\rho} = -\mathrm{i}[\hat{H}, \rho] - \mathrm{i}\sum_{j=1}^{2}\kappa_j\left\{[\mathrm{i}(L_j^{\dagger}\hat{F}_j - \hat{F}_j L_j)/2, \rho] + \mathcal{D}[\mathrm{i}L_j - \mathrm{i}\hat{F}_j]\rho\right\}. \tag{7.82}$$

This allows us to use the same code words, and Eqs. (7.70) and (7.71) suggest using the following feedback operators:

$$\hat{F}_1 = XI + ZX,$$

$$\hat{F}_2 = IX + XZ. \tag{7.83}$$

Using also the same driving Hamiltonian (7.74) as in the jump case, the resulting master equation is

$$\dot{\rho} = \kappa_1\mathcal{D}[YI - \mathrm{i}ZX]\rho + \kappa_2\mathcal{D}[IY - \mathrm{i}XZ]\rho. \tag{7.84}$$

Exercise 7.28 *Verify this, and show that it preserves the above code space.*
Hint: *First show that* $YI - \mathrm{i}ZX = YI(II - XX)$.

7.5.3 Generalization to spontaneous emission of n qubits

We will now demonstrate a simple n-qubit code that allows correction of spontaneous-emission errors, while encoding $n - 1$ qubits. Both of the above calculations (jump and diffusion) generalize. The master equation is the same as (7.68), and the index j runs from 1 to n. Again we need only a single stabilizer generator, namely $X^{\otimes n}$. The number of code words is thus 2^{n-1}, enabling $n - 1$ logical qubits to be encoded. Since it uses only one physical qubit in excess of the number of logical qubits, this is again obviously an optimal code.

First, we consider the jump case. As previously, a spontaneous-emission jump fulfils the error-correction condition (7.69) (this will be shown in an even more general case in the following subsection). Therefore there exists a unitary that will correct for the spontaneous-emission jump. Additionally, it is easy to see by analogy with (7.75) that

$$\hat{H} = \kappa_j \sum_j X^{\otimes j-1} Y X^{\otimes n-j} \tag{7.85}$$

protects against the nontrivial no-emission evolution. Therefore the code space is protected.

Next, for a diffusive unravelling, we again choose homodyne measurement of the Y-quadrature. The same driving Hamiltonian (7.85) is again required, and the feedback operators generalize to

$$\hat{F}_j = I^{\otimes j-1} XI^{\otimes n-j} + X^{\otimes j-1} ZX^{\otimes n-j}. \tag{7.86}$$

The master equation becomes

$$\dot{\rho} = \sum_j \kappa_j \mathcal{D}[I^{\otimes j-1} Y I^{\otimes n-j} (I^{\otimes n} - X^{\otimes n})], \tag{7.87}$$

which manifestly has no effect on states in the code space.

7.5.4 Generalization to arbitrary local measurements on n qubits

In this section, we generalize the above theory to n qubits with arbitrary local (that is, single-qubit) measurements. We find the condition that the stabilizers of the code space must satisfy and show that it is always possible to find an optimal code space (that is, one with a single stabilizer group generator). We give the explicit feedback protocol for a family of unravellings parameterized by a complex number γ, as introduced in Section 4.4. A simple jump unravelling has $\gamma = 0$, while the diffusive unravelling requires $|\gamma| \to \infty$, with the measured quadrature defined by $\phi = \arg(\gamma)$.

Consider a Hilbert space of n qubits obeying (7.68), but with the lowering operators L_j replaced by arbitrary single-qubit operators $\hat{c}_j$ and with $\kappa_j \equiv 1$ (which is always possible since these rates can be absorbed into the definitions of $\hat{c}_j$). Let us consider a single jump operator $\hat{c}$ acting on a single qubit. We may then write $\hat{c}$ in terms of traceless Hermitian operators $\hat{A}$ and $\hat{B}$ as

$$e^{-i\phi}\hat{c} = \chi I + \hat{A} + i\hat{B} \equiv \chi I + \vec{a} \cdot \vec{\sigma} + i\vec{b} \cdot \vec{\sigma}, \tag{7.88}$$

where χ is a complex number, $\vec{a}$ and $\vec{b}$ are real vectors, and $\vec{\sigma} = (X, Y, Z)^\top$.

We now use the standard condition (7.69), where here we take $\hat{E} = \hat{c} + \gamma$ (see Section 4.4). Henceforth, γ is to be understood as real and positive, since the relevant phase ϕ has been taken into account in the definition (7.88). From Eq. (7.69), we need to consider

$$\hat{E}^\dagger \hat{E} = (|\chi + \gamma|^2 + \vec{a}^2 + \vec{b}^2)I + \mathrm{Re}(\chi + \gamma)\hat{A} + \mathrm{Im}(\chi + \gamma)^* i\hat{B} + (\vec{a} \times \vec{b}) \cdot \vec{\sigma}$$

$$\equiv (|\chi + \gamma|^2 + \vec{a}^2 + \vec{b}^2)I + \hat{D}, \tag{7.89}$$

where $\hat{D}$ is Hermitian and traceless.

Now the sufficient condition for error correction for a stabilizer code is that the stabilizer should anticommute with the traceless part of $E^\dagger E$ [Got96]. This condition becomes explicitly

$$0 = \{\hat{S}, \hat{D}\}. \tag{7.90}$$

As long as this is satisfied, there is some feedback unitary e^{-iV} that will correct the error.

As usual, even when the error with measurement operator $\sqrt{dt}\,\hat{E}$ does not occur, there is still non-unitary evolution. As shown in Section 4.4, it is described by the measurement operator

$$\hat{M}_0 = 1 - \hat{E}^\dagger \hat{E} \, dt - \frac{|\gamma|}{2}(e^{-i\phi}\hat{c} - e^{i\phi}\hat{c}^\dagger)dt - i\hat{H} \, dt. \tag{7.91}$$

Now we choose the driving Hamiltonian

$$\hat{H} = i\hat{D}\hat{S} + \frac{i|\gamma|}{2}(e^{-i\phi}\hat{c} - e^{i\phi}\hat{c}^\dagger). \tag{7.92}$$

This is an Hermitian operator because of (7.90).

Exercise 7.29 *Show that, with this choice, $\hat{M}_0$ is proportional to the identity plus a term proportional to $\hat{D}(1 - \hat{S})$, which annihilates the code space.*

Thus, for a state initially in the code space, the condition (7.90) suffices for correction of both the jump and the no-jump evolution.

We now have to show that a single $\hat{S}$ exists for all qubits, even with different operators $\hat{c}_j$. Since $\hat{D}_j$ (the operator associated with $\hat{c}_j$ as defined in (7.89)) is traceless, it is always possible to find some other Hermitian traceless one-qubit operator $\hat{s}_j$, such that $\{\hat{s}_j, \hat{D}_j\} = 0$ and $\hat{s}_j^2 = I$. Then we may choose the single stabilizer generator

$$\hat{S} = \hat{s}_1 \otimes \cdots \otimes \hat{s}_n \tag{7.93}$$

so that the stabilizer group[2] is $\{\hat{I}, \hat{S}\}$. Having chosen $\hat{S}$, choosing $\hat{H}$ as

$$\hat{H} = \sum_j i\hat{D}_j\hat{S} + \frac{i|\gamma_j|}{2}(e^{-i\phi_j}\hat{c}_j - e^{i\phi_j}\hat{c}_j^\dagger) \tag{7.94}$$

will, by our analysis above, provide a total evolution that protects the code space, and the errors will be correctable; furthermore, this code space encodes $n - 1$ qubits in n.

Exercise 7.30 *Show that the n-qubit jump process of Section 7.5.1 follows by choosing, $\gamma = 0$ and $\hat{S} = X^{\otimes n}$, and that $\hat{D}_j = \kappa_j Z_j$.*

Exercise 7.31 *Show that the n-qubit diffusion process in Section 7.5.1 follows by choosing, $\forall j, |\gamma_j| \to \infty$ and $\phi_j = \pi/2$.*
Hint: See Ref. [AWM03].

7.5.5 Other generalizations

In the above we have emphasized that it is always possible to choose one stabilizer, and so encode $n - 1$ qubits in n qubits. However, there are situations in which one might choose a less efficient code with more than one stabilizer. In particular, it is possible to choose a stabilizer $\hat{S}_j$ for each error channel $\hat{c}_j$, with $\hat{S}_j \neq \hat{S}_k$ in general. For example, for the spontaneous emission errors $\hat{c}_j = X_j - iY_j$ one could choose $\hat{S}_j$ as particular stabilizers

---

[2] Strictly, this need not be a stabilizer group, since $\hat{S}$ need not be in the Pauli group, but the algebra is identical, so the analysis is unchanged.

of the universal five-qubit code. This choice is easily made, since the usual generators of the five-qubit code are $\{XZZXI, IXZZX, XIXZZ, ZXIXZ\}$ as discussed above. For each qubit j, we may pick from this set a stabilizer $\hat{S}_j$ that acts as X on that qubit, since X anticommutes with $\hat{D}_j = Z_j$.

In this case, since there are four stabilizer generators, only a single logical qubit can be encoded. However, this procedure would be useful in a system where spontaneous emission is only the dominant error process. If these errors could be detected (with a high degree of efficiency) then they could be corrected using the feedback scheme given above. Then other (rarer) errors, including missed spontaneous emissions, could be corrected using standard canonical error correction, involving measuring the stabilizer generators as explained in Section 7.4.2. The effect of missed emissions from detector inefficiency is discussed in Ref. [AWM03].

Another generalization, which has been investigated in Ref. [AWJ04], is for the case in which there is more than one decoherence channel per qubit, but they are all still able to be monitored with high efficiency. If there are at most *two* error channels per qubit then the encoding can be done with a single stabilizer (and hence $n - 1$ logical qubits) just as above. If there are more than two error channels per qubit then in general two stabilizers are required. That is, one can encode $n - 2$ logical qubits in n physical qubits, requiring just one more physical qubit than in the previous case. The simplest example of this, encoding two logical qubits in four physical qubits, is equivalent to the well-known quantum erasure code [GBP97] which protects against qubit loss.

7.6 QEC using continuous feedback

We turn now, from correction of detected errors by feedback, to correction of undetected errors by conventional error correction. As explained in Section 7.4.2, this usually consists of projective measurement (of the stabilizer generators) at discrete times, with unitary feedback to correct the errors. Here we consider a situation of *continuous* error correction, which may be more applicable in some situations. That is, we consider continual weak measurement of the stabilizer generators, with Hamiltonian feedback to keep the system within the code space. This section is based upon Ref. [SAJM04].

For specificity, we focus on bit-flip errors for which the code words are given in Eq. (7.54), and we assume a diffusive unravelling of the measurement of the stabilizer generators. These measurements will have no effect when the system is in the code space and will give error-specific information when it is not. However, because the measurement currents are noisy, it is impossible to tell from the current in an infinitesimal interval whether or not an error has occurred in that interval. Therefore we do not expect Markovian feedback to be effective. Rather, we must filter the current to obtain information about the error syndrome.

The optimal filter for the currents in this case (and more general cases) has been determined by van Handel and Mabuchi [vHM05]. Since the point of the encoding is to make the quantum information invisible to the measurements, the problem reduces to a classical one of estimating the error syndrome. It is known in classical control theory as the Wonham filter

[Won64]. Here we are using the word 'filter' in the sense of Chapter 6: a way to process the currents in order to obtain information about the system (or, in this case, about the errors). The filtering process actually involves solving nonlinear coupled differential equations in which the currents appear as coefficients for some of the terms. As discussed in Chapter 6, it is difficult to do such processing in real time for quantum systems. This motivates the analysis of Ref. [SAJM04], which considered a non-optimal, but much simpler, form of filtering: a linear low-pass filter for the currents.

In this section we present numerical results from Ref. [SAJM04] showing that, in a suitable parameter regime, a feedback Hamiltonian proportional to the sign of the filtered currents can provide protection from errors. This is perhaps not surprising, because, as seen in Section 7.4, the information about the error syndrome is contained in the signatures of the stabilizer generator measurements (that is, whether they are plus or minus one), a quantity that is fairly robust under the influence of noise.

The general form of this continuous error-correcting scheme is similar to the discrete case. It has four basic elements.

1. Information is encoded using a stabilizer code suited to the errors of concern.
2. The stabilizer generators are monitored and a suitable smoothing of the resulting currents determined.
3. From consideration of the discrete error-correcting unitaries, a suitable feedback Hamiltonian that depends upon the signatures of the smoothed measurement currents is derived.
4. The feedback is added to the system dynamics and the average performance of the QEC scheme is evaluated.

Given m stabilizer generators and d errors possible on our system, the stochastic master equation describing the evolution of a system under this error correction scheme is

$$d\rho_c(t) = \sum_{k=1}^{d} \gamma_k \mathcal{D}[\hat{E}_k]\rho_c(t)dt$$

$$+ \sum_{l=1}^{m} \kappa \mathcal{D}[\hat{M}_l]\rho_c(t)dt + \sqrt{\eta\kappa}\mathcal{H}[\hat{M}_l]\rho_c(t)dW_l(t)$$

$$+ \sum_{k=1}^{d} -iG_k(t)[\hat{F}_k, \rho_c(t)]dt. \tag{7.95}$$

Note that we have set the system Hamiltonian, $\hat{H}$ (which allows for gate operations on the code space) to zero in (7.95). The first line describes the effects of the errors, where $\sqrt{\gamma_k}\hat{E}_k$ is the Lindblad operator for error k, with γ_k a rate and $\hat{E}_k$ dimensionless. The second line describes the measurement of the stabilizers $\hat{M}_l$, with κ the measurement rate (assumed for simplicity to be the same for all measurements). We also assume the same efficiency η for all measurements so that the measurement currents dQ_l/dt can be defined by

$$dQ_l = 2\kappa \, \text{Tr}[\rho_c \hat{M}_l]dt + \sqrt{\kappa/\eta} \, dW_l. \tag{7.96}$$

The third line describes the feedback, with $\hat{F}_k$ a dimensionless Hermitian operator intended to correct error $\hat{E}_k$. Each G_k is the feedback strength (a rate), a function of the smoothed (dimensionless) currents

$$R_l(t) = (1 - e^{-rT})^{-1} \int_{t-T}^{t} r e^{-r(t-t')} \, dQ_l(t')/(2\kappa). \tag{7.97}$$

Here the normalization of this low-pass filter has been defined so that $R_l(t)$ is centred around ± 1. We take T to be moderately large compared with $1/r$.

In a practical situation the γ_ks are outside the experimenters' control (if they could be controlled, they would be set to zero). The other parameters, κ, r and the characteristic size of G_l (which we will denote by λ), can be controlled. The larger the measurement strength κ, the better the performance should be. However, as will be discussed in Section 7.6.1, in practice κ will be set by the best available measurement device. In that case, we expect there to be a region in the parameter space of r and λ where this error-control scheme will perform optimally. This issue can be addressed using simulations.

To undertake numerical simulations, one needs to consider a particular model. The simplest situation to consider is protecting against bit-flips using the three-qubit bit-flip code of Section 7.4.2. We assume the same error rate for the three errors, and efficient measurements. This is described by the above SME (7.95), with $\gamma_k = \gamma$, $\eta = 1$, and

$$\hat{E}_1 = XII, \qquad \hat{E}_2 = IXI, \qquad \hat{E}_3 = IIX, \tag{7.98}$$

$$\hat{M}_1 = ZZI, \qquad \hat{M}_2 = IZZ. \tag{7.99}$$

A suitable choice for $\hat{F}_k$ is to set them equal to $\hat{E}_k$. Because the smoothed currents R_l correspond to the measurement syndrome (the sign of the result of a strong measurement of $\hat{M}_l$), we want G_k to be such that the following apply.

1. If $R_1(t) < 0$ and $R_2(t) > 0$, apply XII.
2. If $R_1(t) > 0$ and $R_2(t) < 0$, apply IIX.
3. If $R_1(t) < 0$ and $R_2(t) < 0$, apply IXI.
4. If $R_1(t) > 0$ and $R_2(t) > 0$, do not apply any feedback.

These conditions can be met by the following (somewhat arbitrary) choice:

$$G_1(t) = \begin{cases} \lambda R_1(t) & \text{if } R_1(t) < 0 \text{ and } R_2(t) > 0, \\ 0 & \text{otherwise,} \end{cases} \tag{7.100}$$

$$G_2(t) = \begin{cases} \lambda R_2(t) & \text{if } R_1(t) > 0 \text{ and } R_2(t) < 0, \\ 0 & \text{otherwise,} \end{cases} \tag{7.101}$$

$$G_3(t) = \begin{cases} \lambda R_1(t) & \text{if } R_1(t) < 0 \text{ and } R_2(t) < 0, \\ 0 & \text{otherwise.} \end{cases} \tag{7.102}$$

Recall that λ is the characteristic strength of the feedback.

A numerical solution of the above SME was presented in Ref. [SAJM04]. As expected, it was found that the performance improved as κ increased. Also it was found that the

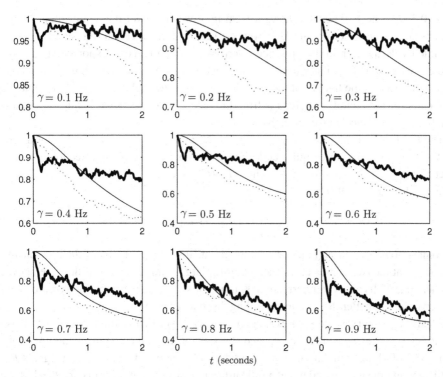

Fig. 7.6 Fidelity curves with and without error correction for several error rates γ. The thick solid curve is the average fidelity $F_3(t)$ of the three-qubit code with continuous error correction. The parameters used were $dt = 10^{-4}$ s, $\kappa = 150\,\mathrm{s}^{-1}$, $\lambda = 150\,\mathrm{s}^{-1}$, $r = 20\,\mathrm{s}^{-1}$ and $T = 0.15$ s. The dotted curve is the average fidelity $F_1(t)$ of one qubit without error correction. The thin solid curve is the fidelity $F_{3d}(t)$ achievable by discrete QEC when the duration between applications is t. Figure 2 adapted with permission from M. Sarovar et al., Phys. Rev. A **69**, 052324, (2004). Copyrighted by the American Physical Society.

optimum values of r and λ increase with κ, for γ fixed. This is as expected, because the limit where κ, r and λ are large compared with γ should approximate that of frequent strong measurements with correction. It was found that the best performance was achieved for $\lambda \geq \kappa$. However, as will be discussed in Section 7.6.1, in practice λ may (like κ) be bounded above by the physical characteristics of the device. This would leave only one parameter (r) to be optimized.

The performance of this error-correction scheme can be gauged by the average fidelity $F_3(t)$ between the initial encoded three-qubit state and the state at time t [SAJM04]. This is shown in Fig. 7.6 for several values of the error rate γ (the time-units used are nominal; a discussion of realistic magnitudes is given in Section 7.6.1). Each plot also shows the fidelity curve $F_1(t)$ for one qubit in the absence of error correction. A comparison of these two curves shows that the fidelity is preserved for a longer period of time by the error-correction scheme for small enough error rates. Furthermore, for small error rates

$(\gamma < 0.3 \text{ s}^{-1})$ the $F_3(t)$ curve shows a great improvement over the exponential decay in the absence of error correction. However, we see that, past a certain threshold error rate, the fidelity decay even in the presence of error correction behaves exponentially, and the two curves look very similar; the error-correcting scheme becomes ineffective. In fact, well past the threshold, the fidelity of the (supposedly) protected qubit becomes lower than that of the unprotected qubit. This results from the feedback 'corrections' being so inaccurate that the feedback mechanism effectively increases the error rate.

The third line in the plots of Fig. 7.6 is of the average fidelity achievable by discrete QEC (using the same three-qubit code) when the time between the detection-correction operations is t. The value of this fidelity ($F_{3d}(t)$) as a function of time was analytically calculated in Ref. [ADL02] as

$$F_{3d} = \frac{1}{4}(2 + 3e^{-2\gamma t} - e^{-6\gamma t}). \qquad (7.103)$$

A comparison between $F_3(t)$ and $F_{3d}(t)$ highlights the relative merits of the two schemes. The fact that the two curves cross each other for large t indicates that, if the time between applications of discrete error correction is sufficiently large, then a continuous protocol will preserve fidelity better than a corresponding discrete scheme.

All the $F_3(t)$ curves show an exponential decay at very early times, $t \lesssim 0.1$ s. This is an artefact of the finite filter length and the specific implementation of the protocol in Ref. [SAJM04]: the simulations did not produce the smoothed measurement signals $R_l(t)$ until enough time had passed to get a full buffer of measurements. That is, feedback started only at $t = T$. We emphasize again that this protocol is by no means optimal.

The effect of non-unit efficiency η was also simulated in Ref. [SAJM04], as summarized by Fig. 7.7. The decay of fidelity with decreasing η indicates that inefficient measurements have a negative effect on the performance of the protocol as expected. However, the curves are quite flat for $1 - \eta$ small. This is in contrast to the correction of detected errors by Markovian feedback as considered in Section 7.5, where the rate of fidelity decay would be proportional to $1 - \eta$. This is because in the present case the measurement of the stabilizer generators has no deleterious effect on the encoded quantum information. Thus a reduced efficiency simply means that it takes a little longer to obtain the information required for the error correction.

7.6.1 Practical considerations for charge qubits

Several schemes for solid-state quantum computing using the charge or spin degree of freedom of single particles as qubits, with measurements to probe this degree of freedom, have been proposed. Here we examine the weak measurement of one such proposed qubit: a single electron that can coherently tunnel between two quantum dots [HDW+04]. The dots are formed by two P donors in Si, separated by a distance of about 50 nm. Surface gates are used to remove one electron from the double-donor system leaving a single electron on the P–P$^+$ system. This system can be regarded as a double-well potential. Surface gates

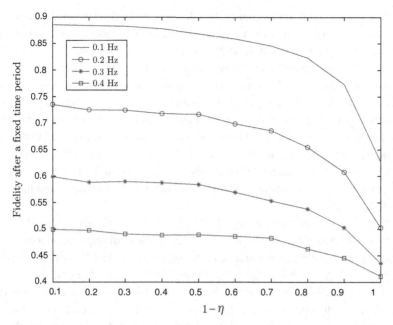

Fig. 7.7 Average fidelity after a fixed amount of time as a function of inefficiency $1 - \eta$ for several error rates. The parameters used were $dt = 10^{-4}$ s, $\kappa = 50\,\text{s}^{-1}$, $\lambda = 50\,\text{s}^{-1}$, $r = 10\,\text{s}^{-1}$ and $T = 0.15\,\text{s}$. Figure 3 adapted with permission from M. Sarovar *et al.*, *Phys. Rev.* A **69**, 052324, (2004). Copyrighted by the American Physical Society.

can then be used to control the barrier between the wells, as well as the relative depth of the two wells. It is possible to design the double-well system so that, when the well depths are equal, there are only two energy eigenstates below the barrier. These states, $|+\rangle$ and $|-\rangle$, with energies E_+ and E_-, are symmetric and antisymmetric, respectively. The localized states describing the electron on the left or right of the barrier can thus be defined as

$$|L\rangle = \frac{1}{\sqrt{2}}(|+\rangle + |-\rangle), \tag{7.104}$$

$$|R\rangle = \frac{1}{\sqrt{2}}(|+\rangle - |-\rangle). \tag{7.105}$$

An initial state localized in one well will then tunnel to the other well at the frequency $\Delta = (E_+ - E_-)$.

Using $|L\rangle$ and $|R\rangle$ as the logical basis states $|0\rangle$ and $|1\rangle$, respectively, we can define Pauli matrices in the usual way. Then the Hamiltonian for the system can be well approximated by

$$\hat{H} = \frac{\omega(t)}{2}Z + \frac{\Delta(t)}{2}X. \tag{7.106}$$

A (time-dependent) bias gate can control the relative well depth $\omega(t)$ and similarly a barrier gate can control the tunnelling rate $\Delta(t)$. Further details on the validity of this Hamiltonian and how well it can be realized in the P–P$^+$ in Si system can be found in Ref. [BM03].

A number of authors have discussed the sources of decoherence in a charge qubit system such as this one [BM03, FF04, HDW$^+$04]. For appropriate donor separation, phonons can be neglected as a source of decoherence. The dominant sources are fluctuations in voltages on the surface gates controlling the Hamiltonian and electrons moving in and out of trap states in the vicinity of the dot. The latter source of decoherence is expected to dominate at low frequencies (long times), as for so-called $1/f$ noise. In any case, both sources can be modelled using the well-known spin–boson model (see Section 3.4.1) The key element of this model for the discussion here is that the coupling between the qubit and the reservoir is proportional to Z.

If the tunnelling term proportional to $\Delta(t)X$ in Eq. (7.106) were not present, decoherence of this kind would lead to pure dephasing. However, in a general single-qubit gate operation, both dephasing and bit-flip errors can arise in the spin–boson model. We use the decoherence rate calculated for this model as indicative for the bit-flip error rate in the toy model used above in which only bit-flips occur. Hollenberg *et al.* [HDW$^+$04] calculated that, for a device operating at $10\,\mathrm{K}$, the error rate would be $\gamma = 1.4 \times 10^6\,\mathrm{s}^{-1}$. This rate could be made a factor of ten smaller by operating at lower temperatures and improving the electronics controlling the gates.

We now turn to estimating the measurement strength κ for the P–P$^+$ system. In order to read out the qubit in the logical basis, we need to determine whether the electron is in the left or the right well quickly and with high probability of success. The technique of choice is currently based on radio-frequency single-electron transistors (RF-SETs) [SWK$^+$98]. A single-electron transistor is a very sensitive transistor whose operation relies upon single-electron tunneling onto and off a small metallic island (hence its name). That is, the differential resistance of the SET can be controlled by a very small bias voltage, which in this case arises from the Coulomb field associated with the qubit electron. Depending on whether the qubit is in the L or R state, this field will be different and hence the SET resistance will be different. In the RF configuration (which enables $1/f$ noise to be filtered from the signal) the SET acts as an Ohmic load in a tuned tank circuit. The two different charge states of the qubit thus produce two levels of power reflected from the tank circuit.

The electronic signal in the RF circuit carries a number of noise components, including amplifier noise, the Johnson noise of the circuit and 'random telegraph' noise in the SET bias conditions due to charges hopping randomly between charge trap states in or near the SET. The quality of the SET is captured by the minimum charge sensitivity per root hertz, S. In Ref. [BRS$^+$05] a value of $S \approx 5 \times 10^{-5} e/\sqrt{\mathrm{Hz}}$ was measured, for the conditions of observing the single-shot response to a charge change $\Delta q = 0.05e$. Here e is the charge on a single electron, and Δq means a change in the bias field for the SET corresponding to moving a charge of Δq from its original position (on the P–P$^+$ system) to infinity. This is of order the field change expected for moving the electron from one P donor to the other. Thus the characteristic rate for measuring the qubit in the charge basis is of order $(\Delta q/S)^2 = 10^6\,\mathrm{Hz}$. Thus we take $\kappa\eta = 10^6\,\mathrm{s}^{-1}$. For definiteness we will say that $\eta = 1$

(that is, a quantum-limited measurement), even though that is almost certainly not the case (see for example Refs. [WUS+01, Goa03]). Note also that we are ignoring the difficulties associated with measuring stabilizers such as ZZI. That is, we simply use the one-qubit measurement rate for this joint multi-qubit measurement.

We next need to estimate typical values for the feedback strength. The feedback Hamiltonian is proportional to an X operator, which corresponds to changing the tunnelling rate Δ for each of the double-dot systems that comprise each qubit. In Ref. [BM03], the maximum tunnelling rate was calculated to be about $10^9\,\mathrm{s}^{-1}$, for a donor separation of 40 nm. We take this to be the upper bound on λ.

To summarize, in the P–P$^+$-based charge qubit, with RF-SET readout, we have $\gamma \approx \kappa \approx 10^6\,\mathrm{s}^{-1}$ and $\lambda \lesssim 10^9\,\mathrm{s}^{-1}$. The fact that the measurement strength and the error rate are of the same order of magnitude for this architecture is a problem for our error-correction scheme. This means that the rate at which we gain information is about the same as the rate at which errors happen, and it is difficult to operate a feedback correction protocol in such a regime. Although it is unlikely that the measurement rate could be made significantly larger in the near future, as mentioned above it is possible that the error rate could be made smaller by improvements in the controlling electronics.

7.7 Continuous QEC without measurement

So far, both for discrete (as discussed in Section 7.4.2) and for continuous (as discussed in the preceding section) QEC, we have treated the measurement and control steps as involving a classical apparatus. However, as discussed in Section 1.3.1, measurement results can be stored in quantum systems and represented by quantum operators. Similarly, as discussed in Section 5.8.1, this information can be used to control the quantum systems that were measured by application of a suitable Hamiltonian. This suggests that it should be possible to implement the QEC process using only a few additional qubits, known as *ancilla* qubits. In other words, the entire process of detection and correction can be done with Hamiltonian dynamics and thus can be implemented with a quantum circuit.

A circuit that implements the three-qubit error-correction protocol without measurement is given in Fig. 7.8. In this circuit, the first three controlled-NOT gates effectively calculate the error syndrome (for the encoded state in the top three qubits), storing the result in the two ancilla qubits. Then the correction is done by direct coupling between the ancillae and the encoded qubits using Toffoli gates (doubly controlled-NOT gates). It is important to note that the ancilla qubits must be reset to the $|0\rangle$ state after each run of the circuit. This is a consequence of the fact that the entropy generated by the errors is moved into the ancilla subsystem and must be carried away before the next run of the circuit.

This circuit illustrates the essential ideas behind implementing error correction without measurement: introduction of ancilla qubits, their direct coupling to the encoded qubits, and the resetting of these ancilla qubits after each cycle. If this cycle comprising detect, correct and reset is performed often enough, and the only errors in our system are independent bit-flip errors at randomly distributed times, then one can preserve the value of a logical

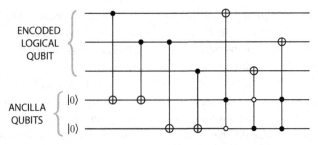

Fig. 7.8 A circuit for implementing error correction using the three-qubit bit-flip code without measurement. The top three qubits form the encoded logical qubit and the bottom two form the ancilla. The first four gates are C-NOT gates as described in Section 7.4.1. The last three are Toffoli gates, which are similar but have two controls (shown by the open or filled circles). The target (large encircled cross) undergoes a bit-flip iff the controls have the appropriate value (zero for an open circle, one for a filled circle). Note that, to repeat the error-correction procedure, the ancilla qubits must be replaced or reset to the $|0\rangle$ state at the end of each run (at the far right of the circuit). Figure 1 adapted with permission from M. Sarovar and G. J. Milburn, *Phys. Rev.* A **72**, 012306, (2005). Copyrighted by the American Physical Society.

qubit indefinitely. Note that we are assuming that the operations involved in the circuit – the unitary gates and the ancilla reset – are instantaneous. In this section we address the obvious question: can we replace these instantaneous discrete operations by continuous processes? That is, can we use a finite apparatus to obtain a continuous version of 'coherent' QEC (see Section 5.8.1) just as there are continuous versions of conventional QEC with measurement as discussed in Section 7.6?

The answer to this question is yes, as shown in Ref. [SM05]. Following that reference, we need to modify two components of the circuit model.

1. The unitary gates which form the system–ancilla coupling are replaced by a finite-strength, time-independent Hamiltonian. This Hamiltonian will perform both the detection and the correction operations continuously and simultaneously.
2. The ancilla reset procedure is replaced by the analogous continuous process of *cooling*. Each ancilla qubit must be independently and continuously cooled to its ground state $|0\rangle$.

These changes lead to a continuous-time description of the process in terms of a Markovian master equation, under the assumption that both open-system components – the errors and the ancilla cooling – are Markovian processes.

We illustrate this continuous-time implementation for the three-qubit bit-flip code example used previously. The continuous time description of the circuit of Fig. 7.8 is

$$\frac{d\rho}{dt} = \gamma(\mathcal{D}[XIIII] + \mathcal{D}[IXIII] + \mathcal{D}[IIXII])\rho$$

$$+ \lambda(\mathcal{D}[IIIL^{\dagger}I] + \mathcal{D}[IIIIL^{\dagger}])\rho - i\kappa[\hat{H}, \rho]. \qquad (7.107)$$

Here, the ordering of the tensor product for all operators in the equation runs down the circuit as shown in Fig. 7.8 (i.e. the first three operators apply to the encoded qubit and the last two to the ancilla), while $L \equiv \frac{1}{2}(X + iY) = |1\rangle\langle0|$ is a qubit lowering operator as before. The parameters are γ, the bit-flip error rate; κ, the strength of the coherent detection and correction (the Hamiltonian operator $\hat{H}$ is dimensionless); and λ, the rate of the cooling applied to the ancilla qubits.

To construct the dimensionless Hamiltonian in Eq. (7.107), we first determine Hamiltonians $\hat{H}_D$ and $\hat{H}_C$ that perform the detection and correction operations, respectively. The detection Hamiltonian is given by

$$\hat{H}_D = \hat{D}_1 \otimes (XI) + \hat{D}_2 \otimes (XX) + \hat{D}_3 \otimes (IX). \qquad (7.108)$$

Here, $\hat{D}_1 = |100\rangle\langle100| + |011\rangle\langle011|$ is the projector onto the subspace where there has been a bit-flip error on the first physical qubit, and $\hat{D}_2$ and $\hat{D}_3$ similarly for the second and third physical qubits. These operators act on the three qubits encoding the logical qubit, while the Pauli operators cause the appropriate bit-flips in the ancilla qubits. Similarly, the correction Hamiltonian is

$$\hat{H}_C = \hat{C}_1 \otimes (PI) + \hat{C}_2 \otimes (PP) + \hat{C}_3 \otimes (IP). \qquad (7.109)$$

Here $P \equiv (1 - Z)/2 = |1\rangle\langle1|$, the projector onto the logical one state of a qubit. We have also defined $\hat{C}_1 = X \otimes (|00\rangle\langle00| + |11\rangle\langle11|)$, an operator that corrects a bit-flip on the first physical qubit (assuming that the second and third remain in the code space), and $\hat{C}_2$ and $\hat{C}_3$ similarly for the second and third physical qubits.

The operation in Fig. 7.8 of detection followed by correction can be realized by the unitary $\hat{U}_{DC} = \exp(-i\hat{H}_C\pi/2)\exp(-i\hat{H}_D\pi/2)$.

Exercise 7.32 *Verify this.*

Now, by the Baker–Campbell–Hausdorff theorem (A.118), it follows that the unitary $\hat{U}_{DC}$ has a generator of the form

$$\hat{H} = \hat{H}_D + \hat{H}_C + i\alpha[\hat{H}_D, \hat{H}_C], \qquad (7.110)$$

for some α. That is, $\ln(\hat{U}_{DC}) \propto -i\hat{H}$.

Exercise 7.33 *Verify this.*
Hint: *Show that $\{\hat{H}_C, \hat{H}_D, i[\hat{H}_D, \hat{H}_C]\}$ form a Lie algebra (see Box 6.2).*

Although it would be possible to determine α from the above argument, it is more fruitful to consider it as a free parameter in $\hat{H}$. That is because the above argument is just a heuristic to derive a suitable $\hat{H}$, since the circuit model does not have the cooling process simultaneous with the detection and correction. It was shown in Ref. [Sar06] that good results were obtained with $\alpha = 1$, and this is the value used in Ref. [SM05].

Note that in Eq. (7.107) the error processes are modelled only on qubits that form the encoded state. One could extend the error dynamics on to the ancilla qubits. However, in the parameter regime where the error correction is effective, $\lambda \gg \gamma$, the cooling will dominate all other ancilla dynamics. Thus we can ignore the error dynamics on the ancilla qubits.

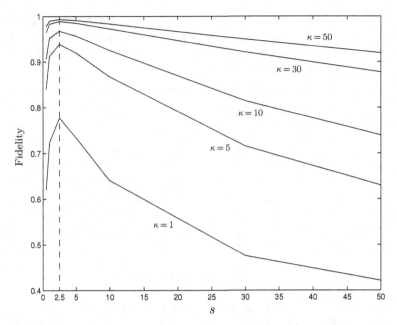

Fig. 7.9 Fidelity, after a fixed period of time ($T = 10$), of an encoded qubit (three-qubit code) undergoing continuous error correction using cooled ancillae. Here time is measured in arbitrary units, with $\gamma = 1/20$. The curves are for different Hamiltonian strengths (κ) and the horizontal axis shows how the cooling rate is scaled with κ; i.e. $\lambda = s\kappa$, where s is varied along the horizontal axis. Figure 2 adapted with permission from M. Sarova and G. J. Milburn, *Phys. Rev.* A **72**, 012306, (2005). Copyrighted by the American Physical Society.

In Ref. [SM05], Eq. (7.107) was solved by numerical integration and the fidelity $F(t) \equiv \langle \psi | \rho(t) | \psi \rangle$ determined. Here $\rho(t)$ is the reduced state of the encoded subsystem and $\rho(0) = | \psi \rangle \langle \psi |$ is the initial logical state. For a given error rate γ we expect there to be an optimal ratio between the Hamiltonian strength κ and the cooling rate λ. Figure 7.9 shows the fidelity after a fixed period of time $1/(2\gamma)$ for several values of these parameters, and it is clear that the best performance is when $\lambda \approx 2.5\kappa$. This optimal point is independent of the ratio of κ to γ and of the initial state of the encoded qubits. The following results were all obtained in this optimal parameter regime.

Figure 7.10 shows the evolution of fidelity with time for a fixed error rate and several values of κ. This clearly shows the expected improvement in performance with an increase in the Hamiltonian strength. Large values of κ and λ are required in order to maintain fidelity at reasonable levels. To maintain the fidelity above 0.95 up to time $T = 1/(2\gamma)$ requires $\kappa/\gamma > 200$. However, a comparison with the unprotected qubit's fidelity curve shows a marked improvement in coherence, due to the error-correction procedure. Therefore, implementing error correction even in the absence of ideal resources is valuable. This was also evident in the scenario of error correction *with* measurement in the preceding section.

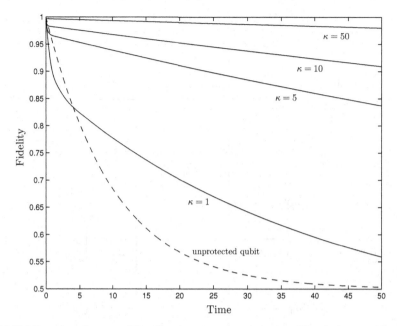

Fig. 7.10 Fidelity curves for several Hamiltonian strengths versus time. Time is measured in arbitrary units, with $\gamma = 1/20$. The solid curves are the fidelity of an encoded qubit (three-qubit code) with continuous error correction. The dashed curve is the fidelity of one qubit undergoing random bit-flips without error correction. Figure 3 adapted with permission from M. Sarovar and G. J. Milburn, *Phys. Rev.* A **72**, 012306, (2005). Copyrighted by the American Physical Society.

Aside from describing a different implementation of error correction, the scheme above casts error correction in terms of the very natural process of cooling; it refines the viewpoint that error correction extracts the entropy that enters the system through errors. Error correction is not cooling to a particular state such as a ground state, but rather a subspace of Hilbert space, and the specially designed coupling Hamiltonian allows us to implement this cooling to a (nontrivial) subspace by a simple cooling of the ancilla qubits to their ground state.

7.8 Linear optical quantum computation

7.8.1 Measurement-induced optical nonlinearity

One of the earliest proposals [Mil89] for implementing quantum computation was based on encoding a single qubit as a single-photon excitation of an optical field mode. The qubits were assumed to interact via a medium with a nonlinear refractive index, such as discussed in Section 5.3. However, it is extremely difficult to implement a significant unitary coupling between two optical modes containing one or two photons. In 2001, Knill *et al.* [KLM01] discovered an alternative approach based on how states change due to measurement. They

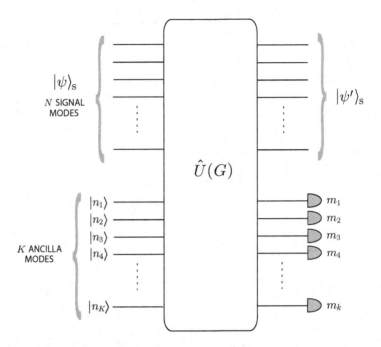

Fig. 7.11 A general conditional linear optical gate.

showed that non-deterministic photonic qubit gates are possible with linear optical networks when some of the input modes (referred to as *ancilla* modes) are prepared in single-photon states before the optical network and directed to photon counters at the network output. The conditional state of all non-ancilla modes (the *signal* modes), conditioned on a particular count on the output ancilla modes, is given by a non-unitary transformation of the input signal state and can simulate a highly nonlinear optical process. This transformation is defined in terms of a conditional measurement operator acting on the signal modes alone.

Consider the situation depicted in Fig. 7.11. In this device $N + K$ modes pass through a linear optical device, comprising only mirrors and beam-splitters. We describe this by a unitary transformation (that is, we ignore losses through absorption etc.) so that the total photon-number is conserved. The K ancilla modes are prepared in photon-number eigenstates. At the output, photon-number measurements are made on the ancilla modes alone. We seek the *conditional* state for the remaining N modes, *given* the ancilla photon-number count.

The linear optical device performs a unitary transformation on all the input states:

$$\hat{U}(G) = \exp[-i\hat{\mathbf{a}}^\dagger G \, \hat{\mathbf{a}}], \qquad (7.111)$$

where $\hat{\mathbf{a}}$ is a vector of annihilation operators,

$$\hat{\mathbf{a}} = \begin{pmatrix} \hat{a}_1 \\ \hat{a}_2 \\ \vdots \\ \hat{a}_N \\ \hat{a}_{N+1} \\ \vdots \\ \hat{a}_{N+K} \end{pmatrix}, \tag{7.112}$$

and G is an Hermitian *matrix* (not an operator). This transformation induces a unitary transform on the vector $\hat{\mathbf{a}}$:

$$\hat{U}^{\dagger}(G)\hat{\mathbf{a}}\,\hat{U}(G) = S(G)\hat{\mathbf{a}}. \tag{7.113}$$

One should not confuse the unitary transformation $\hat{U}(G)$ (an operator) with the induced unitary representation $S(G) = \exp(-iG)$ (a matrix). Because $S(G)$ is unitary, the transformation leaves the total photon number invariant:

$$\hat{U}^{\dagger}(G)\hat{\mathbf{a}}^{\dagger}\hat{\mathbf{a}}\,\hat{U}(G) = \hat{\mathbf{a}}^{\dagger}S(G)S(G)^{\dagger}\hat{\mathbf{a}} = \hat{\mathbf{a}}^{\dagger}\hat{\mathbf{a}}. \tag{7.114}$$

Remember that, as in Chapter 6, $\hat{\mathbf{a}}^{\dagger} = (\hat{a}_1^{\dagger}, \ldots, \hat{a}_{N+K}^{\dagger})$, so that $\hat{\mathbf{a}}^{\dagger}\hat{\mathbf{a}} = \sum_k \hat{a}_k^{\dagger}\hat{a}_k$.

Exercise 7.34 *Verify that $S(G) = \exp(-iG)$.*
Hint: *Show that, for a Hamiltonian $\hat{\mathbf{a}}^{\dagger}G\,\hat{\mathbf{a}}$, $d\hat{\mathbf{a}}/dt = -iG\hat{\mathbf{a}}$.*

The *conditional state* of the signal modes, $|\psi'\rangle_s$, is determined by

$$|\psi'\rangle_s = \frac{1}{\sqrt{\wp(\vec{m})}}\hat{M}(\vec{m}|\vec{n})|\psi\rangle_s. \tag{7.115}$$

Here the observed count is represented by the vector of values $\vec{m}$, and the probability for this event is $\wp(\vec{m})$. The measurement operator is

$$\hat{M}(\vec{n}|\vec{m}) = {}_{\text{anc}}\langle \vec{m}|\hat{U}(G)|\vec{n}\rangle_{\text{anc}} \tag{7.116}$$

with

$$|\vec{m}\rangle_{\text{anc}} = |m_1\rangle_{N+1} \otimes |m_2\rangle_{n+2} \otimes \ldots \otimes |m_k\rangle_{N+K}. \tag{7.117}$$

As an example consider a three-mode model defined by the transformation

$$S(G)\hat{\mathbf{a}} = \begin{pmatrix} s_{11} & s_{12} & s_{13} \\ s_{21} & s_{22} & s_{23} \\ s_{31} & s_{32} & s_{33} \end{pmatrix}\begin{pmatrix} \hat{a}_1 \\ \hat{a}_2 \\ \hat{a}_3 \end{pmatrix}. \tag{7.118}$$

We will regard $\hat{a}_2$ and $\hat{a}_3$ as the ancilla modes, prepared in the single-photon state $|1, 0\rangle$. That is, $n_2 = 1$ and $n_3 = 0$. We will condition on a count of $m_2 = 1$ and $m_3 = 0$. We want the two-mode conditional measurement operator $\hat{M}(1, 0|1, 0)$ acting on mode $\hat{a}_1$. Since photon number is conserved, the only non-zero elements of this operator are

$$\langle n|\hat{M}(1, 0|1, 0)|n\rangle = \langle n, 1, 0|\hat{U}(G)|n, 1, 0\rangle$$
$$= (n!)^{-1}\langle 0, 0, 0|\hat{a}_1^n\hat{a}_2\hat{U}(G)(\hat{a}_1^\dagger)^n\hat{a}_2^\dagger|0, 0, 0\rangle. \tag{7.119}$$

This expression simplifies to

$$\frac{1}{n!}\langle 0, 0, 0|(s_{11}\hat{a}_1 + s_{12}\hat{a}_2 + s_{13}\hat{a}_3)^n(s_{21}\hat{a}_1 + s_{22}\hat{a}_2 + s_{23})\hat{a}_3(\hat{a}_1^\dagger)^n\hat{a}_2^\dagger|0, 0, 0\rangle. \tag{7.120}$$

Exercise 7.35 *Show this, and also show that, since $\hat{a}_3^\dagger$ does not appear in Eq. (7.120), further simplification is possible, namely to*

$$(n!)^{-1}\langle 0, 0, 0|(s_{11}\hat{a}_1 + s_{12}\hat{a}_2)^n(s_{21}\hat{a}_1 + s_{22}\hat{a}_2)(\hat{a}_1^\dagger)^n\hat{a}_2^\dagger|0, 0, 0\rangle. \tag{7.121}$$

Hint: *First show that* $\langle 0, 0, 0|\hat{U}^\dagger(G) = \langle 0, 0, 0|$, *and so replace* $\hat{a}_1^n\hat{a}_2\hat{U}(G)$ *by* $\hat{U}^\dagger(G)\hat{a}_1^n\hat{a}_2\hat{U}(G)$.

From this it can be shown that a formal expression for $\hat{M}(10|10)$ is

$$\hat{M}(1, 0|1, 0) = s_{12}s_{21}\hat{a}_1^\dagger\hat{A}\hat{a}_1 + s_{22}\hat{A}, \tag{7.122}$$

where $\hat{A} = \sum_{n=0}^\infty (s_{11} - 1)^n(\hat{a}_1^\dagger)^n\hat{a}_1^n/n!$. We will not use this expression directly. However, it does serve to emphasize the optical nonlinearity in the measurement, since it contains all powers of the field operator.

7.8.2 Two-qubit gates

The above non-deterministic transformations can be used for universal quantum computation. This scheme and others like it are called linear optical quantum computation (LOQC) schemes. Universal quantum computation [NC00] can be achieved if one can perform arbitrary single-qubit unitaries, and implement a two-qubit entangling gate (such as the C-NOT gate), between any two qubits. In order to see how these one- and two-qubit gates are possible in LOQC, we need to specify a physical encoding of the qubit. In this section we use a 'dual-rail' logic based on two modes and one photon:

$$|0\rangle_L = |1\rangle_1 \otimes |0\rangle_2, \tag{7.123}$$

$$|1\rangle_L = |0\rangle_1 \otimes |1\rangle_2. \tag{7.124}$$

The modes could be distinguished spatially (e.g. a different direction for the wave vector), or they could be distinguished by polarization.

One single-qubit gate that is easily implemented uses a beam-splitter (for spatially distinguished modes) or a wave-plate (for modes distinguished in terms of polarization). These linear optical elements involving two modes can be described by the unitary transformation

$$\hat{U}(\theta) = \exp\left[-i\theta(\hat{a}_1^\dagger\hat{a}_2 + \hat{a}_1\hat{a}_2^\dagger)\right], \tag{7.125}$$

which coherently transfers excitations from one mode to the other. Another simple single-qubit gate is a relative phase shift between the two modes, which can be achieved simply by altering the optical path-length difference. For spatially distinguished modes this can be done by altering the actual length travelled or using a thickness of refractive material (e.g. glass), whereas for polarization-distinguished modes, a thickness of bi-refringent material (e.g. calcite) can be used. This gate can be modelled by the unitary $\hat{U}(\phi) = \exp\left[-\mathrm{i}(\phi/2)(\hat{a}_1^\dagger \hat{a}_1 - \hat{a}_2^\dagger \hat{a}_2)\right]$.

Exercise 7.36 *Show that $X = \hat{a}_1^\dagger \hat{a}_2 + \hat{a}_1 \hat{a}_2^\dagger$ and $Z = \hat{a}_1^\dagger \hat{a}_1 - \hat{a}_2^\dagger \hat{a}_2$ act as the indicated Pauli operators on the logical states defined above.*

By concatenating arbitrary rotations around the X and Z axes of the Bloch sphere of the qubit, one is able to implement arbitrary single-qubit gates.

Exercise 7.37 *Convince yourself of this.*
Hint: *For the mathematically inclined, consider the Lie algebra generated by X and Z (see Section 6.6.2). For the physically inclined, think about rotating an object in three-dimensional space.*

A simple choice for an entangling two-qubit gate is the conditional sign-change (CS) gate. In the logical basis it is defined by

$$|x\rangle_L|y\rangle_L \rightarrow \mathrm{e}^{\mathrm{i}\pi x \cdot y}|x\rangle_L|y\rangle_L. \tag{7.126}$$

This was the sort of interaction considered in Ref. [Mil89], in which the logical basis was the photon-number basis. It is then implementable by a so called mutual-Kerr-effect nonlinear phase shift:

$$\hat{U}_{\mathrm{Kerr}} = \exp[\mathrm{i}\pi a_1^\dagger a_1 a_2^\dagger a_2]. \tag{7.127}$$

This requires the photons to interact via a nonlinear medium. In practice it is not possible to get a single-photon phase shift of π, which this transformation implies, without adding a considerable amount of noise from the medium. However, as we now show, we can realize a CS gate non-deterministically using the dual-rail encoding and the general method introduced in the preceding subsection.

With dual-rail encoding, a linear optical network for a two-qubit gate will have, at most, two photons in any mode. As we will show later, the CS gate can be realized if we can realize the following transformation on a single mode in an arbitrary superposition of no, one and two photons:

$$|\psi\rangle = \alpha_0|0\rangle_1 + \alpha_1|1\rangle_1 + \alpha_2|2\rangle_1 \rightarrow |\psi'\rangle = \alpha_0|0\rangle_1 + \alpha_1|1\rangle_1 - \alpha_2|2\rangle_1, \tag{7.128}$$

with success probability independent of α_n. We will refer to this as a nonlinear sign-change gate (NS gate).

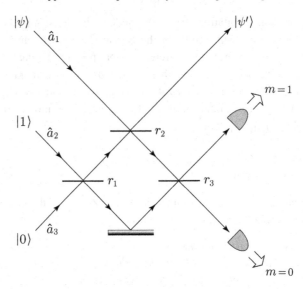

Fig. 7.12 The NS gate $|\psi\rangle \rightarrow |\psi'\rangle$ constructed from three beam-splitters with reflectivities of $r_i = \sin\theta_i$ with $\theta_1 = \theta_3 = 22.5°, \theta_2 = 114.47°$. Adapted by permission from Macmillan Publishers Ltd: *Nature*, E. Knill *et al.*, **409**, 46, Figure 1, copyright 2001.

The NS gate can be achieved using the measurement operator $\hat{M}(10, 10)$ of Eq. (7.122). From Eq. (7.121), for $n \in \{0, 1, 2\}$, we require

$$s_{22} = \lambda,$$

$$s_{22}s_{11} + s_{12}s_{21} = \lambda,$$

$$2s_{11}s_{12}s_{21} + s_{22}s_{11}^2 = -\lambda,$$

for some complex number λ. The phase of λ corresponds to an unobservable global phase shift, while $|\lambda|^2$ is the probability of the measurement outcome under consideration. One solution is easily verified to be

$$S = \begin{pmatrix} 1 - 2^{1/2} & 2^{-1/4} & (3/2^{1/2} - 2)^{1/2} \\ 2^{-1/4} & 1/2 & 1/2 - 1/2^{1/2} \\ (3/2^{1/2} - 2)^{1/2} & 1/2 - 1/2^{1/2} & 2^{1/2} - 1/2 \end{pmatrix}. \qquad (7.129)$$

Here $\lambda = 1/2$, so the success probability is $1/4$. This is the best that can be achieved in a linear optical system via a non-deterministic protocol without some kind of feedforward protocol [Eis05]. An explicit linear optical network to realize this unitary transformation S using three beam-splitters is shown in Fig. 7.12.

In Fig. 7.13, we show how two non-deterministic NS gates can be used to implement a CS gate in dual-rail logic. Here the beam-splitters are all $50:50$ ($\theta = \pi/4$). Since success requires both the NS gates to work, the overall probability of success is $1/16$. A simplification of this scheme that uses only two photons was proposed by Ralph *et al.* [RLBW02].

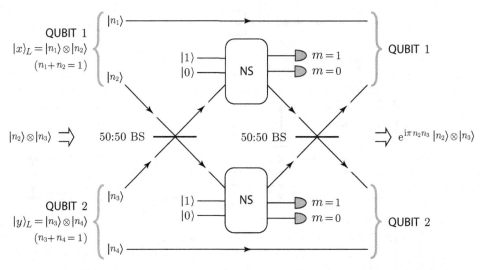

Fig. 7.13 A CS two-qubit gate implemented using two non-deterministic NS gates. This gate has no effect on the two-qubit input state, except to change the sign of the $|1\rangle|1\rangle$ component, as indicated. Adapted by permission from Macmillan Publishers Ltd: *Nature*, E. Knill *et al.*, **409**, 46, Figure 1, copyright 2001.

It is simplified first by setting the beam-splitter parameters r_1 and r_3 to zero in the NS gate implementation (Fig. 7.12), and second by detecting exactly one photon at each logical qubit output. The device is non-deterministic and succeeds with probability of 1/9, but is not scalable insofar as success is heralded by the coincident detection of both photons at distinct detectors: failures are simply not detected at all. It is this simplified gate that was the first to be experimentally realized in 2003 [OPW+03], and it has become the work-horse for LOQC experiments [KMN+07].

7.8.3 Teleporting to determinism

A cascaded sequence of non-deterministic gates is useless for quantum computation because the probability of many gates working in sequence would decrease exponentially. This problem may be avoided by using a protocol based on qubit teleportation as described in Section 7.2. In essence we hold back the gate until we are sure it works and then teleport it on to the required stage of the computation.

The idea that teleportation can be used for universal quantum computation was first proposed by Gottesman and Chuang [GC99]. The idea is to prepare a suitable entangled state for a teleportation protocol with the required gate already applied. We illustrate the idea for teleporting a C-NOT gate in Fig. 7.14. Consider two qubits in an unentangled pure state $|\alpha\rangle \otimes |\beta\rangle$ as shown in Fig. 7.14. Now, to teleport these two qubits one can simply teleport them separately, using two copies of the Bell state $|\psi\rangle$ with measurements and feedforward as introduced in Section 7.2. Now say that, before performing the teleportation protocol,

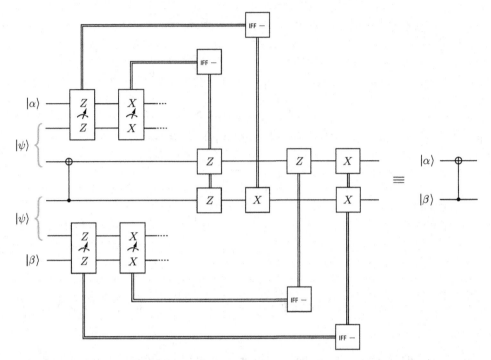

Fig. 7.14 Quantum teleportation of a C-NOT gate on to the state $|\alpha\rangle \otimes |\beta\rangle$ (with $|\beta\rangle$ being the control). The C-NOT at the start of the circuit can be considered part of the preparation of the entangled resource used in the teleportation, which can be discarded and reprepared if this C-NOT fails. Other details are as in Fig. 7.1.

we implement a C-NOT gate between two qubits in the four-qubit state $|\psi\rangle \otimes |\psi\rangle$ – the two qubits that will carry the teleported state. The result is to produce an entangled state (in general) at the output of the dual-rail teleporter, rather than the product state $|\alpha\rangle \otimes |\beta\rangle$. Moreover, by modifying the controls applied in the teleportation protocol the device can be made to output a state that is identical to that which would have been obtained by applying a C-NOT gate directly on the state $|\alpha\rangle \otimes |\beta\rangle$ (with $|\beta\rangle$ being the control).

Exercise 7.38 *Verify that the circuit in Fig. 7.14 works in this way.*

This teleportation of the C-NOT gate works regardless of the initial state of the two qubits. As dicussed above, the C-NOT gate can be realized using a non-deterministic NS gate. The point of the teleportation protocol is that, if it fails, we simply repeat the procedure with another two entangled states $|\psi\rangle \otimes |\psi\rangle$, until the preparation succeeds. When it has succeeded, we perform the protocol in Fig. 7.14. Note that the entangled state $|\psi\rangle$ can also be prepared non-deterministically using a NS gate.

There is one remaining problem with using teleportation to achieve two-qubit gates in LOQC: it requires the measurement of the operators XX and ZZ on a pair of qubits. This

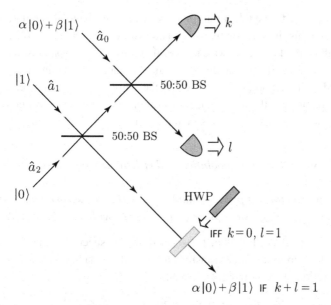

Fig. 7.15 A simple non-deterministic teleportation protocol. The protocol works whenever the total count at the output photon counters is unity.

can be achieved with these two qubits alone, using only single-qubit unitaries, single-qubit measurements in the logical basis and two applications of a C-NOT gate.

Exercise 7.39 *Try to construct the circuit that achieves this using the resources described.*

The problem is that the C-NOT itself is a two-qubit gate! It would seem that this would lead to an infinite regress, with an ever-decreasing probability of success. However, it can be shown that, by using the appropriate entangled resource, the teleportation step can be made near-deterministic. This near-deterministic teleportation protocol requires only photon counting and the ability to perform local quantum control on the basis of these measurement results.

Figure 7.15 shows the basic LOQC quantum teleportation protocol. Note that the states $|0\rangle$ and $|1\rangle$ here are photon-number states, not the dual-rail encoded logical states introduced in Section 7.8.2. That is, the teleporter actually works on a 'single-rail' qubit $|\phi\rangle = \alpha|0\rangle_0 + \beta|1\rangle_0$, transferring it from mode a_0 to mode a_2. A dual-rail qubit can be teleported by teleporting the two single-rail qubits in its two modes. In Section 7.9 we will consider LOQC based on single-rail qubits, for which the teleportation scheme of Fig. 7.15 can be used directly.

The teleportation scheme begins by preparing the ancilla state $|\psi\rangle = (|01\rangle_{12} + |10\rangle_{12})/\sqrt{2}$. In terms of single-rail qubits in modes 1 and 2, this is an entangled state. We denote it $|t_1\rangle_{12}$, because it is a teleportation resource that can be created by sharing one photon between modes 1 and 2, simply using a beam-splitter as shown in Fig. 7.15. A second beam-splitter then mixes modes 0 and 1, and the number of photons in each of these modes is counted. The teleportation works whenever the total count on modes 0 and 1 is unity. To see this, it is instructive to consider what happens in the other cases. If the total count at the output is 0, then we can infer that initially mode a_0 must have been in the vacuum state. Likewise, if the total count is 2 then we can infer that initially mode a_0 must have contained a single photon. In both cases the output photon count serves to measure the number of photons in the input mode, destroying the quantum information there. This is a failure of the teleportation, but a so-called 'heralded' failure because we know it has occurred.

Exercise 7.40 *Show that the probability of this heralded failure is $1/2$, independently of $|\phi\rangle$.*
Hint: *First show that, for this purpose, mode a_1 entering the second beam-splitter can be considered as being in either state $|0\rangle$ or state $|1\rangle$, each with probability $1/2$.*

If the teleporter does not fail as just described, then it succeeds. That is, the input state appears in mode 2 up to a simple transformation without having interacted with mode 2 after the preparation of the initial ancilla state.

Exercise 7.41 *Taking the beam-splitter transformations to be*

$$|01\rangle \rightarrow \frac{1}{\sqrt{2}}(|01\rangle + |10\rangle), \qquad |10\rangle \rightarrow \frac{1}{\sqrt{2}}(|10\rangle - |01\rangle), \qquad (7.130)$$

show that the conditional states at mode a_2 when $k + l = 1$ are given by

$$|\psi\rangle_2 = \begin{cases} \alpha|0\rangle_2 + \beta|1\rangle_2 & \text{for } k = 1, l = 0, \\ \alpha|0\rangle_2 - \beta|1\rangle_2 & \text{for } k = 0, l = 1. \end{cases} \qquad (7.131)$$

In the second case, the state can be corrected back to $|\phi\rangle$ by applying the operator Z. In this single-rail case, this corresponds simply to an optical phase shift of π.

The probability of success of the above teleporter is $1/2$, which is not acceptable. However, we can improve the probability of successful teleportation to $1 - 1/(n + 1)$ by generalizing the initial entangled resource from $|t_1\rangle_{12}$ to an n-photon state

$$|t_n\rangle_{1\cdots(2n)} = \sum_{j=0}^{n} |1\rangle^{\otimes j}|0\rangle^{\otimes(n-j)}|0\rangle^{\otimes j}|1\rangle^{\otimes(n-j)}/\sqrt{n+1}. \qquad (7.132)$$

Here the notation $|a\rangle^{\otimes j}$ means j copies of the state $|a\rangle$: $|a\rangle \otimes |a\rangle \otimes \cdots \otimes |a\rangle$. The modes are labelled 1 to $2n$, from left to right. Note that this could be thought of as a state of $2n$ single-rail qubits, or n dual-rail qubits (with the kth qubit encoded in modes $n + k$ and k). States of this form can be prepared non-deterministically 'off-line' (i.e. prior to being needed in the quantum computation itself).

To teleport the state $\alpha|0\rangle_0 + \beta|1\rangle_0$ using $|t_n\rangle$, we first couple the modes 0 to n by a unitary transformation $\hat{F}_{n+1}$, which implements an $(n + 1)$-point Fourier transform on these modes:

$$\hat{a}_k \to \frac{1}{\sqrt{n+1}} \sum_{l=0}^{n} e^{i2\pi kl/(n+1)} \hat{a}_l. \tag{7.133}$$

Since this is linear, it can be implemented with passive linear optics; for details see Ref. [KLM01]. After applying $\hat{F}_{n+1}$, we measure the number of photons in each of the modes 0 to n.

Suppose this measurement detects k photons altogether. It is possible to show that, if $0 < k < n+1$, then the teleported state appears in mode $n+k$ and only needs to be corrected by applying a phase shift. The modes $2n - l$ are in state $|1\rangle$ for $0 \le l < (n-k)$ and can be reused in future preparations requiring single photons. The remaining modes are in the vacuum state $|0\rangle$. If $k = 0$ the input state is measured and projected to $|0\rangle_0$, whereas if $k = n+1$ it is projected to $|1\rangle_0$. The probability of these two failure events is $1/(n+1)$, regardless of the input. Note that both the necessary correction and which mode we teleported to are unknown until after the measurement.

Exercise 7.42 *Consider the above protocol for $n = 3$. Show that*

$$|t_2\rangle = \frac{1}{\sqrt{3}}(|0011\rangle + |1001\rangle + |1100\rangle). \tag{7.134}$$

Say the results of photon counting on modes 0, 1 and 2 are r, s and t, respectively. Show that the teleportation is successful iff $0 < r + s + t < 3$. Compute the nine distinct conditional states that occur in these instances and verify that success occurs with probability of 2/3.

The problem with the approach presented above is that, for large n, the obvious networks for preparing the required states have very low probabilities of success, but to attain the stringent accuracy requirements for quantum computing [NC00] one does require large n. However, it is possible to make use of the fact that failure is detected and corresponds to measurements in the photon-number basis. This allows exponential improvements in the probability of success for gates and state production with small n, using quantum codes and exploiting the properties of the failure behaviour of the non-deterministic teleportation. For details see Knill *et al.* [KLM01]. Franson [FDF+02] suggested a scheme by which the probability of unsuccessfully teleporting the gate will scale as $1/n^2$ rather than $1/n$ for large n. Unfortunately the price is that gate failure does not simply result in an accidental qubit error, making it difficult to scale.

Some important improvements have been made to the original scheme, making quantum optical computing experimentally viable. Nielsen [Nie04] proposed a scheme based on the cluster-state model of quantum computation. This is an alternative name for the 'one-way' quantum computation introduced in Ref. [RB01], in which quantum measurement and control completely replace the unitary two-qubit gates of the conventional circuit model. A large entangled state (the 'cluster') is prepared, then measurements are performed on individual qubits, and the results are used to control single-qubit unitaries on other qubits in the cluster, which are then measured, and so on. The cluster state does not have to be completely constructed before the computation begins; to an extent it can be assembled on the fly, as described by Browne and Rudoplph [BR05]. This allows LOQC with far fewer physical resources than the original KLM scheme. The assembly of cluster states relies on the basic non-deterministic teleportation protocol introduced above, in which success is heralded. What makes these schemes viable is that failure corresponds to an accidental qubit measurement, as noted above, and the deleterious effect of this is restricted to a single locality within the growing cluster and can be repaired.

To conclude, we summarize the physical requirements for scalable linear optics quantum computation: (i) single-photon sources; (ii) fast, efficient single-photon detectors; (iii) low-loss linear optical networks; and (iv) fast electro-optical control. The specifications for the single-photon source are particularly challenging: it must produce a sequence of identical single-mode pulses containing exactly one photon. The key quality test is that it must be possible to demonstrate very high visibility in the interference of photons in successive pulses (known as Hong–Ou–Mandel interference [HOM87]). Potential single-photon sources have been demonstrated [MD04]. The requirement for fast single-photon detectors that can reliably distinguish zero, one and two photons is also difficult with current technology, but is achievable by a variety of means. Low photon loss is not in principle a problem, and there are LOQC protocols that can correct for loss [KLM01]. The required electro-optical control (which can be considered feedforward) is also not a problem in principle. However, it is certainly challenging because the slowness of electro-optics requires storing photons for microsecond time-scales. For small numbers of gates some of these technical requirements can be ignored, and there is a considerable body of experimental work in this area [KMN⁺07, LZG⁺07].

7.9 Adaptive phase measurement and single-rail LOQC

In Section 7.8.2 we introduced the dual-rail encoding for LOQC in order to show how one- and two-qubit gates could be implemented. However, as noted in Section 7.8.3, the basic unit for two-qubit gates, namely the non-deterministic teleporter of Fig. 7.15, works on single-rail qubits. This suggests that it is worth considering using single-rail encoding, if one could work out a way to do single-qubit unitaries in the two-dimensional subspace spanned by the single-mode Fock states $|0\rangle$ and $|1\rangle$. This idea was first tried in Ref. [LR02], where non-deterministic single-rail single-qubit gates were constructed. However, these gates had low probability of success and, moreover, to obtain high fidelities required commensurately many resources. The resources used were coherent states and photon counting.

More recently, it was shown [RLW05] that, by adding an extra resource, namely dyne detection and the ability to do feedback, the resource consumption of single-rail logic could be dramatically reduced. Specifically, this allowed a *deterministic* protocol to prepare arbitrary single-rail qubits and to convert dual-rail qubits into single-rail qubits. Moreover, the reverse conversion can be done (albeit non-deterministically), thus allowing dual-rail single-qubit gates to be applied to single-rail qubits. The basic idea is to use feedback and dyne measurement to create an *adaptive phase measurement* [Wis95], as was realized experimentally [AAS⁺02]. Since this section ties together the idea of adaptive measurement, from Chapter 2, with continuous quantum measurement and feedback theory, from Chapters 4–6, to find an application in quantum information processing, it seems a fitting topic on which to end this book.

7.9.1 Dyne measurement on single-rail qubits

Consider a single optical mode within a high-quality cavity prepared in state $|\psi(0)\rangle$ at time $t = 0$. Say the cavity has only one output beam, giving rise to an intensity decay rate γ. Say also that this beam is subject to unit-efficiency dyne detection with the local oscillator having phase $\Phi(t)$ (for homodyne detection, this phase is time-independent). Then, working in the frame rotating at the optical frequency, the linear quantum trajectory describing the system evolution is (see Section 4.4.3)

$$d|\bar{\psi}_J(t)\rangle = \left[-\tfrac{1}{2}\gamma\hat{a}^\dagger\hat{a}\,dt + \sqrt{\gamma}e^{-i\Phi(t)}\hat{a}J(t)dt\right]|\bar{\psi}_J(t)\rangle. \tag{7.135}$$

Here $J(t)$ is the dyne current, which is ostensibly white noise in order for $\langle\bar{\psi}_J(t)|\bar{\psi}_J(t)\rangle$ to be the appropriate weight for a particular trajectory (see Section 4.4.3).

Now say that the mode initially contains at most one photon: $|\psi(0)\rangle = c_0|0\rangle + c_1|1\rangle$. Then there is a simple analytical solution for the conditioned state:

$$|\bar{\psi}_J(t)\rangle = (c_0 + c_1 R_t^*)|0\rangle + c_1 e^{-\gamma t/2}|1\rangle, \tag{7.136}$$

where R_t is a functional of the dyne photocurrent record up to time t:

$$R_t = \int_0^t e^{i\Phi(s)}e^{-\gamma s/2}\sqrt{\gamma}\,J(s)ds. \tag{7.137}$$

Exercise 7.43 *Show this.*

The measurement is complete at time $t = \infty$, and the probability of obtaining a particular measurement record $\mathbf{J} = \{J(s)\colon 0 \le s < \infty\}$ is

$$\wp(\mathbf{J}) = \langle\bar{\psi}_\mathbf{J}(t)|\bar{\psi}_\mathbf{J}(t)\rangle\wp_{\text{ost}}(\mathbf{J}). \tag{7.138}$$

Here $\wp_{\text{ost}}(\mathbf{J})$ is the ostensible probability of $\mathbf{J}$; that is, the distribution it would have if $J(t)dt$ were equal to a Wiener increment $dW(t)$. Now, from the above solution (7.136), $\wp(\mathbf{J})$ depends upon the system state only via the single complex functional $A = R_\infty$. That is, all of the information about the system in the complete dyne record $\mathbf{J}$ is contained in the complex number A. We can thus regard the dyne measurement in this case as a measurement yielding the result A, with probability distribution

$$\wp(A)d^2A = |c_0 + c_1 A^*|^2\wp_{\text{ost}}(A)d^2A. \tag{7.139}$$

Here $\wp_{\text{ost}}(A)$ is the distribution for A implied by setting $J(t)dt = dW(t)$. Thus, the measurement can be described by the POM

$$\hat{E}(A)d^2A = (|0\rangle + A|1\rangle)(\langle 0| + A^*\langle 1|)\wp_{\text{ost}}(A)d^2A. \qquad (7.140)$$

In the above the shape of the mode exiting from the cavity is a decaying exponential $u(t) = \gamma e^{-\gamma t}$. The mode-shape $u(t)$ means, for example, that the mean photon number in the part of the output field emitted in the interval $[t, t + dt)$ is $|c_1|^2 u(t)dt$.

Exercise 7.44 *Verify this using the methods of Section 4.7.6.*

We can generalize the above theory to dyne detection upon a mode with an arbitrary mode-shape $u(t)$, such that $u(t) \geq 0$ and $U(\infty) = 1$, where

$$U(t) = \int_0^t u(s)ds. \qquad (7.141)$$

We do this by defining a time-dependent decay rate, $\gamma(t) = u(t)/[U(t) - 1]$. Then we can consider modes with finite duration $[0, T]$, in which case $U(T) = 1$. For a general mode-shape, Eq. (7.140) still holds, but with $A = R_T$ and

$$R_t = \int_0^t e^{i\Phi(s)}\sqrt{u(s)}\, J(s)ds. \qquad (7.142)$$

Exercise 7.45 *Show this by considering $\gamma(t)$ as defined above.*

7.9.2 Adaptive phase measurement on single-rail qubits

For any time $t < T$ the measurement is *incomplete* in the sense defined in Section 1.4.2. Thus one can change the sort of information obtained about the system by *adapting* the measurement at times $0 < t < T$ (see Section 2.5.2). In the present context, the only parameter that can be controlled by a feedback loop is the local oscillator phase $\Phi(t)$. This allows adaptive dyne detection as discussed in Section 2.6. Indeed, here we consider the adaptive scheme introduced in Ref. [Wis95] which was realized in the experiment of Armen *et al.* [AAS$^+$02]. This is to set

$$\Phi(t) = \arg R_t + \pi/2. \qquad (7.143)$$

From Eq. (7.142), we can then write a differential equation for R as

$$dR_t = i\frac{R_t}{|R_t|}\sqrt{u(t)}\, J(t)dt. \qquad (7.144)$$

Bearing in mind that $[J(t)dt]^2 = dt$ (both ostensibly and actually), this has the solution

$$R_t = \sqrt{U(t)}e^{i\varphi(t)}, \qquad (7.145)$$

where

$$\varphi(t) = \int_0^t \sqrt{\frac{u(s)}{U(s)}} J(s) \mathrm{d}s. \tag{7.146}$$

Exercise 7.46 *Verify Eq. (7.145).*
Hint: *Consider first the SDE for $|R|^2$ and then that for $\varphi(t) = [1/(2\mathrm{i})] \ln(R_t / R_t^*)$.*

Now ostensibly $J(s)\mathrm{d}s = \mathrm{d}W(s)$, so $\varphi(t)$ is a Gaussian random variable with mean zero and variance

$$\int_0^t \frac{u(s)}{U(s)} \mathrm{d}s = \log\left(\frac{U(t)}{U(0)}\right). \tag{7.147}$$

But $U(0) = 0$ by definition, so ostensibly the variance of $\varphi(t)$ is infinite. That is, under this adaptive scheme $A = R_T$ describes a variable with a random phase and a modulus of unity. Since the modulus is deterministic, it can contain no information about the system. Thus, under this scheme, all of the information is contained in $\theta = \arg(A)$. Since this is ostensibly random, the POM for this measurement is, from Eq. (7.140),

$$\hat{E}(\theta)\mathrm{d}\theta = |\theta\rangle\langle\theta| \frac{\mathrm{d}\theta}{\pi}. \tag{7.148}$$

Here $|\theta\rangle = (|0\rangle + e^{\mathrm{i}\theta}|1\rangle)/\sqrt{2}$ is a truncated phase state [PB97]. This is precisely Example 2 introduced in Section 1.2.5 to illustrate measurements for which the effect $\hat{E}(\theta)\mathrm{d}\theta$ is not a projector.

This adaptive dyne detection is useful for estimating the unknown phase of an optical pulse [Wis95, WK97, WK98, AAS$^+$02], but for LOQC we are interested only in its role in state preparation when the system mode is entangled with other modes. Say the total state is $|\Psi\rangle$. Then, from Eq. (7.148), the conditioned state of the other modes after the measurement yielding result θ is

$$\langle\theta|\Psi\rangle/\sqrt{\pi}, \tag{7.149}$$

where the squared norm of this state is equal to the probability density for obtaining this outcome. We now discuss applications of this result.

7.9.3 Preparing arbitrary single-rail qubit states

A basic qubit operation is the preparation of superposition states. For dual-rail encoding in LOQC this is trivial to perform by making single-qubit gates act on a one-photon state, as described in Section 7.8.2. Arbitrary single-rail superposition states, $\alpha|0\rangle + e^{-\mathrm{i}\phi}\sqrt{1-\alpha^2}|1\rangle$, with α and ϕ real numbers, are not so easy to produce. Previous suggestions for deterministic production of such states involved nonlinearities significantly larger than is currently feasible. Alternatively, non-deterministic techniques based on photon counting [PPB98, LR02] have been described and experimentally demonstrated [LM02], but these have low probabilities of success. A non-deterministic scheme based on homodyne detection has also been demonstrated [BBL04], but has a vanishing probability of success for high-fidelity preparation. We now show that it is possible deterministically to produce an arbitrary single-rail state from a single-photon state using linear optics and adaptive phase measurements.

We begin by splitting a single photon into two modes at a beam-splitter with intensity reflectivity η, producing $|\Psi\rangle = \sqrt{\eta}|1\rangle|0\rangle + \sqrt{1-\eta}|0\rangle|1\rangle$. If we then carry out an adaptive phase measurement on the first mode we obtain a result θ, which prepares the second mode in state

$$\sqrt{\eta}|0\rangle + e^{-i\theta}\sqrt{1-\eta}|1\rangle. \tag{7.150}$$

Exercise 7.47 *Verify this, and show that the result θ is completely random (actually random, not just ostensibly random).*

Now, by feedforward onto a phase modulator on the second mode, this random phase can be changed into any desired phase θ'. Thus we can deterministically produce the arbitrary state

$$\sqrt{\eta}|0\rangle + e^{-i\theta'}\sqrt{1-\eta}|1\rangle. \tag{7.151}$$

7.9.4 Quantum gates using adaptive phase measurements

We now have to show how to perform single-qubit unitaries on our single-rail qubits. Some of these are easy, such as the phase rotation used above. However, others, such as the Hadamard gate (which is essential for quantum computation) are more difficult. This is defined by the transformations $|0\rangle \rightarrow (|0\rangle + |1\rangle)/\sqrt{2}$ and $|1\rangle \rightarrow (|0\rangle - |1\rangle)/\sqrt{2}$. A Hadamard transformation plus arbitrary phase rotation will allow us to perform arbitrary single-qubit unitaries [LR02]. A non-deterministic Hadamard transformation for single-rail qubits based on photon counting was described in Ref. [LR02], but its success probability was very low. Here we show that a Hadamard transformation based on a combination of photon counting and adaptive phase measurements, whilst still non-deterministic, can have a much higher success probability.

The key observation is that a deterministic mapping of dual-rail encoding into single-rail encoding can be achieved using adaptive phase measurements. Consider the arbitrary dual-rail qubit $\alpha|01\rangle + e^{-i\phi}\sqrt{1-\alpha^2}|10\rangle$. Suppose an adaptive phase measurement is made on the second rail of the qubit, giving the result θ. If a phase shift of $-\theta$ is subsequently imposed on the remaining rail of the qubit, the resulting state is $\alpha|0\rangle + e^{-i\phi}\sqrt{1-\alpha^2}|1\rangle$, which is a single-rail qubit with the same logical value as the original dual-rail qubit.

What about the reverse operation from a single-rail encoded qubit to a dual-rail encoded qubit? It does not appear to be possible to do this deterministically with only linear optics. However, a non-deterministic transformation is possible by teleporting between encodings. Dual-rail teleportation can be achieved using a dual-rail Bell state such as $|01\rangle|10\rangle + |10\rangle|01\rangle$. (We are ignoring normalization for convenience.) Single-rail teleportation can be

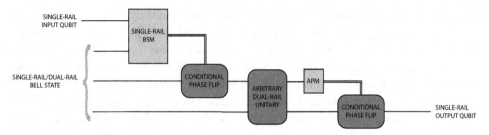

Fig. 7.16 A schematic representation of the application of an arbitrary single-qubit unitary to a single-rail qubit. The single-rail qubit is teleported onto a dual-rail qubit, the unitary is applied, then an adaptive phase measurement is used to convert back to a single-rail qubit. BSM means Bell-state measurement and APM means adaptive phase measurement. All operations are deterministic except the Bell-state measurement, which succeeds 50% of the time. Based on Figure 1 from T. C. Ralph *et al.*, Adaptive Phase Measurements in Linear Optical Quantum Computation, *Journal of Optics B* **7**, S245, (2005), IOP Publishing Ltd.

achieved using a single-rail Bell state such as $|0\rangle|1\rangle + |1\rangle|0\rangle$. In both cases only two of the four Bell states can be identified with linear optics, so the teleportation works 50% of the time, as illustrated in Fig. 7.16.

Now suppose we take a dual-rail Bell state and use an adaptive phase measurement to project one of its arms into a single-rail state. We obtain the state $|0\rangle|10\rangle + |1\rangle|01\rangle$, which is Bell entanglement between dual- and single-rail qubits. If we now perform a Bell measurement between the single-rail half of the entanglement and an arbitrary single-rail qubit then (when successful) the qubit will be teleported onto the dual-rail part of the entanglement, thus converting a single-rail qubit into a dual-rail qubit.

We now have a way of (non-deterministically) performing an arbitrary rotation on an unknown single-rail qubit. The idea is depicted schematically in Fig. 7.16. First we teleport the single-rail qubit onto a dual-rail qubit. Then we perform an arbitrary rotation on the dual-rail qubit. We then use an adaptive phase measurement to transform the dual-rail qubit back into a single-rail qubit. The only non-deterministic step is the Bell measurement in the teleportation, which in this simple scheme has a success probability of 50%. This is a major improvement over previous schemes. As discussed in Section 7.8.3, the success probability for this step can be increased arbitrarily by using larger entangled resources. Also as discussed in that section, the fundamental two-qubit gate, the CS gate, is in fact a single-rail gate. Thus, by employing quantum feedback control we are able to perform universal quantum computation in LOQC using single-rail encoding.

7.10 Further reading

There are many other applications of measurement and control in quantum information processing besides those mentioned in this chapter. Here are a few of them.

Rapid purification of a qubit. A two-level quantum system initially in a completely mixed state will gradually purify to a $\hat{\sigma}_z$ eigenstate under a continuous QND measurement of $\hat{\sigma}_z$ (as in Eq. (5.204) of Section 5.7, but for a single spin). It was shown by Jacobs [Jac03] that, using feedback control to make the state always unbiased with respect to $\hat{\sigma}_z$ (that is, to keep the system Bloch vector in the x–y plane), the information gain from the QND measurement is greater. That is, the rate of increase of the *average purity* of the system can be increased. Moreover, this is achieved by Markovian feedback of the QND current with a time-varying feedback strength, as in Section 5.7. In the asymptotic limit of high purity (long times) the system can be purified to any given degree using measurement and feedback in half the time it would take from the measurement alone. Note, however, that different results are obtained from considering the *average time* required to obtain a given level of purity – see Refs. [WR06, CWJ08, WB08].

Mitigating the effect of a noisy channel. Consider a qubit prepared in one of two non-orthogonal states in the x–z plane of the Bloch sphere, with the same value of x but opposite values of z. Say this qubit is subjected to dephasing noise; that is, a rotation around the z axis by a random angle described by some probability distribution. The task is to use measurement and feedback control to attempt to correct the state of the qubit; that is, to undo the effect of the *noisy channel*. It was demonstrated in Ref. [BMG$^+$07] that projective measurements are not optimal for this task and that there exists a non-projective measurement with an optimum measurement strength that achieves the best trade-off between gaining information about the system (or the noise) and disturbing it through measurement back-action. Moreover, a quantum control scheme that makes use of this weak measurement followed by feedback control is provably optimal for ameliorating the effect of noise on this system.

Controlling decoherence by dynamical decoupling. In this chapter we have discussed methods of controlling decoherence that are based on quantum error correction, both for undetected errors (conventional quantum error correction) and for detected errors. An alternative is to try to prevent the errors from happening in the first place by decoupling the system from its environment. *Dynamical decoupling*, introduced by Viola and co-workers [VLK99], uses open-loop control, without ancillae or measurement. On the basis of the idea of bang-bang control [VL98], the quantum system is subjected to a sequence of impulsive unitary transformations so that the evolution is described on longer time-scales by an effective modified Hamiltonian in which unwanted interactions are suppressed. In later work [SV06] the idea of a randomly generated (but known to the experimenter) sequence of unitary control pulses was shown to overcome some of the limitations for regular dynamical decoupling when there are rapidly fluctuating interactions or when the usual deterministically generated sequence of control pulses would be too long to implement.

Adaptive phase estimation inspired by quantum computing. At the heart of Shor's 1994 factoring algorithm is a routine known as the quantum phase-estimation algorithm [NC00]. It relates to accurately estimating the eigenvalues of an unknown unitary operator and involves an algorithm called the quantum Fourier transform (QFT). As shown in Ref. [GN96], the QFT algorithm can be performed using single-qubit measurement and control, much as in cluster-state quantum computing as discussed above. In fact, the quantum phase-estimation algorithm can be used as an (adaptive) protocol for estimating the phase ϕ in a single-qubit phase gate $\exp(i\phi Z/2)$, using *only* single-qubit operations (preparations, measurements and control), as long as it is possible for the gate to be applied multiple times to a given single qubit between preparation and measurement [GLM06]. The quantum phase-estimation algorithm enables a canonical measurement of phase (see Section 2.4), but the nature of the prepared states means that it does not attain the Heisenberg limit for the phase variance (2.133). However, a simple generalization of the quantum phase-estimation algorithm, using the principles of adaptive phase estimation discussed in Section 2.5, enables a variance scaling at the Heisenberg limit to be achieved, with an overhead factor of less than 2.5 [HBB$^+$07]. Moreover, this was recently demonstrated experimentally by Higgins *et al.* using single-photon multi-pass interferometry [HBB$^+$07] – the first experiment to demonstrate Heisenberg-limited scaling for phase estimation. In this chapter, we have concentrated on showing that quantum computing can benefit from an understanding of quantum measurement and control, but this work demonstrates that the converse is also true.

Appendix A

Quantum mechanics and phase-space

A.1 Fundamentals of quantum mechanics

A.1.1 Pure states and operators

The state of a quantum-mechanical system corresponding to maximal knowledge is known as a *pure* state. It is represented by a vector in Hilbert space $\mathbb{H}$, a complex vector space with an arbitrarily large dimensionality. We use Dirac's notation and write the basis vectors for the Hilbert space as $|\phi_j\rangle$ for $j \in \{0, 1, 2, \ldots, D-1\}$. Strictly the formulation we present below holds only for finite D, but the generalizations for infinite D present few problems, although there are some traps for the unwary. For a fuller discussion, see, for example, the excellent book by Ballentine [Bal98].

We define an inner product on the Hilbert space so that the basis states are orthonormal: $\langle\phi_i|\phi_j\rangle = \delta_{ij}$. Then, for a complete basis, we can write an arbitrary pure state, or *state vector*, or *ket* as

$$|\psi\rangle = \sum_i \psi_i |\phi_i\rangle, \tag{A.1}$$

where, for all i, $\psi_i \in \mathbb{C}$ (the complex numbers). The dual vector, or *bra*, is defined as

$$\langle\psi| = \sum_i \psi_i^* \langle\phi_i|. \tag{A.2}$$

If the state vector is to be normalized we require the inner product, or *bracket*, to satisfy

$$\langle\psi|\psi\rangle = \sum_i |\psi_i|^2 = 1. \tag{A.3}$$

In order to relate the state to quantities of physical interest, we need to introduce *operators*. An operator is an object that maps a Hilbert space onto itself, and so can be written in Dirac notation as

$$\hat{A} = \sum_{ij} A_{ij} |\phi_i\rangle\langle\phi_j|, \tag{A.4}$$

where the A_{ij} are complex numbers. Operators are sometimes called q-numbers, meaning quantum numbers, as opposed to c-numbers, which are ordinary classical or complex numbers. Ignoring some subtle issues to do with infinite-dimensional Hilbert spaces, we can simply state that all physical quantities (commonly called *observables*) are associated with *Hermitian* operators. An Hermitian operator is one that is equal to its Hermitian adjoint, defined as

$$\hat{A}^{\dagger} = \sum_{ij} A_{ji}^{*} |\phi_i\rangle \langle \phi_j|. \tag{A.5}$$

In matrix terms, $A = (A^*)^{\mathsf{T}}$. In Table A.1 we summarize definitions such as these and their relations to linear algebra. A knowledge of the results in this table is assumed in subsequent discussions.

If $\hat{\Lambda}$ is the operator which represents a physical quantity Λ, then the simplest connection we can make with the state of the system $|\psi\rangle$ is that the mean value of Λ is given by

$$\langle \Lambda \rangle = \langle \psi | \hat{\Lambda} | \psi \rangle, \tag{A.6}$$

where, unless otherwise stated, we take the ket to be normalized. We derive this expression from more basic considerations in Section 1.2.2.

Note that Eq. (A.6) shows that the *absolute* phase of a state plays no physical role; $e^{i\phi}|\psi\rangle$ gives the same mean value for all observables as does $|\psi\rangle$. Of course the *relative* phase of states in a superposition *does* matter. That is, for a state such as $e^{i\phi_1}|\psi_1\rangle + e^{i\phi_2}|\psi_2\rangle$, the average value of physical quantities will depend in general upon $\phi_2 - \phi_1$.

Exercise A.1 *Convince yourself of these statements.*

Any Hermitian operator $\hat{\Lambda}$ can be diagonalized as

$$\hat{\Lambda} = \sum_{\lambda} \lambda |\lambda\rangle \langle \lambda|, \tag{A.7}$$

where $\{\lambda\}$ are the eigenvalues of $\hat{\Lambda}$ which are real, while $\{|\lambda\rangle\}$ forms a complete basis. Here for simplicity we have taken the spectrum – the set of eigenvalues – to be discrete and non-degenerate (that is, all eigenvalues are different).

Exercise A.2 *Using this representation, show that $\langle \Lambda \rangle$ is real.*

If we assume that the operator for Λ^2 is $\hat{\Lambda}^2$ (which is justified in Section 1.2.2), then it is not difficult to show that the variance in Λ,

$$\mathrm{Var}[\Lambda] = \langle \psi | \hat{\Lambda}^2 | \psi \rangle - \langle \psi | \hat{\Lambda} | \psi \rangle^2, \tag{A.8}$$

is in general greater than zero. This is the puzzling phenomenon of quantum noise; even though we have a state of maximal knowledge about the system, there is still some uncertainty in the values of physical quantities. Moreover, it is possible to derive so-called *uncertainty relations* of the form

$$\mathrm{Var}[\Lambda]\mathrm{Var}[B] \geq |\langle \psi | [\hat{\Lambda}, \hat{B}] | \psi \rangle|^2 / 4, \tag{A.9}$$

where $[\hat{\Lambda}, \hat{B}] \equiv \hat{\Lambda}\hat{B} - \hat{B}\hat{\Lambda}$ is called the commutator. If the commutator is a c-number (that is, it is proportional to the identity operator), this relation puts an absolute lower bound on the product of the two uncertainties.

Exercise A.3 *The position Q and momentum P of a particle have operators that obey $[\hat{P}, \hat{Q}] = -i\hbar$ (see Section A.3). Using Eq. (A.9), derive Heisenberg's uncertainty relation*

$$\langle (\Delta P)^2 \rangle \langle (\Delta Q)^2 \rangle \geq (\hbar/2)^2, \tag{A.10}$$

where $\Delta P = P - \langle P \rangle$ and similarly for Q.

Table A.1. *Linear algebra and quantum mechanics*

Linear algebra		Quantum operator algebra	
$\mathbb{H}$	Complex vector space	$\mathbb{H}$	Hilbert space
$\dim(\mathbb{H})$	dimension D	$\dim(\mathbb{H})$	dimension D
A	Matrix	$\hat{A}$	Operator
$\vec{v}$	Column vector	$|\psi\rangle$	State vector or ket
$(\vec{v}^*)^\top$	Conjugate row vector	$\langle\psi|$	Bra
$(\vec{v}^*)^\top\vec{u}$	Inner or dot product $\vec{v}\cdot\vec{u}$	$\langle\psi|\theta\rangle$	Inner product
$\vec{v}\cdot\vec{u}=0$	Orthogonality	$\langle\psi|\theta\rangle=0$	Orthogonality
$\vec{v}\cdot\vec{v}=1$	Unit vector	$\langle\psi|\psi\rangle=1$	Normalized state vector
$\vec{u}(\vec{v}^*)^\top$	Outer product (a matrix)	$|\theta\rangle\langle\psi|$	Outer product (an operator)
$\{\vec{\alpha}\}$	Eigenvectors $A\vec{\alpha}=\alpha\vec{\alpha}$	$\{|\alpha\rangle\}$	Eigenstates $\hat{A}|\alpha\rangle=\alpha|\alpha\rangle$
$\{\alpha\}$	Eigenvalues (complex)	$\{\alpha\}$	Eigenvalues (complex)
$(A^*)^\top$	Hermitian adjoint	$\hat{A}^\dagger$	Hermitian adjoint
U	Unitary matrix $(U^*)^\top=U^{-1}$	$\hat{U}$	Unitary operator $\hat{U}^\dagger=\hat{U}^{-1}$
Λ	Hermitian matrix $\Lambda=(\Lambda^*)^\top$	$\hat{\Lambda}$	Hermitian operator $\hat{\Lambda}=\hat{\Lambda}^\dagger$
	$\implies\{\lambda\}$ Real eigenvalues		$\implies\{\lambda\}$ Real eigenvalues
	$\implies\{\vec{\lambda}\}$ Orthogonal eigenvectors		$\implies\{|\lambda\rangle\}$ Orthogonal eigenstates
$\{\vec{e}_j\}_{j=0}^{D-1}$	Orthonormal basis	$\{|\phi_j\rangle\}_{j=0}^{D-1}$	Orthonormal basis
	$\implies\vec{e}_j\cdot\vec{e}_k=\delta_{jk}$		$\implies\langle\phi_j|\phi_k\rangle=\delta_{jk}$
U	Change of basis $\vec{e}_j{}'=\sum_k U_{jk}\vec{e}_k$	$\hat{U}$	Change of basis $|\phi_j'\rangle=\sum_k U_{jk}|\phi_k\rangle$
I	Identity $\sum_j(\vec{e}_j^*)\vec{e}_j^\top$	$\hat{I}$	Identity $\sum_j|\phi_j\rangle\langle\phi_j|$
v_j	Vector component $\vec{e}_j\cdot\vec{v}$	ψ_j	Probability amplitude $\psi_j=\langle\phi_j|\psi\rangle$
A_{jk}	Matrix element $(\vec{e}_j^*)^\top A\vec{e}_k$	A_{jk}	Matrix element $\langle\phi_j|\hat{A}|\phi_k\rangle$
$\text{tr }A$	Trace $\sum_j A_{jj}$	$\text{Tr}[\hat{A}]$	Trace $\sum_j\langle\phi_j|\hat{A}|\phi_j\rangle$
$\mathbb{H}=\mathbb{H}_1\otimes\mathbb{H}_2$	Tensor product	$\mathbb{H}=\mathbb{H}_1\otimes\mathbb{H}_2$	Tensor product
	$\implies D=\dim(\mathbb{H})=D_1\times D_2$		$\implies D=\dim(\mathbb{H})=D_1\times D_2$
$\vec{v}_1\otimes\vec{u}_2$	Tensor product	$|\psi\rangle_1\otimes|\theta\rangle_2$	Tensor product $|\psi\rangle_1|\theta\rangle_2$
	$\implies\vec{E}_{k\times D_2+j}=(\vec{e}_k)_1\otimes(\vec{e}_j)_2$		$\implies|\Phi_{k\times D_2+j}\rangle=|\phi_k\rangle_1\otimes|\phi_j\rangle_2$
$A_1\otimes B_2$	Tensor product	$\hat{A}_1\otimes\hat{B}_2$	Tensor product
	$\implies A_1\otimes B_2[\vec{v}_1\otimes\vec{u}_2]=$		$\implies\hat{A}_1\otimes\hat{B}_2[|\psi\rangle_1\otimes|\theta\rangle_2]=$
	$(A_1\vec{v}_1)\otimes(B_2\vec{u}_2)$		$\hat{A}_1|\psi\rangle_1\otimes\hat{B}_2|\theta\rangle_2$

A.1.2 Mixed states

The pure states considered so far are appropriate only if one knows everything one can know about the system. It is easy to imagine situations in which this is not the case. Suppose one has a physical device that prepares a system in one of N states, $|\psi_j\rangle: j = 1, \ldots, N$, with corresponding probabilities $\wp_j$. These states need not be orthogonal, and N can be greater than the dimension of the Hilbert space of the system. We will call the action of this device a *preparation procedure* and the set $\{\wp_j, |\psi_j\rangle: j = 1, 2, \ldots, N\}$ an *ensemble* of pure states.

If one has no knowledge of which particular state is produced, the expected value of a physical quantity is clearly the weighted average

$$\langle \Lambda \rangle = \sum_j \wp_j \langle \psi_j | \hat{\Lambda} | \psi_j \rangle. \tag{A.11}$$

We can combine the classical and quantum expectations in a single entity by defining a new operator. This is called (for historical reasons) the density operator, and is given by

$$\rho = \sum_{j=1}^{N} \wp_j |\psi_j\rangle\langle\psi_j|. \tag{A.12}$$

We can then write

$$\langle \Lambda \rangle = \text{Tr}\left[\rho \hat{\Lambda}\right], \tag{A.13}$$

where the trace operation is defined in Table A.1.

Exercise A.4 *Show this, by first showing that* $\text{Tr}[|\psi\rangle\langle\theta|] = \langle\theta|\psi\rangle$.

The density operator is also known as the *density matrix*, or (in analogy with the state vector) the *state matrix*. It is the most general representation of a quantum state and encodes all of the physical meaningful information about the preparation of the system. Because of its special role, the state matrix is the one operator which we do not put a hat on.

The state matrix ρ is positive: all of its eigenvalues are non-negative. Strictly, it is a positive semi-definite operator, rather than a positive operator, because some of its eigenvalues may be zero. The eigenvalues also sum to unity, since

$$\text{Tr}[\rho] = \sum_j \wp_j = 1. \tag{A.14}$$

In the case in which the ensemble of state vectors has only one element, ρ represents a pure state. In that case it is easy to verify that $\rho^2 = \rho$. Moreover, this condition is sufficient for ρ to be a pure state, since $\rho^2 = \rho$ means that ρ is a projection operator (these are discussed in Section 1.2.2). Using the normalization condition (A.14), it follows that ρ must be a rank-1 projection operator, which we denote as $\hat{\pi}$. That is to say, it must be of the form

$$\rho = |\psi\rangle\langle\psi| \tag{A.15}$$

for some ket $|\psi\rangle$. A state that cannot be written in this form is often called a *mixed* or *impure* state. The 'mixedness' of ρ can be measured in a number of ways. For instance, the *impurity* is usually defined to be one minus the *purity*, where the latter is $p = \text{Tr}\left[\rho^2\right]$.

Exercise A.5 *Show that* $0 \leq p \leq 1$, *with* $p = 1$, *if and only if* ρ *is pure.*
Hint: *The trace of the matrix* ρ^2 *is most easily evaluated in the diagonal basis for* ρ.

Alternatively, one can define the von Neumann entropy

$$S(\rho) = -\text{Tr}[\rho \log \rho] \geq 0, \tag{A.16}$$

where the equality holds if and only if ρ is pure. To obtain a quantity with the dimensions of thermodynamic entropy, it is necessary to multiply it by Boltzmann's constant k_B.

An interesting point about the definition (A.12) is that it is not possible to go backwards from ρ to the ensemble of state vectors $\{\wp_j, |\psi_j\rangle : j = 1, 2, \ldots, N\}$. Indeed, for any Hilbert space, there is an uncountable infinity of ways in which any impure state matrix ρ can be decomposed into a convex (i.e. positively weighted) ensemble of rank-1 projectors. This is quite different from classical mechanics, in which different ensembles of states of complete knowledge correspond to different states of incomplete knowledge. Physically, we can say that any mixed quantum state admits infinitely many preparation procedures.

The non-unique decomposition of a state matrix can be shown up quite starkly using a two-dimensional Hilbert space: an electron drawn randomly from an ensemble in which half are spin up and half are spin down is *identical* to one drawn from an ensemble in which half are spin left and half spin right. No possible experiment can distinguish between them.

Exercise A.6 *Show this by showing that the state matrix under both of these preparation procedures is proportional to the identity.*
Hint: *If the up and down spin basis states are* $|\uparrow\rangle$ *and* $|\downarrow\rangle$, *the left and right spin states are* $|\rightarrow\rangle = (|\uparrow\rangle + |\downarrow\rangle)/\sqrt{2}$ *and* $|\leftarrow\rangle = (|\uparrow\rangle - |\downarrow\rangle)/\sqrt{2}$.

In this case, it is because the state matrix has degenerate eigenvalues that it is possible for both of these ensembles to comprise orthogonal states. If ρ has no degenerate eigenvalues, it is necessary to consider non-orthogonal ensembles to obtain multiple decompositions.

A.1.3 Time evolution

An isolated quantum system undergoes reversible evolution generated by an Hermitian operator called the Hamiltonian or energy operator $\hat{H}$. There are two basic ways of describing this time evolution, called the Schrödinger picture (SP) and the Heisenberg picture (HP). In the former, the state of the system changes but the operators are constant, whereas in the latter the state is time-independent and the operators are time-dependent.

Using units where $\hbar = 1$ (a convention we use in most places in this book, except for parts of Chapter 6), the SP evolution of the state matrix is

$$\frac{\text{d}}{\text{d}t}\rho(t) = -\text{i}[\hat{H}, \rho(t)]. \tag{A.17}$$

For pure states the corresponding equation (the Schrödinger equation) is

$$\frac{\text{d}}{\text{d}t}|\psi(t)\rangle = -\text{i}\hat{H}|\psi(t)\rangle. \tag{A.18}$$

It is easy to see that the solutions of these equations are

$$\rho(t) = \hat{U}(t, 0)\rho(0)\hat{U}^\dagger(t, 0), \qquad |\psi(t)\rangle = \hat{U}(t, 0)|\psi(0)\rangle, \tag{A.19}$$

where $\hat{U}(t, 0) = \exp(-\mathrm{i}\hat{H}t)$. This is called the unitary evolution operator, because it satisfies the unitarity conditions

$$\hat{U}^{\dagger}\hat{U} = \hat{U}\hat{U}^{\dagger} = \hat{1}. \tag{A.20}$$

If the Hamiltonian is a time-dependent operator $\hat{H}(t)$, as would arise from a classical external modulation of the system, then the evolution is still unitary. The evolution operator is

$$\hat{U}(t, 0) = \hat{1} + \sum_{n=1}^{\infty}(-\mathrm{i})^n \int_0^t \mathrm{d}s_n \, \hat{H}(s_n) \int_0^{s_n} \mathrm{d}s_{n-1} \, \hat{H}(s_{n-1}) \cdots \int_0^{s_2} \mathrm{d}s_1 \, \hat{H}(s_1). \tag{A.21}$$

Exercise A.7 *Show this, and show also that $\hat{U}(t, 0)$ is unitary.*
Hint: *Assuming the solutions (A.19), derive the differential equation and initial conditions for $\hat{U}(t, 0)$ and $\hat{U}^{\dagger}(t, 0)$. Then show that Eq. (A.21) satisfies these, and that the unitarity conditions (A.20) are satisfied at $t = 0$ and are constants of motion.*

In the HP, the equation of motion for an arbitrary operator $\hat{A}$ is

$$\frac{\mathrm{d}}{\mathrm{d}t}\hat{A}(t) = +\mathrm{i}[\hat{H}(t), \hat{A}(t)]. \tag{A.22}$$

Note that, because $\hat{H}(t)$ commutes with itself at a particular time t, the Hamiltonian operator is one operator that is the same in both the HP and the SP. The solution of the HP equation is

$$\hat{A}(t) = \hat{U}^{\dagger}(t, 0)\hat{A}(0)\hat{U}(t, 0). \tag{A.23}$$

The two pictures are equivalent because all expectation values are identical:

$$\mathrm{Tr}\big[\hat{A}(t)\rho(0)\big] = \mathrm{Tr}\big[\hat{U}^{\dagger}(t, 0)\hat{A}(0)\hat{U}(t, 0)\rho(0)\big]$$

$$= \mathrm{Tr}\big[\hat{A}(0)\hat{U}(t, 0)\rho(0)\hat{U}^{\dagger}(t, 0)\big]$$

$$= \mathrm{Tr}\big[\hat{A}(0)\rho(t)\big]. \tag{A.24}$$

Here the placement of the time argument t indicates which picture we are in.

Often it is useful to split a Hamiltonian $\hat{H}$ into $\hat{H}_0 + \hat{V}(t)$, where $\hat{H}_0$ is time-independent and easy to deal with, while $\hat{V}(t)$ (which may be time-dependent) is typically more complicated. Then the unitary operator (A.21) can be written as

$$\hat{U}(t, 0) = \mathrm{e}^{-\mathrm{i}\hat{H}_0 t}\hat{U}_{\mathrm{IF}}(t, 0), \tag{A.25}$$

where $\hat{U}_{\mathrm{IF}}(t, 0)$ is given by

$$\hat{1} + \sum_{n=1}^{\infty}(-\mathrm{i})^n \int_0^t \mathrm{d}s_n \, \hat{V}_{\mathrm{IF}}(s_n) \int_0^{s_n} \mathrm{d}s_{n-1} \, \hat{V}_{\mathrm{IF}}(s_{n-1}) \cdots \int_0^{s_2} \mathrm{d}s_1 \, \hat{V}_{\mathrm{IF}}(s_1). \tag{A.26}$$

Here $\hat{V}_{\mathrm{IF}}(t) = \mathrm{e}^{\mathrm{i}\hat{H}_0 t}\hat{V}(t)\mathrm{e}^{-\mathrm{i}\hat{H}_0 t}$ and IF stands for 'interaction frame'.

Exercise A.8 *Show this, by showing that $\mathrm{e}^{-\mathrm{i}\hat{H}_0 t}\hat{U}_{\mathrm{IF}}(t, 0)$ obeys the same differential equation as $\hat{U}(t, 0)$.*

That is, one can treat $\hat{V}_{\mathrm{IF}}(t)$ as a time-dependent Hamiltonian, and then add the evolution $\mathrm{e}^{-\mathrm{i}\hat{H}_0 t}$ at the end. This can be used to define an *interaction picture* (IP), so called because

$\hat{V}(t)$ is often the 'interaction Hamiltonian' coupling two systems, while $\hat{H}_0$ is the 'free Hamiltonian' of the uncoupled systems. The IP is a sort of half-way house between the SP and HP, usually defined so that operators evolve according to the unitary $e^{-i\hat{H}_0 t}$, while states evolve according to the unitary $\hat{U}_{\mathrm{IF}}(t, 0)$. That is, one breaks up the expectation value for an observable A at time t as follows:

$$\langle A(t) \rangle = \mathrm{Tr}\left[\hat{U}^\dagger(t, 0)\hat{A}(0)\hat{U}(t, 0)\rho(0)\right] \tag{A.27}$$

$$= \mathrm{Tr}\left[\left\{e^{i\hat{H}_0 t}\hat{A}(0)e^{-i\hat{H}_0 t}\right\}\left\{\hat{U}_{\mathrm{IF}}(t, 0)\rho(0)\hat{U}_{\mathrm{IF}}^\dagger(t, 0)\right\}\right]. \tag{A.28}$$

An alternative approach to using the identity $\hat{U}(t, 0) = e^{-i\hat{H}_0 t}\hat{U}_{\mathrm{IF}}(t, 0)$ is simply to *ignore* the final $\exp(-i\hat{H}_0 t)$ altogether, and just use $\hat{U}_{\mathrm{IF}}(t, 0)$ as one's unitary evolution operator. The latter is often simpler, since $\hat{V}_{\mathrm{IF}}$ may often be made time-independent (even if $\hat{H}$ is explicitly time-dependent) by a judicious division into $\hat{H}_0$ and $\hat{V}$. If it cannot, then a *secular* or *rotating-wave* approximation is often used to make it time-independent (see Exercise 1.30).

We refer to the method of just using $\hat{U}_{\mathrm{IF}}$ as 'working in the interaction frame'. This terminology is used in analogy with, for example, 'working in a rotating frame' to calculate projectile trajectories on a rotating Earth. Working in the interaction frame is very common in quantum optics, where it is often (but incorrectly) called 'working in the interaction picture'. The interaction frame is not a 'picture' in the same way as the Heisenberg or Schrödinger picture. The HP or SP (or IP) includes the complete Hamiltonian evolution, whereas working in the interaction frame ignores the 'boring' free evolution. The interaction frame may contain either a Heisenberg or a Schrödinger picture, depending on whether $\hat{U}_{\mathrm{IF}}(t, 0)$ is applied to the system operators or the system state. The HP in the IF has time-independent states and time-dependent operators:

$$\rho(t) = \rho(0); \qquad \hat{A}(t) = \hat{U}_{\mathrm{IF}}^\dagger(t, 0)\hat{A}(0)\hat{U}_{\mathrm{IF}}(t, 0). \tag{A.29}$$

The SP in the IF has time-independent states and time-dependent operators:

$$\rho(t) = \hat{U}_{\mathrm{IF}}(t, 0)\rho(0)\hat{U}_{\mathrm{IF}}^\dagger(t, 0); \qquad \hat{A}(t) = \hat{A}(0). \tag{A.30}$$

Thus the SP state in the IF is the same as the IP state (as usually defined). But the SP operators in the IF are *not* the same as the IP operators, which are evolved by $\hat{U}_0(t, 0)$ as in Eq. (A.28).

We make frequent use of the interaction frame in this book, so it is necessary for the reader to understand the distinctions explained above. In fact, because we use the interaction frame so often, we frequently omit the IF subscript, after warning the reader that we are working in the interaction frame. Thus the reader must be very vigilant, since we often use the terms 'Heisenberg picture' and 'Schrödinger picture' with the phrase 'in the interaction frame' understood.

A.2 Multipartite systems and entanglement

A.2.1 Multipartite systems

Nothing is more important in quantum mechanics than understanding how to describe the state of a large system composed of subsystems. For example, in the context of measurement, we need to be able to describe the composite system composed of the system and the apparatus by which it is measured. The states of composite systems in quantum mechanics are described using the *tensor product*.

Consider two systems A and B prepared in the states $|\psi_A\rangle$ and $|\psi_B\rangle$, respectively. Let the dimension of the Hilbert space for systems A and B be D_A and D_B, respectively. The state of the total system is the tensor product state $|\Psi\rangle = |\psi_A\rangle \otimes |\psi_B\rangle$. More specifically, if we write the state of each component in an orthonormal basis $|\psi_A\rangle = \sum_{j=0}^{D_A-1} a_j |\phi_j^A\rangle$, $|\psi_B\rangle = \sum_{k=0}^{D_B-1} b_k |\phi_k^B\rangle$, then the state of the total system is

$$|\Psi\rangle = \sum_{j=0}^{D_A-1} \sum_{k=0}^{D_B-1} a_j b_k \left(|\phi_j\rangle_A \otimes |\phi_k\rangle_B \right). \tag{A.31}$$

Note that the dimension of the Hilbert space of the composite system C is $D_C = D_A \times D_B$. We can define a composite basis $|\Phi_{l(k,j)}\rangle_C = |\phi_j\rangle_A \otimes |\phi_k\rangle_B$, where $l(k, j)$ is a new index for the composite system. For example, we could have $l = k \times D_A + j$. Then an arbitrary pure state of C can be written as

$$|\Psi\rangle_C = \sum_{l=0}^{D_C-1} c_l |\Phi_l\rangle_C. \tag{A.32}$$

This is a vector in the compound Hilbert space $\mathbb{H}_C = \mathbb{H}_A \otimes \mathbb{H}_B$. Similarly, an arbitrary mixed state for C is represented by a state matrix acting on the D_C-dimensional tensor-product Hilbert space.

If we have n component systems, each of dimension D, the dimension of the total system is D^n, which is exponential in n. The dimension of the Hilbert space for many-body systems is *very* big! It is worth comparing this exponential growth of Hilbert-space dimension in multi-component quantum systems with the description of multi-component classical systems. In classical mechanics, a state of complete knowledge of a single particle in three dimensions is specified by six numbers (a 3-vector each for the position and momentum). For two particles, 12 numbers are needed, and so on. That is, the size of the description increases *linearly*, not exponentially, with the number of subsystems. However, this quantum–classical difference is not present for the case of states of incomplete knowledge. For a single particle these are defined classically as a probability distribution on the configuration space (which here is the phase space) $\mathbb{R}^6$. In general this requires an infinite amount of data to represent, but this can be made finite by restricting the particle to a finite phase-space volume and introducing a minimum resolution to the description. Since the dimensionality of the phase-space increases linearly (as $6n$) with the number of particles n, the amount of data required grows exponentially, just as in the quantum case. This is an example of how quantum states are like classical states of incomplete knowledge – see also Section 1.2.1.

A.2.2 Entanglement

In Section A.1.2 we introduced the state matrix by postulating a classical source of uncertainty (in the preparation procedure). In the context of compound quantum systems, mixedness arises naturally within quantum mechanics itself. This is because of the fundamental feature of quantum mechanics termed *entanglement* by Schrödinger in 1935 [Sch35b]. Entanglement means that, even if the combined state of two systems is pure, the state of either of the subsystems need not be pure. This means that, in contrast to classical systems, maximal knowledge of the whole does not imply maximal knowledge of the parts. Formally, we say that the joint pure state need not factorize. That is, there exist states $|\Psi_{AB}\rangle \in \mathbb{H}_{AB} \equiv \mathbb{H}_A \otimes \mathbb{H}_B$ such that

$$|\Psi_{AB}\rangle \neq |\psi_A\rangle |\psi_B\rangle, \tag{A.33}$$

where $|\psi_A\rangle \in \mathbb{H}_A$ and $|\psi_B\rangle \in \mathbb{H}_B$. Note that we are omitting the tensor-product symbols for kets, as will be done when confusion is not likely to arise.

If we were to calculate the mean of an operator $\hat{\Lambda}_A$, operating on states in $\mathbb{H}_A$, then we would use the procedure

$$\langle \Lambda_A \rangle = \langle \Psi_{AB} | (\hat{\Lambda}_A \otimes \hat{I}_B) | \Psi_{AB} \rangle$$

$$= \sum_j \langle \Psi_{AB} | \phi_j^B \rangle \hat{\Lambda}_A \langle \phi_j^B | \Psi_{AB} \rangle = \sum_j \langle \tilde{\psi}_j^A | \hat{\Lambda}_A | \tilde{\psi}_j^A \rangle, \tag{A.34}$$

where $|\tilde{\psi}_j^A\rangle$ is the unnormalized state $\langle \phi_j^B | \Psi_{AB} \rangle$. Using the fact that $\langle \psi | \hat{\Lambda} | \psi \rangle = \mathrm{Tr}\big[|\psi\rangle \langle \psi | \hat{\Lambda}\big]$, we get

$$\langle \Lambda_A \rangle = \mathrm{Tr}\big[\rho_A \hat{\Lambda}_A\big], \tag{A.35}$$

where

$$\rho_A = \sum_j \langle \phi_j^B | \Psi_{AB} \rangle \langle \Psi_{AB} | \phi_j^B \rangle \equiv \mathrm{Tr}_B[|\Psi_{AB}\rangle \langle \Psi_{AB}|] \tag{A.36}$$

is called the *reduced* state matrix for system A. The operation Tr_B is called the *partial trace* over system B.

It should be noted that the result in Eq. (A.36) also has a converse, namely that any state matrix ρ_A can be constructed as the reduced state of a (non-unique) pure state $|\Psi_{AB}\rangle$ in a larger Hilbert space. This is sometimes called a *purification* of the state matrix ρ_A, and is an example of the Gelfand–Naimark–Segal theorem [Con90].

Exercise A.9 *Construct a $|\Psi_{AB}\rangle$ that is a purification of ρ_A, given that the latter has the preparation procedure $\{\wp_j, |\psi_j\rangle\}$.*

For a bipartite system in a pure state, the entropy of one subsystem is a good measure of the degree of entanglement [NC00]. In particular, the entropy of each subsystem is the same. Note that the von Neumann entropy is *not* an extensive quantity, as is assumed in thermodynamics. As the above analysis shows, the entropy of the subsystems may be positive while the entropy of the combined system is zero. For systems with more than two parts, or for systems in mixed states, quantifying the entanglement is a far more difficult exercise, with many subtleties and as-yet unresolved issues.

The equality of the entropies of the subsystems of a pure bipartite system is known as the Araki–Lieb identity. It follows from an even stronger result: for a pure compound system, the eigenvalues of the reduced states of the subsystems are equal. This can be proven as follows. Let $\{|\phi_\lambda^A\rangle\}$ be the eigenstates of ρ_A:

$$\rho_A |\phi_\lambda^A\rangle = \wp_\lambda |\phi_\lambda^A\rangle. \tag{A.37}$$

Since these form an orthonormal set (see Box 1.1) we can write the state of the compound system using this basis for system A as

$$|\Psi_{AB}\rangle = \sum_\lambda \sqrt{\wp_\lambda} |\phi_\lambda^A\rangle |\phi_\lambda^B\rangle, \tag{A.38}$$

where $|\phi_\lambda^B\rangle \equiv \langle \phi_\lambda^A | \Psi_{AB} \rangle / \sqrt{\wp_\lambda}$.

Exercise A.10 *From this definition of* $|\phi_\lambda^B\rangle$, *show that* $\{|\phi_\lambda^B\rangle\}$ *forms an orthonormal set, and furthermore that*

$$\rho_B|\phi_\lambda^B\rangle = \wp_\lambda|\phi_\lambda^B\rangle. \tag{A.39}$$

Thus the eigenvalues of the reduced states of the two subsystems are equal. The decomposition in Eq. (A.38), using the eigenstates of the reduced states, is known as the Schmidt decomposition.

Note that the orthonormal set $\{|\phi_\lambda^B\rangle\}$ need not be a complete basis for system B, since the dimension of B may be greater than the dimension of A. If the dimension of B is *less* than the dimension of A, then it also follows that the *rank* of ρ_A (that is, the number of *non-zero* eigenvalues it has) is limited to the dimensionality of B. Clearly, for a purification of ρ_A (as defined above), the dimensionality of B can be as low as the rank of ρ_A, but no lower.

A.3 Position and momentum

A.3.1 Position

Consider an operator $\hat{Q}$ having the real line as its spectrum. This could represent the position of a particle, for example. Because of its continuous spectrum, the eigenstates $|q\rangle$ of $\hat{Q}$ are not normalizable. That is, it is not possible to have $\langle q|q\rangle = 1$. Rather, we use *improper states*, normalized such that

$$\int_{-\infty}^{\infty} dq\,|q\rangle\langle q| = \hat{1}. \tag{A.40}$$

Squaring the above equation implies that the normalization for these states is

$$\langle q|q'\rangle = \delta(q - q'). \tag{A.41}$$

The position operator is written as

$$\hat{Q} = \int dq\,|q\rangle q\,\langle q|. \tag{A.42}$$

Here we are using the convention that the limits of integration are $-\infty$ to ∞ unless indicated otherwise.

A pure quantum state $|\psi\rangle$ in the position representation is a function of q,

$$\psi(q) = \langle q|\psi\rangle, \tag{A.43}$$

commonly called the *wavefunction*. The probability density for finding the particle at position q is $|\psi(q)|^2$, and this integrates to unity. The state $|\psi\rangle$ is recovered from the wavefunction as follows:

$$|\psi\rangle = \int dq\,|q\rangle\langle q|\psi\rangle = \int dq\,\psi(q)|q\rangle. \tag{A.44}$$

It is worth remarking more about the nature of the continuum in quantum mechanics. The probability interpretation of the function $\psi(q)$ requires that it belong to the set $L^{(2)}(\mathbb{R})$. That is, the integral (technically, a Lebesgue integral) $\int |\psi(q)|^2\,dq$ must be finite, so that it can be set equal to unity for a normalized wavefunction. Although the space of $L^{(2)}(\mathbb{R})$ functions is infinite-dimensional, it is a countable infinity. That is, the basis states for the Hilbert space $\mathbb{H} = L^{(2)}(\mathbb{R})$ can be labelled by integers; an example basis is the set of harmonic-oscillator eigenstates discussed in Section A.4.1 below. The apparent continuum of the position states $\{|q\rangle\}$ (or the momentum states $\{|p\rangle\}$ defined below) does not contradict this: these 'states' are not normalizable and so are not actually in the Hilbert space. They exist as limits of true states, but the limit lies outside $\mathbb{H}$.

A.3.2 Momentum

It turns out that, if $\hat{Q}$ does represent the position of a particle, then its momentum is represented by another operator with the real line as its spectrum, $\hat{P}$. Using $\hbar = 1$, the eigenstates for $\hat{P}$ are related to those for $\hat{Q}$ by

$$\langle q|p \rangle = (2\pi)^{-1/2} e^{ipq}. \tag{A.45}$$

Here the normalization factor is chosen so that, analogously to Eqs. (A.40) and (A.41), we have

$$\int dp\, |p\rangle\langle p| = \hat{1}, \qquad \langle p|p' \rangle = \delta(p - p'). \tag{A.46}$$

Exercise A.11 *Show Eq. (A.46), using the position representation and the result that* $\int dy\, e^{iyx} = 2\pi\delta(x)$.

The momentum-representation wavefunction is thus simply the Fourier transform of the position-representation wavefunction:

$$\psi(p) = \langle p|\psi \rangle = (2\pi)^{-1/2} \int dq\, e^{-ipq} \psi(q). \tag{A.47}$$

From the above it is easy to show that in the position representation $\hat{P}$ acts on a wavefunction identically to the differential operator $-i\, \partial/\partial q$. First, in the momentum representation,

$$\hat{P} = \int dp\, |p\rangle\, p\, \langle p|. \tag{A.48}$$

Thus,

$$\langle q|\hat{P}|\psi \rangle = \int dp \int dq'\, \langle q|p\rangle\, p\, \langle p|q'\rangle\langle q'|\psi \rangle \tag{A.49}$$

$$= (2\pi)^{-1} \int dp \int dq'\, p\, e^{ip(q-q')} \psi(q'). \tag{A.50}$$

Now $p e^{ip(q-q')} = i\, \partial e^{ip(q-q')}/\partial q'$, so, using integration by parts and the fact (required by normalization) that $\psi(q)$ vanishes at $\pm\infty$, we obtain

$$\langle q|\hat{P}|\psi \rangle = -i(2\pi)^{-1} \int dp \int dq'\, e^{ip(q-q')} \frac{\partial}{\partial q'} \psi(q') \tag{A.51}$$

$$= -i\frac{\partial}{\partial q} \psi(q). \tag{A.52}$$

It is now easy to find the commutator between $\hat{Q}$ and $\hat{P}$:

$$\langle q|[\hat{Q}, \hat{P}]|\psi\rangle = \langle q|[\hat{Q}, \hat{P}] \int dq'\, \psi(q')|q'\rangle \tag{A.53}$$

$$= q(-i)\frac{\partial}{\partial q}\psi(q) - (-i)\frac{\partial}{\partial q}q\psi(q) \tag{A.54}$$

$$= i\psi(q) = i\langle q|\psi\rangle. \tag{A.55}$$

Now $\psi(q)$ here is an arbitrary function, apart from the assumption of differentiability and vanishing at $\pm\infty$. Thus it must be that

$$[\hat{Q}, \hat{P}] = i. \tag{A.56}$$

The fact that the commutator here is a c-number makes this an example of a *canonical commutation relation*.

A.3.3 Minimum-uncertainty states

From the above canonical commutation relation it follows (see Exercise A.3) that the variances in Q and P must satisfy

$$\langle(\Delta P)^2\rangle\langle(\Delta Q)^2\rangle \geq 1/4. \tag{A.57}$$

(Remember that we have set $\hbar = 1$.) The states which saturate this are known as minimum-uncertainty states (MUSs). It can be shown that these are Gaussian pure states. By this we mean that they are states with a Gaussian wavefunction. For a MUS, they are parameterized by three real numbers. Below, we take these to be q_0, p_0 and σ.

The position probability amplitude (i.e. wavefunction) for a MUS takes the form

$$\psi(q) = (\pi\sigma^2)^{-1/4}\exp\left[+ip_0(q - q_0) - (q - q_0)^2/(2\sigma^2)\right]. \tag{A.58}$$

Here we have chosen the overall phase factor to give $\psi(q)$ a real maximum at $q = q_0$. It is then easily verified that the moments for Q are

$$\langle Q\rangle = q_0, \tag{A.59}$$

$$\langle(\Delta Q)^2\rangle = \sigma^2/2. \tag{A.60}$$

Note that the variance does not equal σ^2, as one might expect from Eq. (A.58), because $\wp(q) = |\psi(q)|^2$.

The Fourier transform of a Gaussian is also Gaussian, and in the momentum representation

$$\psi(p) = (\pi/\sigma^2)^{-1/4}\exp\left[-iq_0 p - (p - p_0)^2\sigma^2/2\right]. \tag{A.61}$$

From this it is easy to show that

$$\langle P\rangle = p_0, \tag{A.62}$$

$$\langle(\Delta P)^2\rangle = 1/(2\sigma^2). \tag{A.63}$$

The saturation of the Heisenberg bound (A.57) follows.

Exercise A.12 *Verify Eqs. (A.59)–(A.63).*

A.4 The harmonic oscillator

So far there is nothing that sets a natural length (or, consequently, momentum) scale for the system. The simplest dynamics which does so is that generated by the harmonic oscillator Hamiltonian

$$\hat{H} = \frac{\hat{P}^2}{2m} + \frac{m\omega^2 \hat{Q}^2}{2}. \tag{A.64}$$

Here m is the mass of the particle and ω is the oscillator frequency. This Hamiltonian applies to any mode of harmonic oscillation, such as a mode of a sound wave in a condensed-matter system, or a mode of the electromagnetic field. In the latter case, $\hat{Q}$ is proportional to the magnetic field and $\hat{P}$ to the electric field.

Classically the harmonic oscillator has no characteristic length scale, but quantum mechanically it does, namely

$$\sigma = \sqrt{\hbar/(m\omega)}, \tag{A.65}$$

where we have temporarily restored $\hbar$ to make its role apparent. If we define the (non-Hermitian) operator

$$\hat{a} = \frac{1}{\sqrt{2}}\left(\frac{\hat{Q}}{\sigma} + i\frac{\sigma\hat{P}}{\hbar}\right) \tag{A.66}$$

then we can rewrite the Hamiltonian as

$$H = \hbar\omega(\hat{a}^\dagger\hat{a} + \hat{a}\hat{a}^\dagger)/2 = \hbar\omega(\hat{a}^\dagger\hat{a} + \tfrac{1}{2}). \tag{A.67}$$

Now, from the commutation relations of $\hat{Q}$ and $\hat{P}$ we can show that

$$[\hat{a}, \hat{a}^\dagger] = 1. \tag{A.68}$$

Also, we can show that the state $|\psi_0\rangle$ with wavefunction

$$\psi_0(q) = \langle q|\psi_0\rangle \propto \exp[-q^2/(2\sigma^2)] \tag{A.69}$$

is an eigenstate of $\hat{a}$ with eigenvalue 0.

Exercise A.13 *Show this using the position representation of $\hat{P}$ as $-i\hbar\, \partial/\partial q$.*

Thus it is also an eigenstate of the Hamiltonian (A.67), with eigenvalue $\hbar\omega/2$. Since $\hat{a}^\dagger\hat{a}$ is obviously a positive semi-definite operator, this is the lowest eigenvalue of the Hamiltonian. That is, we have shown that the quantum harmonic oscillator has a *ground state* that is a minimum-uncertainty state with $q_0 = p_0 = 0$ and a characteristic length σ given by Eq. (A.65).

A.4.1 Number states

From the above it is easy to show that the eigenvalues of $\hat{a}^\dagger\hat{a}$ are the non-negative integers, as follows. From the commutation relations (A.68) it follows that (for integer k)

$$[\hat{a}^\dagger\hat{a}, (\hat{a}^\dagger)^k] = \hat{a}^\dagger[\hat{a}, (\hat{a}^\dagger)^k] = k(\hat{a}^\dagger)^k. \tag{A.70}$$

Exercise A.14 *Show this.*
Hint: *Start by showing it for $k = 1$ and $k = 2$ and then find a proof by induction.*

Then, if we define an unnormalized state $|\psi_n\rangle = (\hat{a}^\dagger)^n |\psi_0\rangle$ we can easily show that

$$(\hat{a}^\dagger \hat{a})|\psi_n\rangle = (\hat{a}^\dagger \hat{a})(\hat{a}^\dagger)^n |\psi_0\rangle = \left[n(\hat{a}^\dagger)^n + (\hat{a}^\dagger)^n (\hat{a}^\dagger \hat{a}) \right] |\psi_0\rangle$$

$$= n(\hat{a}^\dagger)^n |\psi_0\rangle = n|\psi_n\rangle, \tag{A.71}$$

which establishes the result and identifies the eigenstates.

Thus we have derived the eigenvalues of the harmonic oscillator as $\hbar\omega \left(n + \frac{1}{2} \right)$. The corresponding unnormalized eigenstates are $|\psi_n\rangle$, which we denote $|n\rangle$ when normalized. If the Hamiltonian (A.64) refers to a particle, these are states with an integer number of elementary excitations of the vibration of the particle. They are therefore sometimes called *vibron* number states, that is, states with a definite number of vibrons. If the harmonic oscillation is that of a sound wave, then these states are called *phonon* number states. If the oscillator is a mode of the electromagnetic field, they are called *photon* number states. Especially in the last case, the ground state $|0\rangle$ is often called the vacuum state.

The operator $\hat{N} = \hat{a}^\dagger \hat{a}$ is called the *number operator*. Because $\hat{a}^\dagger$ raises the number of excitations by one, with

$$|n\rangle \propto (\hat{a}^\dagger)^n |0\rangle, \tag{A.72}$$

it is called the creation operator. Similarly, $\hat{a}$ lowers it by one, and is called the annihilation operator. To find the constants of proportionality, we must require that the number states be normalized, so that

$$\langle n|m\rangle = \delta_{nm}. \tag{A.73}$$

Now, since $|n\rangle$ is an eigenstate of $a^\dagger a$ with eigenvalue n,

$$\langle n|\hat{a}^\dagger \hat{a}|n\rangle = n\langle n|n\rangle = n. \tag{A.74}$$

However, we also have

$$\langle n|\hat{a}^\dagger \hat{a}|n\rangle = \langle \psi|\psi\rangle, \tag{A.75}$$

where $|\psi\rangle = \hat{a}|n\rangle \propto |n-1\rangle$. Thus the constant of proportionality must be such that

$$|\psi\rangle = \hat{a}|n\rangle = e^{i\phi}\sqrt{n}|n-1\rangle \tag{A.76}$$

for some phase ϕ. We choose the convention that $\phi = 0$, so that

$$\hat{a}|n\rangle = \sqrt{n}|n-1\rangle. \tag{A.77}$$

Similarly, it can be shown that

$$\hat{a}^\dagger|n\rangle = \sqrt{n+1}|n+1\rangle. \tag{A.78}$$

Exercise A.15 *Show this, and show that the above two relations are consistent with* $|n\rangle$ *being an eigenstate of* $\hat{a}^\dagger \hat{a}$*. Show also that the normalized number state is given by* $|n\rangle = (n!)^{-1/2}(\hat{a}^\dagger)^n |0\rangle$.

Note that $\hat{a}$ acting on the vacuum state $|0\rangle$ produces nothing, a null state.

A.4.2 Coherent states

No matter how large n is, a number state $|n\rangle$ never approaches the classical limit of an oscillating particle (or oscillating field amplitude). That is because for a system in a number state the average values of Q and P are always zero.

Exercise A.16 *Show this.*

For this reason, it is useful to consider a state for which there is a classical limit, the *coherent* state. This state is defined as an eigenstate of the annihilation operator

$$\hat{a}|\alpha\rangle = \alpha|\alpha\rangle, \tag{A.79}$$

where α is a complex number (because $\hat{a}$ is not an Hermitian operator). There are no such eigenstates of the creation operator $\hat{a}^\dagger$.

Exercise A.17 *Show this.*
Hint: *Assume that there exist states $|\beta\rangle$ such that $\hat{a}^\dagger|\beta\rangle = \beta|\beta\rangle$ and consider the inner product $\langle n|(\hat{a}^\dagger)^{n+1}|\beta\rangle$. Hence show that the inner product of $|\beta\rangle$ with any number state is zero.*

It is easy to find an expression for $|\alpha\rangle$ in terms of the number states as follows. In general we have

$$|\alpha\rangle = \sum_{n=0}^{\infty} c_n|n\rangle. \tag{A.80}$$

Since $\hat{a}|\alpha\rangle = \alpha|\alpha\rangle$ we get

$$\sum_{n=0}^{\infty} \sqrt{n}c_n|n-1\rangle = \sum_{n=0}^{\infty} \alpha c_n|n\rangle. \tag{A.81}$$

By equating the coefficients of the number states on both sides we get the recursion relation

$$c_{n+1} = \frac{\alpha}{\sqrt{n+1}}c_n, \tag{A.82}$$

so that $c_n = (\alpha^n/\sqrt{n!})c_0$. On choosing c_0 real and normalizing the state, we get

$$|\alpha\rangle = \exp(-|\alpha|^2/2) \sum_n \frac{\alpha^n}{\sqrt{n!}}|n\rangle. \tag{A.83}$$

The state $|\alpha := 0\rangle$ is the same state as the state $|n := 0\rangle$. For α finite the coherent state has a non-zero mean photon number:

$$\langle\alpha|\hat{a}^\dagger\hat{a}|\alpha\rangle = \big((\alpha|\alpha^*)(\alpha|\alpha)\big) = |\alpha|^2. \tag{A.84}$$

The number distribution (the probability of measuring a certain excitation number) for a coherent state is a *Poissonian* distribution of mean $|\alpha|^2$:

$$\wp_n = |\langle n|\alpha\rangle|^2 = e^{-|\alpha|^2}\frac{(|\alpha|^2)^n}{n!}. \tag{A.85}$$

This distribution has the property that the variance is equal to the mean. That is,

$$\langle(\hat{a}^\dagger\hat{a})^2\rangle - \langle\hat{a}^\dagger\hat{a}\rangle^2 = |\alpha|^2. \tag{A.86}$$

Exercise A.18 *Verify this, either from the distribution (A.85) or directly from the coherent state using the commutation relations for $\hat{a}$ and $\hat{a}^\dagger$.*

Setting $\hbar = 1$, it is simple to show that

$$\langle \alpha | \hat{Q} | \alpha \rangle = \sqrt{2} \sigma \, \mathrm{Re}[\alpha], \tag{A.87}$$

$$\langle \alpha | \hat{P} | \alpha \rangle = (\sqrt{2}/\sigma) \mathrm{Im}[\alpha], \tag{A.88}$$

$$\langle \alpha | (\Delta \hat{Q})^2 | \alpha \rangle = \sigma^2/2, \tag{A.89}$$

$$\langle \alpha | (\Delta \hat{P})^2 | \alpha \rangle = 1/(2\sigma^2), \tag{A.90}$$

$$\langle \alpha | \Delta \hat{Q} \, \Delta \hat{P} + \Delta \hat{P} \, \Delta \hat{Q} | \alpha \rangle = 0. \tag{A.91}$$

That is, a coherent state is a minimum-uncertainty state, as defined in Section A.3.3.

Because $\hat{a}$ is not Hermitian, the coherent states do not form an orthonormal set. In fact it can be shown that

$$\langle \beta | \alpha \rangle = e^{\beta^* \alpha - (|\alpha|^2 + |\beta|^2)/2}, \tag{A.92}$$

from which it follows that $|\langle \beta | \alpha \rangle|^2 = e^{-|\alpha - \beta|^2}$. If α and β are very different (as they would be if they represent two macroscopically distinct fields) then the two coherent states are very nearly orthogonal. Another consequence of their non-orthogonality is that the coherent states form an *overcomplete* basis. Whereas for number states we have

$$\sum_n |n\rangle \langle n| = \hat{1}, \tag{A.93}$$

the identity, for coherent states we have

$$\int d^2\alpha |\alpha\rangle \langle \alpha| = \pi \hat{1}. \tag{A.94}$$

This has applications in defining the trace, for example

$$\mathrm{Tr}[\rho] = \frac{1}{\pi} \int d^2\alpha \langle \alpha | \rho | \alpha \rangle. \tag{A.95}$$

Exercise A.19 *Show Eq. (A.94) using the expansion (A.83).*
Hint: *Write $\alpha = r e^{i\phi}$ so that $d^2\alpha = r \, dr \, d\phi$. The result $n! = \int_0^\infty dx \, x^n e^{-x}$ may be useful.*

Unlike number states, coherent states are not eigenstates of the Hamiltonian. However, they have the nice property that they remain as coherent states under evolution generated by the harmonic-oscillator Hamiltonian

$$\hat{H} = \omega \hat{a}^\dagger \hat{a}. \tag{A.96}$$

Here we have dropped the $1/2$ from the Hamiltonian (A.67) since it has no physical consequences (at least outside general relativity). The amplitude $|\alpha|$ of the states remains the same; only the phase changes at rate ω (as expected):

$$\exp(-i\hat{H}t)|\alpha\rangle = |e^{-i\omega t}\alpha\rangle. \tag{A.97}$$

Exercise A.20 *Show this, using Eq. (A.83).*

This form-invariance under the harmonic-oscillator evolution is why they are called coherent states.

Coherent states can be generated from the vacuum state as follows:

$$|\alpha\rangle = \hat{D}(\alpha)|0\rangle, \tag{A.98}$$

Appendix A

where

$$\hat{D}(\alpha) = e^{\alpha \hat{a}^\dagger - \alpha^* \hat{a}} = e^{-i(i\alpha \hat{a}^\dagger - i\alpha^* \hat{a})} \tag{A.99}$$

is called the displacement operator. This is easiest to see as follows. First, note that if we define the family of operators $\hat{O}_\mu = \hat{U}_\mu^\dagger \hat{O}_0 \hat{U}_\mu$, where $\hat{U}_\mu = e^{(\alpha \hat{a}^\dagger - \alpha^* \hat{a})\mu}$, then these are solutions to the equation

$$\frac{d}{d\mu} \hat{O}_\mu = -[\alpha \hat{a}^\dagger - \alpha^* \hat{a}, \hat{O}_\mu]. \tag{A.100}$$

Exercise A.21 *Show this, by analogy with the Heisenberg equations of motion.*

Now, applying this to $\hat{O}_0 = \hat{a}$, we see that $\hat{O}_\mu = \hat{a} + \mu\alpha$ is a solution to Eq. (A.100). Then, noting that $\hat{D}(\alpha) = \hat{U}_1 = \hat{D}^\dagger(-\alpha)$, we have

$$\hat{a}\hat{D}(\alpha)|0\rangle = \hat{D}(\alpha)\hat{D}^\dagger(\alpha)\hat{a}\hat{D}(\alpha)|0\rangle = \hat{D}(\alpha)(\hat{a} + \alpha)|0\rangle = \alpha\hat{D}(\alpha)|0\rangle, \tag{A.101}$$

which proves the above result.

A.4.3 Squeezed states

Because the harmonic-oscillator Hamiltonian picks out a particular class of minimum-uncertainty states (the coherent states), the other minimum-uncertainty states are given a special name in this situation: the squeezed states. In fact, any Gaussian pure state other than a coherent state is called a squeezed state. Whereas a coherent state requires one complex parameter α to specify, a general squeezed state requires two additional real parameters:

$$|\alpha, r, \phi\rangle = \hat{D}(\alpha)|r, \phi\rangle, \tag{A.102}$$

where

$$|r, \phi\rangle = \exp\left[r\left(e^{-2i\phi}\hat{a}^2 - e^{2i\phi}\hat{a}^{\dagger 2}\right)/2\right]|0\rangle \tag{A.103}$$

is known as a squeezed vacuum. This is an appropriate name since it is in fact a zero-amplitude coherent state for rotated and rescaled canonical coordinates, $\hat{Q}'$ and $\hat{P}'$, defined by

$$\hat{Q} + i\hat{P} = (\hat{Q}'e^r + i\hat{P}'e^{-r})e^{i\phi}. \tag{A.104}$$

For example, if $\phi = 0$ then $\hat{P} = e^{-r}\hat{P}'$, so the variance of P is smaller by a factor of e^{-2r} than that of a coherent state. By contrast, $\hat{Q} = e^r\hat{Q}'$, so the variance of Q is e^{2r} times larger than that of a coherent state. That is, the reduction in the variance of one coordinate *squeezes* the uncertainty into the conjugate coordinate. The term squeezed state is often applied more broadly, to any state (pure or mixed, Gaussian or not) of a harmonic oscillator in which the uncertainty in one coordinate is below that of the vacuum state.

A.5 Quasiprobability distributions

It is often convenient to represent quantum states as quantum probability distributions, or quasiprobability distributions, over non-commuting observables. Here we consider the three most commonly used distributions, called the P, Q and W distributions (or functions).

A.5.1 Normal order and the P function

Once an annihilation operator has been defined, *normal ordering* can be defined. A normally ordered operator expression is one in which all annihilation operators appear to the right of all creation operators. For example, $\hat{A} = (\hat{a}^\dagger \hat{a})^2$ is not a normally ordered operator expression, but $\hat{A} = \hat{a}^\dagger \hat{a}^\dagger \hat{a} \hat{a} + \hat{a}^\dagger \hat{a}$ *is* a normally ordered expression. Note that this example shows that it does not make sense to speak of a 'normally ordered operator' (although this is common parlance), since it is the *same* operator being represented here, in one case by a normally ordered expression and in the other by a non-normally ordered expression. One advantage of normal ordering is that one can see immediately whether any term will give zero when acting on a vacuum state by seeing whether it has at least one annihilation operator.

Classically, the c-number analogues of $\hat{a}$ and $\hat{a}^\dagger$, which we can denote α and α^*, commute. This means that, regardless of the ordering of an expression $f(\alpha, \alpha^*)$, we have

$$\langle f(\alpha, \alpha^*) \rangle = \int d^2\alpha \, P(\alpha, \alpha^*) f(\alpha, \alpha^*), \tag{A.105}$$

where $P(\alpha, \alpha^*)$ is a probability distribution over phase-space.[1] Quantum mechanically, we do have to worry about ordering, but we could ask the following: for a given ρ, is there a distribution P such that

$$\text{Tr}[\rho f_n(\hat{a}, \hat{a}^\dagger)] = \int d^2\alpha \, P(\alpha, \alpha^*) f_n(\alpha, \alpha^*), \tag{A.106}$$

where f_n is a normally ordered expression? The answer is yes, but in general P is an extremely singular function (i.e. more singular than a δ-function). The relation between the P function (as it is called) and ρ is

$$\rho = \int d^2\alpha \, P(\alpha, \alpha^*) |\alpha\rangle\langle\alpha|. \tag{A.107}$$

Thus, if P is only as singular as a δ-function, then ρ is a mixture of coherent states.

Exercise A.22 *Assuming a non-singular P function, verify Eq. (A.106) from Eq. (A.107).*

A.5.2 Antinormal order and the Q function

Antinormal ordering is, as its name implies, the opposite to normal ordering. For example, $\hat{A} = \hat{a}\hat{a}\hat{a}^\dagger\hat{a}^\dagger - 3\hat{a}\hat{a}^\dagger + 1$ is an antinormally ordered operator expression for the operator $\hat{A}$ defined above. As for normal ordering, one can ask the question, for a given ρ, is there a distribution Q such that

$$\text{Tr}[\rho f_a(\hat{a}, \hat{a}^\dagger)] = \int d^2\alpha \, Q(\alpha, \alpha^*) f_a(\alpha, \alpha^*), \tag{A.108}$$

---

[1] We write, for example, $P(\alpha, \alpha^*)$ rather than $P(\alpha)$, to avoid implying (wrongly) that these functions are analytical functions in the complex plane.

where f_a is an *antinormally* ordered expression? Again the answer is yes. Moreover, the Q function (as it is called) is always smooth and positive, and is given by

$$Q(\alpha, \alpha^*) = \pi^{-1}\langle\alpha|\rho|\alpha\rangle. \tag{A.109}$$

Exercise A.23 *From this definition, verify Eq. (A.108).*

A.5.3 Symmetric order and the Wigner function

A final type of ordering commonly used is symmetric ordering. This can be defined independently of an annihilation operator, as an expression that is symmetric in $\hat{a}$ and $\hat{a}^\dagger$ is also symmetric in position $\hat{Q}$ and momentum $\hat{P}$. A symmetric operator expression is one in which every possible ordering is equally weighted. Using the same example as previously, a symmetric expression is

$$\hat{A} = \frac{1}{6}[\hat{a}^\dagger\hat{a}^\dagger\hat{a}\hat{a} + \hat{a}^\dagger\hat{a}\hat{a}^\dagger\hat{a} + \hat{a}^\dagger\hat{a}\hat{a}\hat{a}^\dagger + \hat{a}\hat{a}^\dagger\hat{a}^\dagger\hat{a} + \hat{a}\hat{a}^\dagger\hat{a}\hat{a}^\dagger + \hat{a}\hat{a}\hat{a}^\dagger\hat{a}^\dagger]$$

$$- \frac{1}{2}[\hat{a}^\dagger\hat{a} + \hat{a}\hat{a}^\dagger]. \tag{A.110}$$

We can ask the same question as in the preceding subsections, namely is there a distribution function W such that

$$\text{Tr}[\rho f_s(\hat{a}, \hat{a}^\dagger)] = \int d^2\alpha \, W(\alpha, \alpha^*) f_s(\alpha, \alpha^*), \tag{A.111}$$

where f_s is a symmetrically ordered expression? The answer again is yes, and this distribution is known as the Wigner function, because it was introduced by Wigner [Wig32]. Its relation to ρ is that

$$W(\alpha, \alpha^*) = \frac{1}{\pi^2}\int d^2\omega \, \text{Tr}\left[\rho \exp[\omega(\hat{a}^\dagger - \alpha^*) - \omega^*(\hat{a} - \alpha)]\right]. \tag{A.112}$$

The Wigner function is always a smooth function, but it can take negative values. It was originally defined by Wigner as a function of position q and momentum p. From Eq. (A.66), with $\hbar = 1$, these are related to α by

$$\alpha = \frac{1}{\sqrt{2}}\left(\frac{q}{\sigma} + i\sigma p\right). \tag{A.113}$$

In terms of these variables (using $\omega = x - ik$),

$$W(q, p) = \frac{1}{(2\pi)^2}\int dk \int dx \, \text{Tr}\left[\rho \exp[ik(\hat{Q} - q) + ix(\hat{P} - p)]\right]. \tag{A.114}$$

Note that the characteristic length σ of the harmonic oscillator does not enter into this expression.

A particularly appealing feature of the Wigner function is that its marginal distributions are the true probability distributions. That is,

$$\int dq \, W(q, p) = \wp(p) = \langle p|\rho|p\rangle, \tag{A.115}$$

$$\int dp \, W(q, p) = \wp(q) = \langle q|\rho|q\rangle. \tag{A.116}$$

Exercise A.24 *Show this.*
Hint: *Recall that $\int dq \, e^{ipq} = 2\pi \delta(p)$ and that $\delta(p - \hat{P}) = |p\rangle\langle p|$.*

The Wigner function thus appears like a joint classical probability distribution, except that in many cases it is not positive definite. Indeed, of the pure states, only states with a Gaussian wavefunction $\psi(q)$ have a positive-definite Wigner function. Another appealing property of the Wigner function is that the overlap between two states is given simply by the integral of the products of their respective Wigner functions:

$$\text{Tr}[\rho_1\rho_2] = 2\pi\hbar \int dq \int dp \, W_1(q, p)W_2(q, p). \tag{A.117}$$

Exercise A.25 *Show this.*

Finally, the Baker–Campbell–Hausdorff theorem states that, for arbitrary operators $\hat{A}$ and $\hat{B}$ satisfying $[\hat{A}, [\hat{A}, \hat{B}]] = 0$ and $[\hat{B}, [\hat{A}, \hat{B}]] = 0$,

$$\exp(\hat{A} + \hat{B}) = \exp(\hat{A})\exp(\hat{B})\exp(-\tfrac{1}{2}[\hat{A}, \hat{B}]). \tag{A.118}$$

Using this, the Wigner function can be rewritten as

$$W(q, p) = \frac{1}{(2\pi)^2} \int dk \int dx \, \text{Tr}\left[\rho e^{ik(\hat{Q}-q)}e^{ix(\hat{P}-p)}e^{-ikx/2}\right] \tag{A.119}$$

$$= \frac{1}{(2\pi)^2} \int dk \int dx \, \text{Tr}\left[\rho e^{ix(\hat{P}-p)}e^{ik(\hat{Q}-q)}e^{+ikx/2}\right]. \tag{A.120}$$

From this, it is easy to prove the following useful operator correspondences:

$$\hat{Q}\rho \leftrightarrow \left(q + \frac{i}{2}\frac{\partial}{\partial p}\right)W(q, p), \tag{A.121}$$

$$\rho\hat{Q} \leftrightarrow \left(q - \frac{i}{2}\frac{\partial}{\partial p}\right)W(q, p), \tag{A.122}$$

$$\hat{P}\rho \leftrightarrow \left(p - \frac{i}{2}\frac{\partial}{\partial q}\right)W(q, p), \tag{A.123}$$

$$\rho\hat{P} \leftrightarrow \left(p + \frac{i}{2}\frac{\partial}{\partial q}\right)W(q, p). \tag{A.124}$$

Exercise A.26 *Show these. This means showing, for example, that*

$$\int dk \int dx \, \text{Tr}\left[e^{ix(\hat{P}-p)}e^{ik(\hat{Q}-q)}e^{+ikx/2}\hat{Q}\rho\right]$$

$$= \int dk \int dx \left(q + \frac{i}{2}\frac{\partial}{\partial p}\right)\text{Tr}\left[e^{ix(\hat{P}-p)}e^{ik(\hat{Q}-q)}e^{+ikx/2}\rho\right]. \tag{A.125}$$

Note that here ρ is not restricted to being a state matrix. It can be an arbitrary operator with Wigner representation $W(q, p)$, provided that the integrals converge and boundary terms can be ignored.

Appendix B

Stochastic differential equations

B.1 Gaussian white noise

Although the description 'stochastic differential equation' (SDE) sounds rather general, it is usually taken to refer only to differential equations with a Gaussian white-noise term. In this appendix we begin by reviewing SDEs of this sort, which are also known as Langevin equations. In the final section, we generalize to other sorts of noise (in particular jumps). This review is intended not to be mathematically rigorous, but rather to build intuition about the physical assumptions behind the formalism. In particular, the concept of stochastic integration will not be introduced at all. A more formal treatment of SDEs and stochastic integrals, still aimed at physical scientists rather than mathematicians, can be found in Ref. [Gar85]. Another more elementary introduction may be found in Refs. [Gil93, Gil96].

Consider the one-dimensional case for simplicity. A SDE for the random variable X may then be written as

$$\dot{X} = \alpha(X) + \beta(X)\xi(t). \tag{B.1}$$

Here, the time argument of X has been omitted, α and β are arbitrary real functions, and $\xi(t)$ is a rapidly varying random process. This process, referred to as noise, is continuous in time, has zero mean and is a stationary process. The last descriptor means that all of its statistics, including in particular its correlation function

$$E[\xi(t)\xi(t+\tau)], \tag{B.2}$$

are independent of t. Here E denotes an ensemble average or expectation value as usual. The noise is normalized such that its correlation function integrates to unity:

$$\int_{-\infty}^{\infty} d\tau \, E[\xi(t)\xi(t+\tau)] = 1. \tag{B.3}$$

Note that Eq. (B.3) implies that $[\alpha] = [X]T^{-1}$ and $[\beta] = [X]T^{-1/2}$, where here $[A]$ denotes the dimensionality of A and T is the time dimension.

We are interested in the case of Markovian SDEs, for which the correlation time of the noise must be zero. That is, we can replace Eq. (B.3) by

$$\int_{-\epsilon}^{\epsilon} d\tau \, E[\xi(t)\xi(t+\tau)] = 1, \tag{B.4}$$

for all $\epsilon > 0$. In this limit, $\xi(t)$ is called *Gaussian white noise*, which is completely characterized by the two moments

$$E[\xi(t)\xi(t')] = \delta(t - t'), \tag{B.5}$$

$$E[\xi(t)] = 0. \tag{B.6}$$

The correlation function contains a singularity at $t = t'$ because of the constraint of Eq. (B.4). Because of this singularity, one has to be very careful in finding the solutions of Eq. (B.1). The noise $\xi(t)$ is called *white* because the spectrum is flat in this limit, just like the spectrum of white light is flat (in the visible range of frequencies anyway). Recall that the spectrum of a noise process is the Fourier transform of the correlation function.

Physically, an equation like (B.1) could be obtained by deriving it for a physical (non-white) noise source $\xi(t)$, and then taking the idealized limit. In that case, Eq. (B.1) is known as a *Stratonovich* SDE. This result is known as the Wong–Zakai theorem [WZ65]. The Stratonovich SDE for some function f of X is found by using the standard rules of differential calculus, that is,

$$\dot{f}(X) = f'(X)[\alpha(X) + \beta(X)\xi(t)], \tag{B.7}$$

where the prime denotes differentiation with respect to X. As stated above, the differences from standard calculus arise when actually solving Eq. (B.1).

Let $X(t)$ be known, and equal to x. If one were to assume that the infinitesimally evolved variable X were given by

$$X(t + dt) = x + [\alpha(x) + \beta(x)\xi(t)]dt \tag{B.8}$$

and, further, that the stochastic term $\xi(t)$ were independent of the system at the same time, then one would derive the expected increment in X from t to $t + dt$ to be

$$E[dX] = \alpha(x)dt. \tag{B.9}$$

The second assumption here seems perfectly reasonable since the noise is not correlated with any of the noise which has interacted with the system in the past, and so would be expected to be uncorrelated with the system. Applying the same arguments to f yields

$$E[df] = f'(x)\alpha(x)dt. \tag{B.10}$$

That is, all expectation values are independent of β. In particular, if we consider $f(X) = X^2$, then the above imply that the infinitesimal increase in the variance of X is

$$E[d(X^2)] - d(E[X])^2 = 2x\alpha(x)dt - 2x\alpha(x)dt = 0. \tag{B.11}$$

That is to say, the stochastic term has not introduced any noise into the variable X.

Obviously this result is completely contrary to what one would wish from a stochastic equation. The lesson is that it is invalid to make simultaneously the following three assumptions.

1. The chain rule of standard calculus applies (Eq. (B.7)).
2. The infinitesimal increment of a quantity is equal to its rate of change multiplied by dt (Eq. (B.8)).
3. The noise and the system at the same time are independent.

With a Stratonovich SDE the first assumption is true, and the usual explanation [Gar85] is that the second is also true but that the third assumption is false. Alternatively (and this is the interpretation we adopt), one can characterize a Stratonovich SDE by saying that the second assumption is false (or true only in an implicit way) and that the third is still true.

In this way of looking at things, the fluxion $\dot{X}$ in a Stratonovich SDE is just a symbol that can be manipulated using the usual rules of calculus. It should not be turned into a ratio of differentials dX/dt. In particular, $E[\dot{X}]$ is not equal to $dE[X]/dt$ in general. This point of view is useful for later generalization to jump processes in Section B.6, where one can still consider starting with an SDE containing non-singular noise, and then taking the singular limit. In the jump case, the third assumption is inapplicable, so the problem must lie with the second assumption. Since the term Stratonovich is restricted to the case of Gaussian white noise, we will also use a more general terminology, referring to any SDE involving $\dot{X}$ as an *implicit* equation.

A different choice of which postulates to relax is that of the Itô stochastic calculus. With an Itô SDE, the first assumption above is false, the second is true in an explicit manner and the third is also true (for Gaussian white noise, but not for jumps). The Itô form has the advantage that it simply allows the increment in a quantity to be calculated, and also allows ensemble averages to be taken easily. It has the disadvantage that one cannot use the usual chain rule.

B.2 Itô stochastic differential calculus

Because different rules of calculus apply to the Itô and Stratonovich forms of a SDE, the equations will appear differently in general. The Itô form of the Stratonovich equation (B.1) is

$$dX = [\alpha(X) + \tfrac{1}{2}\beta(X)\beta'(X)]dt + \beta(X)dW(t). \tag{B.12}$$

Here, the infinitesimal Wiener increment has been introduced, defined by

$$dW(t) = \xi(t)dt. \tag{B.13}$$

This is called a Wiener increment because if we define

$$W(t) = \int_{t_0}^{t} \xi(t')dt' \tag{B.14}$$

then this has all of the properties of a Wiener process. That is, if we define $\Delta W(t) = W(t + \Delta t) - W(t)$, then this is independent of $W(s)$ for $s < t$, and has a Gaussian distribution with zero mean and variance Δt:

$$\Pr[\Delta W(t) \in (w, w + dw)] = [2\pi \, \Delta t]^{-1/2} \exp[-w^2/(2\,\Delta t)]dw. \tag{B.15}$$

It is actually quite easy to see these results. First the independence of $\Delta W(t)$ from $W(s)$ for $s < t$ follows simply from Eq. (B.5). Second, it is easy to show that

$$E[\Delta W(t)^2] = \Delta t, \tag{B.16}$$

$$E[\Delta W(t)] = 0. \tag{B.17}$$

Exercise B.1 *Verify these using Eqs. (B.5) and (B.6).*

To go from these moments to the Gaussian distribution (B.15), note that, for any finite time increment Δt, the Wiener increment ΔW is the sum of an infinite number of independent noises $\xi(t)dt$, which are identically distributed. By the central limit theorem [Gil83], since the sum of the variances is finite, the resulting distribution is Gaussian. We thus see why $\xi(t)$ was called Gaussian white noise: because of the Gaussian probability distribution of the Wiener process. Note that the Wiener process is not differentiable, so strictly $\xi(t)$ does not exist. This is another way of seeing why stochastic calculus is a tricky business and why we have to worry about the Itô versus Stratonovich definitions.

In Eq. (B.12) we have introduced a convention of indicating Itô equations by an explicit representation of an infinitesimal increment (as on the left-hand side of Eq. (B.12)), whereas Stratonovich equations will be indicated by an implicit equation with a fluxion on the left-hand side (as in Eq. (B.1)). If an Itô (or *explicit*) equation is given as

$$dX = a(X)dt + b(X)dW(t), \qquad (B.18)$$

then the corresponding Stratonovich equation is

$$\dot{X} = a(X) - \tfrac{1}{2}b'(X)b(X) + b(X)\xi(t). \qquad (B.19)$$

Here the prime indicates differentiation with respect to X. A simple non-rigorous derivation of this relation will be given in Section B.3.

In the Itô form, the noise is independent of the system, so the expected increment in X from Eq. (B.18) is simply

$$E[dX] = a(X)dt. \qquad (B.20)$$

However, the nonsense result (B.11) is avoided because the chain rule does not apply to calculating $df(X)$. The actual increment in $f(X)$ is simple to calculate by using a Taylor expansion for $f(X + dX)$. The difference from the usual chain rule is that second-order infinitesimal terms cannot necessarily be ignored. This arises because the noise is so singular that second-order noise infinitesimals are as large as first-order deterministic infinitesimals. Specifically, the infinitesimal Wiener increment $dW(t)$ can be assumed to be defined by the following Itô rules:

$$E[dW(t)^2] = dt, \qquad (B.21)$$

$$E[dW(t)] = 0. \qquad (B.22)$$

These can be obtained from Eqs. (B.16) and (B.17) simply by taking the infinitesimal limit $\Delta \to d$.

Note that there is actually no restriction that $dW(t)$ must have a Gaussian distribution. As long as the above moments are satisfied, the increment $\Delta W(t)$ over any finite time will be Gaussian from the central limit theorem. By a similar argument, it is actually possible to omit the expectation value in Eq. (B.21) because, over any finite time, a time average effects an ensemble average of what is primarily a deterministic rather than stochastic quantity. This can be seen as follows. Consider the variable

$$\Delta\tau = \sum_{j=0}^{N-1}[\delta W(t_j)]^2, \qquad (B.23)$$

where $t_j = t_0 + j\,\delta t$, where $\delta t = \Delta t/N$. Then it follows that

$$\langle\Delta\tau\rangle = \Delta t, \qquad (B.24)$$

$$\sqrt{\langle(\Delta\tau)\rangle^2 - \langle\Delta\tau\rangle^2} = \frac{\Delta t}{\sqrt{N/2}}. \qquad (B.25)$$

Exercise B.2 *Show these results.*
Hint: *For the second of these, first show that $\langle[\delta W(t_j)]^2[\delta W(t_k)]^2\rangle = (\delta t)^2(1 + 2\delta_{jk})$. Remember that the $\delta W(t_j)$ and $\delta W(t_k)$ are independent for $k \neq j$, while, for $j = k$, use the fact that, for a Gaussian random variable X of mean 0, $\langle X^4\rangle = 3\langle X^2\rangle^2$.*

In the limit $N \to \infty$, where $\delta t \to dt$, the standard deviation in $\Delta \tau$ vanishes and $\Delta \tau$ converges to Δt in the mean-square sense. Since this is true for any finite time interval, we may as well replace dW^2 by dt.

Using this result and expanding the Taylor series to second order gives the modified chain rule

$$df(X) = f'(X)dX + \tfrac{1}{2}f''(X)(dX)^2. \tag{B.26}$$

Specifically, with dX given by Eq. (B.18), and using the rule $dW(t)^2 \equiv dt$,

$$df(X) = \left[f'(X)a(X) + \tfrac{1}{2}f''(X)b(X)^2\right]dt + f'(X)b(X)dW(t). \tag{B.27}$$

With this definition, and with $f(X) = X^2$, one finds that the expected increase in the variance of X in a time dt is

$$E[dX(t)^2] - d(E[X(t)])^2 = b(x)^2 \, dt. \tag{B.28}$$

That is to say, the effect of the noise is to increase the variance of X. Thus, the correct use of the stochastic calculus evades the absurd result of Eq. (B.11).

B.3 The Itô–Stratonovich relation

Consider again the Stratonovich equation, with the (boring) deterministic term set to zero:

$$\dot{X} = \beta(X)\xi(t). \tag{B.29}$$

Assuming that the chain rule of standard calculus applies, and that the noise at time t is independent of the system at that time, we have shown that naively turning this from an equation for the rate of change of X into an equation for the increment of X,

$$X(t + dt) = X(t) + \beta(X)\xi(t)dt, \tag{B.30}$$

leads to absurd results in general. This is because, when the noise $\xi(t)dt$ is as singular as we are assuming (scaling as $\sqrt{dt}$, rather than dt), even an infinitesimal time increment cannot be assumed to yield a change of size scaling as dt.

Since Eq. (B.30) comes from a first-order Taylor expansion of $X(t + dt)$ in dt, it makes sense from the above arguments that we should use a higher-order expansion. The all-order expansion is

$$X(t + dt) = \exp\left(dt \, \frac{\partial}{\partial s}\right)X(s)\bigg|_{s=t}. \tag{B.31}$$

We can evaluate this by rewriting Eq. (B.29) as

$$\frac{\partial}{\partial s}X(s)\bigg|_{s=t} = \beta(X(s))\xi(t)|_{s=t}. \tag{B.32}$$

Note that $\xi(t)$ is assumed constant while $X(s)$ changes. This is an expression of the fact that the noise $\xi(t)$ cannot in reality be δ-correlated. As emphasized above, equations of the Stratonovich form arise naturally only when $\xi(t)$ is a physical (non-white) noise source, and the idealization to white noise is made later. Thus the physical noise will have some finite correlation time over which it remains relatively constant. This idealization is valid if the physical correlation time is much smaller than the characteristic evolution time of the system.

We now expand Eq. (B.31) to second order in dt. As in the Itô chain rule, this is all that is necessary. The result, using Eq. (B.32), is

$$X(t + dt) = X(t) + dt\,\beta(X(t))\xi(t) + \frac{1}{2}(dt)^2\left[\frac{\partial}{\partial s}\beta(X(s))\xi(t)\right]_{s=t}.$$ (B.33)

Now, using the usual chain rule to expand $\partial\beta(X(s))/\partial s$, again using Eq. (B.32), we get

$$dX = \beta(X)\xi(t)dt + \frac{1}{2}(\xi(t)dt)^2\beta(X)\beta'(X).$$ (B.34)

Replacing $\xi(t)dt$ by $dW(t)$ and using Eq. (B.21) yields the correct Itô equation (B.12).
In cases for which β is linear in X, as in

$$\dot{X}(t) = \lambda X(t)\xi(t),$$ (B.35)

the Itô equation

$$dX(t) = [\lambda\,dW(t) + (\lambda^2/2)dt]X(t)$$ (B.36)

can be found easily since Eq. (B.31) becomes

$$X(t + dt) = \exp(\lambda\,dW(t))X(t).$$ (B.37)

This case is particularly relevant in quantum systems.

B.4 Solutions to SDEs

B.4.1 The meaning of 'solution'

Because the equations we are considering are stochastic, they have no simple solution as for deterministic differential equations, as a single number that changes with time. Rather, there are infinitely many solutions, depending on which noise $\xi(t)$ actually occurs. It might seem that this is more of a *problem* than a *solution*, since it is not easy to characterize such an infinite ensemble in general. However, this infinite ensemble of solutions has definite *statistical* properties, because the noise $\xi(t)$ has definite statistical properties. For example, the *moments* $E[X(t)]$ and $E[X(t)^2]$ are *deterministic functions of time*, as is the correlation function $E[X(t)X(t+\tau)]$.

To find averages such as these, in general a stochastic numerical solution is required. That is, the SDE is solved for one particular realization of the noise and the result $X(t)$ recorded. It is then solved again for a different (and independent) realization of the noise. Any given moment can then be approximated by the finite ensemble average F. For example, the one-time average $E[f(X(t))]$, at any particular time t, can be estimated from

$$F[f(X(t))] \equiv \frac{1}{M}\sum_{j=1}^{M} f(X_j(t)),$$ (B.38)

where $X_j(t)$ is the solution from the jth run and M is the total number of runs. The error in the estimate $F[f(X(t))]$ can be estimated by the usual statistical formula

$$\sigma\{F[f(X(t))]\} = \sqrt{\frac{F[[f(X(t))]^2] - F[f(X(t))]^2}{M}}.$$ (B.39)

Thus M has to be chosen large enough for this to be below some acceptable level. Two-time averages such as correlation functions, and the uncertainties in these estimates, may be determined in a similar way.

B.4.2 *Itô versus Stratonovich*

The existence of two forms of the same SDE, Itô and Stratonovich, may seem problematic at this point. Which one is actually used to solve SDEs? The answer depends on the method of solution.

Using a simple Euler step method, the Itô SDE giving an explicit increment is appropriate. That is, for the one-dimensional example, one has

$$X(t_{j+1}) = X(t_j) + a\big(X(t_j)\big)\delta t + b\big(X(t_j)\big)\sqrt{\delta t}\, S_j, \tag{B.40}$$

where δt is a very small increment, with $t_{j+1} - t_j = \delta t$, and S_j is a random number with a standard normal distribution[1] generated by the computer for this time step. The S_{j+1} for the next time step is a new number, and the numbers in one run should be independent of those in any other run. If one were to use a more sophisticated integration routine than the Euler one, then the Stratonovich equation may be the one needed. See Ref. [KP00] for a discussion.

In some cases, it is possible to obtain analytical solutions to a SDE. By this we mean a closed integral form. Of course, this integral will not evaluate to a number, because it will contain the noise term $\xi(t)$. However, it can be manipulated so as to give moments easily. Again, the question arises, which equation is actually integrated in these cases, the Itô one or the Stratonovich one? Here the answer is that in practice it does not matter. The only cases in which an analytical solution is possible are those in which the Itô equation has been (perhaps by an appropriate change of variable) put in the form

$$dX = a(t)dt + b(t)dW, \tag{B.41}$$

that is, where a and b are not functions of X. In this case the Stratonovich equation is

$$\dot{X} = a(t) + b(t)\xi(t). \tag{B.42}$$

That is, it looks the same as the Itô equation, so one could naively integrate it instead, to obtain the solution

$$X(t) = X(0) + \int_0^t a(s)ds + \int_0^t b(s)\xi(s)ds. \tag{B.43}$$

B.5 The connection to the Fokker–Planck equation

An alternative to describing a stochastic process using a SDE for X is to use a Fokker–Planck equation (FPE). This is an evolution equation for the probability distribution $\wp(x)$ for the variable. In this section we show how the FPE corresponding to a SDE can very easily be derived. In the process we obtain other results that are used in the main text.

First note that the probability density for a continuous variable X is by definition

$$\wp(x) = \mathrm{E}[\delta(X - x)]. \tag{B.44}$$

---

[1] That is, a Gaussian distribution with mean zero and variance unity.

Now $\delta(X - x)$ is just a function of X, so we can consider the SDE it obeys. If X obeys

$$dX = a(X)dt + b(X)dW(t) \tag{B.45}$$

then, using the Itô chain rule, one obtains

$$d\delta(X - x) = \left[\frac{\partial}{\partial X}\delta(X - x)\right][a(X)dt + b(X)dW(t)]$$

$$+ \left[\frac{1}{2}\frac{\partial^2}{(\partial X)^2}\delta(X - x)\right]b(X)^2\,dt, \tag{B.46}$$

$$= \left[-\frac{\partial}{\partial x}\delta(X - x)\right][a(X)dt + b(X)dW(t)]$$

$$+ \left[\frac{1}{2}\frac{\partial^2}{(\partial x)^2}\delta(X - x)\right]b(X)^2 dt. \tag{B.47}$$

Exercise B.3 *Convince yourself that, for an arbitrary smooth function* $f(X)$,

$$\left[\frac{\partial}{\partial x}\delta(X - x)\right]f(X) = \frac{\partial}{\partial x}[\delta(X - x)f(x)]. \tag{B.48}$$

Hint: *Consider the first-principles definition of a differential.*

Using the result of this exercise and its generalization to second derivatives, and then taking the expectation value over X, gives

$$d\wp(x) = \left\{-\frac{\partial}{\partial x}[a(x)dt + b(x)dW(t)] + \frac{1}{2}\frac{\partial^2}{(\partial x)^2}b(x)^2\,dt\right\}\wp(x). \tag{B.49}$$

If $\wp(x; t) = \delta(X(t) - x)$ at some time, then by construction this will remain true for all times by virtue of the stochastic equation (B.49). However, this equation (which we call a stochastic FPE) is more general than the SDE (B.45), insofar as it allows for initial uncertainty about X. Moreover, it allows the usual FPE to be obtained by assuming that we do not know the particular noise process dW driving the stochastic evolution of X and $\wp(x)$. Replacing dW in Eq. (B.49) by its expectation value gives the (deterministic) FPE

$$\dot{\wp}(x) = \left\{-\frac{\partial}{\partial x}a(x) + \frac{1}{2}\frac{\partial^2}{(\partial x)^2}b(x)^2\right\}\wp(x). \tag{B.50}$$

Note that this $\wp(x)$ is *not* the same as that appearing in Eq. (B.49) because we are no longer conditioning the distribution upon knowledge of the noise process. In Eq. (B.50), the term involving first derivatives is called the drift term and that involving second derivatives the diffusion term.

B.6 More general noise

As we have noted, our characterization of the Itô–Stratonovich distinction as an explicit–implicit distinction is not standard. Its advantage becomes evident when one considers point-process noise. Recall that, when one starts with evolution driven by physical noise and then idealizes this as Gaussian white noise, one ends up with a Stratonovich equation, which has to be converted into an Itô equation in order to find an explicit solution. Similarly, if one has an equation driven by physical noise that one then idealizes as a point process (that is, a time-series of δ-functions), one also ends up with an implicit equation that one has to make explicit. The implicit–explicit relation is more general than the Itô–Stratonovich one for two reasons. First, for point-process noise the defining characteristic of an Itô equation, namely that the stochastic increment is independent of the current values of the system variables, need not be true. Secondly, when feedback is considered, this Itô rule fails even for Gaussian white noise. That is because the noise which is fed back is necessarily correlated with the system at the time it is fed back, and cannot be decorrelated by invoking Itô calculus.

Although point-process noise may be non-white (that is, it need not have a flat noise spectrum), it must still have an infinite bandwidth. If the correlation function for the noise were a smooth function of time, then there would be no need to use any sort of stochastic calculus; the normal rules of calculus would apply. But, for any noise with a singular correlation function, it is appropriate to make the implicit–explicit distinction. We write a general explicit equation (in one dimension) as

$$dX = k(X)dM(t). \tag{B.51}$$

Here, deterministic evolution is being ignored, so $dM(t)$ is some stochastic increment. If $dM(t) = dW(t)$ then Eq. (B.51) is an Itô SDE. More generally, $dM(t)$ will have well-defined moments that may depend on the system $X(t)$. A stochastic calculus will be necessary if second- or higher-order moments of $dM(t)$ are not of second or higher order in dt. For Gaussian white noise, only the second-order moments fit this description, with $dW(t)^2 = dt$. In contrast, all moments must be considered for a point-process increment $dM(t) = dN(t)$.

The point-process increment can be defined by

$$E[dN(t)] = \lambda(X)dt, \tag{B.52}$$

$$dN(t)^2 = dN(t). \tag{B.53}$$

Here $\lambda(X)$ is a positive function of the random variable X (here assumed known at time t). Equation (B.52) indicates that the mean of $dN(t)$ is of order dt and may depend on the system. Equation (B.53) simply states that $dN(t)$ equals either zero or one, which is why it is called a point process. From the stochastic evolution it generates it is also known as a jump process. Because dN is infinitesimal (at least in its mean), we can say that all second- and higher-order products containing dt are $o(dt)$. This notation means that such products (like $dN\,dt$, but not dN^2) are negligible compared with dt. Obviously all moments of $dN(t)$ are of the same order as dt, so the chain rule for $f(X)$ will completely fail.

Unlike dW, which is independent of the system at the same time, dN does depend on the system, at least statistically, through Eq. (B.52). In fact, we can use the above equations to show that

$$E[dN(t)f(X)] = E[\lambda(X)f(X)]dt, \tag{B.54}$$

for some arbitrary function f.

Exercise B.4 *Convince yourself of this.*

It turns out [Gar85] that for Markovian processes it is sufficient to consider only the above two cases, $dM = dW$ and $dM = dN$.

Now consider a SDE that, like the Stratonovich equation for Gaussian white noise, arises from a physical process in which the singularity of the noise is an idealization. Such an equation would be written, using our convention, as

$$\dot{X} = \chi(X)\mu(t), \tag{B.55}$$

where $\mu(t)$ is a noisy function of time that is idealized by

$$\mu(t) = dM(t)/dt. \tag{B.56}$$

Equation (B.55) is an *implicit* equation in that it gives the increment in X only implicitly. It has the advantage that $f(X)$ would obey an implicit equation as given by the usual chain rule,

$$\dot{f}(X) = f'(X)\chi(X)\mu(t). \tag{B.57}$$

Notice that the third distinction between Itô and Stratonovich calculus, namely that based on the independence of the noise term and the system at the same time, has not entered this discussion. This is because, even in the explicit equation (B.51), the noise may depend on the system. The independence condition is simply a peculiarity of Gaussian white noise. The implicit–explicit distinction is more general than the Stratonovich–Itô distinction. As we will show below, the relationship between the Stratonovich and Itô SDEs can be easily derived within this more general framework.

The general problem is to find the explicit form of an implicit SDE with arbitrary noise. For implicit equations, the usual chain rule (B.57) applies, and can be rewritten

$$\dot{f} = f'(X)\chi(X)\mu(t) \equiv \phi\big(f(X)\big)\mu(t), \tag{B.58}$$

where this equation defines $\phi(f)$. Now, in order to solve Eq. (B.55), it is necessary to find an explicit expression for the increment in X. The correct answer may be found by expanding the Taylor series to all orders in dM. This can be written formally as

$$X(t + dt) = \exp\left(dt \, \frac{\partial}{\partial s} \right) X(s)|_{s=t} \tag{B.59}$$

$$= \exp\left[\chi(x)dM(t)\frac{\partial}{\partial x} \right] x|_{x=X(t)}. \tag{B.60}$$

Here we have used the relation

$$\left[\frac{d}{ds}X(s) = \chi\big(X(s)\big)\frac{dM(t)}{dt} \right]_{s=t}, \tag{B.61}$$

which is the explicit meaning of the implicit Eq. (B.55). Note that $\mu(t)$ is assumed to be constant, while $X(s)$ is evolved, for the same reasons as explained following Eq. (B.32). If the noise $\mu(t)$ is the limit of a physical process (which is the limit for which Eq. (B.55) is intended to apply), then it must have some finite correlation time over which it remains relatively constant. The noise can be considered δ-correlated if that time can be considered to be infinitesimal compared with the characteristic evolution time of the system X.

The explicit SDE is thus defined to be

$$dX(t) = \left(\exp\left[\chi(X)dM(t)\frac{\partial}{\partial X} \right] - 1 \right) X(t), \tag{B.62}$$

which means

$$dX(t) = \left(\exp\left[\chi(x)dM(t)\frac{\partial}{\partial x} \right] - 1 \right) x|_{x=X(t)}. \tag{B.63}$$

This expression will converge for all $\chi(X)$ for $dM = dN$ or $dM = dW$, and is compatible with the chain-rule requirement (B.58) for the implicit form. This can be seen from calculating the increment in $f(X)$ using the explicit form:

$$
\begin{aligned}
df &= f\big(X(t) + dX(t)\big) - f\big(X(t)\big) \\
&= f\left(\exp\left[\chi(x)dM(t)\frac{\partial}{\partial x}\right]x\big|_{x=X(t)}\right) - f\big(X(t)\big) \\
&= \exp\left[\chi(x)dM(t)\frac{\partial}{\partial x}\right]f(x)\big|_{x=X(t)} - f\big(X(t)\big) \\
&= \left(\exp\left[\phi(f)dM(t)\frac{\partial}{\partial f}\right] - 1\right)f\big|_{f=f(X(t))},
\end{aligned}
\tag{B.64}
$$

as expected from Eq. (B.58). This completes the justification for Eq. (B.62) as the correct explicit form of the implicit Eq. (B.55).

For deterministic processes ($\chi = 0$), there is no distinction between the explicit and implicit forms, since only the first-order expansion of the exponential remains with dt infinitesimal. There is also no distinction if $\chi(x)$ is a constant. For Gaussian white noise, the formula (B.62) is the rule given in Section B.3 for converting from Stratonovich to Itô form. That is, if the Stratonovich SDE is Eq. (B.55) with $dM(t) = dW(t)$, then the Itô SDE is

$$
dX(t) = \chi(X)dW(t) + \tfrac{1}{2}\chi(X)\chi'(X)dt.
\tag{B.65}
$$

Exercise B.5 *Show this, using the Itô rule $dW(t)^2 = dt$.*

This rule implies that it is necessary to expand the exponential only to second order. This fact makes the inverse transformation (Itô to Stratonovich) easy. For the jump process, the rule $dN(t)^2 = dN(t)$ means that the exponential must be expanded to all orders. This gives

$$
dX(t) = dN(t)\left(\exp\left[\chi(x)\frac{\partial}{\partial X}\right] - 1\right)x(t).
\tag{B.66}
$$

In this case, the inverse transformation would not be easy to find in general, but there seems no physical motivation for requiring it.

B.6.1 Multi-dimensional generalization

The multi-dimensional generalization of the above formulae is obvious. Writing X_i for the componenents of the vector $\vec{X}$ and using the Einstein summation convention, if the implicit form is

$$
\dot{X}_i(t) = \chi_{ij}\big(\vec{X}(t)\big)\mu_j(t),
\tag{B.67}
$$

then the explicit form is

$$
dX_i(t) = \left(\exp\left[\chi_{kj}(\vec{X})dM_j(t)\frac{\partial}{\partial X_k}\right] - 1\right)X_i(t).
\tag{B.68}
$$

This is quite complicated in general. Fortunately, when considering quantum feedback processes, the equations for the state are linear. Thus, if one has the implicit equation

$$\dot{\rho}(t) = \mu(t)\mathcal{K}\rho(t), \tag{B.69}$$

where $\mathcal{K}$ is a Liouville superoperator, then the explicit SDE is simply

$$d\rho(t) = (\exp[\mathcal{K}\,dM(t)] - 1)\rho(t). \tag{B.70}$$

Exercise B.6 *Convince yourself that this is consistent with Eq. (B.68).*

References

[AAS+02] M. A. Armen, J. K. Au, J. K. Stockton, A. C. Doherty, and H. Mabuchi, *Adaptive homodyne measurement of optical phase*, Phys. Rev. Lett. **89**, 133602, (2002).

[AB61] Y. Aharonov and D. Bohm, *Time in the quantum theory and the uncertainty relation for time and energy*, Phys. Rev. **122**, 1649, (1961).

[ABC+01] G. Alber, Th. Beth, Ch. Charnes, A. Delgado, M. Grassl, and M. Mussinger, *Stabilizing distinguishable qubits against spontaneous decay by detected-jump correcting quantum codes*, Phys. Rev. Lett. **86**, 4402, (2001).

[ABC+03] ———, *Detected-jump-error-correcting quantum codes, quantum error designs, and quantum computation*, Phys. Rev. A **68**, 012316, (2003).

[ABJW05] D. J. Atkins, Z. Brady, K. Jacobs, and H. M. Wiseman, *Classical robustness of quantum unravellings*, Europhys. Lett. **69**, 163, (2005).

[ACH07] E. Andersson, J. D. Cresser, and M. J. W. Hall, *Finding the Kraus decomposition from a master equation and vice versa*, J. Mod. Opt. **54**, 1695, (2007).

[AD01] F. Albertini and D. D'Alessandro, *Notions of controllability for quantum mechanical systems*, Proceedings of the 40th IEEE Conference on Decision and Control, vol. 2, IEEE, New York (also available as eprint:quant-ph/0106128), p. 1589, 2001.

[ADL02] C. Ahn, A. C. Doherty, and A. J. Landahl, *Continuous quantum error correction via quantum feedback control*, Phys. Rev. A **65**, 042301, (2002).

[AK65] E. Arthurs and J. L. Kelly, *On the simultaneous measurement of a pair of conjugate observables*, Bell. Syst. Tech. J. **44**, 725, (1965).

[AL87] R. Alicki and K. Lendi, *Quantum dynamical semigroups and applications*, Lecture Notes in Physics, vol. 717, Springer, Berlin, 1987.

[Alt02] C. Altafini, *Controllability of quantum mechanical systems by root space decomposition of su(n)*, J. Math. Phys. **43**, 2051, (2002).

[AMW88] P. Alsing, G. J. Milburn, and D. F. Walls, *Quantum nondemolition measurements in optical cavities*, Phys. Rev. A **37**, 2970, (1988).

[Ash70] A. Ashkin, *Atomic-beam deflection by resonance-radiation pressure*, Phys. Rev. Lett. **25**, 1321, (1970).

[Ash90] R. B. Ash, *Information theory*, Dover, New York, 1990.

[AWJ04] C. Ahn, H. M. Wiseman, and K. Jacobs, *Quantum error correction for continuously detected errors with any number of error channels per qubit*, Phys. Rev. A **70**, 024302, (2004).

[AWM03] C. Ahn, H. M. Wiseman, and G. J. Milburn, *Quantum error correction for continuously detected errors*, Phys. Rev. A **67**, 052310, (2003).

[Bal98] L. E. Ballentine, *Quantum mechanics: A modern development*, World Scientific, Singapore, 1998.

[Ban01] K. Banaszek, *Fidelity balance in quantum operations*, Phys. Rev. Lett. **86**, 1366, (2001).

[Bar90] A. Barchielli, *Direct and heterodyne detection and other applications of quantum stochastic calculus to quantum optics*, Quantum Opt. **2**, 423, (1990).

[Bar93] ———, *Stochastic differential equations and a posteriori states in quantum mechanics*, Int. J. Theor. Phys. **32**, 2221, (1993).

[BB00] Ya. M. Blanter and M. Büttiker, *Shot noise in mesoscopic conductors*, Phys. Rep. **336**, 1, (2000).

[BBC+93] C. H. Bennett, G. Brassard, C. Crépeau, R. Jozsa, A. Peres, and W. K. Wootters, *Teleporting an unknown quantum state via dual classical and Einstein–Podolsky–Rosen channels*, Phys. Rev. Lett. **70**, 1895, (1993).

[BBL04] S. A. Babichev, B. Brezger, and A. I. Lvovsky, *Remote preparation of a single-mode photonic qubit by measuring field quadrature noise*, Phys. Rev. Lett. **92**, 047903, (2004).

[BC94] S. L. Braunstein and C. M. Caves, *Statistical distance and the geometry of quantum states*, Phys. Rev. Lett. **72**, 3439, (1994).

[BCM96] S. L. Braunstein, C. M. Caves, and G. J. Milburn, *Generalized uncertainty relations: Theory, examples, and Lorentz invariance*, Annals Phys. **247**, 135, (1996).

[BCS57] J. Bardeen, L. N. Cooper, and J. R. Schrieffer, *Theory of superconductivity*, Phys. Rev. **108**, 1175, (1957).

[BCS+04] M. D. Barrett, J. Chiaverini, T. Schaetz, J. Britton, W. M. Itano, J. D. Jost, E. Knill, C. Langer, D. Leibfried, R. Ozeri, and D. J. Wineland, *Deterministic quantum teleportation of atomic qubits*, Nature **429**, 737, (2004).

[BDD05] J. Bylander, T. Duty, and P. Delsing, *Current measurement by real-time counting of single electrons*, Nature **434**, 361, (2005).

[Bel64] J. S. Bell, *On the Einstein–Podolsy–Rosen paradox*, Physics **1**, 195, (1964), reprinted in Ref. [Bel87].

[Bel83] V. P. Belavkin, *Towards the theory of the control of observable quantum systems*, Autom. Remote Control **44**, 178, (1983) (also available as eprint:quant-ph/0408003).

[Bel87] J. S. Bell, *Speakable and unspeakable in quantum mechanics*, Cambridge University Press, Cambridge, 1987.

[Bel88] V. P. Belavkin, *Nondemolition measurement and nonlinear filtering of quantum stochastic processes*, Lecture Notes in Control and Information Sciences, Springer, Berlin, p. 245, 1988.

[Bel99] ———, *Measurement, filtering and control in quantum open dynamical systems*, Rep. Math. Phys. **43**, 405, (1999).

[Bel02] ———, *Quantum causality, stochastics, trajectories and information*, Rep. Prog. Phys. **65**, 353, (2002).

[Ber94] P. R. Berman (ed.), *Cavity quantum electrodynamics (advances in atomic, molecular and optical physics)*, Academic Press, Boston, MA, 1994.

[BF28] M. Born and V. Fock, *Beweis des Adiabatensatzes*, Z. Phys. **51**, 165–180, (1928).

[BFK00] S. L. Braunstein, C. A. Fuchs, and H. J. Kimble, *Criteria for continuous-variable quantum teleportation*, J. Mod. Opt. **47**, 267, (2000).

[BFM02] T. A. Brun, J. Finkelstein, and N. D. Mermin, *How much state assignments can differ*, Phys. Rev. A **65**, 032315, (2002).

[BGS$^+$99] B. C. Buchler, M. B. Gray, D. A. Shaddock, T. C. Ralph, and D. E. McClelland, *Suppression of classical and quantum radiation pressure noise via electro-optic feedback*, Opt. Lett. **24**, 259, (1999).

[BH93] D. Bohm and B. J. Hiley, *The undivided universe: An ontological interpretation of quantum theory*, Routledge, London, 1993.

[BHD$^+$96] M. Brune, E. Hagley, J. Dreyer, X. Maître, A. Maali, C. Wunderlich, J.-M. Raimond, and S. Haroche, *Observing the progressive decoherence of the 'meter' in a quantum measurement*, Phys. Rev. Lett. **77**, 4887, (1996).

[BHIW86] J. C. Bergquist, R. G. Hulet, W. M. Itano, and D. J. Wineland, *Observation of quantum jumps in a single atom*, Phys. Rev. Lett. **57**, 1699, (1986).

[BK92] V. B. Braginsky and F. Y. Khalili, *Quantum measurement*, Cambridge University Press, Cambridge, 1992.

[BK98] S. L. Braunstein and H. J. Kimble, *Teleportation of continuous quantum variables*, Phys. Rev. Lett. **80**, 869, (1998).

[BM03] S. D. Barrett and G. J. Milburn, *Measuring the decoherence rate in a semiconductor charge qubit*, Phys. Rev. B **68**, 155307, (2003).

[BM04] A. J. Berglund and H. Mabuchi, *Feedback controller design for tracking a single fluorescent molecule*, Appl. Phys. B. **78**, 653, (2004).

[BMG$^+$07] A. M. Branczyk, P. E. M. F. Mendonca, A. Gilchrist, A. C. Doherty, and S. D. Bartlett, *Quantum control of a single qubit*, Phys. Rev. A **75**, 012329, (2007).

[Boh13] N. Bohr, *On the constitution of atoms and molecules*, Phil. Mag. **26**, 1, (1913).

[BP02] H.-P. Breuer and F. Petruccione, *The theory of open quantum systems*, Oxford University Press, Oxford, 2002.

[BR97] S. M. Barnett and E. Riis, *Experimental demonstration of polarization discrimination at the Helstrom bound*, J. Mod. Opt. **44**, 1061, (1997).

[BR05] D. E. Browne and T. Rudolph, *Resource-efficient linear optical quantum computation*, Phys. Rev. Lett. **95**, 010501, (2005).

[Bro73] R. W. Brockett, *Lie theory and control systems defined on spheres*, SIAM J. Appl. Math. **25**, 213, (1973).

[BRS$^+$05] T. M. Buehler, D. J. Reilly, R. P. Starrett, A. D. Greentree, A. R. Hamilton, A. S. Dzurak, and R. G. Clark, *Single-shot readout with the radio-frequency single-electron transistor in the presence of charge noise*, Appl. Phys. Lett. **86**, 143117, (2005).

[BRW$^+$06] P. Bushev, D. Rotter, A. Wilson, F. Dubin, C. Becher, J. Eschner, R. Blatt, V. Steixner, P. Rabl, and P. Zoller, *Feedback cooling of a single trapped ion*, Phys. Rev. Lett. **96**, 043003, (2006).

[BS92] V. P. Belavkin and P. Staszewski, *Nondemolition observation of a free quantum particle*, Phys. Rev. A **45**, 1347, (1992).

[BS94]   J. M. Bernardo and A. F. M. Smith, *Bayesian theory*, Wiley, Chichester, 1994.

[BSSM07] L. Bouten, J. Stockton, G. Sarma, and H. Mabuchi, *Scattering of polarized laser light by an atomic gas in free space: A quantum stochastic differential equation approach*, Phys. Rev. A **75**, 052111, (2007).

[BP86]   S. M. Barnett and D. T. Pegg, *Phase in quantum optics*, J. Phys. A **19**, 3849, (1986).

[Büt88] M. Büttiker, *Symmetry of electrical conduction*, IBM J. Res. Dev. **32**, 317–334, (1988).

[BvH08] L. Bouten and R. van Handel, *On the separation principle of quantum control*, Quantum Stochastics and Information: Statistics, Filtering and Control (V. P. Belavkin and M. I. Guţă, eds.), World Scientific, Singapore, p. 206, 2008 (also available as eprint: math-ph/0511021).

[BW00]   D. W. Berry and H. M. Wiseman, *Optimal states and almost optimal adaptive measurements for quantum interferometry*, Phys. Rev. Lett. **85**, 5098, (2000).

[BWB01] D. W. Berry, H. M. Wiseman, and J. K. Breslin, *Optimal input states and feedback for interferometric phase estimation*, Phys. Rev. A **63**, 053804, (2001).

[Car93] H. J. Carmichael, *An open systems approach to quantum optics*, Springer, Berlin, 1993.

[Car99] ———, *Statistical methods in quantum optics, Vol. 1: Master equations and Fokker–Planck equations*, Springer, Berlin, 1999.

[Car07] ———, *Statistical methods in quantum optics, Vol. 2: Non-classical fields*, Springer, Berlin, 2007.

[CBR91] H. J. Carmichael, R. J. Brecha, and P. R. Rice, *Quantum interference and collapse of the wavefunction in cavity QED*, Opt. Commun. **82**, 73, (1991).

[CCBFO00] H. J. Carmichael, H. M. Castro-Beltran, G. T. Foster, and L. A. Orozco, *Giant violations of classical inequalities through conditional homodyne detection of the quadrature amplitudes of light*, Phys. Rev. Lett. **85**, 1855, (2000).

[CCBR01] R. B. M. Clarke, A. Chefles, S. M. Barnett, and E. Riis, *Experimental demonstration of optimal unambiguous state discrimination*, Phys. Rev. A **63**, 040305(R), (2001).

[CFS02a] C. M. Caves, C. A. Fuchs, and R. Schack, *Conditions for compatibility of quantum-state assignments*, Phys. Rev. A **66**, 062111, (2002).

[CFS02b] ———, *Quantum probabilities as Bayesian probabilities*, Phys. Rev. A **65**, 022305, (2002).

[Che00] A. Chefles, *Quantum state discrimination*, Contemp. Phys. **41**, 401, (2000).

[CHP03] J. M. Courty, A. Heidmann, and M. Pinard, *Back-action cancellation in interferometers by quantum locking*, Europhys. Lett. **63**, 226, (2003).

[CK85]   R. J. Cook and H. J. Kimble, *Possibility of direct observation of quantum jumps*, Phys. Rev. Lett. **54**, 1023, (1985).

[CKT94] H. J. Carmichael, P. Kochan, and L. Tian, *Coherent states and open quantum systems: A comment on the Stern–Gerlach experiment and Schrödinger's cat*, Proceedings of the International Symposium on

Coherent States: Past, Present, and Future (D. H. Feng, J. R. Klauder, and M. R. Strayer, eds.), World Scientific, Singapore, p. 75, 1994.

[CM87]   C. M. Caves and G. J. Milburn, *Quantum-mechanical model for continuous position measurements*, Phys. Rev. A **36**, 5543, (1987).

[CM91]   S. L. Campbell and C. D. Meyer, *Generalized inverses of linear transformations*, Dover Publications, New York, 1991.

[CMG07]   R. L. Cook, P. J. Martin, and J. M. Geremia, *Optical coherent state discrimination using a closed-loop quantum measurement*, Nature **446**, 774, (2007).

[Con90]   J. B. Conway, *A course in functional analysis*, 2nd edn, Springer, New York, 1990.

[Coo56]   L. N. Cooper, *Bound electron pairs in a degenerate Fermi gas*, Phys. Rev. **104**, 1189, (1956).

[CR98]   A. N. Cleland and M. L. Roukes, *Nanostructure-based mechanical electrometry*, Nature **392**, 160, (1998).

[CRG89]   C. Cohen-Tannoudji, J. Dupont Roc, and G. Grynberg, *Photons and atoms: Introduction to quantum electrodynamics*, Wiley-Interscience, New York, 1989.

[CSVR89] H. J. Carmichael, S. Singh, R. Vyas, and P. R. Rice, *Photoelectron waiting times and atomic state reduction in resonance fluorescence*, Phys. Rev. A **39**, 1200, (1989).

[CT06]   T. M. Cover and J. A. Thomas, *Elements of information theory*, 2nd edn, Wiley-Interscience, New York, 2006.

[CW76]   H. J. Carmichael and D. F. Walls, *Proposal for the measurement of the resonant Stark effect by photon correlation techniques*, J. Phys. B **9**, L43, (1976).

[CWJ08]   J. Combes, H. M. Wiseman, and K. Jacobs, *Rapid measurement of quantum systems using feedback control*, Phys. Rev. Lett. **100**, 160503, (2008).

[CWZ84]   M. J. Collett, D. F. Walls, and P. Zoller, *Spectrum of squeezing in resonance fluorescence*, Opt. Commun. **52**, 145, (1984).

[D'A07]   D. D'Alessandro, *Introduction to quantum control and dynamics*, Chapman & Hall, London, 2007.

[Dat95]   S. Datta, *Electronic transport in mesoscopic systems*, Cambridge University Press, Cambridge, 1995.

[Dav76]   E. B. Davies, *Quantum theory of open systems*, Academic Press, London, 1976.

[DCM92]   J. Dalibard, Y. Castin, and K. Mølmer, *Wave-function approach to dissipative processes in quantum optics*, Phys. Rev. Lett. **68**, 580, (1992).

[DDS⁺08] S. Deléglise, I. Dotsenko, C. Sayrin, J. Bernu, M. Brune, J.-M. Raimond, and S. Haroche, *Reconstruction of non-classical cavity field states with snapshots of their decoherence*, Nature **455**, 510, (2008).

[DGKF89] J. C. Doyle, K. Glover, P. P. Khargonekar, and B. A. Francis, *State-space solutions to standard H_2 and H_∞ control problems*, IEEE Trans. Automatic Control **34**, 831, (1989).

[DHJ⁺00] A. C. Doherty, S. Habib, K. Jacobs, H. Mabuchi, and S. M. Tan, *Quantum feedback control and classical control theory*, Phys. Rev. A **62**, 012105, (2000).

[Die88]   D. Dieks, *Overlap and distinguishability of quantum states*, Phys. Lett. A **126**, 303–306, (1988).

[Di6́88] L. Diósi, *Localized solution of a simple nonlinear quantum Langevin equation*, Phys. Lett. A **132**, 233, (1988).

[Diós93] ———, *On high-temperature Markovian equation for quantum Brownian motion*, Europhys. Lett. **22**, 1, (1993).

[Dió08] ———, *Non-Markovian continuous quantum measurement of retarded observables*, Phys. Rev. Lett. **100**, 080401, (2008), Erratum *ibid.* **101**, 149902, (2008).

[Dir27] P. A. M. Dirac, *The quantum theory of the emission and absorption of radiation*, Proc. Roy. Soc. London. A **114**, 243, (1927).

[Dir30] ———, *The principles of quantum mechanics*, Clarendon Press, Oxford, 1930.

[DJ99] A. C. Doherty and K. Jacobs, *Feedback control of quantum systems using continuous state estimation*, Phys. Rev. A **60**, 2700, (1999).

[DOG03] B. D'Urso, B. Odom, and G. Gabrielse, *Feedback cooling of a one-electron oscillator*, Phys. Rev. Lett. **90**, 043001, (2003).

[Dol73] S. J. Dolinar, *An optimum receiver for the binary coherent state quantum channel*, MIT Research Laboratory of Electronics Quarterly Progress Report **111**, 115, (1973).

[DPS02] A. C. Doherty, P. A. Parrilo, and F. M. Spedalieri, *Distinguishing separable and entangled states*, Phys. Rev. Lett. **88**, 187904, (2002).

[DPS03] G. M. D'Ariano, M. G. A. Paris, and M. F. Sacchi, *Quantum tomography*, Adv. Imaging Electron Phys. **128**, 205, (2003).

[DZR92] R. Dum, P. Zoller, and H. Ritsch, *Monte Carlo simulation of the atomic master equation for spontaneous emission*, Phys. Rev. A **45**, 4879, (1992).

[Ein17] A. Einstein, *On the quantum theory of radiation*, Phys. Z. **18**, 121, (1917).

[Eis05] J. Eisert, *Optimizing linear optics quantum gates*, Phys. Rev. Lett. **95**, 040502, (2005).

[EPR35] A. Einstein, B. Podolsky, and N. Rosen, *Can quantum-mechanical description of physical reality be considered complete?*, Phys. Rev. **47**, 777, (1935).

[Fan57] U. Fano, *Description of states in quantum mechanics by density matrix and operator techniques*, Rev. Mod. Phys. **29**, 74, (1957).

[FDF+02] J. D. Franson, M. M. Donegan, M. J. Fitch, B. C. Jacobs, and T. B. Pittman, *High-fidelity quantum logic operations using linear optical elements*, Phys. Rev. Lett. **89**, 137901, (2002).

[FF04] L. Fedichkin and A. Fedorov, *Error rate of a charge qubit coupled to an acoustic phonon reservoir*, Phys. Rev. A **69**, 032311, (2004).

[FHWH07] C. Fricke, F. Hohls, W. Wegscheider, and R. J. Haug, *Bimodal counting statistics in single-electron tunneling through a quantum dot*, Phys. Rev. B **76**, 155307, (2007).

[FJ01] C. A. Fuchs and K. Jacobs, *Information-tradeoff relations for finite-strength quantum measurements*, Phys. Rev. A **63**, 062305, (2001).

[FMO00] G. T. Foster, S. L. Mielke, and L. A. Orozco, *Intensity correlations in cavity QED*, Phys. Rev. A **61**, 053821, (2000).

[FSB+98] A. Furusawa, J. L. Sørensen, S. L. Braunstein, C. A. Fuchs, H. J. Kimble, and E. S. Polzik, *Unconditional quantum teleportation*, Science **282**, 706, (1998).

[FSRO02] G. T. Foster, W. P. Smith, J. E. Reiner, and L. A. Orozco, *Time-dependent electric field fluctuations at the subphoton level*, Phys. Rev. A **66**, 033807, (2002).

[Fuc96] C. A. Fuchs, *Distinguishability and accessible information in quantum theory*, Ph.D. thesis, The University of New Mexico, Albuquerque, NM, 1996 (also available as eprint:quant-ph/9601020).

[Gar85] C. W. Gardiner, *Handbook of stochastic methods for physics, chemistry and the natural sciences*, Springer, Berlin, 1985.

[Gar86] ———, *Inhibition of atomic phase decays by squeezed light: A direct effect of squeezing*, Phys. Rev. Lett. **56**, 1917, (1986).

[Gar04] ———, *Input and output in damped quantum systems III: Formulation of damped systems driven by fermion fields*, Opt. Commun. **243**, 57, (2004).

[GAW04] J. Gambetta, T. Askerud, and H. M. Wiseman, *Jumplike unravelings for non-Markovian open quantum systems*, Phys. Rev. A **69**, 052104, (2004).

[GBD⁺07] C. Guerlin, J. Bernu, S. Deléglise, C. Sayrin, S. Gleyzes, S. Kuhr, M. Brune, J.-M. Raimond, and S. Haroche, *Progressive field-state collapse and quantum non-demolition photon counting*, Nature **448**, 889, (2007).

[GBP97] M. Grassl, Th. Beth, and T. Pellizzari, *Codes for the quantum erasure channel*, Phys. Rev. A **56**, 33, (1997).

[GC85] C. W. Gardiner and M. J. Collett, *Input and output in damped quantum systems: Quantum stochastic differential equations and the master equation*, Phys. Rev. A **31**, 3761, (1985).

[GC99] D. Gottesman and I. L. Chuang, *Demonstrating the viability of universal quantum computation using teleportation and single-qubit operations*, Nature **402**, 390, (1999).

[GG01] F. Grosshans and P. Grangier, *Quantum cloning and teleportation criteria for continuous quantum variables*, Phys. Rev. A **64**, 010301, (2001).

[Gil83] D. T. Gillespie, *A theorem for physicists in the theory of random variables*, Am. J. Phys. **51**, 520, (1983).

[Gil93] ———, *Fluctuation and dissipation in Brownian motion*, Am. J. Phys. **61**, 1077, (1993).

[Gil96] ———, *The mathematics of Brownian motion and Johnson noise*, Am. J. Phys. **64**, 225, (1996).

[Gis89] N. Gisin, *Stochastic quantum dynamics and relativity*, Helv. Phys. Acta **62**, 363, (1989).

[GKG⁺07] S. Gleyzes, S. Kuhr, C. Guerlin, J. Bernu, S. Deléglise, U. B. Hoff, M. Brune, J.-M. Raimond, and S. Haroche, *Quantum jumps of light recording the birth and death of a photon in a cavity*, Nature **446**, 297, (2007).

[Gla63] R. J. Glauber, *The quantum theory of optical coherence*, Phys. Rev. **130**, 2529, (1963).

[Gle57] A. M. Gleason, *Measures on the closed subspaces of a Hilbert space*, Indiana Univ. Math. J. **6**, 885, (1957).

[GLM06] V. Giovannetti, S. Lloyd, and L. Maccone, *Quantum metrology*, Phys. Rev. Lett. **96**, 010401, (2006).

[GM01] H-S. Goan and G. J. Milburn, *Dynamics of a mesoscopic charge quantum bit under continuous quantum measurement*, Phys. Rev. B **64**, 235307, (2001).

[GMWS01] H-S. Goan, G. J. Milburn, H. M. Wiseman, and H.-B. Sun, *Continuous quantum measurement of two coupled quantum dots using a point contact: A quantum trajectory approach*, Phys. Rev. B **63**, 125326, (2001).

[GN96]   R. B. Griffiths and C-S. Niu, *Semiclassical Fourier transform for quantum computation*, Phys. Rev. Lett. **76**, 3228, (1996).

[Goa03] H-S. Goan, *An analysis of reading out the state of a charge quantum bit*, Quantum Information Computation **3**, 121, (2003).

[Got96] D. Gottesman, *Class of quantum error-correcting codes saturating the quantum Hamming bound*, Phys. Rev. A **54**, 1862, (1996).

[GP92a] N. Gisin and I. C. Percival, *The quantum-state diffusion model applied to open systems*, J. Phys. A **25**, 5677, (1992).

[GP92b] ————, *Wave-function approach to dissipative processes: Are there quantum jumps?*, Phys. Lett. A **167**, 315, (1992).

[GPS01] H. Goldstein, C. P. Poole, and J. L. Safko, *Classical mechanics*, 3rd edn, Addison-Wesley, Reading, MA, 2001.

[GPZ92] C. W. Gardiner, A. S. Parkins, and P. Zoller, *Wave-function quantum stochastic differential equations and quantum-jump simulation methods*, Phys. Rev. A **46**, 4363, (1992).

[GSM04] J. M. Geremia, J. K. Stockton, and H. Mabuchi, *Real-time quantum feedback control of atomic spin-squeezing*, Science **304**, 270, (2004).

[GSM08] ————, *Retraction*, Science **321**, 489a, (2008).

[Gur97] S. A. Gurvitz, *Measurements with a noninvasive detector and dephasing mechanism*, Phys. Rev. B **56**, 15215, (1997).

[GW01]   J. Gambetta and H. M. Wiseman, *State and dynamical parameter estimation for open quantum systems*, Phys. Rev. A **64**, 042105, (2001).

[GW02]   ————, *Non-Markovian stochastic Schrödinger equations: Generalization to real-valued noise using quantum-measurement theory*, Phys. Rev. A **66**, 012108, (2002).

[GW03]   ————, *Interpretation of non-Markovian stochastic Schrödinger equations as a hidden-variable theory*, Phys. Rev. A **68**, 062104, (2003).

[GW04]   ————, *Modal dynamics for positive operator measures*, Foundations Phys. **34**, 419, (2004).

[GWM93] M. J. Gagen, H. M. Wiseman, and G. J. Milburn, *Continuous position measurements and the quantum Zeno effect*, Phys. Rev. A **48**, 132, (1993).

[GZ04]   C. W. Gardiner and P. Zoller, *Quantum noise: A handbook of Markovian and non-Markovian quantum stochastic methods with applications to quantum optics*, Springer, Berlin, 2004.

[Har02] L. Hardy, *Why quantum theory?*, Non-Locality and Modality (Tomasz Placek and Jeremy Butterfield, eds.), Nato Science Series: II, vol. 64, Kluwer, Dordrecht, p. 61, 2002.

[Haw71] A. G. Hawkes, *Spectra of some self-exciting and mutually exciting point processes*, Biometrika **58**, 83, (1971).

[HBB⁺07] B. L. Higgins, D. W. Berry, S. D. Bartlett, H. M. Wiseman, and G. J. Pryde, *Entanglement-free Heisenberg-limited phase estimation*, Nature **450**, 393, (2007).

[HDW⁺04] L. C. Hollenberg, A. S. Dzurak, C. Wellard, A. R. Hamilton, D. J. Reilly, G. J. Milburn, and R. G. Clark, *Charge-based quantum computing using single donors in semiconductors*, Phys. Rev. B **69**, 113301, (2004).

[Hei27] W. Heisenberg, *Über den anschaulichen Inhalt der quantentheoretischen Kinematik und Mechanik*, Z. Phys. **43**, 172, (1927), English translation in Ref. [WZ83].

[Hei30]   ——, *The physical principles of quantum mechanics*, University of Chicago, Chicago, IL, 1930.

[Hel76]   C. W. Helstrom, *Quantum detection and estimation theory*, Mathematics in Science and Engineering, vol. 123, Academic Press, New York, 1976.

[HHM98]   H. Hofmann, O. Hess, and G. Mahler, *Quantum control by compensation of quantum fluctuations*, Opt. Express **2**, 339, (1998).

[HJHS03] A. Hopkins, K. Jacobs, S. Habib, and K. Schwab, *Feedback cooling of a nanomechanical resonator*, Phys. Rev. B **68**, 235328, (2003).

[HJW93]   L. Hughston, R. Jozsa, and W. Wootters, *A complete classification of quantum ensembles having a given density matrix*, Phys. Lett. A **183**, 14, (1993).

[HK97]   D. B. Horoshko and S. Ya. Kilin, *Direct detection feedback for preserving quantum coherence in an open cavity*, Phys. Rev. Lett. **78**, 840, (1997).

[HMG⁺96] B. Huttner, A. Muller, J. D. Gautier, H. Zbinden, and N. Gisin, *Unambiguous quantum measurement of nonorthogonal states*, Phys. Rev. A **54**, 3783, (1996).

[HMP⁺96] Z. Hradil, R. Myška, J. Peřina, M. Zawisky, Y. Hasegawa, and H. Rauch, *Quantum phase in interferometry*, Phys. Rev. Lett. **76**, 4295, (1996).

[Hol82]   A. S. Holevo, *Probabilistic and statistical aspects of quantum theory*, Statistics and Probability, vol. 1, North-Holland, Amsterdam, 1982.

[Hol84]   ——, *Covariant measurements and imprimitivity systems*, Lecture Notes in Mathematics, vol. 1055, Springer, Berlin, p. 153, (1984).

[HOM87]   C. K. Hong, Z. Y. Ou, and L. Mandel, *Measurement of subpicosecond time intervals between two photons by interference*, Phys. Rev. Lett. **59**, 2044, (1987).

[HOW⁺05] L. Hollberg, C. W. Oates, G. Wilpers, C. W. Hoyt, Z. W. Barber, S. A. Diddams, W. H. Oskay, and J. C. Bergquist, *Optical frequency/wavelength references*, J. Phys. B **38**, S469, (2005).

[HR85]   F. Haake and R. Reibold, *Strong damping and low-temperature anomalies for the harmonic oscillator*, Phys. Rev. A **32**, 2462, (1985).

[HTC83]   G. M. Huang, T. J. Tarn, and J. W. Clark, *On the controllability of quantum-mechanical systems*, J. Math. Phys. **24**, 2608, (1983).

[Hus40]   K. Husimi, *Some formal properties of the density matrix*, Proc. Phys. Math. Soc. Japan **22**, 264, (1940).

[HWPC05] K. Hammerer, M. M. Wolf, E. S. Polzik, and J. I. Cirac, *Quantum benchmark for storage and transmission of coherent states*, Phys. Rev. Lett. **94**, 150503, (2005).

[HY86]   H. A. Haus and Y. Yamamoto, *Theory of feedback-generated squeezed states*, Phys. Rev. A **34**, 270, (1986).

[Imr97]   Y. Imry, *Introduction to mesoscopic physics*, Oxford University Press USA, New York, 1997.

[Iva87]   I. D. Ivanovic, *How to differentiate between non-orthogonal states*, Phys. Lett. A **123**, 257, (1987).

[IW89]   N. Ikeda and S. Watanabe, *Stochastic differential equations and diffusion processes*, 2nd edn, North-Holland, Amsterdam, 1989.

[Jac93]   O. L. R. Jacobs, *Introduction to control theory*, Oxford University Press, Oxford, 1993.

[Jac03]   K. Jacobs, *How to project qubits faster using quantum feedback*, Phys. Rev. A **67**, 030301(R), (2003).

[Jac07]   ———, *Feedback control for communication with non-orthogonal states*, Quantum Information Computation **7**, 127, (2007).

[Jam04]   M. R. James, *Risk-sensitive optimal control of quantum systems*, Phys. Rev. A **69**, 032108, (2004).

[Jam05]   ———, *A quantum Langevin formulation of risk-sensitive optimal control*, J. Opt. B **7**, S198, (2005).

[Jau68]   J. M. Jauch, *Foundations of quantum mechanics*, Addison Wesley Longman, Reading, MA, 1968.

[JKP01]   B. Julsgaard, A. Kozhekin, and E. S. Polzik, *Experimental long-lived entanglement of two macroscopic objects*, Nature **413**, 400, (2001).

[JW85]   E. Jakeman and J. G. Walker, *Analysis of a method for the generation of light with sub-Poissonian photon statistics*, Opt. Commun. **55**, 219, (1985).

[JWD07]   S. J. Jones, H. M. Wiseman, and A. C. Doherty, *Entanglement, Einstein–Podolsky–Rosen correlations, Bell nonlocality, and steering*, Phys. Rev. A **76**, 052116, (2007).

[Kas93]   M. A. Kastner, *Artificial atoms*, Phys. Today **46**, 24, (1993).

[KDM77]   H. J. Kimble, M. Dagenais, and L. Mandel, *Photon antibunching in resonance fluorescence*, Phys. Rev. Lett. **39**, 691, (1977).

[KGB02]   N. Khaneja, S. J. Glaser, and R. Brockett, *Sub-Riemannian geometry and time optimal control of three spin systems: Quantum gates and coherence transfer*, Phys. Rev. A **65**, 032301, (2002).

[KJ93]   S. Kotz and N. L. Johnson (eds.), *Breakthroughs in statistics Vol. I*, Springer, New York, 1993.

[KL97]   E. Knill and R. Laflamme, *Theory of quantum error-correcting codes*, Phys. Rev. A **55**, 900, (1997).

[KLM01]   E. Knill, R. Laflamme, and G. J. Milburn, *A scheme for efficient quantum computation with linear optics*, Nature **409**, 46, (2001).

[KLRG03]   N. Khaneja, B. Luy, T. Reiss, and S. J. Glaser, *Optimal control of spin dynamics in the presence of relaxation*, J. Magnetic Resonance **162**, 311, (2003).

[KMN+07]   P. Kok, W. J. Munro, K. Nemoto, T. C. Ralph, J. P. Dowling, and G. J. Milburn, *Linear optical quantum computing with photonic qubits*, Rev. Mod. Phys. **79**, 135, (2007).

[Kor99]   A. N. Korotkov, *Continuous quantum measurement of a double dot*, Phys. Rev. B **60**, 5737, (1999).

[Kor01a]   ———, *Output spectrum of a detector measuring quantum oscillations*, Phys. Rev. B **63**, 085312, (2001).

[Kor01b]   ———, *Selective quantum evolution of a qubit state due to continuous measurement*, Phys. Rev. B **63**, 115403, (2001).

[Kor03]   ———, *Noisy quantum measurement of solid-state qubits: Bayesian approach*, Quantum Noise in Mesoscopic Physics (Y. V. Nazarov, ed.), Kluwer, Dordrecht, p. 205.

[Kor05]   ———, *Simple quantum feedback of a solid-state qubit*, Phys. Rev. B **71**, 201305, (2005).

[KP00]   P. E. Kloeden and E. Platen, *Numerical solution of stochastic differential equations*, Springer, New York, 2000.

[Kra83]   K. Kraus, *States, effects, and operations: Fundamental notions of quantum theory*, Lecture Notes in Physics, vol. 190, Springer, Berlin, 1983.

[KU93]   M. Kitagawa and M. Ueda, *Squeezed spin states*, Phys. Rev. A **47**, 5138, (1993).

[Lan08] P. Langevin, *Sur la théorie du mouvement brownien*, Comptes Rendus Acad. Sci. (Paris) **146**, 550, (1908), English translation by D. S. Lemons and A. Gythiel, Am. J. Phys. **65**, 1079, (1997).

[Lan88] R. Landauer, *Spatial variation of currents and fields due to localized scatterers in metallic conduction*, IBM J. Res. Dev. **32**, 306, (1988).

[Lan92] ———, *Conductance from transmission: common sense points*, Phys. Scripta **T42**, 110, (1992).

[LBCS04] M. D. LaHaye, O. Buu, B. Camarota, and K. C. Schwab, *Approaching the quantum limit of a nanomechanical resonator*, Science **304**, 74, (2004).

[LCD⁺87] A. J. Leggett, S. Chakravarty, A. T. Dorsey, M. P. A. Fisher, A. Garg, and W. Zwerger, *Dynamics of the dissipative two-state system*, Rev. Mod. Phys. **59**, 1, (1987).

[Lin76] G. Lindblad, *On the generators of quantum dynamical semigroups*, Commun. Math. Phys **48**, 119, (1976).

[LJP⁺03] W. Lu, Z. Ji, L. Pfeiffer, K. W. West, and A. J. Rimberg, *Real-time detection of electron tunneling in a quantum dot*, Nature **423**, 422, (2003).

[Llo00] S. Lloyd, *Coherent quantum feedback*, Phys. Rev. A **62**, 022108, (2000).

[LM02]   A. I. Lvovsky and J. Mlynek, *Quantum-optical catalysis: Generating nonclassical states of light by means of linear optics*, Phys. Rev. Lett. **88**, 250401, (2002).

[LMPZ96] R. Laflamme, C. Miquel, J. P. Paz, and W. H. Zurek, *Perfect quantum error correcting code*, Phys. Rev. Lett. **77**, 198, (1996).

[LR91]   P. Lancaster and L. Rodman, *Solutions of continuous and discrete time algebraic Riccati equations: A review*, The Riccati Equation (S. Bittanti, A. J. Laub, and J. C. E. Willems, eds.), Springer, Berlin, p. 11, 1991.

[LR02]   A. P. Lund and T. C. Ralph, *Nondeterministic gates for photonic single-rail quantum logic*, Phys. Rev. A **66**, 032307, (2002).

[Lüd51] G. Lüders, *Concerning the state-change due to the measurement process*, Ann. Phys. (Leipzig) **8**, 322, (1951), English translation by K. Kirkpatrick in Ann. Phys. (Leipzig), **15**, 633, (2006).

[LWP⁺05] N. K. Langford, T. J. Weinhold, R. Prevedel, K. J. Resch, A. Gilchrist, J. L. O'Brien, G. J. Pryde, and A. G. White, *Demonstration of a simple entangling optical gate and its use in Bell-state analysis*, Phys. Rev. Lett. **95**, 210504, (2005).

[LZG⁺07] C.-Y. Lu, X.-Q. Zhou, O. Guhne, W.-B. Gao, J. Zhang, Z.-S. Yuan, A. Goebel, T. Yang, and J.-W. Pan, *Experimental entanglement of six photons in graph states*, Nature Phys. **3**, 91, (2007).

[Mab08] H. Mabuchi, *Coherent-feedback quantum control with a dynamic compensator*, Phys. Rev. A **78**, 032323, (2008).

[Maj98] F. G. Major, *The quantum beat: The physical principles of atomic clocks*, Lecture Notes in Physics, vol. 400, Springer, New York, 1998.

[MB99]   G. J. Milburn and S. L. Braunstein, *Quantum teleportation with squeezed vacuum states*, Phys. Rev. A **60**, 937, (1999).

[MCD93] K. Mølmer, Y. Castin, and J. Dalibard, *Monte Carlo wave-function method in quantum optics*, J. Opt. Soc. Am. B **10**, 524, (1993).

[MCSM]   A. E. Miller, O. Crisafulli, A. Silberfarb, and H. Mabuchi, *On the determination of the coherent spin-state uncertainty level*, private communication (2008).

[MD04]   A. Migdall and J. Dowling, *Special issue on single-photon: detectors, applications, and measurement methods (editorial)*, J. Mod. Opt. **51**, 1265, (2004).

[MdM90a] H. Martens and W. M. de Muynck, *The inaccuracy principle*, Foundations Phys. **20**, 357, (1990).

[MdM90b] H. Martens and W. M. de Muynck, *Nonideal quantum measurements*, Foundations Phys. **20**, 255, (1990).

[Mer98]   N. D. Mermin, *What is QM trying to tell us?*, Am. J. Phys. **66**, 753, (1998).

[MHF⁺90] J. Mertz, A. Heidmann, C. Fabre, E. Giacobino, and S. Reynaud, *Observation of high-intensity sub-Poissonian light using an optical parametric oscillator*, Phys. Rev. Lett. **64**, 2897, (1990).

[MHF91]   J. Mertz, A. Heidmann, and C. Fabre, *Generation of sub-Poissonian light using active control with twin beams*, Phys. Rev. A **44**, 3229, (1991).

[Mil89]   G. J. Milburn, *Quantum optical Fredkin gate*, Phys. Rev. Lett. **62**, 2124, (1989).

[Mil93]   P. Milonni, *The quantum vacuum: An introduction to quantum electrodynamics*, Academic Press, Boston, MA, 1993.

[MK05]   H. Mabuchi and N. Khaneja, *Principles and applications of control in quantum systems*, Int. J. Robust Nonlinear Control **15**, 647–667, (2005).

[MMW05]   S. Mancini, V. I. Man'ko, and H. M. Wiseman, *Special issue on quantum control (editorial)*, J. Opt. B **7**, S177, (2005).

[Mol69]   B. R. Mollow, *Power spectrum of light scattered by two-level systems*, Phys. Rev. **188**, 1969, (1969).

[Møl97]   K. Mølmer, *Optical coherence: A convenient fiction*, Phys. Rev. A **55**, 3195, (1997).

[MPV94]   A. V. Masalov, A. A. Putilin, and M. V. Vasilyev, *Sub-Poissonian light and photocurrent shot-noise suppression in closed optoelectronic loop*, J. Mod. Opt. **41**, 1941, (1994).

[MY86]   S. Machida and Y. Yamamoto, *Observation of sub-Poissonian photoelectron statistics in a negative feedback semiconductor laser*, Opt. Commun. **57**, 290, (1986).

[MZ96]   H. Mabuchi and P. Zoller, *Inversion of quantum jumps in quantum optical systems under continuous observation*, Phys. Rev. Lett. **76**, 3108, (1996).

[NC00]   M. A. Nielsen and I. L. Chuang, *Quantum computation and quantum information*, Cambridge University Press, Cambridge, 2000.

[Nie01]   M. A. Nielsen, *Characterizing mixing and measurement in quantum mechanics*, Phys. Rev. A **63**, 022114, (2001).

[Nie04]   ———, *Optical quantum computation using cluster states*, Phys. Rev. Lett. **93**, 040503, (2004).

[NM08]   A. E. B. Nielsen and K. M Mølmer, *Atomic spin squeezing in an optical cavity*, Phys. Rev. A **77**, 063811, (2008).

[NRO⁺99] G. Nogues, A. Rauschenbeutel, S. Osnaghi, M. Brune, J.-M. Raimond, and S. Haroche, *Seeing a single photon without destroying it*, Nature **400**, 239, (1999).

[NSD86]   W. Nagourney, J. Sandberg, and H. Dehmelt, *Shelved optical electron amplifier: Observation of quantum jumps*, Phys. Rev. Lett. **56**, 2797, (1986).

[NWCL00] R. J. Nelson, Y. Weinstein, D. Cory, and S. Lloyd, *Experimental demonstration of fully coherent quantum feedback*, Phys. Rev. Lett. **85**, 3045, (2000).

[Nyq28] H. Nyquist, *Thermal agitation of electric charge in conductors*, Phys. Rev. **32**, 110, (1928).

[OGW08] N. P. Oxtoby, J. Gambetta, and H. M. Wiseman, *Model for monitoring of a charge qubit using a radio-frequency quantum point contact including experimental imperfections*, Phys. Rev. B **77**, 125304, (2008).

[OPW+03] J. L. O'Brien, G. J. Pryde, A. G. White, T. C. Ralph, and D. Branning, *Demonstration of an all-optical quantum controlled-NOT gate*, Nature **426**, 264, (2003).

[OWW+05] N. P. Oxtoby, P. Warszawski, H. M. Wiseman, H.-B. Sun, and R. E. S. Polkinghorne, *Quantum trajectories for the realistic measurement of a solid-state charge qubit*, Phys. Rev. B **71**, 165317, (2005).

[Oxt07] N. Oxtoby, *Keeping it real: A quantum trajectory approach to realistic measurement of solid-state quantum systems*, Ph.D. thesis, Griffith University, Brisbane, 2007.

[Pau80] W. Pauli, *General principles of quantum mechanics*, Springer, Heidelberg, 1980.

[PB97] D. T. Pegg and S. M. Barnett, *Tutorial review: Quantum optical phase*, J. Mod. Opt. **44**, 225, (1997).

[PDR88] A. P. Peirce, M. A. Dahleh, and H. Rabitz, *Optimal control of quantum-mechanical systems: Existence, numerical approximation, and applications*, Phys. Rev. A **37**, 4950, (1988).

[Pea89] P. Pearle, *Combining stochastic dynamical state-vector reduction with spontaneous localization*, Phys. Rev. A **39**, 2277, (1989).

[Per88] A. Peres, *How to differentiate between non-orthogonal states*, Phys. Lett. A **128**, 19, (1988).

[Per95] ———, *Quantum theory: Concepts and methods*, Springer, Berlin, 1995.

[PJM+04] J. R. Petta, A. C. Johnson, C. M. Marcus, M. P. Hanson, and A. C. Gossard, *Manipulation of a single charge in a double quantum dot*, Phys. Rev. Lett. **93**, 186802, (2004).

[PPB98] D. T. Pegg, L. S. Phillips, and S. M. Barnett, *Optical state truncation by projection synthesis*, Phys. Rev. Lett. **81**, 1604, (1998).

[PR04] M. Paris and J. Rehacek (eds.), *Quantum state estimation*, Lecture Notes in Physics vol. 649, Springer, Berlin, 2004.

[Pre97] J. Preskill, *Lecture notes on quantum computation*, Lecture notes produced for the California Institute of Technology, 1997, http://www.theory.caltech.edu/people/preskill/ph229.

[Pre98] ———, *Reliable quantum computers*, Proc. Roy. Soc. London. A **454**, 385, (1998).

[PV05] F. Petruccione and B. Vacchini, *Quantum description of Einstein's Brownian motion*, Phys. Rev. E **71**, 046134, (2005).

[PVK97] M. B. Plenio, V. Vedral, and P. L. Knight, *Quantum error correction in the presence of spontaneous emission*, Phys. Rev. A **55**, 67, (1997).

[PZ01] J. P. Paz and W. H. Zurek, *Environment induced superselection and the transition from quantum to classical*, Coherent Matter Waves, Les Houches Session LXXII (R. Kaiser, C. Westbrook, and F. David, eds.), EDP Sciences, Springer, Berlin, p. 533, 2001.

[RB01]   R. Raussendorf and H. J. Briegel, *A one-way quantum computer*, Phys. Rev. Lett. **86**, 5188, (2001).

[RBH01]   J.-M. Raimond, M. Brune, and S. Haroche, *Manipulating quantum entanglement with atoms and photons in a cavity*, Rev. Mod. Phys. **73**, 565, (2001).

[RDM02]   B. Rahn, A. C. Doherty, and H. Mabuchi, *Exact performance of concatenated quantum codes*, Phys. Rev. A **66**, 032304, (2002).

[RdVRMK00] H. Rabitz, R. de Vivie-Riedle, M. Motzkus, and K. Kompa, *Whither the future of controlling quantum phenomena?*, Science **288**, 824, (2000).

[Red57]   A. G. Redfield, *On the theory of relaxation processes*, IBM J. Res. Dev. **1**, 19–31, (1957).

[RK02]   R. Ruskov and A. N. Korotkov, *Quantum feedback control of a solid-state qubit*, Phys. Rev. B **66**, 041401, (2002).

[RK03]   ———, *Spectrum of qubit oscillations from generalized Bloch equations*, Phys. Rev. B **67**, 075303, (2003).

[RL98]   T. C. Ralph and P. K. Lam, *Teleportation with bright squeezed light*, Phys. Rev. Lett. **81**, 5668, (1998).

[RLBW02] T. C. Ralph, N. K. Langford, T. B. Bell, and A. G. White, *Linear optical controlled-not gate in the coincidence basis*, Phys. Rev. A **65**, 062324, (2002).

[RLW05]   T. C. Ralph, A. P. Lund, and H. M. Wiseman, *Adaptive phase measurements in linear optical quantum computation*, J. Opt. B **7**, S245, (2005).

[RSD+95] V. Ramakrishna, M. V. Salapaka, M. Dahleh, H. Rabitz, and A. Peirce, *Controllability of molecular systems*, Phys. Rev. A **51**, 960, (1995).

[RSO+04] J. E. Reiner, W. P. Smith, L. A. Orozco, H. M. Wiseman, and J. Gambetta, *Quantum feedback in a weakly driven cavity QED system*, Phys. Rev. A **70**, 023819, (2004).

[SAB+06] M. Steffen, M. Ansmann, R. C. Bialczak, N. Katz, E. Lucero, R. McDermott, M. Neeley, E. M. Weig, A. N. Cleland, and J. M. Martinis, *Measurement of the entanglement of two superconducting qubits via state tomography*, Science **313**, 1423, (2006).

[SAJM04] M. Sarovar, C. Ahn, K. Jacobs, and G. J. Milburn, *Practical scheme for error control using feedback*, Phys. Rev. A **69**, 052324, (2004).

[Sar06]   M. Sarovar, *Quantum control and quantum information*, Ph.D. thesis, University of Queensland, Brisbane, 2006.

[Sch30]   E. Schrödinger, *Zum Heisenbergschen Unschärfeprinzip*, Sitzungsber. Preuß. Akad. Wiss. Berlin (Math.-Phys.) **19**, 296, (1930), English translation by A. Angelow, M.-C. Batoni, *On Heisenberg's uncertainty relation*, Bulg. J. Phys. **26**, 193, (1999); also available as eprint:quant-ph/9903100.

[Sch35a] ———, *Discussion of probability relations between separated systems*, Proc. Camb. Phil. Soc. **31**, 553, (1935).

[Sch35b] ———, *Die gegenwärtige Situation in der Quantenmechanik*, Naturwissenschaften **23**, 807, (1935), English translation by J. D. Trimmer, *The present situation in quantum mechanics*, Proc. Am. Phil. Soc. **124**, 323, (1980).

[Sch49]   P. A. Schilpp (ed.), *Albert Einstein: Philosopher-scientist*, Library of Living Philosophers, Evanston, IL, 1949.

[Sch86]   B. L. Schumaker, *Quantum mechanical pure states with Gaussian wave functions*, Phys. Rep. **135**, 317, (1986).

[Sch95]   B. Schumacher, *Quantum coding*, Phys. Rev. A **51**, 2738, (1995).

[Sch04]   M. Schlosshauer, *Decoherence, the measurement problem, and interpretations of quantum mechanics*, Rev. Mod. Phys. **76**, 1267, (2004).

[SD81]   M. D. Srinivas and E. B. Davies, *Photon counting probabilities in quantum optics*, J. Mod. Opt. **28**, 981, (1981).

[SDCZ01] A. Sørensen, L.-M. Duan, I. Cirac, and P. Zoller, *Many-particle entanglement with Bose–Einstein condensates*, Nature **409**, 63, (2001).

[SDG99]   W. T. Strunz, L. Diósi, and N. Gisin, *Open system dynamics with non-Markovian quantum trajectories*, Phys. Rev. Lett. **82**, 1801, (1999).

[Seg77]   A. Segall, *Optimal control of noisy finite-state Markov processes*, IEEE Trans. Automatic Control **22**, 179, (1997).

[SG64]   L. Susskind and J. Glogower, *Quantum mechanical phase and time operator*, Physics **1**, 49, (1964).

[Sha49]   C. E. Shannon, *Communication theory of secrecy systems*, Bell Syst. Tech. J. **28**, 656–715, (1949).

[SHB03]   W. T. Strunz, F. Haake, and D. Braun, *Universality of decoherence for macroscopic quantum superpositions*, Phys. Rev. A **67**, 022101, (2003).

[SJG⁺07] E. V. Sukhorukov, A. N. Jordan, S. Gustavsson, R. Leturcq, T. Ihn, and K. Ensslin, *Conditional statistics of electron transport in interacting nanoscale conductors*, Nature Phys. **3**, 243, (2007).

[SJM⁺04] D. A. Steck, K. Jacobs, H. Mabuchi, T. Bhattacharya, and S. Habib, *Quantum feedback control of atomic motion in an optical cavity*, Phys. Rev. Lett. **92**, 223004, (2004).

[SM95]   B. C. Sanders and G. J. Milburn, *Optimal quantum measurements for phase estimation*, Phys. Rev. Lett. **75**, 2944, (1995).

[SM01]   A. J. Scott and G. J. Milburn, *Quantum nonlinear dynamics of continuously measured systems*, Phys. Rev. A **63**, 042101, (2001).

[SM05]   M. Sarovar and G. J. Milburn, *Continuous quantum error correction by cooling*, Phys. Rev. A **72**, 012306, (2005).

[SR05]   K. C. Schwab and M. L. Roukes, *Putting mechanics into quantum mechanics*, Phys. Today **58**, 36, (2005).

[SRO⁺02] W. P. Smith, J. E. Reiner, L. A. Orozco, S. Kuhr, and H. M. Wiseman, *Capture and release of a conditional state of a cavity QED system by quantum feedback*, Phys. Rev. Lett. **89**, 133601, (2002).

[SRZ05]   V. Steixner, P. Rabl, and P. Zoller, *Quantum feedback cooling of a single trapped ion in front of a mirror*, Phys. Rev. A **72**, 043826, (2005).

[SS98]   A. Shnirman and G. Schön, *Quantum measurements performed with a single-electron transistor*, Phys. Rev. B **57**, 15400, (1998).

[SSH⁺87] J. H. Shapiro, G. Saplakoglu, S. T. Ho, P. Kumar, B. E. A. Saleh, and M. C. Teich, *Theory of light detection in the presence of feedback*, J. Opt. Soc. Am. B **4**, 1604, (1987).

[SSW90]   J. J. Stefano, A. R. Subberud, and I. J. Williams, *Theory and problems of feedback and control systems*, 2nd edn, McGraw-Hill, New York, 1990.

[Ste96]   A. M. Steane, *Multiple-particle interference and quantum error correction*, Proc. Roy. Soc. London A **452**, 2551, (1996).

[Ste07]   ———, *How to build a 300 bit, 1 giga-operation quantum computer*, Quantum Information Computation **7**, 171, (2007).

[Sto06] J. K. Stockton, *Continuous quantum measurement of cold alkali-atom spins*, Ph.D. thesis, California Institute of Technology, Pasadena, CA, 2006.

[SV06] L. F. Santos and L. Viola, *Enhanced convergence and robust performance of randomized dynamical decoupling*, Phys. Rev. Lett. **97**, 150501, (2006).

[SvHM04] J. K. Stockton, R. van Handel, and H. Mabuchi, *Deterministic Dicke-state preparation with continuous measurement and control*, Phys. Rev. A **70**, 022106, (2004).

[SW49] C. E. Shannon and W. Weaver, *The mathematical theory of communication*, University of Illinois Press, Urbana, IL, 1949.

[SW07] R. W. Spekkens and H. M. Wiseman, *Pooling quantum states obtained by indirect measurements*, Phys. Rev. A **75**, 042104, (2007).

[SWB+05] D. I. Schuster, A. Wallraff, A. Blais, L. Frunzio, R.-S. Huang, J. Majer, S. M. Girvin, and R. J. Schoelkopf, *AC Stark shift and dephasing of a superconducting qubit strongly coupled to a cavity field*, Phys. Rev. Lett. **94**, 123602, (2005).

[SWK+98] R. J. Schoelkopf, P. Wahlgren, A. A. Kozhevnikov, P. Delsing, and D. E. Prober, *The radio-frequency single-electron transistor (RF-SET): A fast and ultrasensitive electrometer*, Science **280**, 1238, (1998).

[TC68] M. Tavis and F. W. Cummings, *Exact solution for an N-molecule radiation-field Hamiltonian*, Phys. Rev. **170**, 379, (1968).

[TMB+06] A. M. Tyryshkin, J. J. L. Morton, S. C. Benjamin, A. Ardavan, G. A. D. Briggs, J. W. Ager, and S. A. Lyon, *Coherence of spin qubits in silicon*, J. Phys. C **18**, S783, (2006).

[TMW02a] L. K. Thomsen, S. Mancini, and H. M. Wiseman, *Continuous quantum nondemolition feedback and unconditional atomic spin squeezing*, J. Phys. B **35**, 4937, (2002).

[TMW02b] ———, *Spin squeezing via quantum feedback*, Phys. Rev. A **65**, 061801, (2002).

[TRS88] P. R. Tapster, J. G. Rarity, and J. S. Satchell, *Use of parametric down-conversion to generate sub-Poissonian light*, Phys. Rev. A **37**, 2963, (1988).

[Tur95] R. Turton, *The quantum dot: A journey into the future of microelectronics*, W. H. Freeman, Oxford, 1995.

[TV08] F. Ticozzi and L. Viola, *Quantum Markovian subsystems: Invariance, attractivity, and control*, IEEE Trans. Automatic Control **53**, 2048, (2008).

[TW02] L. K. Thomsen and H. M. Wiseman, *Atom–laser coherence and its control via feedback*, Phys. Rev. A **65**, 063607, (2002).

[TWMB95] M. S. Taubman, H. Wiseman, D. E. McClelland, and H.-A. Bachor, *Intensity feedback effects on quantum-limited noise*, J. Opt. Soc. Am. B **12**, 1792, (1995).

[Vai94] L. Vaidman, *Teleportation of quantum states*, Phys. Rev. A **49**, 1473, (1994).

[VB96] L. Vandenberghe and S. Boyd, *Semidefinite programming*, Soc. Indust. Appl. Math. Rev. **38**, 49, (1996).

[VES+04] L. M. K. Vandersypen, J. M. Elzerman, R. N. Schouten, L. H. Willems van Beveren, R. Hanson, and L. P. Kouwenhoven, *Real-time detection of single-electron tunneling using a quantum point contact*, Appl. Phys. Lett. **85**, 4394, (2004).

[VL98] L. Viola and S. Lloyd, *Dynamical suppression of decoherence in two-state quantum systems*, Phys. Rev. A **58**, 2733, (1998).

[VLK99] L. Viola, S. Lloyd, and E. Knill, *Universal control of decoupled quantum systems*, Phys. Rev. Lett. **83**, 4888, (1999).

[vHM05] R. van Handel and H. Mabuchi, *Optimal error tracking via quantum coding and continuous syndrome measurement*, arXiv e-print:quant-ph/0511221, 2005.

[vN32] J. von Neumann, *Mathematische Grundlagen der Quantenmechanik*, Springer, Berlin, 1932, English translation by E. T. Beyer, as *Mathematical Foundations of Quantum Mechanics* (Princeton University Press, Princeton, NJ, 1955).

[Wan01] X. Wang, *Spin squeezing in nonlinear spin-coherent states*, J. Opt. B **3**, 93, (2001).

[WB00] H. M. Wiseman and Z. Brady, *Robust unravelings for resonance fluorescence*, Phys. Rev. A **62**, 023805, (2000).

[WB08] H. M. Wiseman and L. Bouten, *Optimality of feedback control strategies for qubit purification*, Quantum Information Processing **7**, 71, (2008).

[WBI+92] D. J. Wineland, J. J. Bollinger, W. M. Itano, F. L. Moore, and D. J. Heinzen, *Spin squeezing and reduced quantum noise in spectroscopy*, Phys. Rev. A **46**, R6797, (1992).

[WD01] H. M. Wiseman and L. Diósi, *Complete parametrization, and invariance, of diffusive quantum trajectories for Markovian open systems*, Chem. Phys. **268**, 91, (2001), Erratum *ibid.* **271**, 227 (2001).

[WD05] H. M. Wiseman and A. C. Doherty, *Optimal unravellings for feedback control in linear quantum systems*, Phys. Rev. Lett. **94**, 070405, (2005).

[WG08] H. M. Wiseman and J. M. Gambetta, *Pure-state quantum trajectories for general non-Markovian systems do not exist*, Phys. Rev. Lett. **101**, 140401, (2008).

[Whi81] P. Whittle, *Risk-sensitive linear/quadratic/Gaussian control*, Adv. Appl. Probability **13**, 764, (1981).

[Whi96] ———, *Optimal control: Basics and beyond*, Wiley, Chichester, 1996.

[Wig32] E. Wigner, *On the quantum correction for thermodynamic equilibrium*, Phys. Rev. **40**, 749, (1932).

[Wis94] H. M. Wiseman, *Quantum theory of continuous feedback*, Phys. Rev. A **49**, 2133, (1994), Errata *ibid.*, **49** 5159 (1994) and *ibid.* **50**, 4428 (1994).

[Wis95] ———, *Adaptive phase measurements of optical modes: Going beyond the marginal q distribution*, Phys. Rev. Lett. **75**, 4587, (1995).

[Wis98] ———, *In-loop squeezing is like real squeezing to an in-loop atom*, Phys. Rev. Lett. **81**, 3840, (1998).

[Wis04] ———, *Squeezing and feedback*, Quantum Squeezing (P. D. Drummond and Z. Ficek, eds.), Springer, Berlin, p. 171, 2004.

[Wis06] ———, *From Einstein's theorem to Bell's theorem: A history of quantum non-locality*, Contemp. Phys. **47**, 79–88, (2006).

[Wis07] ———, *Grounding Bohmian mechanics in weak values and Bayesianism*, New J. Phys. **9**, 165, (2007).

[WJ85a] J. G. Walker and E. Jakeman, *Optical dead time effects and sub-Poissonian photo-electron counting statistics*, Proc. Soc. Photo-Opt. Instrum. Eng. **492**, 274, (1985).

[WJ85b] ———, *Photon-antibunching by use of a photoelectron-event-triggered optical shutter*, Opt. Acta **32**, 1303, (1985).

[WJD07] H. M. Wiseman, S. J. Jones, and A. C. Doherty, *Steering, entanglement, nonlocality, and the Einstein–Podolsky–Rosen paradox*, Phys. Rev. Lett. **98**, 140402, (2007).

[WK97] H. M. Wiseman and R. B. Killip, *Adaptive single-shot phase measurements: A semiclassical approach*, Phys. Rev. A **56**, 944, (1997).

[WK98] ———, *Adaptive single-shot phase measurements: The full quantum theory*, Phys. Rev. A **57**, 2169, (1998).

[WM93a] H. M. Wiseman and G. J. Milburn, *Interpretation of quantum jump and diffusion processes illustrated on the Bloch sphere*, Phys. Rev. A **47**, 1652, (1993).

[WM93b] ———, *Quantum theory of field-quadrature measurements*, Phys. Rev. A **47**, 642, (1993).

[WM93c] ———, *Quantum theory of optical feedback via homodyne detection*, Phys. Rev. Lett. **70**, 548, (1993).

[WM94a] D. F. Walls and G. J. Milburn, *Quantum optics*, Springer, Berlin, 1994.

[WM94b] H. M. Wiseman and G. J. Milburn, *All-optical versus electro-optical quantum-limited feedback*, Phys. Rev. A **49**, 4110, (1994).

[WM94c] ———, *Squeezing via feedback*, Phys. Rev. A **49**, 1350, (1994).

[WMW02] H. M. Wiseman, S. Mancini, and J. Wang, *Bayesian feedback versus Markovian feedback in a two-level atom*, Phys. Rev. A **66**, 013807, (2002).

[Won64] W. M. Wonham, *Some applications of stochastic differential equations to optimal nonlinear filtering*, J. Soc. Indust. Appl. Math., A **2**, 347, (1964).

[Woo81] W. K. Wootters, *Statistical distance and Hilbert space*, Phys. Rev. D **23**, 357, (1981).

[WPW99] C. E. Wieman, D. E. Pritchard, and D. J. Wineland, *Atom cooling, trapping, and quantum manipulation*, Rev. Mod. Phys. **71**, S253, (1999).

[WR06] H. M. Wiseman and J. F. Ralph, *Reconsidering rapid qubit purification by feedback*, New J. Phys. **8**, 90, (2006).

[WT99] H. M. Wiseman and G. E. Toombes, *Quantum jumps in a two-level atom: Simple theories versus quantum trajectories*, Phys. Rev. A **60**, 2474, (1999).

[WT01] H. M. Wiseman and L. K. Thomsen, *Reducing the linewidth of an atom laser by feedback*, Phys. Rev. Lett. **86**, 1143, (2001).

[WUS+01] H. M. Wiseman, D. W. Utami, H.-B. Sun, G. J. Milburn, B. E. Kane, A. Dzurak, and R. G. Clark, *Quantum measurement of coherent tunneling between quantum dots*, Phys. Rev. B **63**, 235308, (2001).

[WV98] H. M. Wiseman and J. A. Vaccaro, *Maximally robust unravelings of quantum master equations*, Phys. Lett. A **250**, 241, (28 December 1998).

[WV01] ———, *Inequivalence of pure state ensembles for open quantum systems: The preferred ensembles are those that are physically realizable*, Phys. Rev. Lett. **87**, 240402, (2001).

[WV02a] ———, *Atom lasers, coherent states, and coherence. I. Physically realizable ensembles of pure states*, Phys. Rev. A **65**, 043605, (2002).

[WV02b] ———, *Atom lasers, coherent states, and coherence II. Maximally robust ensembles of pure states*, Phys. Rev. A **65**, 043606, (2002).

[WW30] V. F. Weisskopf and E. P. Wigner, *Berechnung der natürlichen Linienbreite auf Grund der Diracschen Lichttheorie*, Z. Phys. **63**, 54, (1930).

[WW01] J. Wang and H. M. Wiseman, *Feedback-stabilization of an arbitrary pure state of a two-level atom*, Phys. Rev. A **64**, 063810, (2001).

[WW03a] P. Warszawski and H. M. Wiseman, *Quantum trajectories for realistic photodetection: I. General formalism*, J. Opt. B **5**, 1, (2003).

[WW03b] ———, *Quantum trajectories for realistic photodetection: II. Application and analysis*, J. Opt. B **5**, 15, (2003).

[WWM01] J. Wang, H. M. Wiseman, and G. J. Milburn, *Non-Markovian homodyne-mediated feedback on a two-level atom: a quantum trajectory treatment*, Chem. Phys. **268**, 221, (2001).

[WZ65] E. Wong and M. Zakai, *On the relationship between ordinary and stochastic differential equations*, Int. J. Eng. Sci. **3**, 213, (1965).

[WZ83] J. A. Wheeler and W. H. Zurek, *Quantum theory and measurement*, Princeton University Press, Princeton, NJ, 1983.

[Yam06] N. Yamamoto, *Robust observer for uncertain linear quantum systems*, Phys. Rev. A **74**, 032107, (2006).

[YHMS95] A. Yacoby, M. Heiblum, D. Mahalu, and H. Shtrikman, *Coherence and phase sensitive measurements in a quantum dot*, Phys. Rev. Lett. **74**, 4047, (1995).

[YIM86] Y. Yamamoto, N. Imoto, and S. Machida, *Amplitude squeezing in a semiconductor laser using quantum nondemolition measurement and negative feedback*, Phys. Rev. A **33**, 3243, (1986).

[YMK86] B. Yurke, S. L. McCall, and J. R. Klauder, *SU(2) and SU(1,1) interferometers*, Phys. Rev. A **33**, 4033, (1986).

[ZDG96] K. Zhou, J. C. Doyle, and K. Glover, *Robust and optimal control*, Prentice-Hall, Upper Saddle River, NJ, 1996.

[ZHP93] W. H. Zurek, S. Habib, and J. P. Paz, *Coherent states via decoherence*, Phys. Rev. Lett. **70**, 1187, (1993).

[ZRK05] Q. Zhang, R. Ruskov, and A. N. Korotkov, *Continuous quantum feedback of coherent oscillations in a solid-state qubit*, Phys. Rev. B **72**, 245322, (2005).

[Zur81] W. H. Zurek, *Pointer basis of quantum apparatus: Into what mixture does the wave packet collapse?*, Phys. Rev. D **24**, 1516, (1981).

[Zur82] ———, *Environment-induced superselection rules*, Phys. Rev. D **26**, 1862, (1982).

[Zur03] ———, *Decoherence, einselection, and the quantum origins of the classical*, Rev. Mod. Phys. **75**, 715, (2003).

[ZVTR03] S. Zippilli, D. Vitali, P. Tombesi, and J.-M. Raimond, *Scheme for decoherence control in microwave cavities*, Phys. Rev. A **67**, 052101, (2003).

Index

absorption
  stimulated, 107
actuator, 283f, 296f
algebra
  Lie, 318, 319, 324
algorithm
  quantum Fourier transform, 396
  quantum phase estimation, 396
  Shor's, 341, 396
amplifier
  operational, 196
ancillae, 21, 91
  n-photon entangled, 388
  qubit, 375
  single-photon, 380, 387
  vacuum field, 185, 186, 310
anticommutator, 110
apparatus
  classical, 2, 98
  quantum, 15, 25, 97, 98
approximation
  Born, 99, 101, 104, 109, 116
  Markov, 99, 105, 117
  rotating-wave (RWA), 43, 104, 108, 109, 140,
    303, 333, 337
Arthurs and Kelly model, 23
atom, 15
  alkali, 261
  hydrogen, 10
  radiative decay of, 102
  rubidium, 46
  Rydberg, 133
  three-level, 42
  two-level, 16, 102, 128, 172, 259
atom lasers, 267

back-action
  classical, 6, 33, 280, 282, 289

quantum, 27, 30, 32, 34
  surplus quantum, 40
    elimination of, 267
bandwidth
  detector, 156, 195
    effective, 201
  feedback, 223, 225
    trade-off with gain, 224
basis
  angular momentum, 73
  canonically conjugate, 62
  complementary, 32, 121
  diagonal, 125
  logical, 342
  measurement, 17, 25, 121, 342
  momentum, 27
  orthonormal, 10, 398
  overcomplete, 413
  pointer, 122, 342
  position, 19, 62
  preparation, 342
bath, see environment, 97
Bayesian inference, 4, 11, 19, 49, 81
  generalized, 7
beam-splitter, 70, 352, 382
  polarizing, 90, 221
  time-dependent, 221
birefringence, 383
bits, 341
  fungibility of, 342
  quantum, see qubits
Bloch sphere, 103, 105, 129f, 131f, 172,
    262

c-numbers, 398
cavity, 107
  microwave, 43
  radiative decay of, 107

Printed in the United States
By Bookmasters